AF511519

HYDROPONICS
Principles and Practices

The Authors

Sanjay Mishra is a prolific author with a diverse academic and professional background. He holds a degree in Bio Science from Delhi University, after which he transitioned into the field of computer science. His career in the IT industry spans several years, where he worked at renowned organizations such as NIIT, Aptech, Hartron, Zed CA, GENX Technologies and Aldite Co. During this time, he developed expertise in programming languages like Java, C++, VB6.0, ASP and scripting languages, apart from training numerous students. In 2012, he left the IT field to dedicate himself entirely to agricultural pursuits. He even ventured into commercial hydroponics and aquaponics farming in Prayagraj (U. P.), experimenting with soilless crop production. He has authored several books across various disciplines. His contributions to computer science include two published works, while his deep interest in horticulture is reflected in his book *Herbs, Spices, and Medicinal Plants* and five detailed monographs on vegetable crops. He has also written a book titled *Soilless Crop Production*. As of mid-2025, readers can look forward to four more books, each focusing on distinct vegetable groups. Currently, Sanjay Mishra dedicates his time to experimenting with innovative crop production techniques and sharing his extensive knowledge through his books.

Dr. Sanyat Misra is a Forestry graduate from UAS, Bangalore; M.Sc. (Ag.) Hort. from GBPUAT, Pantnagar (Udham Singh Nagar), Uttarakhand; and Ph.D. from Dr. YSP UHF, Nauni, Solan (H.P.). He served ICAR-NBPGR (New Delhi and Shimla) for 3 years and in 2004 joined BAU, Kanke (Ranchi) as Assistant Professor *cum* Jr. Scientist. At present, he is Head, Deparment of Horticulture looking after administrative, extension, teaching and research activities of the Department. He had been Chairman in case of 7 M.Sc. students. To his credit, he has some 45 research publications, 11 books and 3 bulletins.

Dr. R.L. Misra got retired from Division of Floriculture and Landscaping, ICAR-IARI, New Delhi-110012 in May 2008 as Principal Scientist *cum* Project Coordinator (Floriculture), All India Coordinated Research Project on Floriculture. He also worked as Officer-in-Charge, Indo-Israel Preject on Research and Development on Farms, IARI, New Delhi. As Adjunct Faculty he continued further teaching works to M.Sc. and Ph.D. students at IARI from 2010 to 2014. As Chairman, he guided 6 M.Sc. and 16 Ph.D. students in IARI. To his credit, he developed 30 gladiolus varieties, has more than 500 publications: 11 books, 9 technical bulletins, 15 review articles, 169 research publications, 120 technical articles, 57 symposium papers and more than 140 book chapters. He is Fellow of: i) **Indian Society of Genetics and Plant Breeding** (1980), ii) **Indian Society of Ornamental Horticulture (2009),** iii) **Horticultural Society of India** (2018), and (iv) **Promotion of Horticultural Science, Bengaluru** (2022). Also he is HSI Gold Medal Awardee (2013). **Indian Society of Ornamental Horticulture,** New Delhi in 2018 on his outstanding floricultural works, and **The Orissa Horticultural Society,** Bhubaneswar in 2023 on his outstanding horticultural works conferred Lifetime Achievement Awards. In 2007, he was also conferred with Urivi Vikram Charitable Trust Award in New Delhi.

HYDROPONICS
Principles and Practices

— *Authors* —

Sanjay Mishra

Sanyat Misra

R.L. Misra

ASTRAL
Publishing Authors Across the Globe

2025

Daya Publishing House®
A Division of
Astral International Pvt. Ltd.
New Delhi – 110 002

© 2025 AUTHORS

ISBN 978-93-5919-855-2

Publisher's Note:

Every possible effort has been made to ensure that the information contained in this book is accurate at the time of going to press and the publisher and author cannot accept responsibility for any errors or omissions, however caused. No responsibility for loss or damage occasioned to any person acting, or refraining from action, as a result of the material in this publication can be accepted by the editor, the publisher or the author. The Publisher is not associated with any product or vendor mentioned in the book. The contents of this work are intended to further general scientific research, understanding and discussion only. Readers should consult with a specialist where appropriate.

Every effort has been made to trace the owners of copyright material used in this book, if any. The author and the publisher will be grateful for any omission brought to their notice for acknowledgement in the future editions of the book.

All Rights reserved under International Copyright Conventions. No part of this publication may be reproduced, stored in a retrieval system, or transmitted in any form or by any means, electronic, mechanical, photocopying, recording or otherwise without the prior written consent of the publisher and the copyright owner.

Published by : **Daya Publishing House®**
A Division of
Astral International Pvt. Ltd.
– ISO 9001:2015 Certified Company –
4736/23, Ground Floor, Ansari Road,
Darya Ganj, New Delhi-110 002
Ph. 011-43549197, 8130496929
E-mail: info@astralint.com
Website: www.astralint.com

Printed at : **Replika Press Pvt. Ltd.**

Preface

When the book '**Soilless Crop Production**' was brought out in 2017 we were satisfied as the book had wide acceptance by the readers throughout the country but for the last 2-3 years we were requested to write an exclusive book on **Hydroponics**, hence now it is before you entitled **Hydroponics – Principles and Practices.** We hope that readers will like this immensely as before. Here we have also dealt with '**Aquaponics**', *i.e.* Fish culture along with horticultural crops.

In all, the book starts with **Introductory** remarks describing importance of such a culture. In all there are 20 chapters, each chapter dealing with specialized components. Chapter 12 deals **Hydroponics Step by Step** as a guide to a beginner. From chapter 13 to 17 it has details about individual crop cultivation in brief, such as **Tomato**, **Cucumber**, **Lettuce**, **Capsicum** and **Chrysanthemum**. Chapter 18 deals about **Homegrown Hydroponics: Cultivating a Bounty with Advanced Methods**. Chapter 19 deals with **Organic Hydroponics** and chapter 20 with **Aquaponics**. In the end there is a column about the **References** quoted in this book.

The book has been written keeping in mind for those growers who wish to join this enterprise as a hobby so that apart from growing vegetables, fruits, flowers, herbs and spices they may also grow fish for double entertainment.

All those shall be benefitted through this book whether one is hobbyist, a teacher or professor teaching a curriculum having **Hydroponics** in the curriculum, the students of horticulture and those opting honours or advance courses dealing with Hydroponics as well those scientists who wish to work in this line. Moreover, those areas having frequent flooding or where people have so much of ponds, lakes or any other type of water reservoirs so that they may grow desired vegetables, fruits, spices, herbs and flowers of their choice throughout the year and so consuming the fresh harvests. This practice is already going on in Dal Lake of

Kashmir and elsewhere in the country for supplying fresh produce to inhabitants there.

The authors sincerely acknowledge all the sources and references consulted during the preparation of this book. We welcome any constructive feedback or suggestions for improvement. For any contributions or recommendations, please feel free to reach out via email at mishrasanjay.agri@gmail.com or by phone at 7905336922 or 8076235469. Your insights will be highly appreciated.

Sanjay Mishra

Sanyat Misra

R.L. Misra

Contents

Introductory

Overcoming Agricultural Challenges with Soilless Cultivation

In recent years, the global landscape of agriculture has encountered formidable challenges stemming from population growth, urbanization and soil degradation. The decline in arable land availability exacerbated by rapid influx of people into urban areas, poses a significant threat to food security worldwide. Concurrently, traditional soil-based farming confronts a multitude of obstacles including soil compaction, poor drainage, erosion, declining fertility and soil-borne pathogens due to continuous cultivation. These factors collectively diminish crop yields and compromise food quality, highlighting the urgent need for alternative cultivation methods.

Soilless crop production has emerged as a promising solution to address the limitations of traditional agriculture. By circumventing the constraints imposed by soil quality and land availability, soilless cultivation offers a viable pathway to meet sustainably the escalating demand for food and nutrition. Through innovative techniques of hydroponics *cum* aquaponics certain crops can be grown in nutrient-rich environments such as greenhouses, rooftops and balconies and this will enable the individuals to cultivate a variety of vegetables and food crops without requiring additional space. Moreover, soilless crop cultivation extends beyond food production to encompass ornamental plants providing a holistic approach to urban gardening and sustainable living.

At the core of soilless crop production lies the principle that soil itself is not indispensable for plant growth. Instead, plants derive essential nutrients such as calcium, magnesium, potassium and phosphorus from mineral nutrient solutions or from inert media such as gravel, mineral wool or coconut husk. This paradigm shift enables growers to exert precise control over the root environment, optimizing water, nutrient and temperature parameters to maximize crop yield and quality. Advanced technologies including computer-controlled systems facilitate automation and optimization of cultivation practices which enhance efficiency and productivity further.

Hydroponics and Aquaponics

Hydroponics and aquaponics represent two prominent techniques within the realm of soilless crop production. In hydroponic systems, plants are nourished mostly with inorganic nutrient solutions, allowing for efficient nutrient uptake and enhanced crop growth. Hydroponic cultivation has gained traction worldwide due to its ability to conserve water, minimising fertilizer usage and mitigating climate and pest-related risks. Conversely, **aquaponics** integrates fish cultivation with plant production creating symbiotic ecosystems where fish excretions serve as organic nutrients for plants. This closed-loop system promotes resource efficiency and offers a sustainable solution for both food and fish production.

Benefits and Challenges

Soilless crop production offers numerous benefits including increased crop quality, productivity and resource efficiency. By eliminating the need for soil, growers can mitigate soil-borne diseases, optimize nutrient delivery and cultivate crops year-round in controlled environments. Furthermore, soilless cultivation conserves water, minimises fertilizer usage and reduces energy consumption compared to traditional farming practices. However, challenges such as initial investment costs, disease management and technical expertise remain barriers to widespread adoption particularly in large-scale commercial setups.

Evolution of Hydroponics: A Historical Journey

Hydroponics is a practice of growing plants without soil which has a rich and storied history that spans centuries of civilization. From ancient techniques to modern innovations, the evolution of hydroponics reflects humanity's enduring quest for sustainable agriculture and food security.

Floating Gardens

Ancient origins: Hydroponic gardening traces its origins back to ancient civilizations where resourceful farmers devised ingenious methods to cultivate crops in water-based environments. Historical records indicate that cultures such as the Indians, Babylonians, Aztecs, and Chinese experimented with various forms of hydroponic gardening as early as millennia ago. These early pioneers recognized the potential of soilless cultivation to overcome geographic limitations, optimize resource utilization and ensure reliable food production.

Landmarks in soilless cultivation: One of the most iconic examples of ancient hydroponics is the 'Hanging Gardens of Babylon' constructed around 600 BC. Regarded as one of the 'Seven Wonders of the World', these lush gardens utilized innovative irrigation systems to nurture plants suspended in mid-air, showcasing the remarkable ingenuity of ancient engineering. Similarly, the Aztec Indians developed floating gardens known as **chinampas** around AD 1100, employing layers of mud and vegetation to cultivate crops atop freshwater lakes in Central America. This practice continues to thrive in regions like Kashmir (India) where tradition of floating gardens persists in places like Dal Lake.

Contributions of modern science: The scientific foundations of hydroponics began to take shape in the 18th and 19th centuries with groundbreaking discoveries in plant physiology and nutrient solutions. In 1792, English scientist Joseph Priestley's revelation that plants absorb carbon dioxide and release oxygen laid the groundwork for understanding plant nutrition. Building upon this knowledge, German scientists Julius von Sachs and W. Knop formulated a standardized nutrient solution in the mid-19th century which paved the way for development of soilless cultivation techniques.

Pioneers and innovators: 20th century witnessed a surge of interest in hydroponics, fuelled by pioneering researchers and innovative technologies. In 1920s, Dr. William F. Gericke coined the term **'hydroponics'** to describe the practice of growing plants in water without soil, laying the groundwork for modern hydroponic systems. Meanwhile, advancements such as the Nutrient Film Technique and Drip Irrigation Systems revolutionized nutrient delivery and water management in hydroponic agriculture.

Technological advancements: Throughout the latter half of the 20th century and into the 21st century, hydroponics continued to evolve with the introduction of cutting-edge techniques and materials. From aeroponics, pioneered by Italian scientist Dr. Franco Massantini in the 1970s, to substrate-based systems like rockwool slabs and coconut husk fibres, hydroponic growers have embraced a diverse array of methods to optimize plant growth and yield. Innovations such as aeroponics which deliver nutrients to plant roots *via* a fine mist highlight the ongoing quest for efficiency and sustainability in soilless cultivation.

Future directions: As hydroponics enters the digital age, with advancements in automation, precision agriculture and vertical farming, the future of soilless cultivation holds immense promise. With global challenges such as climate change and population growth placing unprecedented pressure on traditional agricultural systems, hydroponics offers a resilient and adaptable solution for meeting the world's growing food needs. By combining ancient wisdom with modern technology, hydroponic agriculture continues to chart a course towards a more sustainable and food-secure future.

In a nutshell, soilless crop production represents a transformative approach to modern agriculture offering sustainable solutions to address the pressing challenges facing food production systems worldwide. By harnessing technology, innovation and ecological principles soilless cultivation empowers communities as well as individuals to cultivate healthy and nutritious crops in diverse environments. As society grapples with the imperative to achieve food security and environmental sustainability, soilless crop production emerges as a beacon of hope, fostering resilience, efficiency, and abundance in the face of adversity.

While soilless cultivation offers numerous advantages for home gardening and small-scale production, scaling up these techniques to meet the demands of commercial agriculture presents unique challenges. By addressing issues such as disease management, nutrient solution management, water sustainability and energy consumption, growers can unlock the full potential of soilless cultivation and establish it as a viable and sustainable approach to modern agriculture.

Sanjay Mishra

Sanyat Misra

R.L. Misra

Chapter 1

Understanding Hydroponics

1.1 Introduction

Hydroponics, a term coined by Dr. Gericke represents a revolutionary approach to plant cultivation, transcending traditional soil-based methods (Gericke, 1937). Derived from the Greek words '*hydro*' for water and '*ponos*' for labour, hydroponics literally means 'water-working'.

Hydroponics is a type of horticulture and a subset of hydroculture which involves the cultivation of plants, especially horticultural and medicinal plants without soil in a nutrient-rich water solution, offering a precise and controlled environment for optimal growth. This technique enables growers to bypass the limitations of conventional soil-based agriculture, providing opportunities for enhanced crop management and productivity. Terrestrial or aquatic plants may grow with their roots exposed to the nutritious liquid or the roots may be mechanically supported by an inert medium such as perlite, gravel or other substrates (Gericke, 1945).

1.2 Key Principles

At the core of hydroponic systems lies the ability to deliver essential nutrients directly to the plant roots, bypassing the intermediary role of soil. By suspending plants in nutrient solutions, growers can exert precise control over the growth environment ~ adjusting factors such as pH, nutrient concentration and water availability to meet specific plant requirements. This targeted approach allows for accelerated growth rates, extended cropping seasons and increased yields, these all making hydroponics a preferred choice for commercial cultivation and urban farming initiatives.

1.3 Substrates or Growing Media

In hydroponic setups, traditional soil is replaced by inert substrates or growing media which serve as a support structure for plant roots. Common substrates include cocopeat, vermiculite, perlite and rockwool ~ each offering unique advantages in terms of water retention, aeration and nutrient delivery. These substrates are typically arranged in trays, troughs or individual containers ~ providing a stable foundation for plant growth while facilitating efficient nutrient uptake and root development. Though hydroponic systems typically utilize inert media, roots can still influence the pH of the rhizosphere and impact the biological dynamics within it. Root exudates which include various released compounds have the potential to alter the chemistry and microbial activity in the rhizosphere, thus influencing the physiological equilibrium of the nutrient solution. This interaction becomes particularly significant when plants produce secondary metabolites as these compounds can further modulate the rhizosphere environment and affect nutrient uptake and utilization (Nye, 1981; Walker *et al.,* 2003; Suryawanshi, 2021). Transgenic plants cultivated hydroponically have the capability to release pharmaceutical proteins directly into the hydroponic solution through their root exudates (Horn *et al.,* 2004).

1.4 Hydroponic Nutrients

In hydroponic systems, plants receive vital nutrients dissolved in water. Nutrient needs particularly for nitrogen, phosphorus and potassium (N-P-K) vary depending on factors such as plant type, growth stage, weather and season. These nutrients are available in powder, liquid, resin or tablet forms ~ each tailored to control growth and flowering. Fertilizer labels prioritize N-P-K percentages (*e.g.,* 24-8-16) with common macronutrients including potassium nitrate, calcium nitrate and magnesium sulphate. Micronutrients such as iron, manganese and boron may be supplemented often accompanied by chelating agents and humic acids to enhance nutrient uptake.

1.5 Environmental Considerations and System Design

Hydroponic systems are often housed within greenhouse structures or contained environment on inert media, adapted to the controlled-environment agriculture (CEA) process (Boylan, 2020), providing insulation against fluctuating temperatures, minimizing water loss through evaporation and reducing the risk of disease and pest infestation. The enclosed nature of greenhouse environments allows growers to maintain optimal growing conditions year-round, maximizing productivity and crop quality. Moreover, advancements in technology have led to development of automated monitoring and control systems ~ further enhancing the efficiency and precision of hydroponic cultivation practices.

Maintaining optimal environmental conditions is essential for nutrient solution management and plant growth in hydroponic systems. By controlling temperature, humidity, light intensity and CO_2 level within appropriate ranges growers can ensure that plants receive the necessary conditions for efficient nutrient uptake, metabolic processes and healthy growth. Regular monitoring and adjustments to environmental factors are key to optimizing plant performance and maximizing yields in hydroponic cultivation.

1.6 Classification of Hydroponic Systems

Hydroponic systems can be classified based on various criteria including the presence or absence of a growing medium, the method of nutrient delivery and overall design and operation. These classifications provide a framework for understanding the diversity of hydroponic systems available each with its unique advantages, applications and considerations. Growers can select the most suitable hydroponic system based on factors such as crop type, available space, budget and level of automation desired. Here's a summary of the classification of hydroponic systems:

1.6.1 Based on Design and Operation

Hydroponic systems can be broadly classified as 'closed hydroponics systems' and 'open hydroponics systems'.

A. Closed Hydroponic System

In closed hydroponic system, nutrient solution is contained within a closed-loop system where surplus nutrient solution is recovered, replenished and recycled ~ the same nutrient solution is re-circulated and the nutrient concentrations are monitored and adjusted as needed to maintain optimal levels for plant growth. Closed hydroponics can be both simple hydroponic systems (such as 'deep water culture') as well as sophisticated hydroponics (such as NFT). Closed systems are more water-efficient as the nutrient solution is recycled, reducing overall water consumption. Moreover, there is greater control over nutrient levels, pH and EC

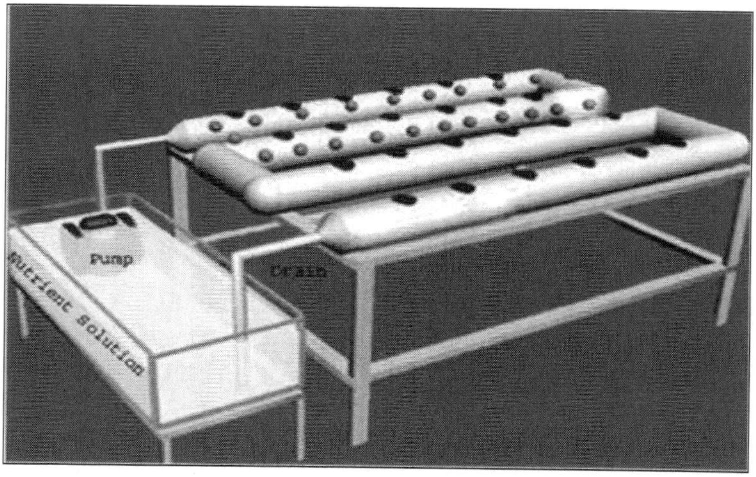

A Closed Hydroponics System.

since the solution is reused, leading to potentially better plant growth and yields. Closed systems are often considered more sustainable as these minimize nutrient wastage and water usage.

B. Open Hydroponic Systems or 'Run to Waste'

In open hydroponic systems, nutrient solution is exposed to the environment and is not recirculated or contained within a closed-loop system. Usually a fresh nutrient solution is introduced for each irrigation cycle and excess nutrient solution is allowed to drain away or evaporate instead of recirculating back into the system. The nutrient solution is usually delivered to the plants using a drip system. While typically considered a closed-loop system, drip irrigation can also be adapted into an open hydroponic system. In this setup, nutrient solution is delivered through drip emitters directly to the base of each plant with excess solution draining away. This method allows for precise control over nutrient delivery and is commonly used in larger-scale hydroponic operations. Open hydroponic systems may require less maintenance as there's no need to monitor or manage the nutrient solution's pH or EC as closely. Generally open systems tend to use more water as the nutrient solution is not recycled and an adequate run-off must also be maintained in order to keep nutrient balance in the root zone. In this case, there is potential nutrient wastage as the excess solution is not reused.

An Open Hydroponics System.

1.6.2 Based on Method of Nutrient Delivery

Closed hydroponics systems can be further classified as 'active hydroponic systems' (recirculated hydroponics systems) and 'passive hydroponic systems' (non-recirculated hydroponics systems).

A. Passive Hydroponic Systems

Passive hydroponics refers to soilless plant cultivation methods where water and nutrients are delivered to the plant roots without use of pumps or active irrigation systems. These systems eliminate the need for mechanical circulation by utilizing natural forces like capillary action or wicking (in passive sub-irrigation) or direct suspension of plant roots in nutrient solution to deliver nutrients.

- ☆ **Examples of water/solution culture (media-less) passive hydroponics:** Static solution culture and Capillary mat system.
- ☆ **Examples of medium culture (media-based) passive hydroponics:** Kratky method, Capillary or wick system, Passive drip systems.
- ☆ **Capillary mat system** can be implemented as both media-less as well as media-based hydroponic systems.

B. Active Hydroponic Systems

Active hydroponic systems are hydroponic setups that utilize mechanical components such as pumps to circulate nutrient solution to the plant roots. Unlike passive hydroponic systems that rely on natural forces like capillary action or wicking, active hydroponic systems actively deliver water and nutrients to the plants providing greater control over nutrient levels, pH and oxygenation.

- ☆ **Examples of media-less active hydroponic systems:** Nutrient film technique (NFT) and Aeroponics.
- ☆ **Examples of media-based active hydroponics:** Drip systems, Ebb and Flow (Flood and Drain) systems.
- ☆ **NFT and Recirculating deep water culture (RDWC)** can be implemented as media-less and media-based hydroponic systems.

C. Hybrid of Acitve and Passive

Deep water culture (DWC) systems can be implemented as solution culture as well as medium culture. In DWC, plant roots are suspended in a nutrient solution reservoir providing oxygen through an air stone or aerating device. Despite technically requiring a pump for aeration, DWC is often viewed as passive due to its simplicity and the lack of visible movement in the nutrient solution. However, it can be considered an active hydroponic system due to its reliance on active aeration and nutrient circulation to support plant growth.

1.6.3 Medium Culture and Solution Culture Hydroponics

Hydroponic techniques can also be classified into those that require growing substrates (media-based) and those that do not (media-less). Both approaches have their advantages and are widely used depending on specific needs, crop types and environmental conditions. Here is a breakdown of both:

A. Medium Culture (Media-based hydroponic or semi-hydroponic)

This encompasses systems where plants are grown utilizing a solid growing medium such as perlite, rockwool, cocopeat or vermiculite to support plant roots where nutrient solution is delivered to the roots, *e.g.* drip systems, ebb and flow systems and wick systems. Though these methods provide benefits like enhanced stability for plant support and improved moisture retention, these may not be classified as true hydroponics as these rely on growing media alongside nutrient solutions.

i) Ebb and Flow (flood and drain)

- Utilizes a growing medium such as perlite, vermiculite, rockwool, coconut coir or gravel.
- Plants are grown in trays or containers filled with growing medium.
- Nutrient solution is periodically flooded into the tray, saturating the medium and providing nutrients to the roots.
- After a set period, the solution drains away, allowing the roots to access oxygen.

ii) Drip System

- Uses a growing medium such as perlite, vermiculite, coconut coir or rockwool.
- Nutrient solution is dripped onto the medium or directly onto the root zone of plants.
- The medium retains moisture and nutrients for the roots to absorb.

iii) Wicking System

- Relies on a growing medium such as vermiculite, perlite, coconut coir or a mix of these.
- Nutrient solution is drawn up through a wick from a reservoir to the growing medium.
- The medium wicks moisture and nutrients to the roots.

iv) Dutch Bucket Systems

- Dutch bucket systems consist of containers filled with a growing medium such as perlite or coconut coir.
- Each container has a drip line or emitter delivering nutrient solution to the base of plant.
- Excess solution drains into a common reservoir which is then recirculated through system.

v) Media Bed Aquaponics

- Media bed aquaponics combines hydroponics with aquaculture where fish waste provides nutrients for plants grown in a media bed.
- Plants are grown in a container filled with a growing medium and nutrient-rich water from the fish tank is pumped into the bed providing nutrients to the plants.

- ☆ The media bed acts as both a support structure for the plants and a biofilter for the aquaponic system.

B. Water/Solution Culture or Media-less or True Hydroponics

Solution culture hydroponics represents a conventional approach to hydroponic systems where plants receive essential nutrients directly from a nutrient solution without use of any growing medium. This method involves structures that support plant roots, ensuring that these remain submerged or continuously exposed to the nutrient solution while facilitating optimal nutrient uptake. Unlike other hydroponic techniques that utilize growing media such as perlite or rockwool, media-less hydroponics relies solely on the nutrient solution to provide plants with all the necessary nutrients for growth and development. True hydroponics can further be classified as **Water/Solution culture** [Static Solution Culture (SSC), Deep Water Culture (DWC), Nutrient Film Technique (NFT) and Membrane System]; **Vapour culture** (Aeroponics) and **Fog culture** (Fogponics). Here are some key characteristics of true hydroponics and how it differs from other hydroponic methods:

- ☆ **Nutrient solution only**: In true hydroponics, plants receive all their essential nutrients directly from a carefully balanced nutrient solution. There is no use of soil, soilless media (such as coco coir or perlite) or any other growing substrate.
- ☆ **Continuous submersion or exposure**: Plant roots in true hydroponic systems are either continuously submerged in the nutrient solution or exposed to it through constant irrigation. This ensures that roots have constant access to water, oxygen and nutrients.
- ☆ **Simple setup**: True hydroponic systems often have a simpler setup compared to other hydroponic methods that utilize growing media. These typically require fewer components and are easier to maintain.
- ☆ **High oxygen availability**: Since there is no growing medium to impede oxygen diffusion to the roots, true hydroponic systems often provide high levels of oxygen to the root zone. This promotes healthy root growth and enhances nutrient uptake efficiency.

Overall, true hydroponics represent a subset of hydroponic techniques that prioritize direct nutrient delivery to plant roots without use of growing media ~ offering simplicity, efficiency, and high oxygen availability for plant growth.

1.6.4 Based on Nutrient Source

Hydroponic systems can also be broadly categorized into organic and inorganic systems based on the type of nutrients used to nourish the plants.

1.7 Crop Selection and Cultivation Techniques

A diverse range of crops can thrive in hydroponic systems including capsicum, cucumber, lettuce, strawberry, tomato, various herbs, cut flowers as well as *Arabidopsis thaliana* which serves as a model organism in plant science and genetics (Zeng *et al.*, 2018). Suitability of hydroponic cultivation for these crops stems from its ability to optimize growing conditions, ensuring consistent nutrient availability, water supply and environmental control. Cultivation techniques vary depending on the crop and system design with growers employing methods such as seedling placement machines or manual seeding to establish plant growth in the substrate.

Each plant species may thrive under different hydroponic methods depending on their specific requirements. Here is a breakdown of some crops and most suitable hydroponic methods for growing them:

i) **Tomatoes, cucumbers and bell peppers:** First two being vine crops and generally do well in nutrient film technique (NFT) systems or Dutch buckets. NFT allows for ample oxygen to the roots and continuous nutrient flow, ideal for these larger plants. Dutch buckets provide stability and support for the heavy fruiting of tomatoes, cucumbers and bell peppers.

ii) **Lettuce and strawberries:** These are well-suited for NFT, deep water culture (DWC) or vertical tower systems. NFT and DWC provide consistent access to nutrients and oxygen, promoting healthy leafy growth in lettuce and runners in strawberries. Vertical tower systems optimize space utilization, making them great for compact crops like lettuce and strawberries.

iii) **Eggplants, broccoli and other cole crops:** These can thrive in various systems including DWC, NFT or even ebb and flow (flood and drain) systems. Systems are chosen that allow for sturdy root support and adequate nutrient availability throughout the growth cycle.

iv) **Roses and chrysanthemums:** These flowering plants can benefit from ebb and flow (flood and drain) systems or drip irrigation systems. Ebb and flow systems provide periodic flooding and draining, mimicking natural watering cycles which can encourage robust root development and flowering. Drip irrigation ensures precise delivery of water and nutrients to the root zone essential for the health and vitality of these ornamental plants.

It is important to note that regardless of the hydroponic method chosen proper monitoring of pH, EC, temperature and humidity is essential for optimal growth of all crops. Moreover, selecting the appropriate nutrient solution tailored to the specific needs of each crop is crucial for success in hydroponic gardening.

1.8 Key Considerations

Achieving success in hydroponic systems hinges on several key factors, each playing a crucial role in fostering optimal plant growth and productivity:

i) **Right nutrient levels**: Tailoring nutrient levels to suit specific requirements of chosen plant varieties is paramount. Specializing in a select few varieties enables growers to fine-tune their nutrient mix, ensuring an ideal balance for robust growth. While finding a versatile nutrient mix that caters to multiple plants is beneficial, adjusting nutrient strength based on individual plant needs can optimize performance.

ii) **Plant selection**: Careful consideration of plant selection based on climate and seasonality is essential particularly in outdoor hydroponic setups. Almost any plant can thrive in a controlled environment such as greenhouse or indoors under artificial lighting, selecting varieties suited to local conditions enhances overall success and yields.

iii) **Nutrient moderation**: Avoiding nutrient excess is crucial as it can be more detrimental than nutrient deficiencies. Opting for a conservative approach to nutrient supplementation not only saves costs but also provides a buffer against potential complications, allowing growers peace of mind even during brief absences.

iv) **Shading**: Implementing appropriate shading measures is vital as excessive sunlight exposure especially during hot months can lead to issues like blossom end rot in susceptible crops such as tomatoes. Balancing light intensity through shading techniques ensures optimal growing conditions and minimizes the risk of plant stress.

v) **Preventing stem rot and suffocation**: Growers must ensure that plant stems are not submerged in the nutrient solution so that suffocation and stem rot are prevented.

In addition to these key considerations, several other factors warrant attention for successful hydroponic cultivation such as:

- ☆ **Water Quality**: Ensuring water quality meeting the standards for hydroponic systems is crucial to prevent contamination and support healthy plant growth. Nutrients appropriate for the plant should be mixed in the proper proportion in the pure irrigation water (water is usually passed through a RO-filter).
- ☆ **pH**: Maintaining the correct pH level in the nutrient solution is essential for nutrient availability and uptake by plants. The growers should maintain optimal nutrient solution composition and pH stability between 5.5 and 6.5.
- ☆ **Temperature and humidity**: Monitoring and controlling temperature and humidity levels create an environment conducive to plant growth while mitigating the risk of stress-related issues.
- ☆ **Optimizing root zone moisture for enhanced plant growth**: Many commercially available hydroponic systems offer the flexibility to adjust moisture levels in the root zone ~ a critical factor influencing plant development. Through careful management of root zone moisture growers can achieve specific outcomes tailored to their crops:

- **Encouraging essential oil production**: Maintaining a drier root zone can stimulate the synthesis of essential oils in aromatic crops such as basil and mint. This adjustment fosters conditions conducive to heightened aroma and flavour profiles so enhancing the market value of these herbs.
- **Promoting vegetative growth**: Conversely, a wetter root zone promotes robust vegetative growth particularly beneficial for large fan-leaved plants. This moisture-rich environment supports increased transpiration and photosynthetic efficiency, fostering vigorous foliage development and overall plant health.

- **Disease and pest control**: Implementing effective strategies for disease and pest management safeguards plants against potential threats, preserving crop health and vitality. Dirty growing conditions, warm water temperatures and high nutrient strengths can create favourable conditions for root diseases such as *Pythium* and *Fusarium*. Implementing proper sanitation practices, monitoring water temperature and adjusting nutrient strength can help prevent the onset of root diseases and ensure healthy plant growth.
- **Preventing algae growth**: Reservoir of nutrient solution should exclude light to prevent algal growth in the nutrient solution.
- **CO_2 levels**: Optimizing CO_2 levels enhances photosynthetic efficiency and promotes vigorous plant growth.
- **Dissolved oxygen**: Ensuring adequate dissolved oxygen levels in the nutrient solution supports root health and prevents anaerobic conditions that can lead to root rot.

These key considerations provide a comprehensive framework for optimizing hydroponic cultivation practices, promoting plant health and maximizing crop yields.

1.9 Hydroponic Components and Equipments

Hydroponic systems consist of various components that work together to support plant growth and provide essential nutrients and oxygen. Here are the key components commonly found in hydroponic setups:

i) **Nutrient reservoir**: The reservoir holds the nutrient solution which is the primary source of water and nutrients for plant growth. Reservoirs come in various sizes and materials such as plastic or stainless steel. It can be anything ~ a large tank or bucket.

ii) **Grow tray or container**: This serves as the foundational support structure for plants in hydroponic systems ~ anchoring their root systems and providing stability. These containers come in various materials such as plastic or metal and may feature channels or holes specifically designed to accommodate plants. The versatility of grow trays allows for customization according to needs of the grower, *vis-a-vis* requirements of the plants to be cultivated. Whether it is a simple pot for individual plants or larger containers for more extensive crops, the design options for grow trays are virtually limitless offering flexibility and adaptability to diverse hydroponic setups and plant varieties.

iii) **Channel:** In hydroponic systems channels serve as conduits for transporting nutrient solution from reservoir to plant roots and back for recirculation. These channels can consist of vinyl tubing or standard PVC pipes having various fittings such as connectors and adapters to ensure a secured and sealed system. Optimal channel materials include Type 400 or class 4 - 100mm PVC pipes chosen for their durability and resistance to sagging which can impede nutrient flow and oxygen delivery to the roots.

PVC pipe system

UV-resistant pipelines are preferred to withstand prolonged exposure to sunlight while painting these pipes white helps mitigate temperature increases in the nutrient solution. Given the minimal flow rate required in hydroponics which is typically 1 to 3 litres per minute, additional features like overflow pipes may be necessary to regulate flow rates effectively.

iv) **Connectors**: Here two commonly used connector types are threaded connectors and slip connectors. Threaded connectors consist of male and female components that screw into each other while slip connectors slide into place without the need for threading. Many connectors are designed as female to maintain pipe diameter allowing the male pipe to fit directly into them. Examples of connectors include 90° elbows, 45° elbows, tee fittings, ball valves, bulkheads, reducers and couplings. When connecting pipes through grow beds or nutrient reservoirs/fish tanks bulkhead fittings or uniseals are typically used.

- **Bulkheads**, available in various sizes and shapes at most home stores, serve as adapters for plumbing through the bottom of grow beds when using a standpipe. These may be purposely designed or standard male and female adapters with rubber washers to ensure a watertight seal.

Bulkheads

- **Uniseals** on the other hand are rubber rings that fit into drilled holes in tanks, clamping around the hole to create a watertight connection. PVC pipes can then be inserted into the seal which usually allows installation in only one direction ensuring a secure seal between the pipe and connector. Uniseals are cost-effective and easy to use making them ideal for plumbing into barrels and other rounded containers including those with irregular surfaces.

Plumbing with uniseal

v) **Water pumps:** Water pumps play crucial role in hydroponic systems by ensuring continuous circulation of nutrient solution from reservoir to grow trays or containers thereby providing plants with a consistent supply of water and essential nutrients. It is essential to select a pump constructed of materials that are resistant to corrosion caused by nutrient salts, typically featuring components like a stainless steel shaft, polycarbonate or stainless steel impeller, pump housing and water seal. While high head pressure is not necessary for hydroponic setups, a domestic water pump with around 0.5 HP is generally sufficient. Safety precautions are paramount when using water pumps in hydroponics due to the conductivity of nutrient solutions; thus, employing a sensitive trip switch to disconnect the electricity supply in case of emergencies is highly recommended. Water pumps can either be submersible, positioned directly within the nutrient solution tank or external depending on the specific system design and requirements.

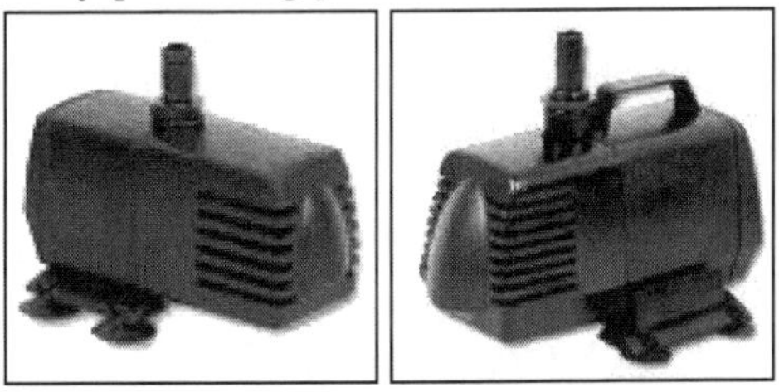

Water pumps

Submersible water pump for large systems

vi) **Growing medium**: Growing medium plays a crucial role in hydroponic systems providing structural support for plant roots, anchoring them in place and aiding in moisture and nutrient retention. While some hydroponic setups opt for media-less systems but many utilize various growing media such as rockwool, perlite, coco coir, expanded clay pebbles, growstones or vermiculite. These media offer stability to the plants and facilitate the uptake of water and nutrients. The choice of growing medium depends on factors like crop type, system design, water availability and personal preference. Experimentation with different media can help determine the most suitable option for a specific hydroponic setup.

vii) **Pots or net pots:** When setting up a hydroponic system the choice of pots or net pots can significantly impact plant growth and system efficiency. While it is possible to add growing media directly into the growing tray, using pots or net pots offers several advantages.

- Net pots, in particular, facilitate rapid flooding and filling with nutrient solution, promoting efficient nutrient uptake by plants. However, when using net pots it is essential to avoid smaller-sized growing media like perlite and sand as these may fall through the holes in the net pots.
- It is worth noting that roots often extend beyond the boundaries of net pots which can sometimes pose challenges when moving or relocating hydroponic plants. Despite this potential inconvenience, net pots are versatile and suitable for various hydroponic systems including NFT, top irrigation, aeroponics and ebb and flow systems.
- Overall, the choice between pots and net pots depends on the specific requirements of grower's hydroponic setup and the types of plants they intend to grow. They must consider factors such as ease of use, nutrient delivery efficiency and root containment when selecting the most suitable option for the hydroponic garden.

Net pot | Net pot filled with medium | Roots emerging from net pot

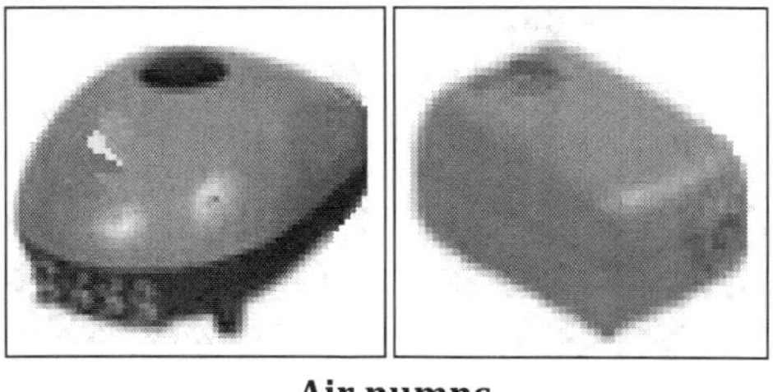

Air pumps

viii) **Aeration System**: Proper aeration is essential to ensure that the nutrient solution remains oxygenated which is crucial for healthy root development. Aeration systems such as air pumps and air stones are often used to introduce oxygen into the nutrient solution. These can be purchased from any aquarium store.

ix) **Timer or controller and oxygen detection sensor**: A timer or controller coupled with an oxygen detection sensor plays a pivotal role in regulating the delivery of nutrient solution in hydroponic systems. These devices ensure precise timing and duration of nutrient solution delivery preventing both overwatering and underwatering while promoting consistent plant growth. Typically, timers control the operation of pumps and if applicable lighting systems by activating them at predefined intervals. During initial growth stages when plants have lower oxygen requirements the circulating time of nutrient solution can be restricted to conserve energy. This can be achieved manually using a timer to set circulating intervals or by incorporating an oxygen concentration detection sensor. The sensor activates the pump whenever the oxygen concentration in the nutrient solution decreases, optimizing oxygen levels and promoting healthy root development.

Timer

x) **Other equipments**

- **pH and EC/ppm meters**: pH meters are used to monitor and adjust the acidity or alkalinity of the nutrient solution while EC or ppm meters measure the concentration of nutrients. Maintaining proper pH and nutrient levels is critical for plant health.
- **Nurtimeter**: In addition to the EC, this meter helps in measuring the nutrient contents of the solution.

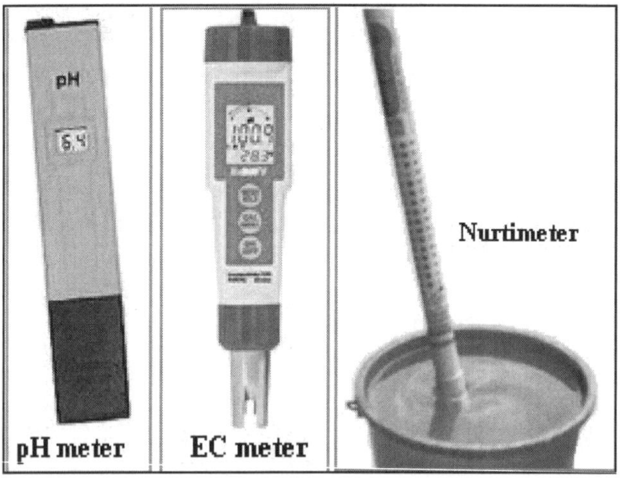

- **Lighting system**: Light plays a crucial role in the growth and development of plants as it serves as a key component in the process of photosynthesis. Various types of grow lights are available including compact fluorescent lighting (CFL's), LED and HID lights, each offering unique advantages and suitability for different applications.

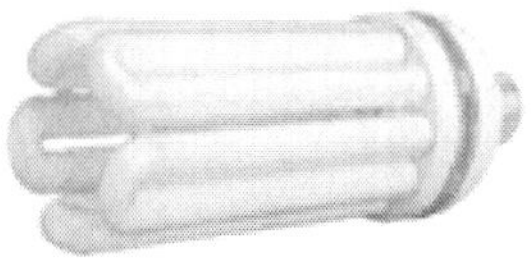

125 watt CFL (Compact Fluorescent Bulb)

- **Light timers**: The feed times and duration along with the grow lights coming on and going off becomes fully automatic after the installation of good timers. Alarms can also be installed to monitor pH and ppm of the nutrient reservoir.
- **Heater:** It is essential for maintaining the optimal temperature of the nutrient solution especially in colder climates or during winter months. Small-scale hydroponic setups can effectively utilize aquarium heaters to regulate the temperature within the nutrient reservoir, ensuring an ideal environment for plant growth and nutrient uptake.
- **Cooling fans or chillers:** Shade cloth, cooling fans or chillers can mitigate overheating issues in summer months.
- **Support structures**: Support structures such as trellises play a crucial role in hydroponic systems due to absence of natural anchorage provided by soil. With the growing medium largely eliminated in liquid cultures, it becomes essential for growers to provide artificial supports to ensure the stability and proper growth of plants. This necessity is particularly pronounced when cultivating tall indeterminate crop varieties like tomatoes and cucumbers or crops with relatively heavy fruits such as bell peppers and eggplants. By implementing trellises and similar structures, growers can train plants to grow along these supports, facilitating optimal growth and fruit development in hydroponic environments.

Cucumber vines trained on trellises

- **Blowers or vibrating pollinators:** Blowers and vibrating pollinators serve as vital tools for promoting pollination within protected structures. Blowers are designed to generate airflow that gently shakes plants aiding in the distribution of pollen and enhancing pollination efficiency. These devices play a crucial role in facilitating pollination processes particularly within enclosed environments where natural pollination may be limited. Meanwhile, vibrating pollinators are compact electrical devices that apply controlled vibrations to individual plants, further promoting pollination within protected structures. By simulating natural pollination mechanisms, these tools contribute to optimal pollination rates and support healthy plant development within greenhouse and indoor farming settings.

Blower

These are the basic components of a hydroponic system, but there can be variations and additional components depending on the specific type of system and the needs of the plants being grown.

1.10 Advantages of Hydroponic Cultivation

Hydroponic cultivation has emerged as a leading method for maximizing biomass and protein production in plants particularly when compared to traditional growth substrates. Its predictable and off-season production is almost free from pests, weeds and soil-borne diseases which results into healthy and hygienic crops produced in a limited area with minimal effort due to well-controlled nutrition levels. This technique offers numerous advantages including enhanced water efficiency which has significant implications for sustainable agriculture and food security. Hydroponics holds promise for addressing global challenges such as climate change and resource scarcity paving the way for innovative solutions in agricultural production both on earth and in space.

Representing a paradigm shift in modern agriculture, hydroponics offers a sustainable and efficient alternative to traditional soil-based farming methods. By harnessing the power of water and nutrients, hydroponic systems empower growers to overcome geographical limitations, optimize resource utilization and cultivate high-quality crops in controlled environments. As the global demand for fresh produce continues to rise hydroponics emerges as a key enabler of food security, environmental stewardship and innovation in the agricultural sector. Through education and adoption of hydroponic techniques we can pave the way for a greener and more resilient future of farming, addressing challenges such as increased biomass and protein production, enhanced water efficiency, resource conservation and climate resilience. By harnessing the power of hydroponics, we can cultivate a greener *cum* more sustainable future for generations to come.

- ☆ **Maximizing biomass and protein production:** Hydroponic cultures consistently yield the highest biomass and protein when compared to plants grown in other substrates under similar environmental conditions (Nagel *et al.,* 2010). By providing plants with precisely calibrated nutrient solutions and optimal growing conditions, hydroponic systems optimize nutrient uptake and metabolic processes resulting in superior growth and productivity.
- ☆ **Faster growth rates**: Plants grown hydroponically often exhibit faster growth rates compared to traditional soil-based cultivation. The precise control over environmental factors such as temperature, humidity and nutrient levels promote optimal growth conditions resulting in shorter crop cycles and increased productivity. It takes a long time to grow seedlings in traditional agriculture but soilless culture can reduce the time substantially as this process can be performed in a concentrated manner.
- ☆ **Water efficiency and resource conservation:** One of the most compelling benefits of hydroponic cultivation is its remarkable water efficiency. Traditional soil-based intensive farming methods consume vast quantities of water with 214 litres required to produce just 1 kilogram of tomatoes (The Guardian, 2024). In contrast, the recirculation of nutrient solutions in hydroponics reduces water consumption so making hydroponics a more sustainable option especially in regions with water scarcity. Hydroponics reduces water usage to 70 liters per kilogram of tomatoes while aeroponics achieves an astonishingly low usage of only 20 liters (Zhang *et al.,* 2018). This significant reduction in water consumption not only conserves precious resources but also mitigates the environmental impact of agricultural practices which makes hydroponics a sustainable alternative for food production.
- ☆ **Optimized nutrient delivery**: In hydroponics, nutrients are delivered directly to the plant roots in a soluble form ensuring efficient uptake and utilization by plants. More equal supply of nutrient solution can be achieved and so more homogeneous crops can be obtained. This targeted delivery minimizes nutrient waste and reduces the need for excess fertilization, resulting in cost savings. All macro- and micro-nutrients can be used efficiently.
- ☆ **Space utilization**: Hydroponic systems can be designed to maximize space utilization allowing for higher plant densities per square metre area. Vertical farming techniques further increase growing area enabling more crops to be grown in a smaller footprint.

- **Reduced labour costs**: Automation and control systems in hydroponics can minimize labour requirements such as watering, fertilizing, weeding, digging or ploughing. This reduces the need for manual labour leading to lower operational expenses.
- **Year-round production**: Hydroponic systems can be implemented indoors or in controlled environments allowing for year-round crop production regardless of seasonal variations. This continuous production can lead to higher yields and increased revenue streams.
- **No crop rotation**
- **Addressing climate change and resource scarcity:** As climate change threatens agricultural yields and exacerbates water scarcity in many regions, hydroponic cultivation offers a promising solution. By requiring less water and nutrients to produce healthy crops, hydroponics enables cultivation in harsh environments with limited access to water. Hydroponics has the potential to empower communities to grow their own plant-based food sustainably even in resource-constrained settings (Gericke, 1938; Boylan, 2020).
- **Reduced pest and disease pressure**: Hydroponic systems can be designed to minimize pest and disease pressures which reduces the reliance on pesticides and fungicides. This not only lowers input costs but also improves crop quality and marketability.
- **Reduced weed pressure**: Without presence of soil, hydroponic systems experience minimal weed pressure which reduces the need for herbicides and manual weed control methods. This not only saves time and labour but also minimizes soil disturbance and erosion.
- **Crop diversity and specialty markets**: Hydroponics allows for cultivation of a wide variety of crops including specialty or high-value crops that may command premium prices in niche markets. Growers can capitalize on market demand for specific crops, further enhancing profitability.
- **Soil conservation**: Traditional agriculture often leads to soil degradation and erosion due to intensive tillage and chemical usage. Organic hydroponics circumvents these issues by eliminating soil cultivation altogether, thus preserving soil structure, fertility and biodiversity.
- **Applications beyond earth:** Hydroponic cultivation is not confined to terrestrial environments; it has also demonstrated its efficacy in space exploration. Hydroponic systems can support plant production experiments in space, offering astronauts a renewable source of fresh food during long-duration missions. This versatility underscores the adaptability and resilience of hydroponic technology, positioning it as a

vital component of future space habitats and extraterrestrial colonization efforts (Stutte *et al.*, 2012).

1.11 Challenges in Scaling Soilless Crop Production

Soilless cultivation has emerged as a promising solution for home gardening and small-scale commercial operations, offering efficient resource utilization and enhanced crop yields. However, when transitioning to large-scale commercial setups, several challenges arise that need to be addressed:

i) **High initial investment costs:** Implementing soilless cultivation on a large scale requires significant upfront investment in infrastructure, equipment and technology. The costs associated with constructing greenhouses, installing hydroponic or aquaponic systems and implementing automated controls can be prohibitive for many growers. This financial barrier may limit the adoption of soilless techniques, particularly for small and medium-sized farmers.

ii) **Disease management and root health:** Large-scale soilless cultivation is susceptible to root diseases caused by pathogens such as *Phytophthora*, *Pythium* and *Fusarium* spp. Unlike traditional soil-based farming where diseases are often confined to specific areas, soilless systems circulate nutrient solutions throughout the entire growing area. This means that once root rot or other diseases occur, they can quickly spread jeopardizing entire crops and resulting in significant financial losses. Effective disease management strategies such as maintaining strict sanitation protocols, implementing crop rotation and utilizing disease-resistant cultivars are essential for mitigating these risks.

iii) **Nutrient solution management:** Maintaining optimal nutrient levels and pH balance in soilless systems presents a continuous challenge for growers. The composition and concentration of nutrient solutions can fluctuate over time leading to imbalances that negatively impact plant health and productivity. While growers typically monitor nutrient levels using pH meters and electric conductivity meters, these tools do not provide insight into changes in solution composition. Moreover, the practice of replacing nutrient solutions periodically either partially or entirely to maintain uniformity is costly and inefficient. Innovations in nutrient monitoring technology and the development of tailored nutrient formulations can help address these challenges, allowing growers to optimize nutrient management and maximize crop yields.

iv) **Water management and sustainability:** Large-scale soilless cultivation requires significant water inputs to sustain crop growth and maintain optimal environmental conditions. Efficient water management practices such as recirculating irrigation systems and water-saving technologies are essential for minimizing water usage and reducing environmental impact.

Sustainable water sourcing strategies such as rainwater harvesting and wastewater recycling can help alleviate pressure on freshwater resources and promote long-term environmental sustainability.

v) **Energy consumption and environmental impact:** The energy requirements associated with heating, lighting and powering large-scale soilless cultivation facilities can be substantial, contributing to environmental emissions and operating costs. Implementing energy-efficient technologies such as LED lighting, solar panels and heat recovery systems can help reduce energy consumption and mitigate environmental impact. Furthermore, adopting sustainable cultivation practices such as organic nutrient formulations and integrated pest management can minimize the use of synthetic inputs and promote ecological balance.

vi) **Failure to the hydroponic system leads to rapid plant death**: Failure of the hydroponic system refers to any malfunction or breakdown in the components or processes involved in hydroponic cultivation. This could include issues such as pump failure, nutrient imbalance, clogging of irrigation lines or environmental factors such as temperature fluctuations or inadequate lighting. When the hydroponic system fails it disrupts the supply of essential nutrients, water and oxygen to the plants ~ leading to rapid deterioration in plant health and ultimate death. Since hydroponic plants rely entirely on the system for their survival, any failure can have immediate and severe consequences, resulting in the loss of the entire crop if not addressed promptly.

☆ **Resource intensiveness**: The requirement for different fertilizers and containment systems in hydroponic agriculture presents a notable challenge adding complexity to nutrient management and cultivation practices. With various plant species exhibiting diverse nutrient needs at different growth stages, maintaining optimal nutrition levels becomes intricate, demanding meticulous monitoring and adjustment of nutrient solution compositions. Certain hydroponic plants may necessitate customized fertilizer formulations tailored to their specific requirements, escalating the sophistication and expertise required in nutrient management compared to traditional farming methods. Accommodating these diverse nutritional needs often entails investing in specialized fertilizers and additives, further intensifying resource demands. Moreover, adapting containment systems to suit unique requirements of each plant species adds another layer of complexity, necessitating careful consideration of system configurations and structural support. As a result, while hydroponic systems offer numerous benefits including water efficiency and year-round production, the need for tailored fertilizers and containment systems poses challenges in terms of management complexity, resource allocation and operational expertise.

1.12 What can be grown Hydroponically?

Hydroponics offers a versatile and efficient method of growing a wide range of crops without soil, utilizing nutrient-rich water solutions to deliver essential elements directly to plant roots. Almost any plant that can be grown in soil can also be cultivated hydroponically, making it suitable for various climates, urban environments and limited spaces. Here's a detailed exploration of what can be grown hydroponically:

i) **Leafy greens**: Leafy greens such as lettuce, spinach, kale, arugula, Swiss chard and mustard greens are among the most popular crops for hydroponic cultivation. Their shallow root systems and rapid growth rates make them well-suited for hydroponic systems where these can quickly absorb nutrients and water for lush foliage production.

ii) **Herbs**: A diverse array of herbs including arugula, basil, chervil, chives, coriander, dill, lemon balm, mache, mint, oregano, rosemary, watercress, parsley, sage, sorrel, sweet marjoram, tarragon and thyme thrive exceptionally well in hydroponic systems, often exhibiting robust growth even in densely planted environments. Known for their rapid maturation, low maintenance requirements and versatile culinary and medicinal applications, herbs are ideally suited for hydroponic cultivation. With minimal care, hydroponic herb gardens yield impressive harvests, making them well-suited for both commercial and home-based hydroponic setups. Herbs that naturally prefer moist conditions tend to fare particularly well in hydroponic environments contrasting with those adapted to drier climates like sage.

iii) **Fruiting vegetables**: Many fruiting vegetables can be grown hydroponically, including tomatoes, peppers, cucumbers, eggplants, zucchinis and strawberries. These crops require adequate support structures and careful management of nutrients and environmental conditions to achieve optimal yield and fruit quality.

iv) **Fruits**: Small and light fruiting plants thrive well in a hydroponics system. Water-loving fruits such as watermelon, cantaloupe, tomatoes, strawberries, blueberries, blackberries, raspberries and grapes are the best choices to be grown hydroponically.

v) **Fruit trees**: Small fruit trees such as dwarf banana, papaya, citrus and lemons may be grown well in a hydroponic greenhouse. A large and heavy fruit tree can be started in a hydroponics system and when it outgrows its container it can be transplanted to the soil. To grow a large tree hydroponically it requires a very well built-up support system.

vi) **Root vegetables and other geophytes**: Root vegetables and other geophytes such as carrots, beets and radishes which are traditionally cultivated in soil can also be grown hydroponically using methods like

DWC or NFT. However, cultivating these crops in hydroponic systems may require specialized setups and careful management to ensure optimal root zone oxygenation. For instance, root crops like beets, carrots and radishes demand ample space for root development necessitating deep containers or channels. The depth of channels should be adjusted based on the size of the crop with larger crops such as potatoes requiring deeper channels (20-25cm) compared to smaller ones such as carrots (7.5-10cm). Moreover, root crops typically have higher potassium requirements. However, it is worth noting that succulent plants which thrive in dry conditions may not be suitable for hydroponic gardening due to their specific moisture needs.

vii) **Microgreens**: Microgreens are young tender greens harvested at an early stage of growth, typically within 7-21 days after germination. Popular microgreens for hydroponic cultivation include radish, broccoli, mustard, sunflower and pea shoots. Hydroponic systems provide optimal conditions for rapid and uniform growth of microgreens which are prized for their intense flavours, vibrant colours and high nutritional content.

viii) **Flowers**: Certain flowers lend themselves well to hydroponic cultivation serving both ornamental and commercial purposes such as orchids, African violets, chrysanthemums, gerbera, daisies and roses. Hydroponic systems offer precise control over nutrient delivery, water availability and environmental conditions ~ fostering optimal conditions for healthy growth and abundant flowering. Seedlings of various flower varieties can be initiated within hydroponic units allowing for controlled development until maturity. Once matured, these plants can be transplanted into other hydroponic units or even transferred to soil-based cultivation. While some ornamentals may be perennials, flowering is often seasonal, however, hydroponic methods enable year-round production of flowers in significant quantities ~ ensuring a steady supply regardless of external conditions.

ix) **Fruit trees and vines**: Though less common but certain fruit trees and vines can be grown hydroponically on a small scale or in controlled environments such as dwarf citrus trees, strawberries, grapes and passion fruit vines. Hydroponic systems can be tailored to root support, trellising and nutrient supplementation to accommodate the specific needs of these crops.

x) **Specialty crops**: Hydroponics enables the cultivation of specialty crops such as exotic salad greens, culinary herbs, medicinal plants and rare or endangered species that may have unique environmental requirements or limited availability in certain regions. Hydroponic techniques can be adapted to suit the precise needs of these crops allowing for sustainable production and conservation efforts.

1.13 Considerations for Selecting Plants

When selecting plants to grow hydroponically, several considerations should be taken into account to ensure successful cultivation and optimal yields. Here are some key factors to consider:

i) **Environmental requirements**: Only those plants are selected that match the environmental conditions of the hydroponic setup, *i.e.* temperature, humidity, light intensity and pH levels. Different plants have varying requirements so it is essential to select species that thrive within your available parameters.

ii) **Space and growth habit**: Available space in the hydroponic system and the growth habits of the plants we intend to grow should be taken into consideration before growing. Some vining plants such as tomatoes or cucumbers may require trellises or vertical support structures while bushy or compact plants such as lettuce or herbs may be grown in smaller spaces.

iii) **Nutrient requirements**: Different plants have distinct nutrient requirements so it is crucial to select species that are compatible with the chosen hydroponic nutrient solution. Some plants may require higher levels of specific nutrients though others lower requirements or be more sensitive to nutrient imbalances.

iv) **Harvesting and yield**: Harvesting requirements and yield potential of the plants should be considered for the plants to be grown hydroponically. The crops such as leafy greens or herbs can be harvested multiple times throughout their growing season while other fruit-bearing crops may have longer maturation periods with offering of higher yields per plant.

v) **Growth cycle and time to maturity**: Growth cycle and time to maturity of the plants should also be taken into account while growing hydroponically. Some crops such as microgreens or lettuce have rapid growth rates and can be harvested within a few weeks while others such as tomatoes or peppers may have longer growing seasons requiring more time to reach maturity.

vi) **Disease and pest resistance**: Choose plant varieties that are known for their disease and pest resistance to minimize the risk of problems in your hydroponic system. Selecting resistant cultivars can help reducing the need for chemical treatments so ensuring healthier *cum* more productive plants.

vii) **Culinary or market demand**: Culinary or market demand of the plants one intends to grow should be taken into consideration while growing the crops hydroponically. One should focus on popular or high-value crops that have a consistent demand and are well-suited for hydroponic cultivation. Local market trends and consumer preferences require to be investigated as a guide for selection of plants.

viii) **Personal preferences and experience**: Lastly, personal preferences and the level of experiences should also be considered with hydroponic gardening. Only those plants are chosen that one enjoys growing and consuming, *vis-a-vis* that aligns with ones skill level and available resources so one should first, experiment with different plant varieties and hydroponic techniques to find what works best.

By considering these factors when selecting plants to grow hydroponically, one can create a diverse and productive hydroponic garden that meets ones needs and preferences while maximizing yields and success.

1.14 Optimizing through Plant-Specific Requirements

Grouping plants with comparable nutritional demands and environmental conditions fosters streamlined hydroponic systems, enhancing efficiency and plant health. To optimize hydroponic systems for diverse crops, it is essential to tailor setups to meet the specific needs of each species. Mixing plants with differing requirements within the same unit is at all not advisable:

- Roses require elevated potassium levels unlike most other plants.
- Leafy greens thrive with higher nitrogen content while root crops benefit from increased potassium.
- Conversely, fruit-bearing plants such as tomatoes, cucumbers, eggplants and peppers prefer lower nitrogen levels, *vis-a-vis* similar light requirements. All these thrive in warm temperatures with ample sunlight/ artificial lighting.
- Broccoli, cabbage and cauliflower have similar nutrient needs and prefer cooler temperatures so allowing them to be grown together. However, cabbage requires slightly higher nitrogen and lower light levels compared to broccoli and cauliflower.

Overall, the potential for hydroponic cultivation is vast and continually expanding as new technologies and crop varieties are coming up day by day. Whether for commercial farming, urban agriculture, home gardening or research purposes, hydroponics offers a flexible and efficient solution for growing a diverse array of crops year-round regardless of climate or soil limitations.

Chapter 2

Passive Hydroponics Systems

Passive hydroponics systems operate without any need of mechanical components for circulation of nutrient solution. Instead, these harness natural forces such as capillary action or wicking as is particularly evident in **passive sub-irrigation** systems. In these setups, water and nutrients are drawn upward through the growing medium or *via* a separate wicking material such as rope or fabric which transports the nutrient solution from reservoir to plant roots. In some passive hydroponic designs, plant roots are directly suspended in the nutrient solution eliminating the need for a growing medium entirely. Nutrient solution for plants is kept in the container which acts as a plant reservoir and the plants are either suspended directly through the holes on the lid of the container or nets/net pots are suspended in the holes that would hold the plants.

By leveraging these natural processes passive hydroponics offers a low-maintenance and efficient method of providing plants with essential nutrients. These systems are simpler and often require less maintenance but these offer less control over the growing environment.

2.1 Classification of Passive Hydroponic Systems

Passive hydroponic systems can be classified based on various factors such as method of nutrient delivery, use of growing media and presence or absence of mechanical components. Here is a classification based on these factors:

2.1.1 Use of Growing Media

i) Media-based

Plants are grown in a solid medium such as perlite, vermiculite, coconut coir or rockwool which provides support to the roots and holds moisture and nutrients such as wick systems and the Kratky method using a growing medium.

- ☆ **Advantages**: Passive medium culture hydroponic systems offer ease of maintenance with watering and feeding involving simply topping up the reservoir of growing solution as water requirement of the plant is easily observable. These systems provide more precise control over nutrient levels and pH compared to water culture systems as the growing medium acts as a buffer and reservoir for nutrients. Moreover, these accommodate a wide range of growing media including perlite, vermiculite, coconut coir and rockwool ~ offering flexibility for different plant species and growth requirements. The use of a growing medium also enhances stability and support for plant roots, reducing the risk of damage or instability especially for larger plants. Moreover, hydroponic media are typically resistant to soil-borne insects and their open aggregate structure facilitates air circulation around the roots so providing increased oxygenation. The absorbent properties of the aggregate help maintaining humidity around the plant and reducing the occurrence of root rot. Overall, passive medium culture systems can significantly reduce the labour required to maintain a large collection of plants.
- ☆ **Disadvantages**: Passive medium culture systems often entail higher initial costs due to need for additional materials and equipments including growing media and containers as compared to simpler water culture setups. Although these require less maintenance than some active hydroponic systems, passive medium culture setups still demand periodic monitoring and adjustment of nutrient levels and pH to ensure optimal plant growth. Over time, the growing media in passive medium culture systems may degrade or compact, diminishing their effectiveness and necessitating replacement or replenishment. However, plants that require intervals between watering and a dry dormant period may struggle in environments with constant moisture. While over-irrigation combined with well-draining media prevents the growing media from drying out, it can lead to insufficient air porosity, potentially depriving roots of oxygen. To mitigate salt accumulation, periodic washing of the system is recommended.

ii) Media-less or Water Culture

Plants are grown without a solid medium with roots of plants suspended in a nutrient solution such as **Static Solution Culture**.

- ☆ **Advantages:** Passive water culture hydroponics systems are straightforward and easy to set up, making them ideal for beginners or those with limited experience in hydroponic gardening. These systems typically require minimal equipment and materials, resulting in lower initial investment costs compared to more complex hydroponic systems. With no moving parts or complex mechanisms, passive water culture systems have minimal maintenance requirements, reducing the need for regular monitoring and adjustments. Plant roots are continuously submerged in oxygen-rich nutrient solution promoting vigorous root growth and nutrient uptake.

☆ **Disadvantages:** If the water level is not properly maintained there is a risk of oxygen depletion in the root zone which can lead to root rot and plant health issues. Passive water culture systems may have limited control over nutrient levels and pH as the nutrient solution is not actively circulated or aerated. Larger or heavier plants may require additional support to prevent them from sinking into the nutrient solution which can be challenging to manage in some setups.

2.1.2 Nutrient Delivery Method

Sub-irrigation and top-irrigation represent two main variations of medium culture passive hydroponics ~ each offering unique advantages and considerations. Whether utilizing the simplicity of sub-irrigation or the precision of top-irrigation both the methods provide an effective means of cultivating plants without need for complex irrigation systems or active components. With proper planning and maintenance, medium culture passive hydroponics systems can yield healthy and vibrant crops in a variety of growing environments.

i) Sub-irrigation

In sub-irrigation passive hydroponics (submersion or bottom-irrigation), nutrient solution is delivered to the plant roots *via* capillary action or wicking from a reservoir below the root zone. It involves placing the plant roots in a growing medium and allowing them to draw water and nutrients from a reservoir located beneath the medium. This method capitalizes on capillary action or wicking to transport the nutrient solution upward into the root zone which ensures continuous hydration and nourishment for the plants as in case of wick systems and capillary mat systems.

☆ **Growing media**: In sub-irrigation systems the growing medium acts as a conduit for water and nutrients facilitating their upward movement through the force of capillary action. Common growing media used in sub-irrigation setups include perlite, vermiculite, coconut coir or a combination thereof. These media are chosen for their ability to efficiently transfer moisture and nutrients to plant roots while maintaining adequate aeration.

☆ **Advantages**: One of the key advantages of sub-irrigation passive hydroponics is its simplicity and ease of implementation. With no need for complex irrigation systems or active components, sub-irrigation setups are straightforward to set up and maintain, making them suitable for novice growers or those seeking a low-maintenance gardening solution.

- ☆ **Disadvantages**: However, sub-irrigation systems also present certain considerations. Over time, the growing medium may become compacted or degraded so reducing its effectiveness in transporting water and nutrients to the roots. Sub-irrigation systems require periodic monitoring to ensure proper hydration levels and nutrient balance as stagnant water in the reservoir can lead to issues such as root rot or nutrient imbalances.

ii) Top Irrigation

Top-irrigation passive hydroponics (drip or trickle irrigation) involves delivering water and nutrients directly to the surface of the growing medium, allowing them to percolate downward to the plant roots. In this method, a reservoir containing the nutrient solution is positioned above the growing medium and a network of tubing or channels disperses the solution evenly across the medium's surface.

- ☆ **Advantages**: Top-irrigation systems offer several advantages such as precise control over water and nutrient delivery, reduced risk of overwatering and improved aeration of the root zone. By delivering water directly to the top of the medium these systems ensure efficient hydration of the plant roots while minimizing water wastage. Top-irrigation systems are highly adaptable and can accommodate a wide range of growing media and plant species. From soilless mixes like perlite or vermiculite to organic substrates such as coco coir or peat moss, top-irrigation systems provide flexibility in cultivation methods and growing environments.
- ☆ **Disadvantages**: However, top-irrigation passive hydroponics also presents certain challenges. Maintaining uniform moisture levels across the growing medium can be tricky especially in larger systems or those with uneven topography. The constant exposure of the growing medium to water and nutrients can lead to nutrient leaching or salt buildup over time, necessitating regular monitoring and adjustment of nutrient concentrations.

2.1.2 Presence of Mechanical Components

- i) **Passive:** Nutrient solution is delivered to roots of the plants without use of mechanical components such as wick systems and the Kratky method.
- ii) **Semi-passive:** Nutrient solution is delivered using simple mechanical components like timers or gravity but without active pumps or aerators, *e.g.* some variants of drip systems or certain top irrigation setups.

These classifications provide a framework for understanding the diversity of passive hydroponic systems, each offering unique advantages and limitations suited to different plants, growing environments and preferences of the grower.

2.2 Static Solution Culture or Root Dipping Hydroponics

Static solution culture [root dipping technique (RDT)] is a non-circulating or passive hydroponic method that offers a straightforward and efficient method for growing plants in a stationary nutrient solution. With proper attention to nutrient management, aeration and system maintenance, growers can achieve successful harvests and enjoy the benefits of hydroponic gardening. This technique offers simplicity and versatility, making it suitable for both novice and experienced hydroponic enthusiasts. In its simplest form, plants are positioned on the surface of the nutrient solution and are left to float akin to lily pads. For larger or heavier plants it is advisable to utilize a support structure or floating device to prevent them from sinking to the bottom of the reservoir.

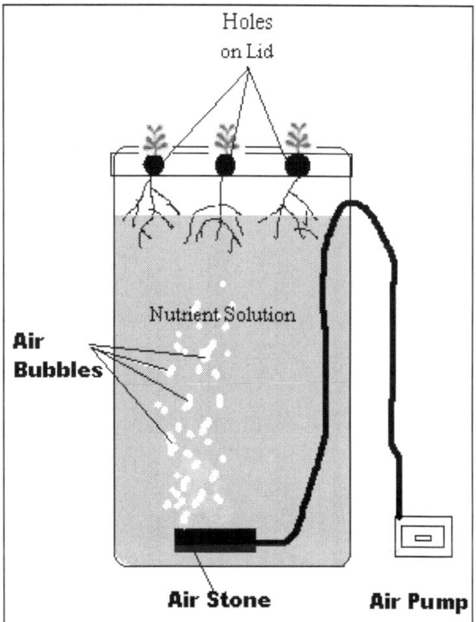

Static Solution Culture Hydroponics.

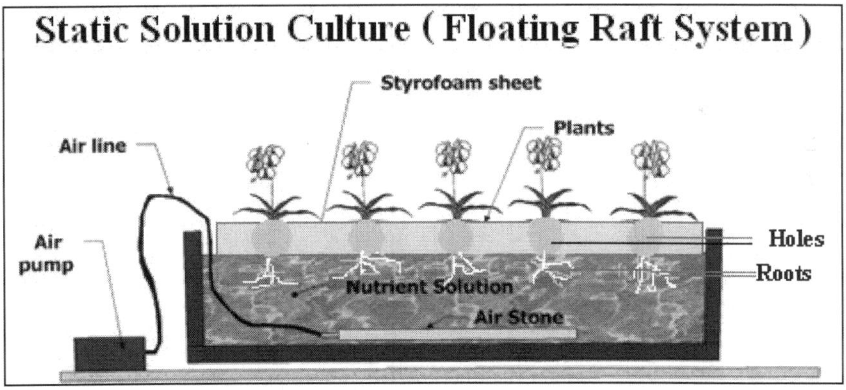

2.2.1 Key Components and Operation

2.2.1.1 Container Selection

Various containers can serve as reservoirs for nutrient solution including aquarium tanks, concrete tanks, buckets, tubs or even household jars. Each plant in the system requires a hole to be precisely cut or drilled into the top of the reservoir. Alternatively, nets or net-pots are suspended in the holes to support plants. In case of a jar or tub, the lid may serve as the top but for other container materials such as cardboard, foil, paper, wood or metal can be utilized to cover the opening securely. Depending on preference and space availability, a single reservoir can cater to a solitary plant or accommodate multiple plants simultaneously. As the plants grow and their nutrient requirements increase, the size of the reservoir can be adjusted accordingly to sustain optimal growth.

The depth of the container should be sufficient (at least 25-30cm) to provide an adequate nutrient solution level while leaving space above for oxygen-absorbing roots. Light exposure can promote algal growth which can compete with plants for nutrients and oxygen. Clear containers should be painted black or shielded using aluminum foil, butcher paper, black plastic or alternative materials to prevent light from reaching the nutrient solution and to mitigate the impacts of negative phototropism.

2.2.1.2 Aeration

The typical approach involves gently aerating the solution though it can alternatively remain un-aerated (Nagel *et al.*, 2010). Air stones and diffusers are commonly used to introduce oxygen into the nutrient solution. These devices release tiny bubbles of air into the nutrient solution, promoting oxygen exchange. Moreover, a natural biofilm of green algae on the inner surface of the container can aid in oxygenation through photosynthesis ~ contributing to the overall health of the plants. Proper aeration ensures healthy root development and nutrient uptake.

In systems where nutrient solution is not aerated, it is advisable to maintain a lower level of solution to ensure that an adequate portion of the roots remains above it, *i.e.* the roots may touch the solution but not fully submerged. This positioning allows the roots to access sufficient oxygen from the air. This approach contrasts with deep water culture (DWC) where plant roots are consistently submerged in a nutrient solution that is highly oxygenated.

2.2.1.3 Diffusers

Diffusers in static solution culture hydroponics work by introducing air into the nutrient solution in the form of microbubbles which oxygenate the water and provide a vital source of oxygen to the plant roots thus promoting healthy growth and nutrient uptake. Here is how diffusers work in this context:

i) **Introduction of oxygen**: Diffusers are typically porous stones or tubes placed at the bottom of the nutrient solution reservoir. These are connected to an air pump located outside the reservoir. When air pump is turned on it pumps air through the diffuser.

ii) **Creation of microbubbles**: As air is forced through the porous material of the diffuser, it breaks into tiny bubbles. These bubbles rise through the nutrient solution carrying oxygen from the air with them.

iii) **Oxygenation of nutrient solution**: As the microbubbles rise through nutrient solution these release oxygen which dissolves into the water.

This oxygenation process ensures that nutrient solution remains well-aerated thus providing a constant supply of oxygen to plant roots.

iv) **Prevention of anaerobic conditions**: Oxygen is essential for root respiration and nutrient uptake. By continuously oxygenating the nutrient solution, diffusers help prevent the development of anaerobic conditions in which harmful microorganisms can thrive. This promotes healthier root growth and overall plant vigour.

v) **Even distribution of oxygen**: Diffusers help ensure that oxygen is evenly distributed throughout the nutrient solution reservoir. This is important for maintaining consistent oxygen levels across all parts of the root system regardless of the size or shape of the container.

vi) **Maintenance**: Regular cleaning and maintenance of diffusers are essential to prevent clogging and ensuring optimal performance. Over time, mineral deposits and organic matter may accumulate on the diffuser so reducing its effectiveness. Cleaning the diffuser with a mild detergent or vinegar solution can help remove these deposits so maintaining proper oxygenation.

2.2.2 Nutrient Solution Management

As plants utilize nutrients from the solution, nutrient concentrations gradually decline over time. Regular monitoring of nutrient levels, pH and EC is essential for maintaining optimal growing conditions. Nutrient solution should be changed or supplemented when levels drop below a certain threshold as indicated by EC readings or plant progress.

Management of nutrient solution involves a systematic approach primarily governed by a predetermined schedule ~ typically weekly or by monitoring its concentration using an electrical conductivity meter. When nutrient concentration falls below a specified threshold, corrective measures are taken by supplementing either water or fresh nutrient solution. Automation can be achieved through utilization of devices such as Mariotte's bottles or float valves which ensure the maintenance of the solution at the desired level without manual intervention.

Note: The principle behind Mariotte's bottle in static solution culture hydroponics is based on hydrostatic pressure. Here's how it works:

i) **Constant water flow**: Mariotte's bottle (constant head reservoir) utilizes the principle of hydrostatic pressure to maintain a consistent water level within the hydroponic system. The bottle consists of a reservoir divided into two chambers ~ a larger upper chamber and a smaller lower chamber connected by a narrow tube or pinhole.

ii) **Hydrostatic pressure**: When reservoir is filled with nutrient solution the weight of liquid creates pressure at the bottom of the container. This pressure pushes the solution up through the narrow tube into the upper chamber.

iii) **Maintaining water level**: As water is consumed by the plants or evaporates from the system, the level of nutrient solution in the main reservoir decreases. However, due to the constant pressure exerted by the column of liquid in the Mariotte's bottle, a steady flow of solution is maintained into the upper chamber which ensures a consistent water level regardless of fluctuations in the main reservoir.

iv) **Stability and consistency**: By keeping the water level constant, Mariotte's bottle helps to stabilize the hydroponic environment. This stability is crucial for promoting optimal nutrient uptake by the plants as well as preventing fluctuations in pH and nutrient concentrations that can negatively impact plant growth.

v) **Simple design**: Mariotte's bottle is a simple and effective solution for maintaining water levels in static solution culture hydroponics. It requires minimal maintenance and can be easily implemented into hydroponic systems of various scales.

In raft solution culture ~ a method employed in hydroponics, plants are strategically positioned on a buoyant plastic sheet which is floated atop the nutrient solution's surface. This arrangement guarantees that the solution remains consistently at a level adequate to sustain root systems of plants thus fostering optimal growth conditions and minimizing the risk of nutrient deprivation (Suryawanshi, 2021).

2.2.3 Temperature Regulation

While plants can thrive at room temperature in most conditions, an aquarium heater may be used to adjust the temperature of the nutrient solution especially in colder climates or during winter months.

- ☆ In static solution culture, the nutrient solution remains relatively still which can make temperature control more challenging compared to systems with active circulation.
- ☆ To manage temperature in static solution culture, it is important to maintain overall ambient temperature of the growing environment within the optimal range for plant growth ~ typically between 18°C and 24°C.
- ☆ Temperature of the nutrient solution should be controlled by regulating temperature of the room or greenhouse where hydroponic setup is located. This can be done using heating or cooling devices such as heaters, fans or air conditioning units.
- ☆ The reservoir of the nutrient solution should be insulated to help stabilize its temperature and prevent fluctuations caused by external environmental conditions.

- Regular monitoring of temperature is essential using a thermometer placed in the nutrient solution to ensure that it remains within the optimal range for plant growth.

2.2.4 pH management

pH management is essential in almost all types of hydroponic systems to ensure optimal nutrient uptake and plant growth. It is important to maintain pH within the optimal range for nutrient uptake typically between 5.5 and 6.5 for most hydroponic crops. However, the methods for managing pH may differ slightly between different systems:

- In static solution culture, pH management typically involves periodic monitoring and adjustment of the pH level of the nutrient solution.
- Since the nutrient solution remains relatively still in static solution culture systems, pH fluctuations may occur more slowly compared to systems with active circulation.
- pH adjustment in static solution culture can be achieved by using pH testing kits or meters to monitor the pH level of the nutrient solution.
- pH can be adjusted by adding pH-up or pH-down solutions to raise or lower the pH, respectively to desired range. It is essential to follow manufacturer instructions and make gradual adjustments to avoid sudden pH swings that could shock the plants.
- Using pH-buffering agents or adjusting the composition of the nutrient solution can help stabilize pH levels over time.

2.2.5 System Maintenance

Stagnant water increases the risk of algae growth and root rot. Regular maintenance including removal of rotting roots is essential for system health. Plant support structures or floating devices may be necessary for heavier plants to prevent sinking and to ensure proper nutrient uptake.

2.2.6 Challenges and Considerations

Providing uniform aeration rates, pH, EC and nutrient levels in each container are the challenging tasks particularly as plants grow larger. Careful monitoring and adjustment are required to maintain optimal growing conditions and preventing nutrient imbalances.

2.3 Kratky Method Hydroponics

Kratky method pioneered by Dr. Bernard Kratky revolutionizes hydroponic gardening with its simplicity and efficiency. This passive hydroponic technique eliminates the need for pumps, electricity or wicks thus making it accessible and cost-effective for home growers and small-scale setups.

2.3.1 Components

Kratky method offers a cost-effective entry into hydroponic gardening requiring only a few essential components to begin cultivation:

i) **Container**: Starting with a container as small as a mason jar or scaling up to a 5-gallon bucket for larger plants, the choice depends on plant size and desired scale of production. For commercial operations, a larger tank may be necessary.

ii) **Lid**: It is a crucial component for minimizing nutrient solution evaporation and providing support for the plant above the water level. Common materials for lids include plastic or Styrofoam.

iii) **Net pots**: These plastic hydroponic containers feature webbed designs and bottom holes facilitating nutrient access to the roots while allowing drainage. These are indispensable for successful hydroponic plant growth.

iv) **Growing medium**: Selecting a pH-neutral growing medium is essential. Options include coconut fibre (a sustainable organic medium derived from coconut husks), rockwool (porous material spun from basalt rock fibres), lightweight expanded clay aggregate (also known as clay pebbles) or perlite (a lightweight volcanic rock).

v) **Hydroponic fertilizer**: Kratky method fertilizer comprises a specific blend of Masterblend fertilizer, calcium nitrate and magnesium sulphate in a 4-18-38 ratio. It is crucial to calculate the appropriate amount based on the container size and water level.

vi) **Additional supplies**: It is advisable to have a pH control kit and a ppm meter on hand for monitoring and adjusting nutrient solution parameters effectively.

By assembling these basic components, growers can embark on their Kratky method hydroponic journey, enjoying the simplicity and accessibility of this innovative gardening approach.

2.3.2 Implementation

In Kratky method, plants are placed on rafts, *i.e.* netted pots filled with a growing medium fixed to the top of a container filled with nutrient solution. Initially, the container is filled with enough solution to submerge the roots completely. At initial setup, the bottom $1/3^{rd}$ of the net pot should be covered in water and Kratky hydroponics nutrients to keep the growing medium wet. As the plants grow the solution level gradually decreases. Unlike traditional hydroponic systems, the nutrient solution is not continuously replenished or aerated. Instead, the gradual reduction in solution level creates an air gap between the roots and the remaining solution which ensures adequate oxygenation. The plant will take up the water as it grows but as the water level declines the roots of the plants will grow and continue to reach the nutrient solution. Growers must carefully calculate the amount of water and nutrients to account for the size of the container for effective growth.

Harvesting in the Kratky method is straightforward: once the plant is ready where nearly depleted nutrient solution does not require replenishment. Gardeners simply calculate the required amount of solution for the plants and fill the tank only once. After harvest the tank can be refilled and new transplants can be placed on top to start another crop. To maintain optimal performance it is recommended to clean the tank after 3-5 growing cycles.

2.3.3 Optimal Plant Selection

When cultivating plants in a Kratky hydroponic system it is crucial to consider the unique growth rates and nutritional needs of each plant species. To maximize growth and prevent competition for resources, it is advisable to grow a single type of plant in each container ~ a practice known as monocropping. It ensures that plants with similar growth rates and nutrient requirements are grouped together minimizing the risk of one plant overpowering another. For instance, deep-rooted and fast-growing varieties like bok choy may outcompete tender leaf lettuce if grown together. In such cases, the bok choy's rapid water uptake can lead to nutrient depletion which may leave the lettuce at a disadvantage.

Even when growing similar plants together caution must be exercised as slight variation in container conditions can affect nutrient availability. In this case if lid supporting the net pots is misaligned one plant may receive more access to the nutrient solution than others resulting in uneven growth and potential competition.

By carefully selecting compatible plant varieties and ensuring proper container management the growers can optimize nutrient uptake and foster healthy growth in their hydroponic systems.

2.3.4 Suitability of the Kratky Method for different Plant Types

Kratky method is particularly well-suited for cultivating leafy greens and herbs, making it a preferred option for small-scale and home hydroponic setups due to its straightforward approach and minimal maintenance demands. For instance, lettuce with its rapid growth and generous yields thrives exceptionally well in Kratky hydroponics. Similarly, spinach and celery also demonstrate robust growth and flourish in Kratky systems. However, herbs like basil and oregano are among the easier varieties to cultivate hydroponically using the Kratky method.

However, the Kratky method may not be the most suitable choice for flowering and fruiting plants such as potatoes and tomatoes with heightened nutrient requirements and greater water needs. These plants often necessitate more active nutrient management and aeration to support their growth. Tomatoes, in particular, are notorious for their high water demands and energy-intensive fruit production process. These typically require spacious root systems and frequent replenishment of nutrient solutions, given the extended duration until harvest. Moreover, achieving precise pH levels and water temperatures is crucial for optimal tomato growth along with ample exposure to light.

Root crops like carrots, turnips, potatoes and beets, which develop below the water line in hydroponic systems, are not well-suited for Kratky setups. Potatoes, in particular, are susceptible to root rot in stagnant water conditions. Carrots, on the other hand, face constraints in net cups due to potential growth restrictions.

2.2.5 Comparison with SSC

While 'static solution culture' and the 'Kratky method' are similar in that they both involve growing plants in containers filled with a stationary nutrient solution, there are differences in nutrient solution management, container design, oxygenation methods and nutrient solution depth. Differences between the two techniques are given below:

i) **Nutrient solution management**: In both static solution culture and the Kratky method plants are grown in containers filled with a nutrient solution. However, in static solution culture the nutrient solution may need to be replenished or adjusted periodically to maintain optimal nutrient levels for plant growth. In contrast, the Kratky method relies on a single initial filling of the nutrient solution, and the solution is not replenished as the plants grow. Instead, the solution level gradually decreases as the plants consume water and nutrients. Therefore, the grower must fill the container with enough water and nutrients to last the lifespan of the plants.

ii) **Container design**: Static solution culture systems can vary in design and complexity ~ ranging from simple containers with a reservoir for the nutrient solution to more elaborate setups with pumps and irrigation systems. On the other hand, the Kratky method is known for its simplicity and requires minimal equipment. Typically, plants are grown in containers such as buckets or jars filled with nutrient solution.

iii) **Oxygenation**: Oxygenation of the nutrient solution is important in both static solution culture and the Kratky method to prevent root suffocation, *vis-a-vis* to promote healthy plant growth. However, the methods of oxygenation may differ between the two techniques. In static solution culture, air stones or diffusers are commonly used to introduce oxygen into the solution. In the Kratky method, oxygenation occurs naturally as the solution level decreases and roots are exposed to air.

iv) **Nutrient solution depth**: In both the techniques, the depth of nutrient solution relative to the plant roots is an important factor. In static solution culture, the nutrient solution typically covers only the lower portion of the roots, allowing the upper roots to access oxygen from the air. In the Kratky method, the initial nutrient solution level is higher but with consumption of plant solution the roots are gradually exposed to air.

2.4 Capillary and Wick System

Capillary and wick systems often used interchangeably are passive hydroponic methods that utilize capillary action to deliver nutrient solution to plant roots. Both capillary and wick systems offer several advantages including simplicity, low cost and ease of setup. However, these may have limitations in larger-scale operations or with certain types of plants that require precise control over nutrient delivery. Moreover, the rate of nutrient uptake in these systems can be slower compared to active hydroponic methods such as drip or nutrient film technique (NFT) systems. Nonetheless, capillary and wick systems remain popular choices for hobbyists, beginners and to those seeking a straightforward approach to hydroponic gardening.

2.4.1 Comparison of Capillary and Wick Systems

While they share similarities, there are subtle differences between the two as defined below:

2.4.1.1 Capillary Action

Both capillary and wick systems rely on capillary action which is the ability of a liquid to flow in narrow spaces without the assistance of external forces like gravity. Capillary action occurs when liquid is drawn into small spaces such as the pores of a material due to adhesive and cohesive forces between the liquid and the material. In both the systems, the wicking material serves as the medium through which capillary action transports the nutrient solution from the reservoir to the plant roots.

2.4.1.2 Key Differences

While both capillary and wick systems are passive hydroponic methods that rely on the movement of water through a medium to deliver nutrients to plant roots, there are subtle differences between the two:

Nature of Water Movement

- In a capillary system water moves through the medium due to capillary action which is the tendency of water to move through narrow spaces in opposition to gravity. In a broader sense a capillary system may include setups where entire growing medium, rather than just a wicking material facilitates capillary action. This could involve using a porous or absorbent growing medium that naturally draws up moisture from the reservoir to the plant roots.
- In a wick system, the movement of water is facilitated by a wick which is a porous material that absorbs water from a reservoir and transports it to the root zone. The wick serves as a conduit for water movement allowing it to move upward against gravity.

Mechanism

- ☆ Capillary action occurs naturally as a result of the interaction between water and the surface of the medium. The medium itself acts as a capillary ~ drawing water upward. The movement is driven by the cohesive and adhesive properties of water and the surface tension between water and the medium.
- ☆ In a wick system, the movement of water is facilitated by physical presence of the wick which actively absorbs and transports water from the reservoir to the root zone. The choice of wicking material and its placement within the growing medium are crucial factors in the functionality of a wick system.

Implementation

- ☆ Capillary systems can involve various media such as soil, perlite or other porous materials where water moves upward through the medium by capillary action. In its simplest form the pot (containing porous medium and plants) is kept in a shallow solution of fertilizer and water or on a capillary mat saturated with nutrient solution.
- ☆ Wick systems specifically use a wick which can be made of materials like cotton rope, felt or nylon to facilitate water movement from the reservoir to the root zone.

Control and Regulation

- ☆ Capillary action relies on the inherent properties of the medium and the environmental conditions such as humidity to regulate water movement. It may be less controllable compared to wick systems.
- ☆ Wick systems offer more control over water delivery since the rate of water uptake can be influenced by factors such as type of wick material, the thickness of the wick and the size of the reservoir. This allows for more precise regulation of moisture levels in root zone.

2.4.2 Wick System

While both capillary and wick systems utilize capillary action to transport nutrient solution to plant roots, capillary systems rely on the properties of the medium itself while wick systems utilize a separate wick material to facilitate water movement. Wick systems offer greater control and regulation over water delivery compared to capillary systems.

The wick system stands out as one of the simplest and most cost-effective passive hydroponic systems available. It requires no moving parts, making it easy to set up and maintain. By utilizing capillary action, nutrient solution is delivered from the reservoir to the plant roots through one or more wicks.

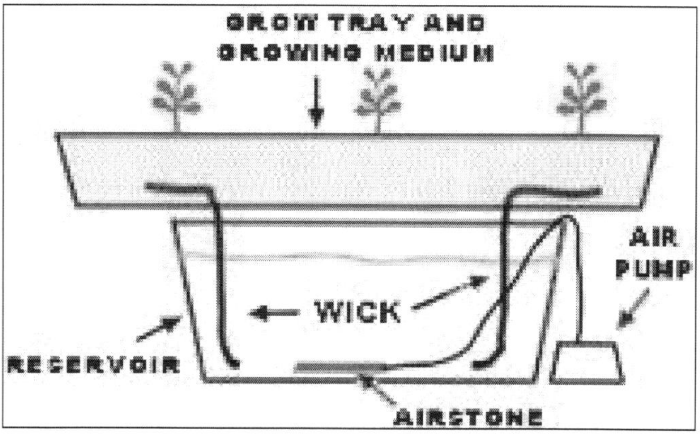

Wick Hydroponic System.

This system is particularly suitable for small-scale or hobbyist growers who may not have access to more complex hydroponic systems or who prefer a hands-off approach to gardening. However, it may not be as suitable for larger plants or those with high water and nutrient requirements as the wicking action may not be able to deliver sufficient moisture and nutrients to meet their needs.

2.4.2.1 Requirements

- ✰ **Wicking materials**: In Wick system a wicking material such as cotton or nylon rope, felt or other porous materials are used to draw up nutrient solution from a reservoir to the plant roots. While cotton ropes are readily available and inexpensive but nylon ropes offer greater durability ensuring long-lasting performance.
- ✰ **Grow medium**: The wicking material is inserted into a growing medium such as perlite, vermiculite or coconut coir which provides support for the plants and helps distributing the nutrient solution evenly.
- ✰ **Nutrient reservoir**: A reservoir containing nutrient solution is placed below the growing medium.

2.4.2.2 Implementation

The implementation of the Wick system involves inserting wicks into a grow tray filled with suitable growing media such as perlite, vermiculture or coconut fibre. The wicking material extends from the nutrient solution in the reservoir up into the growing medium. The number of wicks used depends on factors like system size, plant species and the wicking ability of the chosen material. To maximize growth results it is advisable to add one wick per plant ensuring these are positioned near the roots for efficient nutrient uptake. The wicking material absorbs the nutrient solution from the reservoir and transports it upward to the plant roots through capillary action. The plant roots grow into the moistened growing medium where these absorb water and nutrients from nutrient solution delivered *via* the wicking material.

2.4.2.3 Testing Wicking Ability

Before use, the wicking ability of a rope can be tested by dipping one end into a glass of coloured liquid. Observing its ability to draw the liquid upward provides insight into its effectiveness. Comparing washed and unwashed ropes can further assess wicking performance.

2.4.2.4 Nutrient Solution Management

The reservoir beneath the grow tray is filled with nutrient solution which is drawn up to the roots *via* capillary action. It is essential to refresh the nutrient solution regularly to maintain its strength as plants absorb nutrients. However, pH levels should be monitored and adjusted as needed to ensure optimal nutrient uptake.

2.4.2.5 Aeration

Aeration of the nutrient solution can be achieved by incorporating an air stone and pump into the system. The air stone similar to those used in aquariums is placed in the reservoir and connected to an air pump outside. This setup facilitates the distribution of oxygen through the water *via* tiny bubbles so promoting healthy root development.

While the Wick system is effective for growing leafy vegetables and herbs such as lettuce and rosemary, it may not be suitable for larger plants or those with high water requirements such as tomatoes. In such cases, faster nutrient depletion may occur compared to the rate of supply through the wicks.

Despite its limitations, the Wick system remains a practical choice for beginners and those seeking a low-maintenance hydroponic option. With proper management and attention to plant requirements, it can yield successful results for a variety of crops.

2.5 Capillary Mat System

'Capillary Mat System' is a passive media-based hydroponic method that utilizes capillary action to deliver water and nutrients to plant roots. It offers a simple and efficient way to grow plants hydroponically making it an attractive option for hobbyists, home gardeners and those seeking a low-maintenance hydroponic solution. Here's a comprehensive overview:

2.5.1 Principle of Operation

Capillary action is the phenomenon where a liquid such as water moves upward against gravity in narrow spaces or porous materials due to adhesive and cohesive forces. In 'Capillary Mat System' a mat made of a highly absorbent material such as felt or capillary matting is used to facilitate capillary action. The mat is placed in contact with a reservoir containing nutrient solution. The solution is drawn upward through the mat by capillary action thus moistening entire surface of the mat.

A capillary mat hydroponic system can be implemented as both a media-based and media-less hydroponic system depending on the specific setup and design.

- **Media-based capillary mat system**: In a media-based capillary mat system, plants are typically grown in a soilless growing medium such as perlite, vermiculite or coconut coir. The capillary mat is placed underneath the growing medium and extends into a reservoir of nutrient solution. Growing medium acts as a medium for plant roots to anchor and grow while the capillary mat draws up the nutrient solution from the reservoir through capillary action. This setup provides a continuous supply of water and nutrients to the plant roots so promoting healthy growth and development.
- **Media-less capillary mat system**: In a media-less capillary mat system plants are grown without use of any growing medium. Instead, the plant roots are directly exposed to the capillary mat as these are soaked in a nutrient solution. The capillary mat draws up the nutrient solution through capillary action delivering water and nutrients directly to the plant roots, ensuring continuous hydration and nutrient supply to the plants. This setup eliminates the need for a growing medium allowing for a simpler and potentially more efficient hydroponic system.

2.5.2 Advantages

- **Simple setup**: 'Capillary Mat System' is easy to set up and requires minimal equipment. It is suitable for both indoor and outdoor gardening applications.
- **Water efficiency**: Since the system relies on capillary action, it uses water efficiently by delivering moisture directly to the roots where it is needed most.
- **Low maintenance**: Once set up, the system requires minimal maintenance. Refilling the reservoir with nutrient solution as needed is typically the only maintenance required.

2.5.3 Plant Compatibility

'Capillary Mat System' is well-suited for growing a wide range of plants including herbs, leafy greens and ornamental plants. It may not be suitable for plants with very high water requirements or those that prefer dry conditions as the constant moisture provided by the capillary mat may lead to root rot or other issues.

2.5.4 Monitoring and Management

Regular monitoring of the moisture level in the capillary mat and the nutrient solution reservoir is important to ensure that plants receiving an adequate water and nutrient supply. Adjustments to nutrient solution strength and pH may be necessary over time to maintain optimal growing conditions.

2.5.5 Limitations

While the Capillary Mat System is effective for smaller-scale gardening applications, it may not be suitable for large-scale commercial production due to its reliance on passive capillary action.

2.6 Passive Drip Systems (PDS)

Passive drip systems (drip irrigation hydroponics), are a form of passive hydroponic system where nutrient solution is delivered directly to plant roots in a controlled manner through drip emitters or tubing. PDS hydroponics offer an efficient and precise method of delivering nutrient solution to plant roots, making them a popular choice for growers seeking water-efficient and low-maintenance hydroponic solutions. Here's a comprehensive overview of passive drip systems hydroponics:

2.6.1 System Overview

In passive drip systems, a nutrient reservoir is placed at a higher elevation than the plants being grown. Nutrient solution flows from the reservoir through a network of tubing or piping equipped with drip emitters which deliver precise amounts of solution to root zone of each plant. Gravity acts as the driving force allowing the nutrient solution to passively flow through the system without need for pumps or active circulation mechanisms.

2.6.2 Components

- ☆ **Reservoir**: A container holds the nutrient solution at an elevated position to facilitate gravity-driven flow.
- ☆ **Tubing/piping**: Durable tubing or piping carries nutrient solution from reservoir to plants.
- ☆ **Drip emitters (drippers)**: These devices regulate the flow of nutrient solution and deliver it directly to the root zone of each plant.
- ☆ **Growing medium**: Plants are typically grown in a medium such as perlite, vermiculite, coconut coir or rockwool which provides support and retains moisture while allowing for good drainage.
- ☆ **Trays/beds**: Plants are arranged in trays/beds with each plant receiving its own drip emitter for precise nutrient delivery.

2.6.3 Advantages

- ☆ **Water efficiency**: Passive drip systems are highly water-efficient which deliver nutrient solution directly to plant roots with minimal waste.

- ☆ **Nutrient precision**: Drip emitters allow for precise control over nutrient delivery ensuring each plant receives the appropriate amount of water and nutrients.
- ☆ **Scalability**: These systems can easily be scaled up or down to accommodate different numbers of plants or growing spaces.
- ☆ **Low maintenance**: With no moving parts or complex systems passive drip systems require minimal maintenance once set up properly.

2.6.4 Considerations

- ☆ **pH and nutrient monitoring**: Regular monitoring of pH and nutrient levels is essential to maintain optimal growing conditions.
- ☆ **Clogging**: Drip emitters and tubing may become clogged over time which necessitates periodic inspection and cleaning.
- ☆ **Medium selection**: Choosing the appropriate growing medium is crucial to ensure proper drainage and root aeration while retaining moisture.

2.6.5 Plant Selection

Passive drip systems are suitable for a wide range of plants including vegetables, herbs, flowers, and certain fruiting crops. Plants with similar water and nutrient requirements are typically grouped together to optimize nutrient delivery and together ensuring uniform growth.

2.6.6 Application

Passive drip systems are commonly used in both small-scale and commercial hydroponic operations including greenhouse cultivation and indoor farming. These are particularly well-suited for crops like tomatoes, peppers, cucumbers, leafy greens that benefit from precise nutrient delivery.

Chapter 3

Hybrid of Active and Passive Hydroponics

3.1 Deep Water Culture (DWC)

In 'Deep Water Culture' (DWC) hydroponic systems the roots are submerged in the nutrient solution, and a recirculating pump is typically used to circulate and aerate the solution in the reservoir. Even though the nutrient solution surrounds the roots directly, aeration provided by the pump is essential to ensure adequate oxygenation of the solution, preventing root suffocation and growth of harmful anaerobic pathogens. Therefore, while the nutrient solution is not pumped directly to the roots in DWC systems, the recirculating pump serves the crucial function of maintaining proper aeration and circulation of the solution for optimal plant growth.

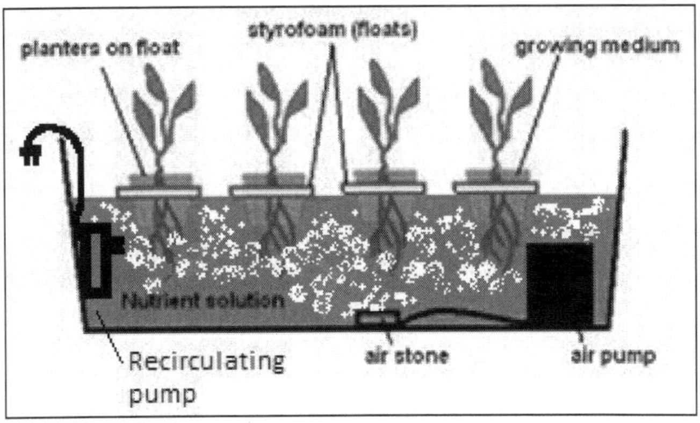

DWC

Moreover, an air pump can also be utilized to provide additional oxygen to the nutrient solution in DWC hydroponic systems. By introducing bubbles into the solution, the air pump enhances oxygen levels so further promoting root health and overall plant vitality. This supplementary oxygenation contributes to an optimal growing environment, supporting vigorous growth and maximizing plant yields. At its core, DWC facilitates rapid plant growth by directly supplying ample oxygen and nutrients to the roots. This direct access optimizes nutrient uptake and metabolic processes, promoting vigorous growth and high yields.

While DWC is primarily categorized as a solution culture hydroponic system, it does offer some versatility in implementation akin to a medium culture system. In traditional DWC setups, plants are suspended directly in the nutrient solution with their roots freely hanging and absorbing nutrients. However, some variations of DWC systems incorporate the use of growing media to support plant roots or provide additional stability. This adaptation introduces elements of medium culture within the DWC framework.

In some DWC systems growers may place net pots filled with growing media such as clay pellets or rockwool into the lids of the nutrient reservoirs. The plants are then positioned in these net pots allowing their roots to extend into the nutrient solution below while being supported by the growing medium. This hybrid approach combines aspects of both solution culture and medium culture hydroponics.

Moreover, some DWC setups may incorporate trays or rafts filled with growing media for providing a platform for plant growth while still allowing their roots to access the nutrient solution below. This design variation allows for greater stability and support for larger plants while maintaining the benefits of a solution culture system.

i) **Solution culture hydroponics system**: The solution culture variant of DWC relies solely on a highly oxygenated nutrient solution to support plant growth. Without use of a medium the plant roots are directly suspended in nutrient solution where these absorb essential elements and oxygen directly. This streamlined approach eliminates the need for a supporting medium, simplifying setup and maintenance.

ii) **Medium culture hydroponics system**: In this variant, a porous inert medium such as perlite, vermiculite or expanded clay pellets serve as a support structure for roots of the plants. The medium is placed within net pots or baskets allowing roots to grow freely while being submerged in the oxygenated nutrient solution. This method provides stability to the plants while ensuring continuous access to oxygen and nutrients. The Kratky Method is similar to deep water culture but uses a non-circulating water reservoir.

In both implementations, DWC epitomizes efficiency and effectiveness in hydroponic cultivation offering unparalleled control over plant nutrition and growth. Its versatility and adaptability make it a preferred choice for growers seeking high yields and superior plant health. It is important to note that while DWC is highly efficient and offers numerous advantages, it requires careful monitoring and management to maintain optimal conditions for plant growth. Inaccurate or inconsistent oxygenation, nutrient levels or pH can compromise the health and productivity of the plants.

3.1.1 Is DWC an Active or Passive Hydroponics System?

DWC systems suspend plant roots in a nutrient solution reservoir. Oxygen is provided to the roots through an air stone or other aerating device. While technically having a pump for aeration, DWC is often considered passive by many growers due to its simplicity. Moreover, DWC may appear passive due to lack of visible movement in nutrient solution. It can be considered an active hydroponic system due to its reliance on active aeration and nutrient circulation to support plant growth. Let's break down why it's primarily active but has passive elements:

i) Active Hydroponic System

- ☆ **Nutrient circulation**: While nutrient solution in DWC systems may appear static on the surface, it is continuously circulated and aerated by the action of air bubbles from the air stones. This circulation ensures that nutrients are evenly distributed and available to the roots throughout the growing container.
- ☆ **Active aeration**: In DWC systems, air pumps and air stones are commonly used to aerate the nutrient solution actively. Oxygenation of nutrient solution is crucial for root health and nutrient uptake in plants. The air stones release bubbles into the nutrient solution, providing a constant supply of oxygen to the roots.
- ☆ **Maintenance**: DWC systems require active management, including monitoring and adjustment of nutrient levels, pH levels and oxygenation. This active involvement in maintaining optimal growing conditions distinguishes DWC as an active hydroponic system.
- ☆ **Comparison with passive hydroponic system**: In contrast, passive hydroponic systems such as wick systems or static solution culture rely on capillary action or gravity to deliver nutrient solution to the plants without need for external pumps or aeration devices. These systems are generally simpler and require less maintenance compared to active systems like DWC.

ii) Passive Hydroponic Elements

- ☆ **Static nutrient solution**: The nutrient solution in a DWC system remains relatively still with the roots suspended directly in the solution. This static nature of the solution resembles passive hydroponic systems where solution does not actively circulate.

- ☆ **Fewer moving parts**: Compared to other active hydroponic systems such as NFT or aeroponics, DWC systems have fewer moving parts and require less ongoing maintenance.
- ☆ Despite the active aeration component, DWC systems also incorporate passive elements. Once the air pump is turned on and the system is set up, there is minimal ongoing intervention required.

In a nutshell while DWC hydroponics primarily operates as an active hydroponic system due to use of air pumps and active aeration, it also incorporates passive elements such as static nature of the nutrient solution and reduced maintenance requirements once set up. Therefore, it can be considered a hybrid system that combines aspects of both active and passive hydroponics.

3.1.2 Traditional DWC Setup

In the traditional DWC setup, plastic buckets or large containers are used with plants housed in net pots suspended from the centre of the lid. The roots dangle into the oxygenated nutrient solution below. An air pump coupled with porous stones oxygenates the solution ensuring a steady supply of oxygen to the roots.

3.1.3 Components of a Simple DWC System

To set up a basic DWC system, growers need a container of around 10-15 liters capacity, air stones for oxygenation, a lid capable of supporting net pots, a pump for oxygenating the solution and net pots (with or without porous medium). Plant roots extend from the net pot into the oxygenated nutrient solution below.

3.1.4 Media-based DWC

In media-based DWC, porous growing media such as expanded clay pellets are used to absorb nutrients and water from reservoir initially. Base of the net pot containing porous media and a plant must be slightly submerged in the solution so that porous medium is able to absorb nutrient solution for the roots. As the root system develops, it gradually immerses in the nutrient solution.

Maintaining the nutrient solution at the same level as the pot base is not necessary as lower levels suffice for optimal growth.

3.1.5 Nutrient Delivery and Oxygenation

The nutrient solution in DWC systems is carefully balanced to provide plants with essential macro- and micro-nutrients required for healthy growth. The oxygen-rich nutrient solution in DWC nourishes plants for fast growth. This solution is typically oxygenated using air pumps and air stones to ensure adequate oxygen levels for root respiration. While plants typically absorb more oxygen from the air than from water, DWC allows roots to absorb large quantities of both oxygen and nutrients from the solution. This leads to sustained and vigorous growth throughout life cycle of the plant.

3.1.6 Differences from 'Static Solution Culture'

DWC shares similarities with 'Static Solution Culture' except that it can function as both solution culture and medium culture hydroponic systems. Unlike static solution culture where nutrient solution remains stationary, DWC keeps roots submerged in a highly oxygenated solution, promoting optimal growth conditions.

3.1.7 Maintenance and Care

As the plants enter the flowering stage it is important to adjust the nutrient solution level gradually, allowing the roots to be gradually exposed to the air. Depending on factors such as plant size, reservoir volume and pH fluctuations the nutrient solution should be either replaced or supplemented with a new half-strength solution approximately once a week or every fortnight.

Regular maintenance tasks are essential to ensure the health and productivity of the hydroponic system. This includes thorough cleaning of the container on a weekly basis to eliminate any algae, mould or salt deposits that may accumulate. Timely monitoring and adjustment of pH levels particularly when replacing the nutrient solution are crucial to maintain an optimal pH range of 5.5 to 6.5.

For transparent containers it is advisable to wrap them with black cloth or plastic to prevent algal growth which can thrive in presence of light. By implementing these maintenance practices growers can create an optimal growing environment for their plants and maximize the success of their hydroponic operation.

3.1.8 Germination and Planting

For optimal results it would be better to germinate the seeds using rockwool cubes before transferring these to DWC pots filled with expanded clay pellets. This method ensures healthy plant development and maximizes growth potential in the DWC system.

3.1.9 Some Crucial Considerations

Here are some crucial considerations to keep in mind for successful hydroponic cultivation:

i) Nutrient Buffering and pH Stability

- ☆ In DWC systems, pH management is similar to static solution culture but may require more frequent monitoring and adjustment due to the active circulation of the nutrient solution. The use of air pumps and air stones in DWC systems promotes oxygenation of the nutrient solution but can also lead to more rapid pH fluctuations compared to static systems. pH levels in DWC systems can be monitored using pH testing kits or meters, and adjustments can be made as needed using pH-up or pH-down solutions. Regular monitoring and adjustment of pH in DWC systems help ensuring stable growing conditions and prevent nutrienting imbalances that can affect plant health and growth.

- A well-buffered nutrient solution should be utilized to ensure pH stability throughout the hydroponic system. One should aim for a pH range of 5.5-6.5 with slight variations depending on the growth stage:
- For vegetative growth, pH level is maintained within 6.0-6.3.
- During flowering and fruiting stages, pH is adjusted in such a way to fall within 5.7-5.9 for optimal nutrient uptake and plant health.

ii) Nutrient Solution Temperature Management

- In DWC systems, active aeration is typically used to oxygenate the nutrient solution which can also help regulate its temperature to some extent. The movement of air bubbles generated by air pump can help to dissipate excess heat from nutrient solution particularly in warmer environments.
- Similar to static solution culture, maintaining the ambient temperature of the growing environment within the optimal range (18°C and 24°C) is important for DWC systems. High temperatures above 24°C can lead to a decrease in dissolved oxygen (DO) levels, potentially impacting root health and nutrient absorption. Conversely, temperatures below 18°C may slow plant metabolism particularly in warm-season crops. Some growers strategically lower nutrient temperatures towards the end of the flowering stage to aid in ripening.
- Monitoring the temperature of the nutrient solution directly using a thermometer and adjusting the operation of the air pump or using other cooling or heating methods as needed can help maintaining the solution within the optimal temperature range.
- Insulating the nutrient solution reservoir and shading the system from direct sunlight can also help prevent overheating of the nutrient solution.

iii) Preventing Stem Submersion

It should be ensured that plant stems are not submerged in nutrient solution to prevent rot and disease. Proper support and spacing of plants can help maintain an optimal growing environment.

iv) Moisture Management in DWC Systems

- In DWC systems, manipulation of moisture levels in the root zone should be done to optimize plant growth and development.
- Adjusting moisture levels can influence essential oil production in aromatic crops such as basil and mint. Dryer root zones may stimulate oil production.
- Conversely, wetter root zones can promote vegetative growth especially in large fan-leaved plants, enhancing transpiration and photosynthetic potential.

v) Preventing Root Diseases

- Maintaining cleanliness and hygiene in the hydroponic system prevents the development of root diseases such as *Pythium* and *Fusarium*.
- Avoiding warm water and high nutrient concentrations can create favourable conditions for growth of pathogens. Regular monitoring and maintenance can help mitigate the risk of root diseases.

By adhering to these essential guidelines hydroponic growers can create an optimal environment for plant growth. These precautions will minimize the risk of nutrient imbalances and diseases, *vis-a-vis* will maximize the overall success of their cultivation endeavours.

Chapter 4

Active Hydroponics Systems

Active hydroponic systems require a pump or other mechanical device to actively delivering the nutrient solution to the root zones of the plants either continuously or intermittently. In active hydroponics since nutrient solution is circulated through the system, these systems are typically more complex and involve more maintenance but offer greater control over the growing environment. Nutrient solution in the reservoir is pumped to the growbed and then allowed to drain back to the reservoir. The main issue of system management for this type of system is increase in nutrient strength and pH as the nutrients are fed and then returned through the system. This increase is more apparent in media-based systems than water-based systems.

In an active hydroponic system, there is a mechanism, such as a pump or irrigation system.

4.1 Classification of Active Hydroponic Systems

Active hydroponic systems can be classified based on various criteria such as nutrient delivery method, growing medium, system design and management approach. Here are some common classifications of active hydroponic systems based on use of growing media, nutrient delivery method, management approach and system design:

4.1.1 Use of Growing Media

Growing medium: Like passive hydroponic systems, an active hydroponic system can be implemented as both medium culture and solution culture systems. Let's delve into each:

i) **Media-based systems:** Within the realm of active hydroponics, medium culture systems are a popular choice for growers seeking a balance between simplicity and control. In an active medium culture hydroponic system a growing medium such as perlite, vermiculite, rockwool, coco coir or clay pellets is used in support of plant roots. Nutrient solution is actively circulated through growing medium using pumps allowing for precise control over nutrient delivery and oxygenation. There are two main variations of medium culture active hydroponics: sub-irrigation (ebb and flow) and top-irrigation (drip irrigation).

ii) **Media-less systems:** Plants are grown without a traditional growing medium with their roots suspended in air or in a nutrient solution. Nutrient solution is actively aerated and circulated using pumps to ensure that roots have access to oxygen and nutrients such as 'Nutrient Film Technique (NFT)', 'Aeroponics' and 'Recirculating Deep Water Culture (RDWC)'.

4.1.2 Nutrient Delivery Method

i) Continuous Flow and Interval-Based Systems

- ☆ In continuous flow systems, nutrient solution is continuously pumped or circulated to the root zones of the plants such as 'nutrient film technique (NFT)' and 'aeroponics'.
- ☆ In interval-based systems, nutrient solution is delivered to the plants periodically or at specific intervals such as 'ebb and flow (flood and drain) systems' and 'drip irrigation systems'.

ii) Sub-Irrigation and Top-Irrigation

Both sub-irrigation and top-irrigation are viable options for medium culture active hydroponics too offering growers flexibility, control and efficiency in nutrient delivery. By understanding the characteristics and requirements of each system growers can choose the method that best suits their needs and preferences, ultimately optimizing plant growth and maximizing yields.

- ☆ **Sub-irrigation in active hydroponics:** Sub-irrigation hydroponic systems (ebb and flow or flood and drain systems) involve periodical flooding the growing medium with nutrient solution and then allowing it to drain back into a reservoir. This cycle of flooding and draining ensures that the roots have access to water and nutrients while also allowing for sufficient aeration between irrigation events.
- ☆ **Top-irrigation in active hydroponics**: Top-irrigation hydroponic systems (drip systems or drip irrigation) involve delivering nutrient solution directly to the base of each plant through drip emitters or tubing. This targeted approach allows for precise control over water and nutrient

delivery, minimizing waste and ensuring that plants receive exact amount of moisture they need. In a top-irrigation system, plants are typically grown in individual pots or containers filled with a growing medium. A pump delivers nutrient solution from a reservoir to a network of tubing or drip lines positioned above the plants. Drip emitters or micro-sprayers attached to the tubing release small amounts of nutrient solution directly onto the growing medium around each plant.

4.1.3 Management Approach

i) **Automated systems:** Systems that incorporate sensors, timers and controllers to automate nutrient delivery, pH adjustment and other environmental factors.

ii) **Manual systems:** Systems that rely on manual intervention for nutrient mixing, pH adjustment and irrigation scheduling.

4.1.4 System Design (Closed or Open Systems)

i) Closed Hydroponic Systems

In a closed hydroponic system the nutrient solution operates within a self-contained loop where excess solution is recovered, replenished and recycled. This means that the same nutrient-rich solution continuously circulates with adjustments made as necessary to maintain optimal nutrient concentrations for plant growth. Closed hydroponic setups encompass both straightforward systems like deep water culture and more intricate setups such as nutrient film technique (NFT).

Advantages: One of the primary advantages of closed hydroponic systems lies in their water efficiency. By recycling the nutrient solution, these systems significantly reduce overall water consumption compared to traditional soil-based agriculture. Moreover, the closed-loop design allows for meticulous control over crucial factors like nutrient levels, pH balance and EC. This precise control fosters an environment where plants can thrive, potentially leading to superior growth rates and increased yields. Another key benefit of closed hydroponic systems is their sustainability. By minimizing nutrient wastage and water usage these setups align with environmentally conscious practices. The ability to reuse the nutrient solution not only conserves resources but also reduces the environmental impact associated with conventional farming methods. Overall, closed hydroponic systems represent a sustainable and efficient approach to cultivation offering growers the opportunity to optimize resource utilization while maximizing crop productivity.

Drawbacks: The drawback of this recovery system is that the pH and nutrient strength require periodic checking and adjustments because the pH and nutrient concentration of the nutrient solution gets changed as the plants use up the nutrients in the nutrient solution.

ii) Open Hydroponic Systems

Open hydroponic systems (run-to-waste systems) represent a dynamic approach to hydroponic cultivation where nutrient solution is exposed to the environment and not recirculated or contained within a closed-loop system. In this method, fresh nutrient solution is introduced for each irrigation cycle and any excess solution is allowed to drain away or evaporate rather than being recycled back into the system. One common method of nutrient delivery in open hydroponic systems is through drip irrigation. In this setup, nutrient solution is delivered *via* drip emitters directly to the base of each plant so ensuring precise control over nutrient delivery while allowing any surplus solution to drain away. This approach is particularly favoured in larger-scale hydroponic operations due to its efficiency and effectiveness. While typically regarded as a closed-loop system, drip irrigation can seamlessly transition into an open hydroponic system thus offering flexibility and adaptability to various growing setups.

Advantage: With open hydroponics often there is less maintenance involved as there is no need for rigorous monitoring or management of the pH or EC of the nutrient solution compared to closed-loop systems.

Disadvantages: It is important to note that open hydroponic systems may consume more water as the nutrient solution is not recycled which necessitates a constant supply of fresh solution. Moreover, maintaining adequate run-off is essential to preserve the nutrient balance in the root zone and prevent potential nutrient buildup. Moreover, there is risk of nutrient wastage since excess solution is not reused which may impact the overall sustainability of the system. Despite these considerations, open hydroponic systems offer a viable and efficient method for hydroponic cultivation, especially in scenarios where precise nutrient delivery and scalability are paramount.

These classifications provide a framework for understanding the diversity of active hydroponic systems available to growers and help in selecting the most suitable system based on factors such as crop type, space constraints, budget and level of automation desired.

4.2 Recirculating Deep Water Culture (RDWC) Hydroponics

It is also called as 'Modular Deep Water Culture (MDWC) hydroponics' which represents a sophisticated evolution of traditional Deep Water Culture (DWC) systems, offering enhanced nutrient delivery and oxygenation for improved plant growth. RDWC hydroponics offers a dynamic and efficient solution for cultivating plants in a soilless environment. By harnessing the benefits of continuous nutrient circulation and oxygenation, RDWC systems empower growers to achieve impressive yields and maximize the potential of their hydroponic operations.

Recirculating Deep Water Culture.

4.2.1 Key Components and Operation

- **System configuration**: RDWC systems feature a central reservoir that supplies nutrient solution to multiple interconnected containers or buckets. These containers are equipped with plant baskets or net pots which allows the roots to be submerged in nutrient solution.
- **Active recirculation**: Unlike traditional DWC systems, RDWC is an active system where nutrient solution continuously circulates from central reservoir to each container and back again. This recirculation is facilitated by pumps and plumbing, ensuring a constant flow of oxygenated nutrient solution throughout the system.
- **Aeration and mixing**: As the nutrient solution returns to the reservoir, it is aerated and mixed using spray nozzles or similar devices. This continuous agitation helps oxygenate the water and maintain a uniform distribution of nutrients and pH levels throughout entire system.

4.2.2 Advantages and Drawbacks over Stand-Alone DWC

- **Advantages**: One of the primary advantages of RDWC over stand-alone DWC systems is the centralized nutrient management. Testing and adjustments need to be performed only at the central reservoir, simplifying maintenance and monitoring compared to individual DWC containers. RDWC systems also facilitate better nutrient uptake and plant health due to continuous circulation and oxygenation of nutrient solution.
- **Drawbacks and considerations**: Despite its advantages, RDWC systems may be more susceptible to spread of root diseases compared to stand-alone DWC setups. The interconnected nature of the system can facilitate the rapid transmission of pathogens from plant to plant. Proper sanitation

practices and vigilant monitoring of plant health are essential to mitigate the risk of disease outbreaks in RDWC systems.

4.2.3 Oxygenation and Plant Growth

RDWC systems are known for their superior oxygenation capabilities achieved through the use of air pumps and porous stones. The increased oxygen levels in the nutrient solution promote faster and healthier plant growth ~ leading to higher yields and overall improved plant vitality.

4.2.4 Maintenance and Management

Regular maintenance of RDWC systems involves monitoring of nutrient levels, pH and oxygenation as well as cleaning and sanitizing equipment to prevent the buildup of algae and pathogens. Proper plant spacing and hygiene practices can help minimize the risk of disease spread and optimize the performance of the system.

4.3 Continuous-Flow Solution Culture or NFT

Continuous-flow solution culture and NFT are similar in that they both involve a continuous flow of nutrient solution over plant roots but these are not exactly the same. While both these techniques involve a continuous flow of nutrient solution, NFT is a specific subset of 'Continuous-Flow Solution Culture' that utilizes channels or gutters to deliver the solution in a thin film over the roots. On the other hand, 'Continuous-Flow Solution Culture' encompasses a broader range of systems where nutrient solution is continuously circulated but may not necessarily involve the use of channels or a thin film of solution. Here is an overview of each method and their differences:

4.3.1 Continuous-Flow Solution Culture

Continuous-Flow Solution Culture is a hydroponic system where plants are grown in a continuously flowing nutrient solution. It can be implemented in a large storage tank of desired size where thousands of plants can be grown. In this system continuously a pump circulates nutrient solution from a reservoir through a network of tubes or channels to deliver it to the plant roots. The excess nutrient solution not absorbed by the plants is collected and returned to the reservoir for recirculation. This method ensures a constant supply of nutrients to the roots and provides ample aeration which promotes healthy plant growth.

4.3.2 Nutrient Film Technique (NFT)

NFT represents a specialized hydroponic system categorized under 'Continuous-Flow Solution Culture'. In NFT setups, plants typically inhabit channels or gullies that are gently sloped, facilitating the passage of a thin film of nutrient solution over their roots. This precise application of solution ensures that upper segments of roots remain consistently moist without being submerged thereby granting access to both water and air. This strategic design optimizes oxygen availability to the roots while providing essential nutrients.

Operational mechanics involve a continuous pumping of nutrient solution from a reservoir to the highest point of the channel, allowing it to cascade over the roots before returning to the reservoir. The shallow depth of the nutrient film ensures that roots remain exposed to air as well as solution, fostering ideal oxygenation and nutrient uptake for plant vitality.

4.3.2.1 Components

In a nutrient film technique hydroponic system several components work together to facilitate growth of plants by providing a continuous flow of nutrient solution to the roots. Here are the key components of an NFT system:

i) **Growing Channels or Gullies**

- ❖ In NFT hydroponic systems, plants thrive within channels or gullies meticulously crafted to cradle their roots while ensuring a seamless flow of nutrient solution. These channels crafted from sturdy materials like PVC, ABS, plastic or metal come in various lengths and widths tailored to the specific needs of the cultivated crop. Engineered with a gentle incline, these facilitate the uninterrupted movement of nutrient solution from one end to the other.
- ❖ For efficient plant placement, a readily available cost-effective round tube or PVC pipe is often employed. Equidistant holes or planting sites drilled along the top of these channels accommodate net pots or seedlings which enable the growth of multiple plants within a single system. However, the use of PVC may introduce challenges as roots in the centre may access deeper solution depths while those near the edges encounter shallower levels, potentially resulting in uneven growth. This issue is mitigated by adopting flat-bottomed channels.

ii) **Planting sites**: Planting sites within NFT systems vary accommodating small net pots, baskets or net cups. Remarkably no additional growing medium is necessitated beyond the initial rockwool cubes used for seedling germination, exemplifying the simplicity and efficiency of the NFT hydroponic approach. With roots extending into the nutrient-rich solution coursing through the channels, plants absorb water and nutrients directly thus fostering robust growth and development.

iii) **Nutrient solution reservoir**: The nutrient solution reservoir stores the nutrient solution that will be continuously recirculated through the NFT system. Reservoirs can vary in size depending on the scale of the operation. These are typically made of plastic or other non-reactive materials to prevent contamination of the nutrient solution.

iv) **Submersible pump**: A submersible pump is used to circulate the nutrient solution from the reservoir to the top of the growing channels or gullies. The pump ensures a steady flow of solution allowing it to cascade continuously down the channels in a thin film over the roots of the plants.

v) **Tubing and fittings**: Tubing and fittings are used to connect the submersible pump to the growing channels or gullies, creating a closed-loop system for nutrient solution. Flexible tubing is typically used to transport the solution and various fittings such as connectors, elbows and tees may be used to route the tubing as needed.

vi) **Drip emitters or spreader mat**: Drip emitters or a spreader mat are used to distribute evenly the nutrient solution across the bottom of the growing channels or gullies. Drip emitters release the solution in a controlled manner while a spreader mat helps to create a thin film of solution that flows over the roots of the plants.

vii) **Support structure**: A support structure is used to elevate the growing channels or gullies and provide stability for the plants. This structure may consist of racks, stands or frames made of metal or sturdy plastic. It should be designed to withstand the weight of the growing medium and plants.

viii) **Growing medium**: While not always necessary, a growing medium such as rockwool, coconut coir or perlite may be used to support the plants and provide stability for the root system. The growing medium in net pots or net cups helps to anchor the plants in place and retain moisture around the roots.

ix) **Monitoring and control equipment**: Monitoring and control equipment such as pH meters, EC meters and timers may be used to monitor the pH of nutrient solution and nutrient levels and to control the timing and duration of nutrient solution delivery.

These components work together to create a controlled hydroponic environment where plants can thrive by receiving a continuous supply of water, nutrients and oxygen directly to their roots. Proper maintenance and monitoring of the system are essential to ensure optimal plant growth and productivity.

4.3.2.2 Plant Spacing and Arrangement

Plants in NFT systems are spaced according to their growth requirements with adequate distance between each plant to prevent overcrowding and competition for resources. Depending on the size of the system and the type of crop being grown, plants may be arranged in single rows or multiple rows within the channels or gullies.

4.3.2.3 Implementation

In NFT, a channel often referred to as a grow tray is positioned above a nutrient solution reservoir. A pump is utilized to convey the nutrient solution to upper end of the grow tray while a drain pipe facilitates the return of unused solution to the reservoir *via* gravity. This feeding cycle can adopt either an intermittent or continuous approach. In intermittent feeding regulated by a timer and repeated multiple times per hour, roots experience exposure to air between each feed so enhancing aeration. This variation is commonly termed '**Pulsed NFT**'. Conversely, a continuous feed cycle runs continuously and is predominantly favoured in commercial operations.

To ensure proper oxygenation of nutrient solution an air stone can be inserted into the reservoir connected to an external air pump. The channel is strategically inclined to facilitate downward flow of water towards the drain pipe. At the lower end of the channels, the nutrient solution is collected and directed back to the reservoir for recycling.

Monitoring the salt concentration of solution before recycling is imperative to maintain optimal nutrient balance. The flow rate of the nutrient solution should be meticulously adjusted typically ranging from 2 to 3 liters per minute, contingent on the length of the channel.

A well-designed NFT system encompasses the appropriate channel slope, flow rate, channel length and nutrient solution composition. Planting in this hydroponic system typically begins with seeds germinated in rockwool cubes which are later transferred into the channel holes at a certain stage of growth where these mature to fruition.

4.3.2.4 Design Parameters for NFT

<table>
<tr><th></th><th>Recommended</th><th>Most Effective</th><th>Remarks</th></tr>
<tr><td>Channel slope</td><td>Gradient 1:30 to 1:40</td><td>1:100, but difficult to implement</td><td>Even with these slopes waterlogging may occur</td></tr>
<tr><td>Flow rate</td><td>1 to 1.5 litres per minute</td><td>1 litre per minute</td><td>At planting, rates may be half of this (maximum flow rate should not exceed 2 litres per minute)</td></tr>
<tr><td>Channel length</td><td>8 to 12 metres</td><td>8 to 10 metres</td><td></td></tr>
<tr><td colspan="4">For rapidly growing crops, oxygen levels typically remain sufficient within the nutrient solution yet nitrogen depletion may occur along the length of extended channels. To counteract this, channels exceeding 12 meters in length necessitate an additional nutrient feed placed midway to replenish nitrogen levels. It is advisable to reduce the flow rate to 1 litre per minute through each outlet to ensure optimal nutrient distribution.

In longer channels, plants positioned at the beginning typically exhibit superior health compared to those located at the outlet-end. The ideal channel length varies depending on the specific plant species being cultivated and the nutrient concentration within the solution. Adjustments to channel length should be made accordingly to maintain optimal growing conditions and maximize crop productivity.</td></tr>
</table>

	Recommended	*Most Effective*	*Remarks*
Channel depth and width	The dimensions of the channel should be carefully considered to provide ample space for mature root systems of plants. Channels that are too narrow or shallow risk becoming obstructed by roots which can result in blockages, overflowing, or the formation of stagnant areas that impede proper drainage. Positioning the last plant sufficiently above the drain outlet ensures that roots do not obstruct the drainage system.		
Nutrient solution	Balanced nutrient solution depending on the type and age of plants.		

NFT hydroponics shines brightest when cultivating lightweight fast-growing plants of small to medium size ideally suited for rapid growth cycles and streamlined harvesting. The efficiency of this method thrives with crops like leafy greens, herbs and certain varieties of vegetables that exhibit prolific growth without necessitating extensive structural support. However, for heavier specimens such as tomatoes or squash which boast substantial biomass and require robust support system throughout their growth stages, supplemental support systems become imperative. Trellises, cages or other structural frameworks are indispensable aids, ensuring the plants remaining upright and stable as they mature, preventing potential damage or collapse under their own weight.

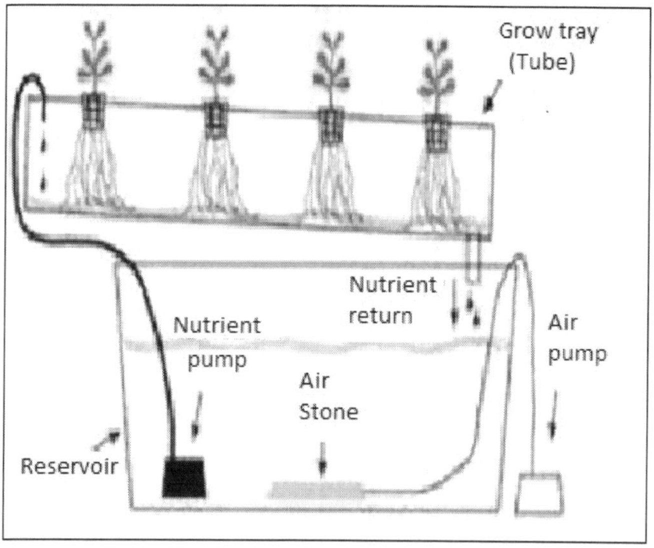

Beyond the need for support, it is essential to consider specific requirements of each crop when designing a NFT system. Factors such as root depth, spacing and nutrient demands must be meticulously calibrated to optimize growth and yield. Maintaining proper pH and nutrient levels along with regular monitoring for signs

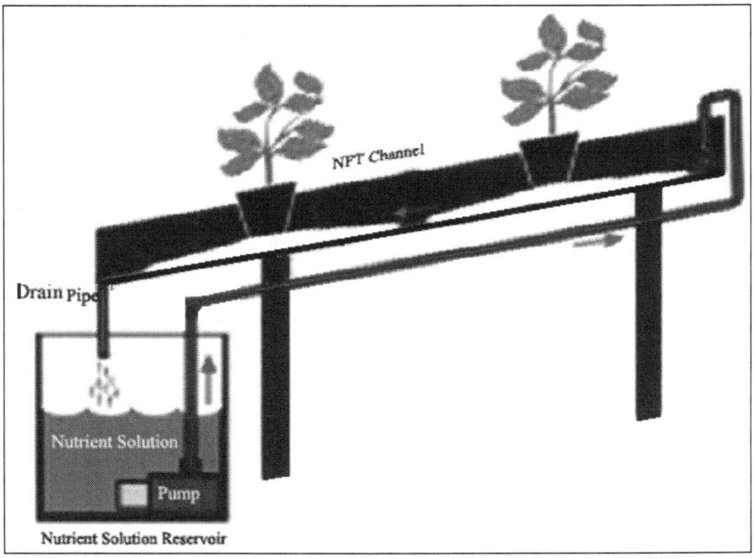

of nutrient deficiencies or imbalances are paramount for sustaining healthy plant development in NFT setups.

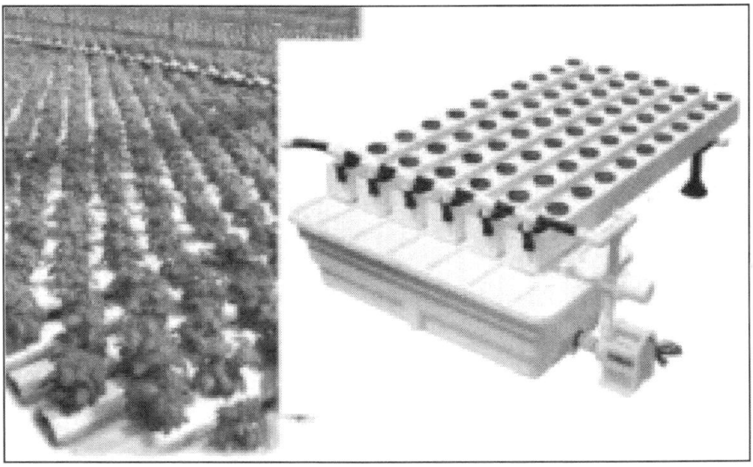

NFT Hydroponic System.

Moreover, the versatility of NFT hydroponics extends beyond traditional horticultural applications, finding resonance in research settings, educational environments and urban farming initiatives. Its compact footprint, resource efficiency and adaptability make it an attractive choice for growers seeking sustainable solutions to modern agricultural challenges.

4.3.2.5 Advantages of NFT

i) **Precise control of environment**: NFT systems feature a separate storage tank for nutrient solution allowing for easy monitoring and adjustment of temperature and nutrient concentrations. This precise control contributes to optimal plant growth and health.

ii) **Optimal root environment**: Plant roots in NFT systems are exposed to abundant water, oxygen and nutrients ~ fostering vigorous growth and high-quality produce. This optimized root environment results in higher yields over extended cropping periods while conserving water and nutrients.

iii) **Root inspection and disease management**: Without presence of growing media it is easier to inspect plant roots for signs of disease or nutrient deficiencies. This facilitates timely intervention and disease management practices ~ contributing to healthier plants and improved crop yields.

iv) **Prevention of salt build-up**: The flood and drain mechanism inherent in NFT helps preventing localized salt build-up in the root zone. It promotes uniform pH and conductivity levels so ensuring optimal nutrient uptake by plant roots.

v) **Environmental sustainability**: NFT is environmentally friendly with minimal risk of groundwater contamination compared to other hydroponic systems. This makes it a sustainable choice for modern agricultural practices.

4.3.2.6 Disadvantages of NFT

i) **Risk of system failure**: NFT systems are susceptible to failure in the event of power outage, pump failure or blockage in tubes. Backup electricity and regular checks of pumps and tubes are necessary to mitigate this risk.

ii) **Limitations for certain plants**: NFT may not be suitable for plants with large tap-root systems such as carrots as these require more space and support than provided in the channels.

iii) **Water and temperature fluctuations**: The reduced volume of the storage tank in NFT systems can lead to rapid fluctuations in water temperature and quality particularly in smaller setups. Careful monitoring and management are required to maintain optimal growing conditions.

iv) **Limited scalability**: While NFT systems are suitable for small to medium-scale operations, larger setups may require additional measures to address issues such as system failure and scalability.

Overall, NFT offers numerous advantages in terms of precise environmental control, optimal root health and sustainability. However, it is essential to address potential challenges such as system failure and limitations for certain plant types to ensure successful implementation and operation.

4.3.2.7 Variations and Adaptations of Continuous-Flow Solution Culture

While nutrient film technique is the most common type of continuous-flow solution culture, there are a few variations and adaptations of this method. By combining elements of NFT with other hydroponic techniques or modifications these systems can optimize nutrient delivery, enhance root health and maximize productivity in various growing conditions. Here are some hybrid types of continuous-flow solution culture that offer versatility and customization options to meet the specific needs and preferences of growers.

- ☆ **NFT with substrate addition**: In this hybrid approach a thin layer of substrate such as perlite, coconut coir or vermiculite is added to the bottom of the NFT channels or gullies. The substrate provides additional support for plant roots and helps retaining the moisture, enhancing root growth and nutrient uptake. This adaptation combines the benefits of NFT with the improved stability and resilience offered by substrate-based systems.
- ☆ **NFT with vertical farming**: Vertical NFT is a variation of the traditional NFT hydroponic system where plants are arranged vertically in stacked layers or towers. NFT channels or gullies are arranged vertically allowing for high-density planting and efficient use of space. This adaptation maximizes growing area while minimizing footprint so making it suitable for urban or space-limited environments. Here's an overview of vertical NFT:
    - ❖ **Design and configuration**: In vertical NFT systems plants are arranged in vertical channels or towers which allows for high-density planting in a limited space. The channels or towers are typically constructed from durable materials such as PVC, plastic or metal and these may have multiple levels or tiers to accommodate several layers of plants.
    - ❖ **Planting method**: Plants are inserted into openings or holes in the channels or towers with their roots suspended within the nutrient film flowing through the system. The channels or towers may be designed with a slight slope to facilitate the continuous flow of nutrient solution from top to bottom which ensures all the plants receiving adequate moisture and nutrients.

❖ **Advantages**:

Space efficiency: Vertical NFT systems maximize growing space by utilizing vertical space, making them ideal for urban environments or areas with limited land availability.

High density planting: By stacking plants vertically, vertical NFT systems allow for high-density planting so increasing the potential yield per square metre of growing area.

Efficient nutrient delivery: The continuous flow of nutrient solution over plant roots ensures efficient nutrient uptake so promoting healthy growth and higher yields.

❖ **Suitable crops**: Vertical NFT systems are suitable for a wide range of crops including leafy greens, herbs, strawberries and some other fruiting crops. Crops with shallow root systems or those that can be grown hydroponically are particularly well-suited for vertical NFT.

❖ **Maintenance**: Like traditional NFT systems, vertical NFT requires regular monitoring of nutrient levels, pH and system maintenance to ensure optimal growing conditions. Proper spacing between channels or towers is important to prevent shading and allowing for adequate air circulation and light penetration.

❖ **Applications**: Vertical NFT systems are commonly used in commercial hydroponic operations, indoor farms, vertical farms and urban agriculture projects. These offer a space-efficient and sustainable solution for growing fresh produce in urban areas or locations where land space is limited.

☆ **NFT with aeroponics**: Aeroponic NFT combines elements of aeroponics with nutrient film technique where plant roots are suspended in the air within enclosed channels or pipes. Nutrient solution is sprayed or misted onto the roots at regular intervals so providing both moisture and nutrients in a highly oxygenated environment. Here is an explanation of how these two systems can be integrated:

❖ **Aeroponics**: It is a hydroponic method where plant roots are suspended in the air and nutrient solution is misted or sprayed directly onto the roots. In aeroponic systems, plants are typically grown in enclosed chambers or containers with roots suspended in a nutrient-rich mist or fog. This method provides plants with ample oxygen and nutrient uptake efficiency as the roots are exposed to both ~ air and nutrient solution droplets.

❖ **Nutrient film technique (NFT)**: NFT is a hydroponic method where a thin film of nutrient solution flows continuously over the roots of plants so providing water and nutrients while maintaining oxygenation. In NFT systems, plants are typically grown in channels or gullies with a constant flow of nutrient solution passing over the roots.

- ❖ **Aeroponic NFT**: Combining elements of both aeroponics and NFT known as ~ "Aeroponic NFT" could refer to a hybrid hydroponic system that incorporates aspects of both the methods. In such a system, plants may be suspended in channels or gullies similar to traditional NFT setups but instead of a continuous flow of nutrient solution these are intermittently misted or sprayed with nutrient solution in a manner more akin to aeroponics. The integration of aeroponic principles into NFT systems may offer benefits such as increased oxygenation of the root zone, enhanced nutrient uptake efficiency and potentially improved plant growth *cum* yield. However, the specific design and implementation of an 'Aeroponic NFT' system would depend on various factors including the type of plants being grown, environmental conditions and grower preferences.

- ☆ **NFT with drip irrigation**: Hybrid NFT systems may integrate drip irrigation technology to supplement nutrient delivery or to provide additional water to the root zone. Drip emitters or tubing are installed alongside NFT channels or gullies to deliver controlled amounts of water and nutrients to specific plants or areas within the system. This combination allows for more precise control over nutrient delivery and can be particularly useful for crops with varying water and nutrient requirements.

- ☆ **NFT with deep water culture (DWC)**: In some hybrid setups, NFT channels or gullies are connected to a larger reservoir containing a DWC system. Plant roots are suspended in the nutrient solution within the DWC reservoir while NFT channels or gullies provide additional support and aeration for the plants. This hybrid approach combines the benefits of both NFT and DWC systems so providing optimal nutrient delivery, root oxygenation and plant support.

4.4 Deep Flow Technique (DFT)

The 'Deep Flow Technique' (DFT) is a hydroponic system where plants are grown in a continuous flow of nutrient-rich water. The system typically begins with a reservoir or tank where nutrient solution consisting of water and essential plant nutrients is stored. From there, a pump continuously delivers the solution to the growing area through a network of pipes or channels. These channels often made of plastic or other waterproof materials are long and shallow allowing the nutrient solution to flow through them.

Plants are placed in holes or slots in these channels with their roots submerged directly into the flowing nutrient solution. As the solution passes over the roots, these absorb water and nutrients so facilitating their growth. To ensure proper oxygenation of the roots, an air pump and air stones are commonly used to aerate the nutrient solution as it flows through the channels. This prevents the roots from becoming waterlogged so promotes healthy growth.

Regular monitoring of nutrient solution levels, pH and nutrient concentrations is essential to maintain optimal conditions for plant growth. Periodic cleaning and maintenance of the system are necessary to prevent clogs and algal buildup. Overall, the 'Deep Flow Technique' offers several advantages including efficient nutrient delivery, ease of setup and suitability for a wide range of plant types. It is commonly used in both commercial and hobbyist hydroponic operations for growing leafy greens, herbs and some fruiting plants.

4.4.1 Comparison with DWC

While both DWC and DFT systems involve plants being grown in a deep reservoir of nutrient solution, DWC systems do not have a continuous flow of solution. Instead, the plants are submerged in a static reservoir of nutrient solution, which is aerated to ensure proper oxygen levels for root health.

4.4.2 Comparision with NFT

Both NFT and DFT are hydroponic systems used for growing plants without soil. While these share similarities but these also have distinct differences:

i) Flow of Nutrient Solution

- ☆ **NFT**: In NFT systems a thin film of nutrient solution flows continuously along the bottom of narrow sloping channels. Plant roots are suspended above the flowing solution allowing these to absorb nutrients and oxygen directly from the film.
- ☆ **DFT**: In DFT systems plants are submerged in a deep layer of nutrient solution that flows continuously through shallow channels or trays. The roots are fully immersed in the solution, which provides both nutrients and oxygen.

ii) Root Environment

- ☆ **NFT**: Roots in NFT systems are exposed to the air and nutrient film, allowing for good oxygenation. However, roots can dry out quickly if the flow of nutrient solution is interrupted.
- ☆ **DFT**: Roots in DFT systems are fully submerged in the nutrient solution, providing ample access to both water and nutrients. Oxygenation of the roots is maintained through the flow of aerated solution.

iii) Channel Design

- ☆ **NFT**: NFT channels are typically narrow and sloping with a thin film of nutrient solution flowing along the bottom. These channels are often made of PVC or other materials.

- ☆ **DFT**: DFT channels or trays are wider and shallower than NFT channels, allowing for a deeper layer of nutrient solution. These are often made of plastic or other waterproof materials.

iv) Depth of Nutrient Solution

- ☆ **NFT**: In NFT systems, the nutrient solution typically flows in a thin film along the bottom of narrow sloping channels. The depth of the nutrient film in NFT systems is usually quite shallow, often just a few millimeters deep.
- ☆ **DFT**: In DFT systems (Deep Water Culture), the nutrient solution forms a deeper layer that completely submerges the plant roots. The depth of the nutrient solution in DFT systems can vary but is usually deeper than the thin film in NFT systems. DFT systems often have channels or trays that are several centimeters deep, providing a larger volume of nutrient solution for plant roots to access.

v) Application

- ☆ **NFT**: NFT systems are commonly used for growing leafy greens, herbs and small fruiting plants such as strawberries.
- ☆ **DFT**: DFT systems are also used for growing leafy greens and herbs but these can also accommodate larger fruiting plants such as tomatoes and peppers due to the deeper nutrient solution.

In a nutshell, while both NFT and DFT hydroponic systems involve continuous flow of nutrient solution, these differ in depth of the solution, the design of the channels and the root environment provided to the plants. Each system has its advantages and is suitable for different types of crops and growing conditions.

4.4.3 Deep Flow Technique (DFT) - Pipe System

In DFT - pipe system a shallow layer (usually 2-3cm deep) of nutrient solution flows through 10cm diameter PVC pipes. Plastic net pots containing plants are affixed to these pipes with the bottom of the pots submerged in the nutrient solution. Arrangement of PVC pipes can vary either in a single plane or in a zigzag pattern depending on the specific crops being grown. Planting materials such as old coir dust, carbonized rice husk or a mixture of both are commonly used to fill the net pots. To prevent the planting material from entering the nutrient solution a small piece of netting can be inserted as a lining in the pots. Alternatively, small plastic cups with holes on sides and bottom may be utilized instead of net pots.

As the recycled solution descends into the reservoir it becomes aerated enriching the nutrient solution with oxygen. To ensure proper drainage and nutrient flow, the PVC pipes must have a slope or drop of 1:30 to 1:40. Painting the PVC pipes white can help mitigate the heating of the nutrient solution, maintaining optimal growing conditions. This versatile system can be implemented in both open spaces and protected structures, offering flexibility in hydroponic cultivation methods.

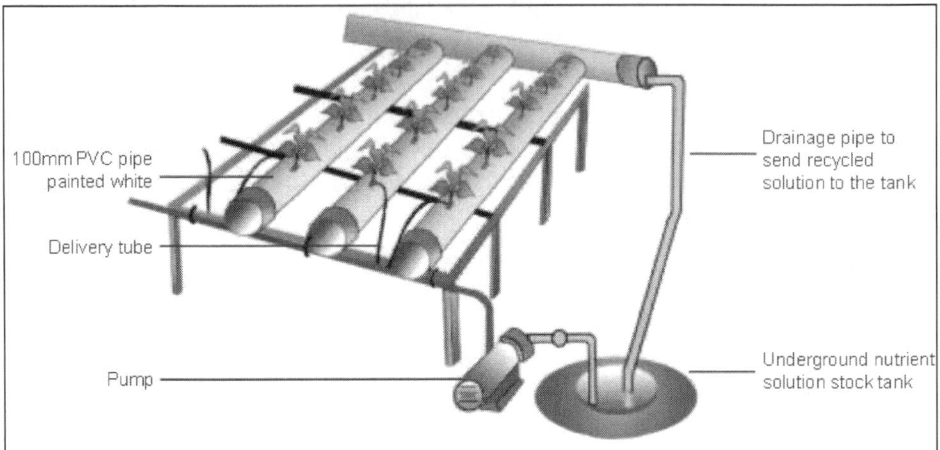

DFT system

Source: **http://ruaf-asia.iwmi.org/Data/Sites/6/PDFs/H_Eng.pdf**

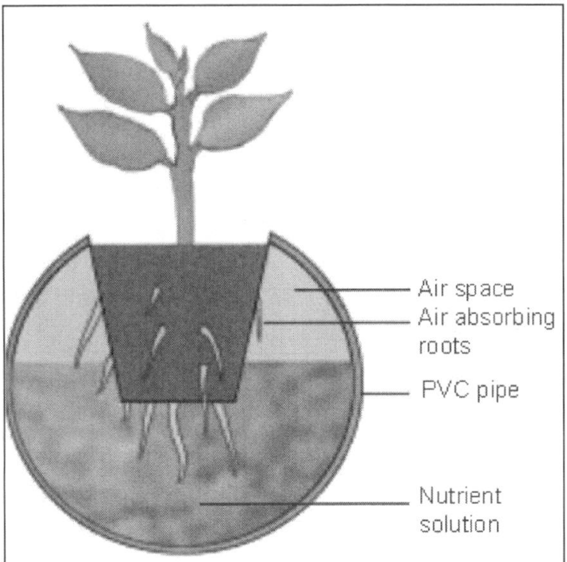

Cross section of a PVC pipe system DFT

Source: http://ruaf-asia.iwmi.org/Data/Sites/6/PDFs/H_Eng.pdf

4.5 Aeroponics

Aeroponics stands at the forefront of modern agricultural innovation offering a paradigm shift in the way we cultivate plants. This cutting-edge technique derived from the Greek words '*aero*' (air) and '*ponos*' (labour) represents a departure from traditional soil-based farming methods, instead suspending plant roots in a nutrient-rich mist or air environment.

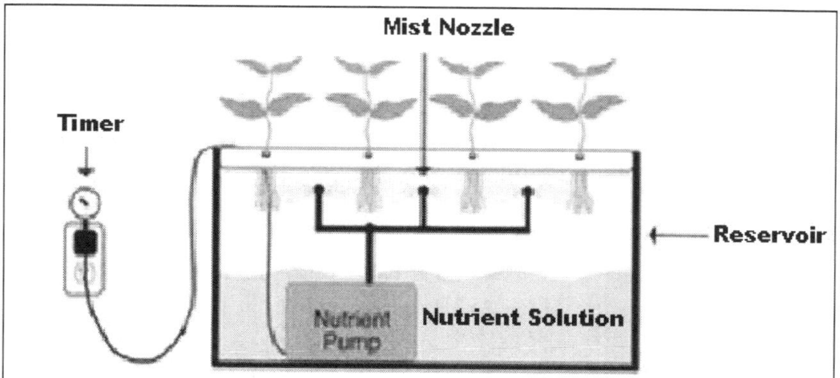

An Aeroponics system displaying roots suspended in root chamber.

In today's market, a wide range of aeroponics systems are available catering to various scales of cultivation from small-scale in-house crop production to large commercial operations. These systems offer growers the flexibility to adapt aeroponic technology to their specific needs and production goals. In a true aeroponics apparatus plants enjoy unrestricted access to atmospheric CO_2 for photosynthesis while their roots have full access to oxygen present in the air.

Lettuce Grown in an Aeroponics Apparatus.

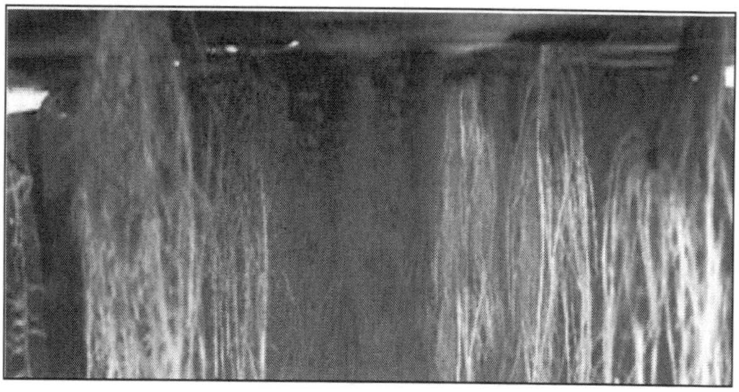

Roots Suspended in Aeroponics Chamber.

To deliver essential nutrients to plant roots aeroponics equipment employs a variety of precision delivery mechanisms such as sprayers, misters, foggers and other devices. For instance, hydro-atomizing sprays operating at 360° can effectively cover large areas of roots ensuring even nutrient distribution and uptake. Alternatively, ultrasonic foggers can be utilized in low-pressure aeroponics setups to finely mist nutrient solutions, promoting optimal nutrient absorption while minimizing waste. These innovative delivery systems not only provide plants with the vital nutrients they need for growth and development but also contribute to the efficient utilization of resources, making aeroponics a sustainable and environmentally friendly cultivation method. Also see "**Fogponics**".

4.5.1 History and Evolution

Carter (1942) conducted groundbreaking research on air culture growing, introducing a method for cultivating plants in water vapour to enable thorough root examination. Since 2006, aeroponics has gained widespread adoption in agriculture globally as highlighted in a report by NASA Spinoff (2006). In 1944, L.J. Klotz made a significant breakthrough by implementing misting techniques for citrus plants which proved instrumental in his investigations into diseases affecting citrus and avocado roots. In 1952, G.F. Trowel achieved success in cultivating apple trees using a spray culture technique (Stoner, 1983).

The term 'aeroponics' was coined by Dr. F. W. Went (1957) ~ a Dutch-American plant physiologist. Dr. Went is widely recognized for his pioneering work in plant physiology including his research on the effects of misting nutrient solutions on plant roots (Stoner, 1983). Dr. Went's contributions laid the foundation for further research and development in the field of aeroponics.

Stoner in 1983 initiated his groundbreaking venture into aeroponic technology by filing his inaugural patent. This marked the inception of a series of enterprises dedicated to researching and propelling the advancement of aeroponics. Among these ventures, 'Genesis Technology Incorporated' (GTi) emerged as a pioneer spearheading the introduction of large-scale aeroponic systems into greenhouse environments for commercial crop cultivation.

Genesis machine (1983): In 1983, GTi introduced the first commercially available aeroponic system ~ the Genesis Machine, inspired by futuristic technology from Star Trek II: The Wrath of Khan. This innovative device, initially named the 'Genesis Rooting System', featured an open-loop water-driven system regulated by a microchip. It delivered a high-pressure hydro-atomized nutrient spray within an aeroponic chamber seamlessly integrating with standard water faucets and electrical outlets for effortless operation. Meanwhile, Stoner at Colorado State University developed 'Organic Disease Control' (ODC) ~ a pioneering formula tailored to boost plant immunity in aeroponic environments. His groundbreaking work attracted attention from NASA and earned widespread acclaim for its innovative approach to plant growth and disease prevention.

Aeroponic propagation (cloning): Aeroponics revolutionizes plant cloning, simplifying the propagation of previously challenging species. Delicate hardwoods and cacti prone to bacterial infections in traditional methods thrive in aeroponic systems. Enhanced root hair development in aeroponics promotes nutrient absorption and growth (Soffer and Burger, 1988). Continuous nutrient supply boosts root development and plant health (Santos and Fisher, 2009) while absence of traditional media reduces root disease risks (Mehandru *et al.*, 2014). Aeroponics is pivotal for propagating plants with historically low success rates, including medicinal species like *Leptadenia reticulata* (Mehandru *et al.,* 2014). It aids the propagation of elm trees, combating Dutch elm disease (Oakes *et al.,* 2009). Compared to misters, aeroponics boasts higher success rates and addresses water consumption, sanitation and nutrient leaching concerns (Peterson *et al.,* 2018).

Air-rooted transplants: Aeroponics revolutionizes tissue culture, streamlining labour-intensive processes and bypassing conventional planting stages. Directly by cloning and transplanting air-rooted plants into open soil it eliminates the need for intermediate steps, reducing labour and risks of disease. Aeroponic root systems exhibit enhanced resilience, leading to improved plant health and resistance to infections. GTi's all-plastic aeroponic system, patented in 1985, marks a significant advancement, enabling plants to thrive without soil or traditional hydroponics. Aeroponics emerges as a time-saving cost-effective technique with vast potential to revolutionize agriculture and promote sustainability.

Genesis growing system (1985): In 1985, GTi introduced the 'Genesis Growing System' ~ a significant advancement in aeroponics. This second-generation hardware featured a closed-loop system ~ a groundbreaking innovation. Unlike its predecessors, the 'Genesis Growing System' recycled effluent regulated by a microprocessor for precise nutrient delivery and environmental control. This system pioneered seed germination support, distinguishing it as the world's first aeroponic system capable of both growth and harvest. Remarkably, many early aeroponic systems including the 'Genesis' still operate effectively today, showcasing the durability and effectiveness of aeroponics technology.

Commercialization: Aeroponics transitioned from laboratories to commercial cultivation in the mid-20th century. B. Briggs pioneered commercial aeroponics in 1966, demonstrating robust root growth in air-rooted cuttings. L. Nir patented an aeroponic apparatus in 1982, utilizing compressed low-pressure air for nutrient delivery. John Prewer, a British researcher conducted successful aeroponic experiments in 1976 producing mature lettuces in just 22 days. R. Stoner patented a microprocessor interface for aeroponic systems in 1983, advancing commercial applications. 'General Hydroponics Europe' (GHE) introduced the Aerogarden system in 1990s, popularizing aeroponic growing in the hobby market. Nutriculture developed scalable droplet-aeroponic systems like the Amazon, showcasing superior results compared to traditional techniques.

Aeroponics in space: NASA funded Stoner's aeroponics research, collaborating with him in 1997 to develop an aeroponic experiment for Mir Space Station. The significance of aeroponics to NASA revolves around efficient resource utilization. In long-term space missions, self-sufficiency in food production is essential to reduce payload weight and conserve water. Aeroponics offers faster growth rates, reduced equipment weight (soil-less) and minimized water usage ~ making it an attractive solution for space travel. Beyond NASA, the EDEN ISS project in Antarctica focuses on safe food production for space stations and planetary outposts with 'Future Exploration Greenhouse' investigating aeroponics among other cultivation techniques.

4.5.2 Components of Aeroponic System

A typical aeroponics system comprises several key components:

i) **Plant support structure**: In aeroponics systems, plants are ingeniously supported through strategically positioned apertures in a panel cover typically fashioned from robust materials like expanded polystyrene, PVC or similar durable substances. This meticulous arrangement optimally exposes the roots to the nutrient-rich mist, fostering optimal growth conditions. Within the confines of an aeroponics setup, plants are nurtured within a controlled or partially enclosed environment. Here, the lower stems and suspended roots of the plants are delicately misted or sprayed with a nutrient-rich water solution. This innovative approach involves suspending the plant roots within a spacious growth chamber while foliage and canopy extend above. To facilitate this unique setup a sophisticated plant support structure is employed. This structure not only segregates the roots but also provides essential stability to entire plant. In instances where larger plants bearing substantial weight are cultivated, specialized trellising mechanisms are utilized. These mechanisms ensure proper support for the plants, effectively mitigating any potential structural strain and promoting healthy growth.

ii) **Growth chamber**: Plants are housed within a controlled environment such as a greenhouse or enclosed chamber where environmental factors such as temperature, humidity and light intensity can be carefully regulated.

iii) **Root zone**: Plant roots are suspended in the air or supported by a structure, allowing for ample oxygenation and nutrient absorption.

iv) **Nutrient delivery system**: Nutrient solution is atomized or sprayed onto the roots at regular intervals, ensuring continuous nourishment.

v) **Reservoir**: A reservoir stores the nutrient solution which is then pumped or misted into the root zone as needed.

vi) **Monitoring and control system**: Sensors and controllers monitor environmental parameters and adjust nutrient delivery accordingly, optimizing plant growth conditions.

4.5.3 Principles of Aeroponic System

This innovative approach relies on several key principles:

1. Root Zone Oxygenation

Plant roots are suspended in the air or supported by a structure allowing for ample oxygenation and nutrient absorption. Root zone oxygenation in aeroponic systems plays a pivotal role in facilitating optimal plant growth and health. Unlike traditional hydroponic systems where roots are submerged or traditional soil-based cultivation methods where roots naturally receive oxygen from the air pockets in the soil, aeroponic systems require specialized mechanisms to ensure roots receiving an adequate oxygen supply while suspended in the air or mist environment. By suspending roots in the air, aeroponics maximizes oxygen uptake so promoting robust root development and enhancing overall plant health. Here's a comprehensive overview of root zone oxygenation in aeroponic systems:

- **Importance of oxygenation**: Oxygen is essential for cellular respiration in plant roots facilitating nutrient uptake, energy production and metabolic processes. In aeroponic systems where roots are exposed to air rather than submerged in water, adequate oxygenation is critical to prevent suffocation and to promote healthy root development.
- **Oxygen delivery mechanisms**: Aeroponic systems utilize various methods to deliver oxygen to plant roots:
  - **Misting or spraying**: Nutrient solution is atomized or sprayed onto the roots at regular intervals enveloping them in a fine mist that not only delivers nutrients but also carries oxygen.
  - **Aeration devices**: Air pumps or aerators may be incorporated into the system to inject oxygen directly into the nutrient solution or create turbulence so increasing oxygen levels.
  - **Root zone design**: Design considerations such as root chamber size, spacing and orientation can impact oxygen diffusion to the roots. Ensuring adequate airspace around roots and optimizing root distribution can enhance oxygen uptake.
- **Optimizing oxygen levels**: Monitoring and controlling oxygen levels in the root zone are essential for maximizing plant growth and preventing oxygen deprivation. Factors to consider include:
  - **Nutrient solution composition**: Balancing nutrient concentrations and pH levels can optimize oxygen solubility in the nutrient solution.
  - **Aeration timing**: Implementing a suitable feed/pause cycle ensures roots receiving intermittent nutrient solution while allowing for periods of oxygenation during pause phases.
  - **Water droplet size**: Fine water droplets produced during misting should be optimized to strike a balance between nutrient delivery and oxygenation. Excessively fine droplets may impede oxygen diffusion while larger droplets may provide inadequate coverage.
- **Benefits of adequate oxygenation**: Proper root zone oxygenation offers numerous benefits, including:
  - Enhanced nutrient uptake and assimilation.
  - Stimulated root growth and development.

- ❖ Improved resistance to stressors such as drought and disease.
- ❖ Increased overall plant vigour and yield potential.

☆ **Challenges and considerations**: Despite its importance, maintaining optimal oxygen levels in aeroponic systems can pose challenges such as:

- ❖ System malfunctions or power outages disrupting oxygen delivery.
- ❖ Root zone saturation or drying out due to improper misting or feeding cycles.
- ❖ Root damage or stunting resulting from inadequate oxygenation.

☆ **Precise nutrient delivery**: Aeroponic systems utilize misting or spraying technology to deliver nutrients directly to root zone. Precise nutrient delivery in aeroponic systems is achieved through a combination of advanced technology, careful monitoring and strategic management of environmental factors. By optimizing nutrient uptake and minimizing waste, aeroponic systems maximize plant productivity and contribute to sustainable agriculture practices. This precise nutrient delivery method ensures plants receiving essential macronutrients, micronutrients and water vapour in optimal concentrations, supporting accelerated growth and higher yields. Several key factors contribute to achieving precise nutrient delivery in aeroponic systems:

- ❖ **Atomization technology**: Aeroponic systems utilize specialized atomization technology to break down the nutrient solution into ultra-fine droplets. This ensures uniform coverage of root zone and facilitates efficient nutrient absorption by plant roots.
- ❖ **Delivery timing and duration**: Nutrient solution is delivered to roots at regular intervals typically through short bursts or pulses. The timing and duration of nutrient delivery cycles are carefully calibrated to meet specific requirements of plants being grown, ensuring optimal nutrient uptake while preventing over-saturation or dehydration of roots. Here is a detailed overview of nutrient delivery intervals in aeroponics:
- ❖ **Continuous nutrient delivery**: Some aeroponic systems employ continuous nutrient delivery where a constant flow of nutrient solution is sprayed or misted onto the plant roots throughout the day. Continuous nutrient delivery ensures that plants have continuous access to essential nutrients and water so promoting steady growth and nutrient uptake.
- ❖ **Intermittent nutrient delivery**: Many aeroponic systems utilize intermittent nutrient delivery cycles where nutrient solution is delivered to the roots at regular intervals followed by periods of rest. The frequency and duration of nutrient delivery cycles vary depending

on factors such as plant species, growth stage, environmental conditions and system design. A common feed/pause cycle involves short bursts of nutrient delivery lasting a few seconds (1.5 to 2 seconds) followed by longer pause periods ranging from 1.5-2 minutes to hours. Typical feed/pause intervals may range from several seconds to several minutes depending on specific requirements of the plants and the system. This cycle strikes a balance allowing roots to uptake essential nutrients while ensuring sufficient oxygenation during the pause phase.

Maintaining the nutrient cycle for aeroponics crop production at home could be a hectic task if a properly designed aeroponics apparatus is not being used. Special care should be taken to maintain the nutrient cycle. Too long or too short pause cycle discourages both lateral root growth and root hair development. Ideally, roots should never be more than slightly moist or excessively dry. A typical feed/pause cycle is 1.5 to 2 seconds on, followed by nearly 1.5 to 2 minutes pause. However, when an accumulator system is incorporated, cycle times can be further reduced to nearly 1 second on and nearly 1 minute pause.

- ❖ **Accumulator systems**: Some advanced aeroponic systems incorporate accumulator or buffering tanks to optimize nutrient delivery intervals and minimize fluctuations in nutrient availability. Accumulator systems store a reservoir of nutrient solution that can be gradually released into the root zone at precise intervals, ensuring consistent nutrient delivery and uptake by the plants. By buffering nutrient delivery, accumulator systems help maintain stable growing conditions and promote uniform plant growth throughout the growth cycle.
- ❖ **Automation and control**: Nutrient delivery intervals in aeroponics systems are often controlled and regulated by automated systems equipped with sensors, timers and controllers. These systems monitor key parameters such as nutrient concentration, pH levels, root moisture content and environmental conditions adjusting nutrient delivery accordingly. Advanced aeroponic systems may utilize feedback loops and algorithms to optimize nutrient delivery intervals based on real-time data and plant responses.
- ❖ **Optimization and fine-tuning**: Nutrient delivery intervals should be optimized and fine-tuned based on factors such as plant growth stage, nutrient requirements, root development and environmental conditions. Regular monitoring and observation of plant health and performance can help identify any signs of nutrient deficiencies or imbalances, allowing for adjustments to be made to nutrient delivery intervals as needed.

- ❖ **Nutrient composition and concentration**: The composition and concentration of the nutrient solution are tailored to the nutritional needs of the plants at different stages of growth. This precise control allows growers to optimize plant performance and maximize yield while minimizing nutrient waste.
- ❖ **Monitoring and control systems**: Aeroponic systems are equipped with sensors and controllers that monitor environmental parameters such as temperature, humidity, pH and nutrient levels in real-time. Automated control systems adjust nutrient delivery based on these parameters ensuring that plants receive the right nutrients in the right amounts at all the times.
- ❖ **Root zone environment**: Maintaining an optimal root zone environment is essential for precise nutrient delivery in aeroponic systems. Factors such as root temperature, oxygenation and moisture levels are closely monitored and controlled to promote healthy root growth and nutrient absorption.
- ❖ **System redundancy and backup**: To mitigate the risk of system failures or malfunctions aeroponic systems may incorporate redundancy measures and backup systems. Backup power sources, nutrient reservoirs and nutrient delivery mechanisms ensure continuity of nutrient supply, safeguarding against disruptions in plant growth.

☆ **Size of water droplets**: The size of water droplets in aeroponics systems is a pivotal factor influencing plant health and growth. Large water droplets while delivering adequate moisture can pose a challenge by limiting oxygen availability to the roots thereby hindering essential growth processes and nutrient uptake. On the other hand, ultra-fine droplets typically generated by ultrasonic misters may trigger excessive root hair growth without promoting the development of a robust lateral root system. This imbalance can undermine the sustained growth and overall vitality of plants within the aeroponics system. Thus, achieving an optimal droplet size is crucial for maximizing nutrient absorption and ensuring the flourishing success of plants in aeroponic cultivation. An apparatus that produces fine mist of nutrient solution is a **true aeroponics system**. This fine mist of solution is generated using a special pump. Whereas the system that produces droplets of nutrient solution is named as **droplet aeroponics system**. Although specially designed droplet aeroponics growing systems are also available but for small scale production at home a timer controlled low pressure pump that delivers nutrient solution *via* jets can be used.

☆ **Controlled environment of growth chamber**: In aeroponic systems, the growth chamber plays a pivotal role in creating an optimal environment for plant growth and development. Aeroponics growers can create ideal growing conditions for a wide range of crops whether under indoor facilities, greenhouses or controlled environment agriculture (CEA) setups where factors like temperature, humidity, lighting, CO_2 levels, air circulation and pest *cum* disease control can be meticulously regulated. This controlled environment provides plants with ideal growing conditions year-round regardless of external weather fluctuations. Here is an elaboration on the controlled environment aspects of aeroponic systems:

- **Temperature control**: Aeroponic systems maintain a controlled temperature range to promote optimal plant growth. Temperature sensors and climate control systems regulate temperatures within the growing environment, preventing extremes that could stress or damage plants. Typically, temperatures are maintained within a range conducive to plant growth which varies depending on the specific crop being cultivated. Aeroponic systems typically operate within a specific temperature range conducive to plant growth usually between 18-25°C for most crops. The growth chamber may be equipped with heating or cooling systems such as heaters, air conditioners or ventilation fans to maintain optimal temperatures year-round.
- **Humidity regulation**: Adequate humidity levels are crucial for preventing excessive moisture loss from plant tissues and promoting transpiration. Humidity levels are closely monitored and controlled in aeroponic systems to prevent excessive moisture loss from plant tissues and to minimize the risk of fungal diseases. The growth chamber may incorporate humidity controllers or misting systems to regulate humidity levels and prevent fluctuations that could stress plants. Humidity sensors and humidifiers or dehumidifiers adjust moisture levels within the growing environment to maintain an optimal balance for plant health.
- **Lighting management**: Adequate lighting is essential for photosynthesis and plant growth. In aeroponic systems, artificial lighting sources such as LED or fluorescent lights are often used to supplement natural sunlight or to provide illumination in indoor growing environments. Light intensity, duration and spectrum can be adjusted to mimic natural sunlight and promote optimal plant development throughout the growth cycle. Light timers and dimmers may be used to simulate natural day-night cycles and optimize light exposure for different stages of plant growth.
- **Carbon dioxide (CO_2) enrichment**: Carbon dioxide is a vital component of photosynthesis so maintaining optimal CO_2 levels enhances plant growth and productivity. CO_2 sensors and CO_2

supplementation systems (CO_2 generators or tanks) are employed in aeroponic environments to ensure that plants have access to sufficient CO_2 for photosynthetic processes particularly in enclosed or indoor growing spaces where natural CO_2 exchange may be limited.

- ❖ **Air circulation and ventilation**: Proper air circulation and ventilation are crucial for maintaining a healthy growing environment in aeroponic systems. Fans and ventilation systems help regulate airflow, prevent the buildup of heat and humidity and distribute fresh air throughout the growing space. Adequate air circulation also promotes exchange of gases and prevents the development of stagnant air pockets so reducing the risk of fungal and bacterial infections.
- ❖ **Pest and disease management**: The growth chamber should be designed to minimize the risk of pest and disease outbreaks which can be detrimental to aeroponic crops. In aeroponics systems, maintaining a controlled environment is essential to nurture robust plant growth and mitigate the risks posed by pests and diseases. Notably, the minimal contact between plants significantly reduces the likelihood of disease transmission while each precisely timed spray pulse can be thoroughly sterilized bolstering the system's defenses against harmful pathogens. Moreover, the system's design allows for swift removal of any diseased plants without compromising the health of neighbouring ones, underscoring its effectiveness in disease management and ensuring the overall vitality of the crop. Controlled environments in aeroponic systems minimize the risk of pest infestations and diseases by creating inhospitable conditions for pathogens and pests. Integrated pest management strategies including regular monitoring, sanitation practices like regular cleaning and disinfection of equipment and surfaces and biological control agents help prevent and mitigate pest and disease outbreaks without need for chemical pesticides.

☆ **Water efficiency**: Aeroponic systems offer a highly efficient way to grow crops with minimal water usage. By delivering water and nutrients in a mist or spray form directly to the roots and minimizing evaporation losses, aeroponics helps conserve water resources and promotes sustainable agriculture practices. This targeted nutrient delivery method minimizes water usage and reduces the risk of water waste or runoff, making aeroponics an environmentally sustainable option for crop cultivation. Here's how aeroponics achieves water efficiency:

- ❖ **Precise nutrient delivery**: Aeroponic systems deliver water and nutrients directly to plant roots in the form of a fine mist or spray. This targeted approach ensures that plants receive only the amount of water they need so minimizing wastage.

- **Reduced evaporation losses**: Unlike traditional soil-based agriculture where water can be lost through evaporation from soil surface, aeroponic systems minimize evaporation losses by delivering water directly to the roots. This helps conserve water and ensures efficient use of resources.
- **Recirculation of nutrient solution**: In many aeroponic systems, the nutrient solution is recirculated ~ meaning that any excess water not absorbed by the plants is collected and reused. This closed-loop system minimizes water wastage and reduces the overall water footprint of the operation.
- **Optimal moisture levels**: Aeroponic systems maintain optimal moisture levels around the plant roots, preventing both overwatering and underwatering. This ensures that plants have access to water when needed without the risk of waterlogging or drought stress.

☆ **Vertical farming potential**: Aeroponic systems are well-suited for vertical farming where crops are grown in stacked layers. This vertical arrangement allows for efficient use of space and enables multiple crops to be grown in a smaller footprint ~ further enhancing water efficiency.

☆ **Space optimization**: In commercial aeroponic operations, the growth chamber may be configured to maximize space utilization and crop density thereby increasing overall production efficiency. Vertical farming techniques, shelving systems or customized rack structures may be employed to accommodate multiple tiers of plants within the limited footprint of the growth chamber. This space-efficient design allows for higher crop yields per square foot making aeroponics particularly well-suited for urban farming or indoor cultivation. Here's an in-depth exploration of how aeroponic systems optimize space:

- **Vertical farming**: Aeroponic systems excel in vertical farming setups where plants are stacked in multiple layers to maximize space utilization. By stacking growing modules vertically aeroponic systems can significantly increase the number of plants grown per square meter compared to traditional horizontal farming methods. This vertical arrangement allows for cultivation of a larger crop volume within a smaller footprint, making aeroponics ideal for urban agriculture and indoor farming operations where space is limited.
- **Compact footprint**: Unlike soil-based agriculture which requires large tracts of arable land aeroponic systems can be implemented in compact indoor or greenhouse environments. Absence of soil and efficient use of vertical space enable aeroponic systems to achieve high crop densities within a relatively small footprint. This compact design makes aeroponics suitable for urban farming initiatives,

rooftop gardens and other space-constrained settings where land availability is limited.

- ❖ **Customizable layouts**: Aeroponic systems offer flexibility in layout and design, allowing growers to tailor their setups to fit specific spatial constraints and production goals. Modular components can be configured and rearranged to optimize space utilization and accommodate different crop types, growth stages and environmental conditions. Growers can adjust the number and size of growing modules, the spacing between plants and the height of vertical towers to maximize growing area and efficiency.
- ❖ **High plant density**: Aeroponic systems support high plant densities with multiple plants grown closely together in a controlled environment. By minimizing the space between plants and optimizing nutrient delivery aeroponic systems can achieve greater crop yields per unit area compared to traditional farming methods. This high plant density maximizes resource efficiency allowing growers to produce more food using less land, water and nutrients.
- ❖ **Year-round production**: Aeroponic systems enable year-round production of fresh fruits, vegetables and herbs regardless of seasonal or climatic limitations. By providing a controlled environment with optimal light, temperature and humidity levels aeroponic systems allow for continuous cultivation and harvesting throughout the year. This consistent production cycle further enhances space utilization as growers can maximize the productivity of their growing areas without being restricted by external factors.

☆ **Pest and disease management**: With plants suspended in mid-air and minimal contact between individuals aeroponic systems are inherently less susceptible to pests and diseases compared to soil-based agriculture. Controlled environment and sterile nutrient delivery system reduce the risk of pathogen transmission further enhancing plant health and productivity. While aeroponics inherently reduces the risk of certain pests and diseases due to absence of soil and reduced contact between plants proactive measures are still necessary to prevent and mitigate potential issues. Here is a comprehensive guide to pest and disease management in aeroponic systems:

- ❖ **Preventative measures**

  Hygiene: Ensure that all the equipments including reservoirs, pipes and nozzles are thoroughly cleaned and sanitized before use to prevent the introduction of pathogens.

  Quarantine: Introduce new plants or cuttings into the system only after careful inspection to avoid introducing pests or diseases.

Plant health: Start with healthy disease-free plant material and maintain optimal growing conditions to promote vigorous plant growth and resilience against pests and diseases.

- **Monitoring and early detection**: Regularly inspect plants for signs of pests such as aphids, mites or whiteflies as well as symptoms of diseases including leaf spots, wilting or abnormal growth. Install monitoring systems such as sticky traps or automated sensors to detect pests or changes in environmental conditions early.
- **Integrated pest management (IPM)**: Implement an integrated approach to pest management incorporating cultural, mechanical, biological and chemical control methods as needed. Cultural methods include proper spacing of plants, pruning of affected plant parts and maintaining optimal growing conditions to minimize pest and disease pressure. Mechanical methods such as using barriers or traps can physically prevent pests from accessing plants or disrupting their lifecycle. Biological control involves introducing beneficial insects or microbes that prey on or compete with pests, effectively reducing their populations. Chemical control should be used judiciously as a last resort opting for least-toxic pesticides and following label instructions carefully to minimize environmental impact.
- **Sanitation and sterilization**: Regularly clean and sterilize all components of the aeroponic system including reservoirs, piping and nozzles to prevent buildup and spread of pathogens. Dispose of any plant debris or fallen leaves promptly to eliminate potential breeding grounds for pests and diseases.
- **Isolation and treatment**: Immediately isolate any plant showing signs of pests or diseases to prevent further spread within the system. Treat affected plants with appropriate remedies such as insecticidal soap for pests or fungicides for fungal diseases following label instructions carefully.
- **Environmental controls**: Maintain optimal environmental conditions including temperature, humidity and airflow to create an inhospitable environment for pests and diseases. Adequate ventilation and air circulation can help reduce humidity levels and prevent the buildup of fungal pathogens.
- **Regular maintenance and monitoring**: Continuously monitor the aeroponic system for any signs of pest infestation or disease outbreaks, taking prompt action to address any issues that arise. Conduct routine maintenance tasks such as cleaning filters, checking nutrient levels and adjusting pH to ensure optimal system performance and plant health.

4.5.4 Research and Innovation

Aeroponics serves as a valuable platform for scientific research and experimentation in plant physiology, genetics and crop breeding. Its controlled environment and precise nutrient delivery system enable researchers to study plant responses to various environmental stimuli and stressors, paving the way for innovative advancements in agriculture. Here are some key areas of research and innovation in aeroponics:

- **Optimizing nutrient delivery**: Researchers are continually investigating ways to enhance the nutrient delivery system in aeroponics to maximize plant growth and health. This includes studying the composition of nutrient solutions, optimizing spray techniques and developing advanced sensors *cum* monitoring systems to precisely control nutrient levels.
- **Root zone management**: Understanding the dynamics of the root zone in aeroponic systems is crucial for optimizing plant growth. Researchers are exploring strategies to improve root health such as optimizing root spacing, preventing root zone diseases and for promoting root branching *cum* development.
- **Environmental control**: Maintaining optimal environmental conditions within aeroponic systems is essential for maximizing crop productivity. Research is focused on developing innovative technologies for controlling factors such as temperature, humidity, light intensity and CO_2 levels to create ideal growing conditions for plants.
- **Crop selection and genetics**: Researchers are exploring the potential of aeroponics for growing a wide range of crops including fruits, vegetables, herbs and specialty crops. There is growing interest in utilizing aeroponics for crop breeding and genetic research to develop varieties that are well-suited for aeroponic cultivation.
- **Automation and robotics**: Automation plays a crucial role in modern aeroponic systems enabling precise control of nutrient delivery, environmental conditions and monitoring. Research in this area focuses on developing advanced robotics and automation technologies to streamline operations, reducing labour costs and for improving overall system efficiency.
- **Space agriculture**: Aeroponics holds promise for space exploration missions as a sustainable method for growing food in extraterrestrial environments. Research in space agriculture may involve studying of plant responses to microgravity conditions, developing compact and efficient aeroponic systems for use in space habitats and exploring novel approaches to nutrient delivery and environmental control in space.
- **Sustainability and resource efficiency**: One of the primary goals of aeroponic research is to enhance sustainability and resource efficiency

in agriculture. This includes developing strategies for reducing water usage, minimizing nutrient waste and for optimizing energy consumption in aeroponic systems.

☆ **Integration with other technologies**: Researchers are exploring how aeroponics can be integrated with other innovative technologies such as artificial intelligence, machine learning and renewable energy systems to further enhance system performance, productivity and sustainability.

4.5.5 Applications of Aeroponics

Aeroponics has a wide range of applications across various sectors:

i) **Agriculture**: In commercial agriculture, aeroponics systems are used to grow a diverse array of crops including leafy greens, herbs, tomatoes, strawberries and more. Its ability to maximize space and conserve resources makes it particularly well-suited for urban farming and vertical cultivation.

ii) **Research**: Aeroponics is widely employed in scientific research and experimentation so facilitating studies on plant physiology and pathology, genetics and crop breeding. This method of plant cultivation has gained significant traction in commercial applications and has demonstrated remarkable success across various sectors. It has proven highly effective for tasks such as propagation, seed germination as well as the cultivation of leafy greens, tomatoes, seed potatoes and more. Its controlled environment and precise nutrient delivery system enable researchers to explore plant responses to various environmental stimuli and stressors.

iii) **Space agriculture**: With its minimal resource requirements and high efficiency, aeroponics holds promise for future space exploration missions. NASA and other space agencies are exploring aeroponics as a means of sustainably growing food crops in extraterrestrial environments such as Mars or lunar bases.

4.5.6 Benefits of Aeroponics

Aeroponics offers several distinct advantages over traditional farming methods:

i) **Water efficiency**: Aeroponics demonstrates exceptional water efficiency by delivering nutrients directly to the roots in a mist or spray form so conserving water significantly compared to conventional soil-based agriculture. Unlike traditional methods prone to water loss through evaporation or runoff, aeroponically grown plants optimize water usage showcasing an impressive reduction of approximately 65 per cent in water consumption when compared to plants cultivated in 'Static Solution Culture' or NFT hydroponic systems.

ii) **Optimizing growth ~ the role of oxygen and CO_2**: In aeroponic systems utilization of fine mist sprayers serves as a pivotal mechanism for ensuring robust root zone oxygenation which in turn catalyzes accelerated plant growth. These sprayers not only deliver essential nutrients but also effectively capture copious amounts of oxygen, directing it directly to the roots. This influx of oxygen fosters an environment conducive to rapid growth, enabling plants to thrive and flourish. Moreover, the unique design of aeroponic systems minimizes contact between the plant and any support structure. With nearly 99 per cent of the stem, leaves and roots suspended in the air, plants in aeroponics enjoy unparallel access to oxygen. This abundant oxygen supply fuels metabolic processes, enhances nutrient uptake and promotes vigorous growth. The elevated position of the plant ensures optimal exposure to atmospheric carbon dioxide which is utilized by leaves during photosynthesis to drive further growth and productivity.

iii) **Space optimization**: Aeroponic systems can be vertically stacked or arranged in tight configurations thus maximizing growing space and increasing crop yield per unit area.

iv) **Nutrient precision**: Another noteworthy advantage of aeroponic cultivation is its efficiency in nutrient utilization. Precise control afforded by aeroponics systems allows for tailored nutrient formulations, optimizing plant growth and health while minimizing waste. Aeroponically grown plants require approximately $1/4^{th}$ of the nutrient input compared to traditional soil-based agriculture or hydroponic systems. This not only conserves resources but also minimizes the risk of nutrient runoff and environmental contamination.

v) **Enhanced mineral absorption**: Aeroponically grown plants boast significantly higher levels of essential minerals compared to their counterparts cultivated through conventional methods. Research indicates that aeroponic plants may contain up to 80 per cent more essential minerals, contributing to superior nutritional quality and potential health benefits for consumers.

vi) **Versatility in plant species**: One of the key strengths of true aeroponics is its adaptability to accommodate diverse plant species. Whether it is leafy greens, herbs, vegetables or even fruit-bearing plants aeroponics provides a conducive environment for their growth and development.

vii) **Reduced pest and disease risk**: In aeroponics systems where plants are suspended in mid-air with minimal contact between individuals, the risk of pests and diseases is significantly reduced compared to traditional soil-based cultivation methods. The controlled environment of aeroponics allows for precise management of factors such as temperature, humidity, and nutrient levels. Consequently, aeroponic plants can effectively be shielded from pests and diseases which may lead to higher yields and allowing for cultivation at higher densities.

viii) **Prevention of waterlogging**: Unlike soil-based or hydroponic systems where roots can become waterlogged, aeroponics ensures that plant roots receive an optimal balance of air and moisture. This prevents the risk of overwatering and root suffocation thus promoting healthier root systems and overall plant vitality.

ix) **Mitigation of transplant shock**: Aeroponically grown plants exhibit remarkable resilience during transplantation with minimal risk of transplant shock. Unlike plants accustomed to soil or hydroponic environments, aeroponic plants seamlessly adapt to soil conditions without experiencing the stress typically associated with transplanting. This ensures smoother transitions and promotes uninterrupted growth and productivity.

4.5.7 Challenges and Considerations

While aeroponics offers numerous advantages, it is important to acknowledge and address its potential drawbacks to ensure successful implementation. Here is a more detailed exploration of the disadvantages of aeroponics:

i) **Initial investment**: The upfront cost of setting up an aeroponics system may be higher compared to soil-based agriculture or simpler hydroponic systems.

ii) **System complexity**: Aeroponic systems can be more complex and technologically advanced than traditional farming methods requiring careful monitoring and maintenance.

- ❖ **Complex nutrient cycle management**: Aeroponics systems require precise management of the nutrient cycle to sustain healthy plant growth and development. Maintaining the correct nutrient balance and pH levels in the nutrient solution can be challenging particularly for novice growers. Proper nutrient management involves monitoring nutrient concentrations, adjusting pH levels and replenishing depleted nutrients to meet the demands of growing plants. Without careful attention to the nutrient cycle plants may experience deficiencies or excesses, resulting in stunted growth, nutrient lockout or other physiological disorders.

- ❖ **Hectic maintenance requirements**: The meticulous nature of nutrient cycle maintenance in aeroponics systems can be time-consuming and labour-intensive particularly for home growers. Regular monitoring of nutrient levels, pH and system components is necessary to prevent nutrient imbalances, equipment malfunctions and other issues that could compromise plant health. Home growers may need to invest additional time and effort in learning and implementing proper maintenance practices to ensure success of their aeroponic cultivation endeavours.

iii) **Risk of system failures**: Aeroponic systems are reliant on various components including pumps, timers and sensors to maintain optimal growing conditions. Power outages and equipment failures or malfunctions can disrupt nutrient delivery leading to plant stress, nutrient deficiencies or even crop failure. To mitigate the risk of system failures growers should implement redundancy measures such as backup power sources or redundant components and conduct regular maintenance checks to identify and address potential issues promptly. A hybrid approach integrating conventional hydroponics can be employed as a backup nutrition and water supply system. This strategic combination of aeroponics and conventional hydroponics provides a safety-net against potential system failures, safeguarding plant health and promoting steady *cum* uniform growth. In the event of an aeroponics system malfunction or power outage, the backup hydroponic system steps in to seamlessly sustain nutrient delivery and hydration to the plants. This continuity of care ensures that plants receive essential resources without interruption, minimizing stress and maximizing resilience.

iv) **Nozzle clogging**: One significant drawback of aeroponics systems is the potential for fine mist nozzles to become clogged over time with mineral deposits or organic matter. Clogged nozzles can impede the delivery of nutrient solution to plant roots leading to uneven nutrient distribution and compromised plant health. Regular maintenance and cleaning of nozzles are essential to prevent clogging and ensuring consistent nutrient delivery.

In response to the issue of nozzle clogging, innovative solutions have emerged. One such solution involves the utilization of special low-mass polymer materials in the design and construction of misting and spray jets. These advanced materials have been specifically engineered to resist mineralization and organic buildup, thereby maintaining the efficiency and effectiveness of the misting system over extended periods. By integrating these next-generation materials into aeroponic systems, growers can significantly reduce the frequency of maintenance tasks associated with nozzle cleaning and replacement. This not only streamlines operations but also ensures that plants receive the optimal nutrient solution delivery required for healthy growth and development.

4.5.8 Aeroponics and Tissue Culture

Aeroponics technology has emerged as a game-changer in the realm of tissue culture offering a revolutionary approach to plant propagation that significantly enhances efficiency, reduces labour and mitigates risks associated with traditional tissue culture techniques. The integration of aeroponics with tissue culture represents a transformative paradigm shift in plant propagation methodologies.

By leveraging the efficiency, precision and resilience of aeroponics, growers and researchers alike can revolutionize the way plants are cloned, propagated and transplanted thus ushering in a new era of sustainable agriculture and botanical innovation. Let us explore the synergies between aeroponics and tissue culture and the transformative impact they have on plant propagation methodologies.

4.5.8.1 Enhanced Cloning and Transplantation

Aeroponics streamlines the cloning process by providing an ideal environment for root development. Unlike traditional tissue culture methods that involve tedious steps and prolonged incubation periods aeroponics accelerates root growth thus allowing for faster cloning of plants.

Furthermore, aeroponics eliminates the need for stage I and stage II plantings into soil. In traditional tissue culture, stage I involves planting tissue culture plants in sterile media followed by stage II where these plants are expanded for eventual transfer into sterile soil. Aeroponics bypasses these intermediary steps by enabling direct transplantation of air-rooted plants into field soil thus saving time, labour and resources.

4.5.8.2 Resilient Root Systems

Aeroponic roots exhibit exceptional resilience and vigour so rendering them less susceptible to wilting, leaf loss and transplant shock compared to conventional tissue culture methods. The nutrient-rich mist environment of aeroponics fosters robust root growth resulting in healthier and more resilient plants that thrive upon transplantation.

4.5.8.3 Reduction of Labour and Disease Risks

Traditional tissue culture techniques entail labour-intensive processes and carry inherent risks of disease infection and failure. By streamlining the cloning and transplantation process aeroponics significantly reduces labour requirements and minimizes the likelihood of disease transmission.

4.5.8.4 Precision and Efficiency

Aeroponics offers unparalleled precision and efficiency in plant propagation with precise control over nutrient delivery, moisture levels and environmental conditions. This level of control ensures optimal growth conditions for plants leading to higher success rates and improved overall efficiency in tissue culture operations.

4.5.8.5 Integration with Tissue Culture Laboratories

Aeroponics systems can seamlessly integrate with tissue culture laboratories providing a continuum of propagation techniques from initial cloning to field transplantation. This integration streamlines the production pipeline, enhances workflow efficiency and facilitates the transition from laboratory-scale tissue culture to large-scale commercial production.

4.5.8.6 Applications and Future Directions

The synergistic combination of aeroponics and tissue culture holds immense promise across a wide range of agricultural sectors including horticulture, forestry and pharmaceuticals. Beyond traditional crop propagation, these technologies pave the way for innovative applications such as rare plant conservation, endangered species preservation and biopharmaceutical production.

The future of aeroponics holds exciting prospects as technology continues to advance and the global demand for sustainable agriculture grows. Here are some future directions and potential advancements in aeroponics:

i) **Automation and robotics**: Advancements in automation and robotics are expected to revolutionize aeroponic farming thus reducing labour requirements and increasing efficiency. Automated systems could monitor plant health, adjust nutrient delivery and perform maintenance tasks thus optimizing growing conditions and maximizing yields.

ii) **Sensor technology**: Integration of advanced sensor technology will enable real-time monitoring of environmental parameters such as temperature, humidity, nutrient levels and plant health. Sensor data can be used to fine-tune nutrient formulations, optimize irrigation schedules and detect early signs of plant stress or disease which warrants for proactive intervention.

iii) **Artificial intelligence (AI)**: AI-driven algorithms can analyze large datasets collected from aeroponic systems to identify trends, patterns and optimal growing conditions. AI-powered decision support systems could provide growers with actionable insights and recommendations for optimizing crop production, resource utilization and energy efficiency.

iv) **Vertical farming and urban agriculture**: Aeroponics is well-suited for vertical farming and urban agriculture initiatives where space is limited and demand for locally grown produce is high. Future developments may focus on scalable modular aeroponic systems designed for high-density vertical farms in urban environments enabling year-round production of fresh nutritious crops.

v) **Nutrient formulations and delivery systems**: Research into innovative nutrient formulations and delivery systems will continue to enhance plant nutrition and health in aeroponic systems. Customized nutrient blends tailored to specific crop requirements and growth stages could improve yields, flavour profiles and nutritional quality of harvested produce.

vi) **Integration with renewable energy sources**: Aeroponic farms may increasingly incorporate renewable energy sources such as solar, wind or hydroelectric power to reduce reliance on conventional energy sources and minimize environmental impact. Off-grid aeroponic systems powered by renewable energy could be deployed in remote or resource-constrained regions, providing sustainable food production solutions.

vii) **Space agriculture and beyond**: Aeroponics has garnered interest from space agencies for its potential to sustainably grow food crops in extraterrestrial environments. Future research may explore the adaptation of aeroponic systems for space colonization missions, lunar or Martian bases and long-duration space flights.

viii) **Education and Outreach**: Education and outreach efforts will play a crucial role in promoting widespread adoption of aeroponic technology and empowering growers with knowledge and skills needed for successful implementation. Training programmes, workshops and educational resources will support growers in mastering aeroponic techniques and best practices for sustainable crop production.

In conclusion, the future of aeroponics is characterized by innovation, sustainability and the pursuit of global food security. By embracing technological advancements, harnessing renewable resources and fostering collaboration across disciplines, aeroponics has the potential to revolutionize agriculture and shape the future of food production on this planet and beyond.

4.5.9 Configurations and Designs of Aeroponics Systems

Aeroponics systems come in various configurations and designs each tailored to specific needs and applications. Here are some common types of aeroponics systems:

4.5.9.1 Low-Pressure Aeroponics System (LPA)

Low-pressure aeroponics systems represent a versatile and accessible approach to aeroponic cultivation particularly suited for home and small-scale setups. These systems operate at lower pressure levels compared to their high-pressure counterparts utilizing gravity-fed or low-pressure pumps to deliver nutrient solution to plant roots suspended in the air. Despite their lower pressure, low-pressure aeroponics (LPA) systems are highly effective in providing adequate oxygenation and nutrient delivery for robust plant growth.

Advantages

i) **Cost-effectiveness**: LPA systems are more affordable and easier to set up compared to high-pressure aeroponics (HPA) systems, making them ideal for hobbyists, educators and small-scale growers operating on a limited budget.

ii) **Accessibility**: With simpler components and lower maintenance requirements LPA systems are accessible to a wider range of growers including beginners and those with limited technical expertise.

iii) **Space efficiency**: LPA systems are well-suited for benchtop or compact growing spaces, making them suitable for urban environments or indoor gardening applications where space is limited.

iv) **Versatility**: These systems can easily be customized and scaled to accommodate different plant types, growth stages and experimental setups, allowing for versatility in experimentation and cultivation.

Components

In a typical low-pressure aeroponics system plant roots are suspended above a nutrient solution reservoir or inside a channel connected to the reservoir. A low-pressure pump powered by electricity or gravity delivers nutrient solution through jets or ultrasonic transducers creating a fine mist that envelops the roots. Misted nutrient solution then drips or drains back into the reservoir for recirculation. In low-pressure aeroponics systems the components and operation are crucial aspects that determine the efficiency and success of plant cultivation. Here is a comprehensive overview:

i) **Plant growth chambers or channels**: Plants are typically housed in growth chambers/channels where their roots are suspended above a reservoir containing nutrient solution. These channels may be constructed from various materials such as PVC, ABS or plastic. These are designed to support the plants while allowing easy access to root zone.

ii) **Nutrient solution reservoir**: The reservoir holds the nutrient solution that is used to nourish plant roots. It may be situated beneath the growth chambers and connected to them *via* tubing or piping.

iii) **Low-pressure pump**: A low-pressure pump powered by electricity or gravity is responsible for delivering nutrient solution to roots. Pump generates pressure to propel the solution through jets or ultrasonic transducers, creating a fine mist that surrounds the roots.

iv) **Jets or ultrasonic transducers**: Jets or ultrasonic transducers are the mechanisms through which the nutrient solution is dispersed into a fine mist. Jets release pressurized streams of solution while ultrasonic transducers use high-frequency vibrations to atomize the solution into tiny droplets.

v) **Recirculation system**: After misting the roots the nutrient solution drips or drains back into the reservoir for recirculation. This closed-loop recirculation system minimizes water and nutrient waste while ensuring continuous nourishment for the plants.

Operation

i) **Nutrient solution delivery**: The low-pressure pump draws nutrient solution from the reservoir and delivers it to the jets or ultrasonic transducers. The solution is then atomized into a fine mist, enveloping the suspended roots in a nutrient-rich fog.

ii) **Root nourishment**: The misted nutrient solution provides direct and efficient nourishment to the plant roots, ensuring optimal nutrient uptake and absorption. This method promotes vigorous root growth and enhances nutrient efficiency leading to healthier and more productive plants.

iii) **Recirculation and replenishment**: After misting the roots any excess solution not absorbed by the roots drips or drains back into the reservoir. The recirculation system continuously replenishes the reservoir with fresh solution, maintaining the desired nutrient concentrations and pH levels.

Challenges and Considerations

i) **Root dryness**: As plants grow and mature in LPA systems, the root systems may encounter dry sections leading to inadequate nutrient uptake and potential stress. Proper monitoring and adjustment of misting frequency and duration can help mitigate this issue.

ii) **Nutrient solution management**: LPA systems may struggle to effectively purify the nutrient solution, leading to accumulation of debris and unwanted pathogens. Regular maintenance including periodic cleaning and sterilization of components is essential to prevent contamination and ensure optimal plant health.

Future Developments and Research Opportunities

As interest in aeroponic cultivation continues to grow ongoing research and innovation in low-pressure aeroponics hold promise for further advancements. Future developments may focus on:

- ☆ Enhancing nutrient delivery efficiency and root zone oxygenation through improved misting technology and nutrient formulations.
- ☆ Developing automated monitoring and control systems to optimize nutrient solution management and environmental conditions.
- ☆ Exploring novel applications of low-pressure aeroponics in areas such as vertical farming, space agriculture and sustainable urban food production.

Additional Considerations

- ☆ **Nutrient solution monitoring**: Regular monitoring of nutrient solution parameters such as pH and nutrient concentrations is essential to ensure optimal plant nutrition.
- ☆ **Root zone temperature control**: Maintaining the appropriate root zone temperature is crucial for promoting root health and nutrient uptake.

- ☆ **Sterilization and sanitation**: Preventing contamination and pathogens in the nutrient solution and root zone is critical for preventing plant diseases and ensuring crop health.
- ☆ **Automation and control systems**: Advanced automation and control systems can optimize nutrient delivery, monitor environmental conditions and adjust system parameters for maximum efficiency and yield.

4.5.9.2 High-Pressure Aeroponics System (HPA)

Advantages

High-pressure aeroponics (HPA) systems represent a pinnacle of precision and efficiency in soilless cultivation employing advanced technologies to optimize plant growth and maximize yields. Here is a more detailed exploration of HPA systems and their key features:

i) **Advanced technology integration**: HPA systems are characterized by their integration of cutting-edge technologies for air and water purification, nutrient sterilization and pressurized nutrient delivery systems. These systems often incorporate sophisticated filtration and sterilization mechanisms to ensure purity and sterility of nutrient solution thus minimizing the risk of contamination and promoting optimal plant health.

ii) **Precision nutrient delivery**: HPA systems utilize high-pressure pumps and specialized misting nozzles to generate a fine mist of nutrient solution at elevated pressures. The high-pressure mist enables precise and uniform coverage of plant roots suspended in the air thus facilitating efficient nutrient absorption and oxygenation. By delivering nutrients directly to root zone with unparalleled accuracy, HPA systems optimize nutrient uptake and minimize waste resulting in superior plant growth and development.

iii) **Thorough root oxygenation**: The high-pressure mist produced by HPA systems ensures thorough root oxygenation thus creating an ideal environment for root development and metabolic activity. Adequate oxygenation of the root zone promotes healthy root growth, enhances nutrient assimilation and improves overall plant vigour and resilience to environmental stressors.

iv) **Suitability for high-value crops**: HPA systems are particularly well-suited for cultivation of high-value crops and plant specimens in commercial and research settings. Their precise nutrient delivery and high level of control over growing conditions make them ideal for producing premium-quality crops with superior flavour, aroma and nutritional content.

v) **Applications in research and innovation**: HPA systems play a vital role in advancing agricultural research and innovation thus providing researchers with a sophisticated platform for studying plant physiology, genetics and crop breeding. These systems enable scientists to manipulate and optimize growing conditions with unparalleled precision thus facilitating experiments and studies that contribute to the development of new crop varieties and cultivation techniques.

vi) **Cost considerations**: While HPA systems may entail higher initial costs compared to low-pressure aeroponics or other hydroponic systems, their long-term benefits in terms of crop quality, yield potential and research capabilities often justify the investment.

Components

i) **Pressure chamber**: The pressure chamber is the heart of the system where pressurized nutrient solution is stored. It typically consists of a reservoir tank equipped with a high-pressure pump and misting nozzles.

ii) **Misting nozzles**: Misting nozzles are strategically positioned within the pressure chamber to atomize nutrient solution into a fine mist. These nozzles ensure uniform coverage of plant roots suspended within the chamber.

iii) **Root chamber**: The root chamber houses plant roots and provides support for the plants. It may consist of trays, baskets or panels where plants are placed thus allowing their roots to hang freely in the mist.

iv) **Nutrient reservoir**: The nutrient reservoir stores the nutrient solution before it is pressurized and delivered to the pressure chamber. It may include a mix tank where nutrients are mixed with water before being pumped into the pressure chamber.

v) **Pump and control system**: High-pressure aeroponics systems rely on powerful pumps capable of generating pressures ranging from 60 to 150 psi. A control system regulates pump operation, misting intervals and nutrient solution concentrations thus ensuring optimal growing conditions.

Operations

i) **Nutrient solution delivery**: The system pumps nutrient solution from the reservoir into the pressure chamber where it is pressurized to high levels. Misting nozzles then atomize the pressurized solution into a fine mist thus enveloping plant roots suspended within the root chamber.

ii) **Misting cycles**: High-pressure aeroponics systems typically operate on misting cycles with intervals ranging from a few seconds to several minutes. These cycles ensure consistent nutrient delivery while allowing roots to access oxygen during brief pause periods.

iii) **Root zone environment**: The misting process creates a highly oxygenated environment around plant roots, promoting robust growth and nutrient absorption. Plants benefit from direct access to both water and oxygen thus resulting in accelerated growth rates and enhanced nutrient uptake efficiency.

Additional Information

i) **Nutrient solution composition**: High-pressure aeroponics systems require precise nutrient formulations to meet nutritional needs of plants. Nutrient solutions may include essential macro and micronutrients, pH buffers and supplements to promote healthy growth.

ii) **Environmental control**: Temperature, humidity and light levels must be carefully monitored and controlled to optimize plant growth and prevent stress. Automated climate control systems may be integrated with high-pressure aeroponics setups to maintain ideal environmental conditions.

iii) **Crop selection and optimization**: High-pressure aeroponics systems are suitable for a wide range of crops including leafy greens, herbs, vegetables and flowering plants. Crop selection, spacing and nutrient requirements should be carefully considered to maximize yields and quality.

iv) **Maintenance and troubleshooting**: Regular maintenance of pumps, misting nozzles and control systems is essential to ensure system reliability and performance. Growers should be familiar with troubleshooting procedures to address issues such as clogged nozzles, leaks or nutrient imbalances promptly.

4.5.9.3 Vertical Aeroponics Towers

Vertical aeroponics towers represent an innovative and space-efficient approach to soilless cultivation thus allowing growers to maximize crop production in limited urban and indoor environments. These towers are designed to maximize growing space by stacking multiple growing levels vertically. Plants are typically arranged in pockets or trays attached to the sides of the tower with roots suspended in the central nutrient chamber.

Vertical aeroponics towers are vertical structures designed to accommodate multiple levels of plant growth in a compact footprint. These towers utilize aeroponic technology where plant roots are suspended in air where these are periodically misted with a nutrient-rich solution. By stacking growing levels vertically, vertical aeroponics towers optimize space utilization and enable high-density cultivation thus making them ideal for urban farming, indoor gardening and small-scale commercial production.

Key Components

i) **Tower structure:** Vertical aeroponics towers are typically constructed from durable materials such as PVC, HDPE (High Density Poly Ethylene) or food-grade plastic. The tower structure consists of stacked growing levels, each level equipped with plant support structures and misting systems.

ii) **Growing medium:** Instead of soil, plants in vertical aeroponics towers are grown using inert substrates such as rockwool cubes, coco coir or foam inserts. These growing media provide support for plant roots while allowing for ample aeration and nutrient uptake.

iii) **Misting system:** Misting systems deliver a fine mist of nutrient solution to plant roots at regular intervals. Misting intervals are carefully controlled to ensure optimal nutrient absorption and oxygenation of root zones.

iv) **Nutrient reservoir:** A centralized nutrient reservoir supplies nutrient solution to misting system. Nutrient solution may be recirculated within the system to minimize waste and maintain nutrient concentrations.

Benefits of Vertical Aeroponics Towers

i) **Maximized space utilization:** Vertical aeroponics towers enable growers to cultivate a large number of plants in a small footprint, making them ideal for urban environments, balconies and indoor spaces with limited space availability.

ii) **High yield potential:** By stacking growing levels vertically, vertical aeroponics towers maximize the use of available space thus allowing for increased crop yields per square foot compared to traditional cultivation methods.

iii) **Water and resource efficiency:** Aeroponic systems are highly water-efficient as these utilize nutrient solution more efficiently than soil-based systems. Vertical aeroponics towers require minimal soil or growing medium thus reducing resource consumption and environmental impact.

iv) **Improved plant health:** The aeroponic growing environment provides plants with direct access to oxygen so promoting healthy root development *cum* proper nutrient absorption. This results in faster growth rates, higher nutrient uptake efficiency and improved overall plant health.

Considerations for Vertical Aeroponics Towers

i) **Maintenance requirements:** Vertical aeroponics towers require regular maintenance to ensure proper functioning of misting systems, nutrient reservoirs and plant support structures. Routine inspection and cleaning are essential to prevent clogs, algal growth and nutrient imbalances.

ii) **Environmental control:** Temperature, humidity and light levels must be carefully monitored and controlled to create an optimal growing environment within vertical aeroponics towers. Climate control systems may be necessary to maintain stable growing conditions, especially in indoor or greenhouse settings.

iii) **Crop selection:** While vertical aeroponics towers can accommodate a wide range of crops, growers should consider factors such as plant size, root depth and light requirements when selecting suitable crops for cultivation. Leafy greens, herbs and compact vegetables are well-suited for vertical aeroponics towers.

4.5.9.4 Rotating Aeroponics Systems (Revolving/rotating drum aeroponics)

Rotating aeroponics systems represent a cutting-edge approach to soilless cultivation, offering unique advantages for plant growth and productivity. Unlike traditional stationary aeroponic systems, rotating aeroponics systems employ rotating platforms or chambers to expose plant roots to nutrient mist from multiple angles. Rotating motion ensures uniform coverage of roots and facilitates even nutrient absorption and oxygenation. These systems are often used in research settings to study plant responses to varying nutrient delivery patterns and environmental conditions.

Components

i) **Rotating chamber or platform**: The central component of a rotating aeroponics system is the rotating chamber or platform which houses plant roots. This chamber rotates at regular intervals ensuring uniform misting of plant roots from all the sides.

ii) **Misting system:** Misting nozzles or sprayers are strategically positioned within the rotating chamber to deliver a fine mist of nutrient solution to plant roots. The misting system operates on a timer or sensor-controlled mechanism to ensure precise nutrient delivery and oxygenation.

iii) **Nutrient reservoir and pump**: A nutrient reservoir stores the nutrient solution before it is pumped into the misting system. A high- or low-pressure pump pressurizes the nutrient solution thus facilitating atomization into a fine mist for root absorption.

iv) **Root support structure**: Roots of plants are suspended within the rotating chamber using a support structure such as mesh baskets, trays or panels. This structure allows roots to hang freely in the misted environment so promoting efficient nutrient absorption and oxygenation.

v) **Control system**: A control system regulates the rotation of the chamber, misting intervals, nutrient solution concentration and other parameters to optimize plant growth.

Operations

i) **Rotation cycle**: The rotating chamber or platform rotates at predetermined intervals typically ranging from a few minutes to several hours. This rotation cycle ensures that plant roots are exposed to nutrient mist from all the angles thus promoting uniform nutrient absorption and root zone oxygenation.

ii) **Misting cycle**: The misting system activates during each rotation cycle delivering a fine mist of nutrient solution to plant roots suspended within the chamber. Misting intervals are carefully controlled to provide plants with adequate hydration and nutrient uptake without waterlogging or suffocating roots.

iii) **Environmental control**: Rotating aeroponics systems may incorporate environmental control systems to monitor and adjust factors such as temperature, humidity and light levels. These controls help maintain optimal growing conditions and maximizing plant growth and productivity.

Benefits

i) **Uniform nutrient delivery**: The rotating design of aeroponics systems ensures that plant roots receive consistent misting from all the angles thus promoting uniform nutrient absorption and growth.

ii) **Enhanced oxygenation**: Rotating chambers facilitate optimal root zone oxygenation by exposing roots to fresh air during each rotation cycle thus preventing suffocation and promoting healthy root development.

iii) **Increased nutrient efficiency**: By delivering nutrients directly to plant roots in a misted form, rotating aeroponics systems maximize nutrient uptake efficiency thus leading to accelerated growth and higher yields.

iv) **Space optimization**: Rotating aeroponics systems are well-suited for vertical farming and space-limited environments thus allowing growers to maximize production per square foot of growing space.

Applications

i) **Commercial agriculture**: Rotating aeroponics systems are increasingly used in commercial agriculture for production of high-value crops such as leafy greens, herbs and specialty vegetables.

ii) **Urban farming**: In urban farming settings, rotating aeroponics systems offer a compact and efficient solution for growing fresh produce on rooftops, in greenhouses and as vertical farming facilities.

iii) **Research and development**: Rotating aeroponics systems are valuable tools for research institutions and universities studying plant physiology, nutrient uptake mechanisms and hydroponic cultivation methods.

4.5.9.5 Hybrid Aeroponics Systems

Hybrid aeroponics systems represent an innovative approach to soilless cultivation ~ combining elements of aeroponics with other hydroponic or soilless techniques to create versatile and efficient growing systems such as some hybrid systems may incorporate aeroponic misting chambers alongside nutrient film technique (NFT) channels or deep water culture (DWC) reservoirs to diversify nutrient delivery methods and optimize plant growth. These systems leverage the strengths of each method to optimize nutrient delivery, root oxygenation and overall plant health. Let us delve into a comprehensive exploration of hybrid aeroponics systems:

Components

i) **Aeroponic misting chamber**: The core component of a hybrid aeroponics system is the misting chamber where plant roots are suspended and exposed to a nutrient-rich mist. Misting chambers typically feature high-pressure pumps and misting nozzles to atomize nutrient solution enveloping roots in a fine mist for optimal nutrient absorption.

ii) **Supplemental growing medium**: Many hybrid aeroponics systems incorporate supplemental growing media such as perlite, vermiculite or coconut coir to provide additional support for plant roots. Growing medium helps anchoring the plants in place within the misting chamber and can retain moisture and nutrients between misting cycles.

iii) **Nutrient delivery system**: Nutrient solution is delivered to the misting chamber either *via* high-pressure pumps or gravity-fed systems, ensuring consistent nutrient availability for plant roots. Growers may customize nutrient formulations based on crop requirements, growth stage and environmental conditions to optimize plant health and productivity.

iv) **Root support structure**: Hybrid aeroponics systems utilize root support structures such as trays, baskets or panels to suspend and protect plant roots within the misting chamber. These structures provide stability for plants and facilitate efficient nutrient uptake while allowing roots to access oxygen and grow freely.

Operational Considerations

i) **Misting intervals and duration**: Hybrid aeroponics systems employ misting cycles to deliver nutrient solution to plant roots while allowing for periods of oxygenation. Misting intervals and duration can be adjusted based on crop requirements, environmental conditions and system design to optimize nutrient delivery and root health.

ii) **Nutrient solution management**: Proper management of nutrient solution is critical for success of hybrid aeroponics systems thus ensuring balanced nutrient concentrations and pH levels. Growers should monitor nutrient solution parameters regularly and make adjustments as needed to prevent nutrient deficiencies or imbalances.

iii) **Environmental control**: Maintaining optimal environmental conditions within the misting chamber is essential for maximizing plant growth and productivity. Temperature, humidity and airflow should be carefully regulated to prevent stress and promote vigorous root development and nutrient absorption.

iv) **Maintenance and troubleshooting**: Regular maintenance of system components including pumps, nozzles and root support structures is necessary to prevent clogs, leaks or malfunctions. Growers should be prepared to troubleshoot issues such as nutrient deficiencies, root rot or pest infestations promptly to minimize crop losses and for ensuring long-term system performance.

Advantages

i) **Versatility**: Hybrid aeroponics systems offer flexibility to adapt to diverse crop types, growth stages and environmental conditions.

ii) **Optimized nutrient delivery**: By combining aeroponic misting technology with supplemental growing media these systems ensure efficient nutrient absorption and root oxygenation.

iii) **Enhanced root health**: Plant roots in hybrid aeroponics systems benefit from direct access to both water and oxygen thus promoting vigorous root growth and nutrient uptake.

iv) **Increased yields and quality**: By providing an ideal growing environment hybrid aeroponics systems can result in higher yields, improved crop quality and faster growth rates compared to traditional cultivation methods.

Challenges and Considerations

i) **Complexity**: Hybrid aeroponics systems may require more advanced setup and management compared to single-method systems thus requiring growers to invest time and resources in system design, operation and maintenance.

ii) **Cost**: The initial investment and ongoing operational costs associated with hybrid aeroponics systems may be higher than those of conventional hydroponic or aeroponic setups, potentially limiting accessibility for some growers.

iii) **Learning curve**: Growers new to hybrid aeroponics may face a learning curve in understanding system dynamics, nutrient management and environmental control which necessitates education and training to optimize system performance.

Future Directions and Applications

The future of hybrid aeroponics systems holds promising opportunities for innovation and advancement in soilless cultivation. Continued research and development efforts are expected to focus on:

i) **Technology integration**: Integrating advanced sensors, automation and data analytics to optimize system performance and resource utilization.

ii) **Crop-specific solutions**: Tailoring hybrid aeroponics systems to meet the unique needs of different crops including fruiting vegetables, herbs, berries and ornamentals.

iii) **Sustainability**: Exploring sustainable practices such as water and energy conservation, nutrient recycling and renewable energy integration to reduce environmental impact and enhance system resilience.

iv) **Vertical farming and urban agriculture**: Adapting hybrid aeroponics systems for vertical farming, rooftop gardens and urban agriculture initiatives to maximize space efficiency and promote local food production.

4.5.9.6 Modular Aeroponics Systems

Modular aeroponics systems represent a versatile and scalable approach to soilless cultivation thus offering growers the flexibility to customize and expand their setups according to specific needs and constraints. Unlike traditional farming methods that rely on soil and modular aeroponics systems suspend plant roots in a nutrient-rich mist or air environment, providing optimal conditions for growth and nutrient absorption.

Components

i) **Modular growing units**: Modular aeroponics systems consist of individual growing units or modules each equipped with its own root chamber, misting mechanism and nutrient delivery system. These modules can be arranged and interconnected in various configurations to accommodate different crop types, growing spaces and production scales.

ii) **Central control system**: A central control system governs the operation of modular aeroponics systems regulating nutrient delivery, misting cycles, environmental parameters and other critical variables. Advanced control systems may feature automation capabilities, allowing for remote monitoring, data analysis and precision control of growing conditions.

iii) **Nutrient reservoir and mixing station**: A nutrient reservoir stores and supplies the nutrient solution to individual modules, ensuring consistent and uniform nutrient delivery. A mixing station or dosing system is responsible for preparing and maintaining the nutrient solution at optimal concentrations, pH levels and nutrient ratios.

iv) **Support structures and growing medium**: Modular aeroponics systems utilize support structures such as racks, trellises or frames to suspend growing modules and provide structural stability. Growing medium options may include inert substrates such as rockwool, coco coir or perlite which support plant roots and facilitate nutrient absorption.

Advantages

i) **Scalability and flexibility**: Modular aeroponics systems can be easily scaled up or down to accommodate changing production needs from small-scale home gardens to large commercial operations. Growers have the flexibility to add or remove modules, adjust planting densities and experiment with different crop varieties without significant infrastructure modifications.

ii) **Space efficiency and vertical farming**: The vertical orientation of modular aeroponics systems maximizes growing space utilization thus making them ideal for vertical farming and urban agriculture initiatives. By stacking modules vertically, growers can achieve higher crop densities and optimize land use in space-constrained environments.

iii) **Resource efficiency and water conservation**: Aeroponics systems are inherently water-efficient as these require less water compared to traditional soil-based farming methods. By delivering nutrients directly to plant roots in a mist form modular aeroponics systems minimize water usage while maximizing nutrient uptake efficiency, reducing overall water consumption and runoff.

iv) **Crop quality and yield potential**: The controlled environment of modular aeroponics systems promotes rapid and vigorous plant growth resulting in higher yields and superior crop quality. Plants grown in aeroponic systems often exhibit increased nutrient density, enhanced flavour profiles and improved resistance to pests and diseases.

Applications

i) **Commercial agriculture**: Modular aeroponics systems are increasingly being adopted in commercial agriculture for production of high-value crops such as leafy greens, herbs, strawberries and microgreens. These systems offer growers a competitive edge by enabling year-round cultivation, faster crop cycles and consistent product quality.

ii) **Urban farming and vertical agriculture**: In urban environments where land is limited and expensive modular aeroponics systems provide a viable solution for localized food production. Vertical farming initiatives leverage the space-saving design of modular aeroponics systems to establish indoor farms in urban centres thus supplying fresh produce to nearby communities.

iii) **Research and education**: Modular aeroponics systems serve as valuable tools for research institutions, universities and educational facilities seeking to study plant physiology, hydroponic techniques and sustainable agriculture practices. These systems offer hands-on learning opportunities for students and researchers to explore innovative growing methods and address pressing agricultural challenges.

Future Prospects and Innovations

i) **Integration of smart technologies**: The integration of smart sensors, data analytics and internet of things (IoT) technologies will enhance the functionality and performance of modular aeroponics systems. Real-time monitoring, predictive analytics and remote control capabilities will empower growers to optimize crop management practices and maximize resource efficiency.

ii) **Advancements in nutrient delivery systems**: Ongoing research and development efforts are focused on improving nutrient delivery systems within modular aeroponics setups. Innovations such as precision dosing, automated nutrient adjustments and dynamic nutrient formulations will enable growers to fine-tune nutrient delivery based on plant requirements and environmental conditions.

iii) **Expansion of vertical farming solutions**: Vertical farming initiatives will continue to expand and driving innovation in modular aeroponics systems for vertical agriculture applications. Technological advancements in vertical farming infrastructure, lighting systems and automation will further optimize space utilization and production efficiency in urban farming environments.

4.5.9.7 Commercial Aeroponics Systems

Commercial aeroponics systems represent a sophisticated fusion of cutting-edge technology and biological expertise designed to optimize plant growth and productivity while minimizing maintenance requirements and environmental impact. Let us delve into a more detailed and elaborate exploration of these systems:

Components

i) **High-pressure devices**: Commercial aeroponics systems often incorporate high-pressure devices for nutrient delivery which facilitate the cultivation of high-value crops and enable multiple crop rotations within the same system. These devices ensure precise misting of nutrient solutions so creating an ideal environment for plant roots to thrive and flourish.

ii) **Biological systems matrix**: The biological systems matrix enhances plant life and crop maturation within commercial aeroponics systems so contributing to extended growth cycles and optimal harvest yields.

By providing a supportive matrix for root development and nutrient absorption the biological systems matrix plays a crucial role in maximizing the productivity and longevity of cultivated crops.

Advantages

i) **Pathogen-free environment**: Commercial aeroponics systems are renowned for their exceptional reliability and low maintenance requirements largely due to their nearly pathogen-free growing environment. By minimizing the risk of pathogen contamination, these systems offer a level of crop safety and consistency unmatched by traditional soil-based cultivation methods.

ii) **Precision subsystems**: The subsystems of commercial aeroponics systems encompass a wide range of precision technologies and controls including effluent control systems, precision timing and nutrient solution pressurization. These subsystems ensure optimal nutrient delivery, temperature regulation and environmental monitoring allowing growers to fine-tune growing conditions for maximum efficiency and crop quality.

iii) **Advanced features**: More advanced commercial aeroponics systems may incorporate features such as data gathering, monitoring, analytical feedback and internet connectivity to various subsystems. These capabilities enable growers to remotely monitor and adjust system parameters, optimize resource utilization and troubleshoot issues in real-time thus enhancing overall operational efficiency and productivity.

Applications and Scalability

i) **Crop diversity**: Commercial aeroponics systems are highly versatile and can accommodate a wide range of crops from leafy greens and herbs to fruits and vegetables. The scalability of these systems allows growers to customize their setups to suit the specific needs and demands of their target markets whether for small-scale urban farms or large-scale commercial operations.

ii) **Market accessibility**: With growing demand for locally grown sustainably produced food, commercial aeroponics systems offer a compelling solution for meeting consumer preferences and market demands. By providing fresh nutrient-rich produce year-round these systems help address food security challenges while promoting environmental stewardship and economic viability.

Future Directions and Innovation

i) **Technological advancements**: Future advancements in commercial aeroponics systems may focus on integrating emerging technologies such as artificial intelligence, machine learning and robotics to further enhance system efficiency and productivity. By leveraging advanced analytics and automation growers can optimize resource allocation, minimize waste and maximize crop yields in a rapidly evolving agricultural landscape.

ii) **Sustainability and resource efficiency**: Ongoing research and development efforts will continue to prioritize sustainability and resource efficiency within the commercial aeroponics industry. Innovations in energy management, water recycling and waste reduction will enable growers to minimize their environmental footprint while maximizing the long-term viability of their operations.

4.6 Fogponics

Fogponics, sometimes referred to as atmoponics can be regarded as the pinnacle of aeroponic technology offering a sophisticated approach to nutrient delivery and plant growth. Derived from the Ancient Greek words '*atmós*' meaning 'vapour' or 'steam' and aeroponics, fogponics represents an innovative approach to plant cultivation. Fogponics harnesses the power of vaporized nutrient solution to envelop plant roots in a cloud-like mist.

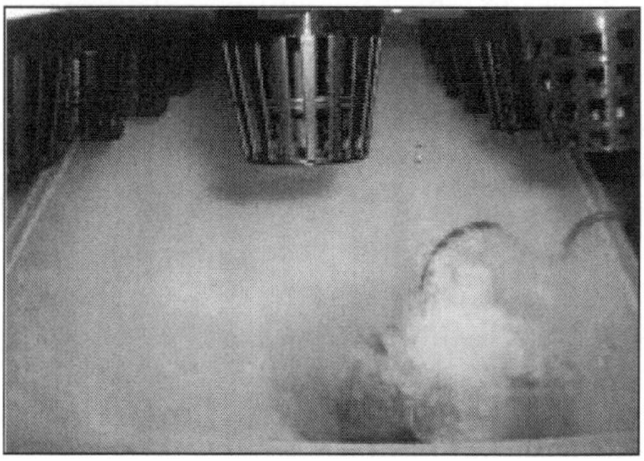

Fogponics Apparatus.

4.6.1 Comparison with Aeroponics

Fogponics stands out from other forms of aeroponics due to its distinctive method of nutrient delivery which involves creation of an ultra-fine mist or suspension of nutrient-rich water particles. Unlike conventional aeroponic systems which utilize sprays to deliver relatively large water droplets to the roots, fogponics employs advanced mechanisms such as ultrasonic technology, compressed air or heating elements to generate a fine mist of micron-sized water particles ranging from 5μm to 30μm or even as a vapour within low-pressure aeroponic devices. This finely atomized mist ensures rapid and efficient absorption by the root system leading to accelerated growth and enhanced nutrient uptake.

i) **Particle size and nutrient absorption**: In aeroponics, the larger water droplets may provide effective nutrient delivery but these can also lead to uneven distribution and slower absorption through roots. In contrast, fogponics ultra-fine particles ensure rapid and uniform nutrient absorption as the micron-sized mist envelops the roots thus maximizing surface area contact and nutrient uptake efficiency.

ii) **Root oxygenation and moisture**: While both aeroponics and fogponics systems offer superior root oxygenation compared to traditional soil-based methods, fogponics finer particles provide enhanced oxygenation and moisture levels to root zone. The micron-sized mist in fogponics creates an oxygen-rich environment around the roots promoting vigorous root growth and overall plant health.

iii) **Mechanisms of particle generation**: Aeroponic systems typically rely on spray nozzles or misting devices to deliver water to the roots generating relatively larger droplets dispersed in the air. In contrast, fogponics systems utilize specialized mechanisms such as ultrasonic transducers, compressed air pumps or heating elements to atomize water into ultra-fine particles or vapour creating a dense fog-like environment around the roots.

iv) **Versatility and efficiency**: Fogponics ability to produce ultra-fine water particles offers greater versatility and efficiency in nutrient delivery thus making it ideal for a wide range of plant species and growth stages. The fine mist produced by fogponics systems ensures minimal water waste and maximum nutrient utilization leading to accelerated plant growth and higher yields.

In a nutshell, while both aeroponics and fogponics share the common principle of soilless cultivation and root zone oxygenation, fogponic's unique approach to nutrient delivery through ultra-fine water particles sets it apart as a highly efficient and innovative hydroponic method. By harnessing advanced particle generation technologies, fogponics systems offer growers unprecedented control over nutrient absorption and plant growth, paving the way for sustainable and high-yielding crop production.

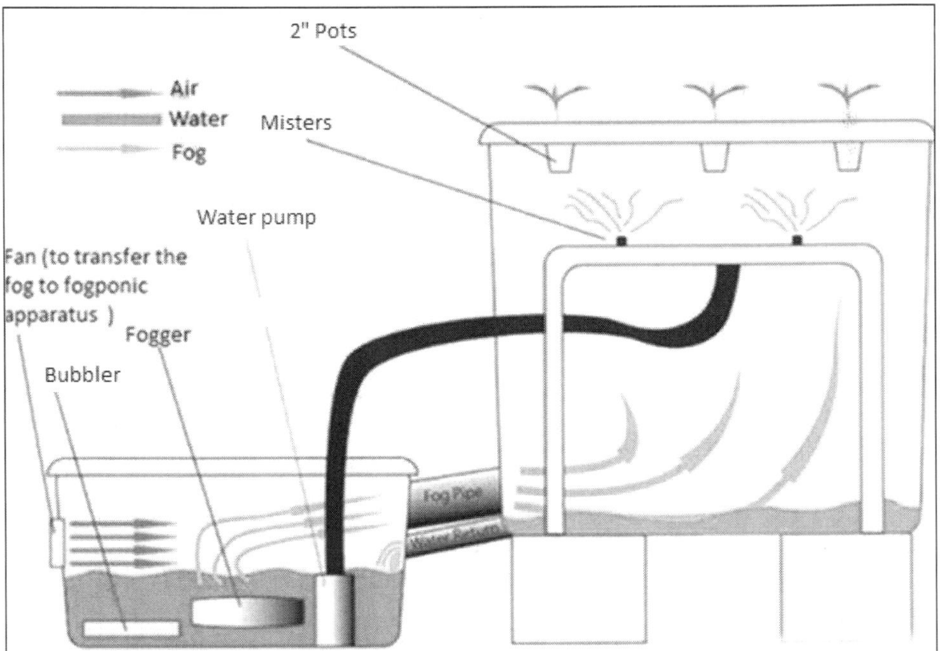

Fogponics system

4.6.2 Essential Components of Fogponics

i) **Nutrient reservoir**: The nutrient reservoir stores the nutrient solution used in fogponics systems. It typically contains a carefully balanced mixture of water and nutrients essential for plant growth.

ii) **Ultrasonic nebulizer or fogger**: The ultrasonic fogger is the core component responsible for converting nutrient solution into a fine mist or fog.

iii) **Low-pressure aeroponic devices**: Fogponics systems typically utilize low-pressure aeroponic devices which maintain a controlled environment for plant roots suspended in the air. These devices create a nutrient-rich fog that surrounds the roots so ensuring continuous hydration and nutrient uptake without need for a growing medium.

iv) **Pressure chamber or rooting chamber**: This chamber houses the plant roots suspended in the nutrient fog. It may consist of trays, racks or panels where plants are placed allowing their roots to access the misted nutrient solution.

v) **Misting lines or nozzles**: Misting lines or nozzles are used to distribute nutrient fog evenly within the pressure chamber. These components ensure uniform coverage of plant roots so facilitating efficient nutrient absorption and oxygenation.

vi) **Pump and tubing system**: A pump and tubing system are employed to circulate the nutrient solution from the reservoir to the ultrasonic nebulizer or fogger. The pump pressurizes the nutrient solution allowing it to be efficiently atomized by the fogging device.

vii) **Control and monitoring system**: A control and monitoring system regulates the operation of fogponics components and tracks key parameters such as nutrient concentrations, pH levels, temperature and humidity. This system ensures that growing conditions remain optimal for plant growth and health.

viii) **Root support structure**: A root support structure provides stability for plant roots within the pressure chamber. It may consist of trays, mesh screens or specialized fixtures designed to securely hold plant roots in place while allowing them to access the nutrient fog.

ix) **Environmental control equipment**: Environmental control equipment such as fans, heaters and humidifiers may be incorporated into fogponics systems to maintain stable growing conditions. These components help regulate temperature, humidity and airflow within the growing environment so ensuring optimal plant growth and health.

By integrating these components into a cohesive system, fogponics enables precise nutrient delivery, enhanced oxygenation and accelerated plant growth making it a promising technique for sustainable agriculture and controlled environment cultivation.

4.6.3 Key features and advantages

i) **Ultrasonic fogging technology**: Ultrasonic foggers utilize high-frequency vibrations to atomize liquid nutrient solutions producing a cloud-like mist with droplets ranging in size from a few microns to several microns. This fine mist provides an ideal medium for nutrient delivery as it enables direct absorption by plant roots while maximizing surface area contact.

ii) **Enhanced nutrient absorption**: The minute particulate size of the nutrient fog in fogponics facilitates rapid absorption by plant roots, maximizing nutrient uptake efficiency. This results in faster growth rates, healthier plants and ultimately higher yields compared to traditional hydroponics or aeroponics systems.

iii) **Oxygenation and aeration**: In addition to delivering nutrients, fogponics systems provide ample oxygenation to the root zone so promoting aerobic conditions essential for healthy root development. The fine mist of nutrient solution ensures that roots are adequately oxygenated, preventing the formation of anaerobic zones and promoting optimal plant growth.

iv) **Precision and control**: Fogponics systems offer precise control over nutrient concentrations, pH levels and environmental conditions allowing growers to tailor the growing environment to the specific needs of their plants. Advanced automation and monitoring systems further enhance control, ensuring optimal growing conditions and maximizing crop productivity.

v) **Energy efficiency**: Fogponics systems require less energy for root growth and maintenance compared to conventional aeroponics or hydroponics setups. The reduced energy demand translates to lower operational costs and environmental impact, making fogponics a sustainable choice for modern agriculture.

vi) **Water conservation**: By utilizing a fog-like mist rather than a continuous stream of nutrient solution, fogponics systems minimize water usage while maintaining optimal hydration for plant roots. This water-saving feature makes fogponics particularly well-suited for regions facing water scarcity or drought conditions.

vii) **Higher crop yields**: Studies have shown that fogponics systems consistently yield higher crop yields compared to traditional aeroponics or hydroponics methods. The precise nutrient delivery and oxygenation provided by fogponics contribute to robust plant growth and bountiful harvests, making it an attractive option for commercial growers seeking to maximize productivity.

Overall, fogponics represents a revolutionary approach to aeroponic cultivation, combining cutting-edge technology with proven growing principles to achieve exceptional results. By harnessing the power of ultrasonic fogging technology, fogponics systems offer growers a highly efficient and versatile solution for producing healthy and vibrant crops in a controlled indoor environment.

4.6.4 Operational Considerations

i) **Particle size optimization**: In fogponics systems, the size of the nutrient particles plays a crucial role in ensuring optimal nutrient absorption by plant roots. Particle sizes ranging from 1-25μm are most effectively absorbed by plants, necessitating careful calibration of misting equipment to achieve desired particle size distribution.

ii) **Nutrient formulation and delivery**: Customizing nutrient formulations tailored to the specific requirements of plant species and growth stages is essential for maximizing yields in fogponics systems. Nutrient solutions should be well-balanced and regularly monitored to prevent deficiencies or excesses that could adversely affect plant health. Employing automated dosing systems can streamline nutrient delivery and ensuring precise control over nutrient concentrations.

iii) **Environmental control**: Maintaining optimal and stable environmental conditions including temperature, humidity and light levels is essential for maximizing efficacy of fogponics systems. Automated climate control systems and sensor technologies can help growers monitor and adjust environmental parameters in real-time, ensuring optimal growing conditions.

iv) **Root zone management**: Monitoring root health and growth is essential in fogponics systems, as the root zone is where nutrient absorption takes place. Regular inspection of root systems for signs of disease, nutrient deficiencies or root rot is crucial for early intervention and preventing crop loss. Proper aeration of the root zone ensures adequate oxygenation and prevents anaerobic conditions that can inhibit root growth.

v) **System maintenance and cleaning**: Regular maintenance and cleaning of fogponics systems are essential to prevent clogging of misting nozzles and ensuring uninterrupted nutrient delivery. Misting nozzles should be inspected and cleaned periodically to remove mineral deposits or debris that could impair their function. Reservoirs, tubing and other system components should be sanitized regularly to prevent the buildup of algae, bacteria or other pathogens.

vi) **Monitoring and data management**: Continuous monitoring of key parameters such as pH levels, nutrient concentrations and environmental conditions is essential for optimizing system performance. Data logging systems can record and analyze data over time, providing insights into trends, patterns and potential issues that may arise. Growers should establish protocols for data interpretation and decision-making based on real-time information gathered from the fogponics system.

By addressing these operational considerations, growers can ensure the efficient operation and management of fogponics systems, leading to optimal plant growth, higher yields and overall success in soilless cultivation.

4.6.5 Future Directions and Research

i) **Integration of smart technologies**: Future advancements in fogponics may involve the integration of smart technologies such as artificial intelligence and machine learning algorithms. These technologies can optimize nutrient formulations, automate system management and provide actionable insights for growers leading to further improvements in efficiency and productivity.

ii) **Crop diversity and adaptation**: Ongoing research efforts are focused on expanding the range of crops suitable for cultivation in fogponics systems. By tailoring nutrient formulations and growing parameters to specific crop requirements, fogponics has the potential to revolutionize the production of a diverse array of fruits, vegetables and specialty crops.

4.6.6 Conclusion

Fogponics represents a paradigm shift in soilless cultivation, offering unparalleled precision, efficiency and sustainability. By harnessing the power of vaporized nutrient solutions, fogponics systems provide an ideal growing environment for plants resulting in higher yields, reduced resource consumption and enhanced crop quality. As technology continues to advance and research progresses, fogponics is poised to play a pivotal role in meeting the challenges of global food security and sustainability in the 21st century.

i) **Drip system**: A common method where a timer-controlled pump delivers nutrient solution to the base of each plant through a network of tubes or drippers. Excess solution is collected and recirculated.

ii) **Ebb and flow (flood and drain)**: Plants are periodically flooded with nutrient solution from a reservoir and then drained. This cyclic flooding and draining help oxygenate the roots.

iii) **Continuous flow**: Similar to Ebb and Flow but with a constant flow of nutrient solution rather than periodic flooding and draining.

iv) **Aquaponics**: Combines hydroponics with aquaculture (fish farming). Fish waste provides nutrients for plants and plants filter the water which is then recirculated back to the fish tanks.

v) **Vertical hydroponics**: Utilizes vertical space by stacking plant containers or using towers, allowing for high plant density and efficient use of space.

vi) **Dutch bucket system**: Plants are grown in individual containers (buckets) filled with growing medium, and a drip system delivers nutrient solution to each bucket.

These hydroponic techniques vary in complexity, cost and resource requirements allowing growers to choose the method that best suits their needs, space and expertise level. Each technique has its advantages and disadvantages, and selection often depends on factors such as the type of plants being grown, available space, budget and desired level of automation.

4.7 Membrane System Hydroponics

Membrane system hydroponics, also known as membrane-based hydroponics or membrane-mediated hydroponics, is a relatively newer approach to hydroponic cultivation. In this system, semi-permeable membranes are utilized to control the exchange of water and nutrients between the plant roots and the nutrient solution. This innovative system offers several advantages over traditional hydroponic methods including improved water and nutrient efficiency, enhanced root zone oxygenation and greater control over nutrient uptake.

4.7.1 Components of Membrane System Hydroponics

i) **Membrane modules**: Membrane modules are the core components of the system consisting of semi-permeable membranes that regulate flow of water and nutrients. These membranes are engineered to allow the passage of water molecules while selectively retaining dissolved nutrients and minerals thus ensuring precise control over nutrient delivery to plant roots.

ii) **Nutrient reservoir**: A nutrient reservoir stores the nutrient solution that is circulated through the membrane modules. The reservoir may contain a carefully balanced nutrient solution composed of essential macro- and micronutrients required for plant growth.

iii) **Pump and circulation system**: A pump and circulation system facilitate the movement of nutrient solution from the reservoir through the membrane modules and back into the reservoir. This continuous circulation ensures uniform nutrient distribution and prevents stagnation thus promoting healthy root growth and optimal nutrient uptake.

iv) **Root zone environment control**: Environmental control measures such as temperature regulation, humidity management and light exposure help maintaining optimal growing conditions within the root zone. These factors contribute to overall plant health and productivity, allowing growers to achieve consistent and reliable crop yields.

v) **Monitoring and automation**: Monitoring sensors and automation systems enable growers to closely monitor key parameters such as nutrient concentrations, pH levels and water quality. Automated controls can adjust nutrient delivery rates, circulation flow rates and environmental conditions in real-time thus optimizing plant growth and resource utilization.

4.7.2 Working of Membrane System Hydroponics

Here's how membrane system hydroponics typically works:

i) **Membrane selection**: The system employs specialized semi-permeable membranes that allow the passage of water and certain nutrients while restricting the movement of larger molecules, ions or contaminants. These membranes are chosen based on their pore size and selectivity to facilitate the desired exchange of substances.

ii) **Root encapsulation**: Plant roots are encapsulated within a membrane-based structure such as a pouch, bag or compartment thus allowing them to interact with the nutrient solution while maintaining separation from the growing medium.

iii) **Nutrient delivery**: A nutrient solution is continuously circulated or periodically replenished within the system. The semi-permeable membranes regulate the transfer of water, ions and other essential nutrients to the roots while preventing the ingress of contaminants or pathogens.

iv) **Root zone environment**: The controlled exchange facilitated by the membranes creates a highly controlled root zone environment. This precise regulation of nutrient availability, pH and oxygen levels can promote optimal plant growth and development.

v) **Water and nutrient efficiency**: Membrane system hydroponics can enhance water and nutrient use efficiency by minimizing losses due to leaching, runoff or evaporation. The controlled exchange also reduces the risk of nutrient imbalances or fluctuations, thereby potentially improving crop yields and quality.

vi) **Modularity and scalability**: The modular nature of membrane-based hydroponic systems allows for scalability and adaptability to various plant species, growth stages and environmental conditions. Growers can easily adjust the system parameters to optimize plant performance and resource utilization.

vii) **Potential applications**: Membrane system hydroponics has the potential for use in various settings including controlled environment agriculture (CEA), urban farming, vertical farming and space agriculture. It offers opportunities for sustainable food production in resource-limited or environmentally challenging environments.

4.7.2 Benefits of Membrane System Hydroponics

i) **Enhanced nutrient efficiency**: Membrane-based filtration allows for precise control over nutrient delivery, minimizing waste and maximizing nutrient uptake by plants.

ii) **Improved root zone oxygenation**: Semi-permeable membranes facilitate the exchange of gases, ensuring adequate oxygenation of the root zone and promoting vigorous root growth.

iii) **Water conservation**: The closed-loop circulation system reduces water consumption by recycling nutrient solution, making membrane hydroponics an environmentally sustainable option.

iv) **Greater control and flexibility**: Growers have the ability to fine-tune nutrient concentrations and environmental parameters, optimizing growing conditions for specific plant varieties and growth stages.

Overall, membrane system hydroponics represents an innovative approach to hydroponic cultivation, leveraging membrane technology to optimize water and nutrient management while providing a favourable environment for plant growth. Further research and development in this field may lead to advancements in agricultural sustainability, productivity and resilience.

4.8 'Ebb and Flow' or 'Flood and Drain' Sub-irrigation

Ebb and Flow hydroponics also known as flood and drain or ebb and flood is a versatile, inexpensive, simple, reliable and widely-used hydroponic system that provides an effective way to grow a variety of plants without soil. In this system, plants are grown in a tray or container filled with an inert growing medium such as gravel, perlite or clay pellets. The tray is periodically flooded with nutrient-rich water from a reservoir below and then allowed to drain back into the reservoir. This cyclical flooding and draining provide plants with water, nutrients and oxygen while preventing water stagnation and root suffocation. This system is great for hobby growers and for indoor growing as these are easy to build and can be made in any number or sizes.

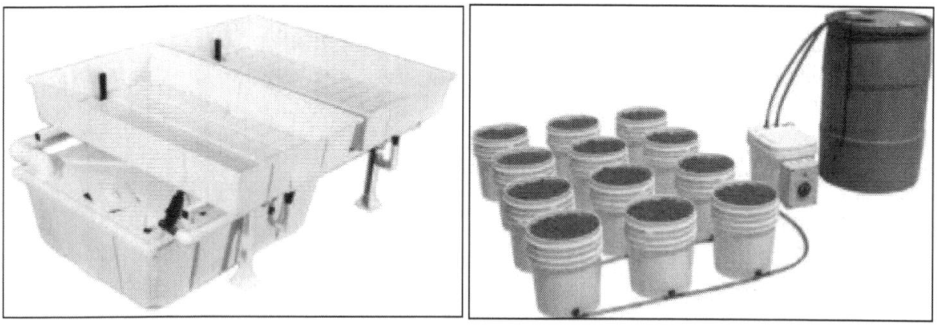

Two Simple Designs of 'Ebb and Flow' System.

Ebb and Flow hydroponics is best suited for smaller crops such as lettuce or other leafy vegetables though larger plants can also be grown in this type of hydroponic system but it would just need to set up things a bit different for the larger plants to deal with the harvesting and replacing of the plants. On a commercial level, 'ebb and flow' hydroponics systems are usually used to grow smaller size plants so as not to have to deal with the added work involved with taller plants.

4.8.1 Key Components

The key components of an Ebb and Flow hydroponic system include:

i) **Grow tray:** This is where the plants are placed and where the growing medium is housed. The tray is designed to hold the plants securely while allowing water to flow freely.

ii) **Reservoir:** A container or tank that holds the nutrient solution. It is usually placed below the grow tray to allow for easy flooding and draining.

iii) **Pump:** A submersible pump (*e.g.* pumps used in aquarium, koi ponds, *etc.*) is used to flood the grow tray periodically with nutrient solution for a short period of 5 to 10 minutes or until the grow tray gets filled. The pump is typically controlled by a timer to ensure that flooding occurs at regular intervals.

iv) **Timer:** An essential component that controls operation of pump, dictating when to flood the grow tray and when to stop. This ensures that the plants receive water and nutrients on a consistent schedule.

v) **Overflow drain:** Allows excess nutrient solution to drain back into the reservoir, preventing waterlogging and ensuring proper aeration of the roots.

4.8.2 Operation

The operation of an Ebb and Flow system is straightforward and effective. A nutrient solution reservoir is positioned below a water-tight grow tray. The grow tray is typically filled with an inert medium such as gravel or coarse sand or it accommodates pots of medium where plants are rooted. This inert medium not only anchors the roots but also serves as a temporary reserve of water and nutrients.

At scheduled intervals a pump is activated, flooding the grow tray with nutrient solution. Adequate drainage is essential to prevent water stagnation which can lead to fungal growth and root rot. To ensure proper drainage the grow tray may be slightly tilted. The duration and frequency of flooding are determined by the plant type and growth stage as well as the water retention capacity of the rooting medium. Some media may require flooding once or twice a day while others may necessitate multiple floodings throughout the day.

The aerated nutrient solution delivers both nutrition and oxygen to the plant roots. Oxygenation is achieved through thin-filming and positive displacement of deoxygenated air as it is pushed out of the root zone by water. Since solution does not remain in constant contact with roots there is no need to oxygenate it in the reservoir. When solution drains back the roots are re-exposed to air promoting passive oxygenation and suppressing pathogen growth.

Maintaining a dry top layer of the media (approximately 5.0 to 6.25cm) helps preventing the weed, fungal and harmful bacterial growth. Media beds should ideally be 30 to 45cm deep or even deeper depending on plant species. The grow bed shell should extend 5cm above the media bed and include an overflow pipe to return excess water to the reservoir thus ensuring optimal system function and plant health.

4.8.3 Designs for Flood/Drain of Nutrient Solution to/from Growbed

- ☆ In a single two-directional path of the solution flow tube the grow-bed is flooded with nutrient solution and then drained back to the reservoir. Once the roots are briefly submerged during flooding the pump is deactivated using a timer-operated switch and the solution is drained back using the same tube. Typically, pumping occurs from the bottom of the grow bed and when the pump switches off the nutrient solution drains back through it.

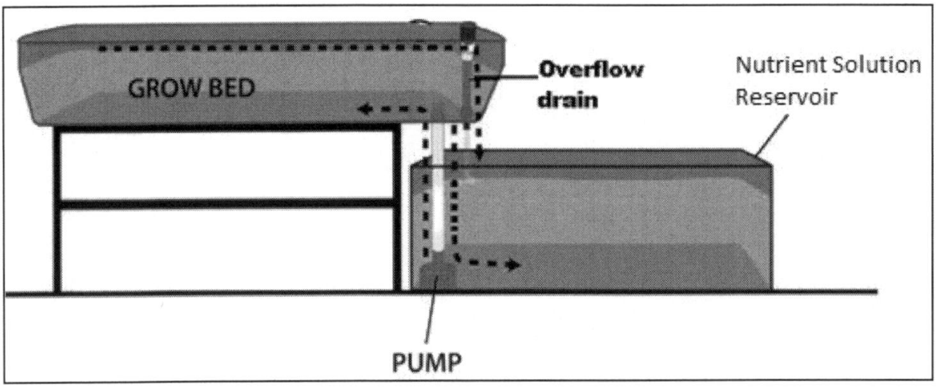

Ebb and Flow System with Single Two-Directional Path for Both, Flood and Drain.

- ☆ Alternatively, the grow bed space can be flooded and drained by either using a timer-controlled pump to fill the bed and then allowing it to drain while the pump is off or by employing a siphon or other intermittent outflow device where the bed continuously fills and then the siphon rapidly drains the bed.

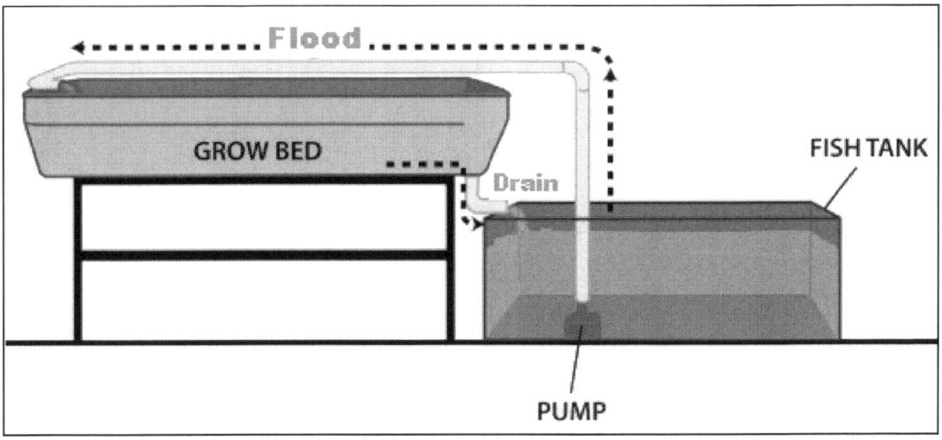

'Ebb and Flow' System with different Paths for Flooding and Draining.

☆ For smaller systems grow beds can be manually flooded one or more times a day depending on the type of inert media and the growth stage of the plants.

Many experts advocate for flood and drain media-based systems utilizing a Bell Siphon to drain considering these more reliable and simpler methods of hydroponics, particularly for beginners. These systems offer better temperature control compared to plastic pipe systems (NFT) and can be constructed using a variety of different containers.

4.8.4 Water Cycling

In a media-filled grow bed system periodic flooding and draining of the media are essential for optimal plant growth. Various methods can achieve this water cycling, ensuring adequate oxygen delivery to the roots. Techniques such as siphons, spill-over mechanisms, toilet valves or timed pumps facilitate the movement of water from the reservoir to the grow bed and back.

The key factor is maintaining a flow rate that effectively cycles water throughout the system. The gradual flooding followed by rapid draining of the grow beds promotes efficient nutrient uptake by the plants and enhances oxygenation of the root zone. The swift drainage draws oxygen deep into the roots crucial for robust growth and overall plant health.

4.8.5 Siphon

A siphon serves as a mechanism for transferring water from one reservoir to another typically from a higher to a lower reservoir and has the capability to lift water over barriers. In flood and drain hydroponic systems siphons play a crucial role in efficiently draining grow beds into the reservoir, employing a simple mechanical method devoid of any moving or electrical components. This elegant solution eliminates the need for timers to regulate the on and off cycle of pumps.

Auto-siphons offer the convenience of self-starting and stopping in response to changes in water levels. Among the commonly used types are the Loop Siphon or U-Bend Siphon and the Bell Siphon. In any grow bed and siphon setup, it is essential to establish a means of water passage through the grow bed. Typically, this involves cutting a hole through the side or bottom of the tank and installing a drain pipe to facilitate efficient water flow.

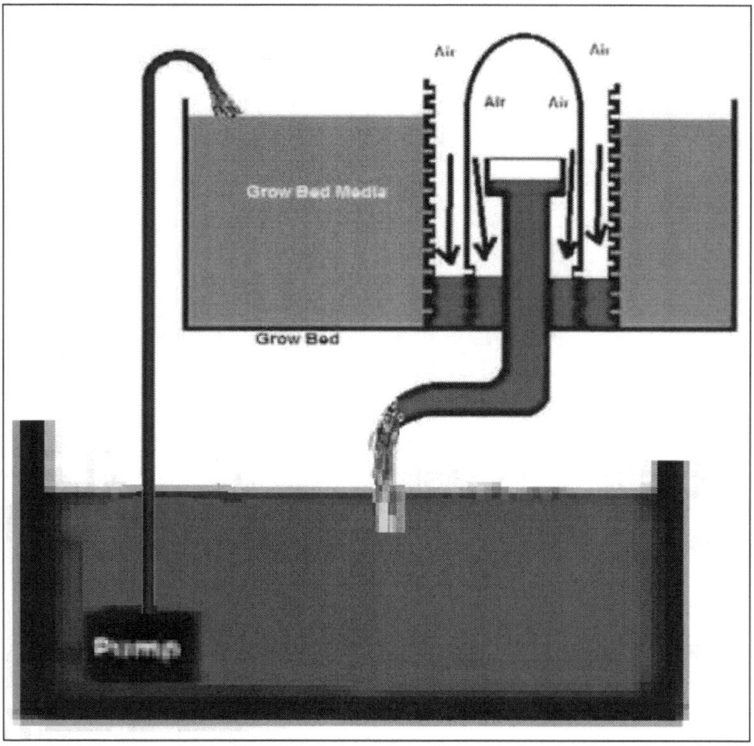

Ebb and Flow Hydroponics with Auto-siphon.

4.8.5.1 Bell Siphon

A bell siphon is a key component of many flood and drain (or ebb and flow) hydroponic systems. It operates on the principle of creating a siphon effect to efficiently drain the nutrient solution from the grow tray back into the reservoir once it reaches a certain level. Here is how a bell siphon works:

- ☆ **Bell and standpipe setup:** The bell siphon consists of two main components: the bell and the standpipe. The bell is a wide hollow tube or cylinder placed in the grow tray while standpipe is a narrower tube connected to the bottom of the bell and extends down into the reservoir.

☆ **Initial filling:** At the start of the flooding cycle the grow tray is filled with nutrient solution. As the level of solution rises it eventually reaches bottom of bell and begins to flow into it.

☆ **Air trapping:** As the nutrient solution fills the bell it displaces the air inside creating an air pocket at the top of the bell. This air pocket prevents the water from flowing out of the bell.

☆ **Critical point reached:** Once the nutrient solution reaches a critical level inside the bell it begins to flow over the top and into the standpipe. This critical level is determined by height of the standpipe relative to the bottom of the bell.

☆ **Siphon activation:** As the nutrient solution flows into the standpipe it creates a suction force due to difference in pressure between top and bottom of standpipe. This suction force initiates the siphon effect.

☆ **Continuous drainage:** Once the siphon effect is established the nutrient solution rapidly drains out of grow tray and into reservoir through standpipe. The siphon continues to operate until level of nutrient solution in the grow tray drops below the bottom of the standpipe.

☆ **Air re-entry:** As the nutrient solution drains out of the grow tray air is drawn into the bell through the standpipe breaking the siphon and allowing the grow tray to refill with fresh nutrient solution for the next flooding cycle.

Overall, the bell siphon provides a simple yet effective method for automating the drainage process in flood and drain hydroponic systems so ensuring that plants receive adequate hydration while preventing waterlogging and root suffocation.

4.8.6 Configurations and Designs of Ebb and Flow Systems

Ebb and Flow hydroponic systems can be designed in various ways to suit different needs and preferences. Here are some common design variations:

i) **Basic flood tray design:** This design consists of a simple flood tray positioned above a reservoir. The flood tray is filled with a growing medium and plants are placed in it. A submersible pump located in the reservoir periodically floods the tray with nutrient solution which then drains back into the reservoir through gravity.

ii) **Multi-tray system:** In larger setups or for growing multiple crops, multiple flood trays can be connected to a single reservoir. Each tray may have its own pump and timer or these can be connected to a central pump and timer system.

iii) **Vertical ebb and flow:** This design utilizes vertical growing space by stacking flood trays on top of each other. Nutrient solution is pumped into the top tray and allowed to overflow into the trays below creating a cascading effect. This maximizes space utilization and allows for higher plant densities.

iv) **Modular systems:** Modular ebb and flow systems consist of individual modules or containers each with its own flood tray and reservoir. These modules can be arranged and stacked in various configurations to create custom layouts based on available space and specific crop requirements.

v) **Customized grow beds:** Ebb and flow systems can be built using customized grow beds made from materials such as PVC, wood or plastic containers. These beds can be designed to fit specific growing areas or to accommodate irregular spaces.

vi) **Integrated systems:** Ebb and flow hydroponics can also be integrated with other hydroponic techniques such as NFT or DWC. This allows growers to combine the benefits of different systems so optimizing plant growth.

vii) **Automated systems:** Advanced 'Ebb and Flow systems' can be equipped with sensors, controllers and automation technology to monitor and control parameters such as nutrient levels, pH and temperature. This ensures precise nutrient delivery and optimal growing conditions.

When designing an 'Ebb and Flow hydroponic system' factors such as available space, budget, crop type and level of automation should be considered to create an efficient and effective growing environment.

4.8.7 Advantage

'Ebb and Flow hydroponics' is an effective and accessible way to grow plants hydroponically offering a balance of simplicity and efficiency for both hobbyists and commercial growers.

Ebb and Flow hydroponics offer several advantages such as:

☆ **Versatility**: Suitable for a wide range of plants ~ from leafy greens to flowering crops.

☆ **Ease of setup**: Relatively simple to assemble and operate, making it ideal for beginners.

☆ **Reduced water usage**: Water is recycled within the system, minimizing waste.

☆ **Inexpensive**: It is simple to design as excessive plumbing is not required. Power consumption is also low.

☆ **Enhanced nutrient uptake**: Flooding and draining promote efficient nutrient absorption by roots.

☆ **Oxygenation**: There is no need to oxygenate the solution because when solution is drained back the roots get re-exposed to air. As many plants prefer some dry time especially if the water is not super aerated.

- **Suppresses the pathogen growth**: The root system provides passive oxygenation at a high level and thus to a great extent suppresses the pathogen growth.

4.8.8 Disadvantages Consideration

These require a reliable pump and timer to automate the flooding and draining cycles thus adding to the initial cost and complexity of the system. If drainage system becomes clogged or pump malfunctions there is risk of overwatering and root suffocation. Regular monitoring and maintenance are necessary to prevent such issues.

- **Maintenance**: Regular monitoring and cleaning of system are required to prevent clogs and to maintain optimal nutrient levels. Removal of harvested or damaged plants can be quite difficult for plants having well developed root system. The rooting medium should be properly sterilized using hydrogen peroxide or chlorine solution and washed to remove root debris or other things.
- **Root health**: Proper drainage is essential to prevent waterlogging and root rot. To prevent growth of pathogen the temperature of nutrient solution can be made slightly below the temperature at which the pathogen growth begins but should not be made so cool that the activities of the roots get suppressed. If there is a suspicion that anaerobic *Pythium* root rot mould started to proliferate on the surface of the root, hydrogen peroxide should be added to nutrient solution. The oxygen liberated from the hydrogen peroxide destroys single-celled organisms. The dosage of hydrogen peroxide may vary with the concentration of peroxide.
- **Nutrient management**: Repeated use of nutrient solution may result in root diseases and nutrient element insufficiencies. Nutrient solution must be regularly monitored and adjusted to meet plant needs. It will require reconstitution, filtering, sterilization and may even require to be replaced within the growing period. This could be a hectic task for large systems.
- **Nutrient solution cycle and PPM**: Nutrient solution should be timely checked for its ppm concentration of all the essential elements. If plants are being fed several times a day, lower ppm nutrients (600-800) are sufficient and the higher ppm nutrient may burn the plants from inside. The ppm concentration may increase due to water evaporation and absorption by the plant roots. For rooting media such as lava rocks the number of feedings can be increased to every 50 to 60 minutes during lights on, and nutrient ppm should be kept between 600 and 800. In lava rocks, the solution drains quickly and a small amount of oxygen and nutrients get trapped in its rough texture.

- ☆ **pH management**: Over a period of time, the pH of the nutrient solution may fluctuate to a range which is unhealthy for the plant. Usually the pH should be adjusted each day during the flowering period.
- ☆ Proper control requires routine checking and replacing with fresh nutrients nearly 5-6 days to avoid toxicity.

4.9 Bubbleponics

Bubbleponics [top-fed deep water culture (DWC)] represents an innovative and efficient method of hydroponic gardening suitable for a wide range of plants from leafy greens to flowering crops like tomatoes and peppers. It is a variation of hydroponic gardening that combines elements of DWC and NFT. In a bubbleponics system, plants are typically grown in individual containers or net pots filled with an inert growing medium such as perlite or rockwool. These containers are suspended above a reservoir filled with a nutrient solution. The key feature of bubbleponics is the use of air pumps and air stones that continuously deliver oxygen to the solution, creating bubbles that rise up and come into contact with the roots of the plants. This oxygenation promotes healthy root development and efficient nutrient absorption similar to DWC systems. Nutrient solution here is pumped from the reservoir to the top of the containers where it is distributed evenly across the roots of the plants. Any excess solution drips back into the reservoir which ensures a constant flow of nutrients to the plants akin to NFT systems.

The growth of plants during the first few weeks is greater in bubbleponics systems as compared with the DWC. This is because as in bubbleponics system the roots get easy access to water and nutrients from the beginning and will quickly grow to reach the oxygenated nutrient solution in the reservoir. When the roots get immersed in the nutrient solution plant growth rate in both ~ DWC and bubbleponics is almost same. Bubbleponics is a popular choice among hydroponic enthusiasts due to its simplicity, efficiency and ability to produce high yields of healthy plants.

Sometimes the term, 'static solution sulture', 'deep water culture' and 'bubbleponics' may be used interchangeably.

4.9.1 Components of a Bubbleponics System

i) **Reservoir:** The reservoir holds the nutrient solution which is continuously oxygenated and circulated throughout the system.

ii) **Grow tray:** Plants are typically placed in net pots or containers filled with an inert growing medium such as hydroton or rockwool which is positioned above the reservoir.

iii) **Airstones and air pump:** An air pump supplies oxygen to the nutrient solution through airstones or air diffusers submerged in the reservoir. These create bubbles that rise through the water thus oxygenating it and preventing stagnation.

iv) **Tubing and connectors:** Tubing connects the air pump to the airstones allowing for the delivery of oxygen to the nutrient solution. Connectors and valves may be used to regulate airflow and ensure even distribution of bubbles.

4.9.2 Advantages of Bubbleponics

Bubbleponics offers several advantages including faster growth rates, increased yields and reduced risk of root rot compared to traditional soil-based gardening. It also provides precise control over nutrient levels and pH leading to optimal plant health and productivity.

4.9.3 Challenges and Considerations

While bubbleponics systems are relatively low-maintenance, regular monitoring of nutrient levels, pH and oxygenation is essential to ensure optimal plant health. Setting up a bubbleponics system may require some initial investment in equipment and materials including air pumps, airstones and nutrient solutions. Despite the benefits of increased oxygenation, over- or under-oxygenation of the nutrient solution can still pose risks to root health. Careful attention to aeration levels is necessary.

4.9.4 Applications of Bubbleponics

Bubbleponics is suitable for a wide range of plants including leafy greens, herbs and flowering crops. It is particularly well-suited for fast-growing plants such as lettuce, spinach, basil and strawberries. Its versatility makes it an excellent choice for hobbyists, commercial growers and researchers alike.

4.9.5 Conclusion

Bubbleponics represents a highly efficient and effective hydroponic system that offers numerous benefits for plant growth and cultivation. By combining the best features of DWC and NFT systems with enhanced oxygenation, bubbleponics provides an ideal environment for robust root development, nutrient absorption and healthy plant growth.

4.10 Drip Hydroponics System (Top Feed or Top Irrigation)

Drip hydroponic systems (drip irrigation hydroponics) are the popular and efficient methods of growing plants hydroponically. This system operates by delivering a controlled amount of nutrient solution directly to the plant roots through a network of drip emitters or tubing. Here is a comprehensive overview of drip hydroponics ~ covering its components, operation, advantages, and applications.

4.10.1 Components of Drip Hydroponic System

i) **Reservoir**: The reservoir holds the nutrient solution which is typically mixed according to specific plant requirements.

ii) **Pump**: A pump is used to circulate the nutrient solution from reservoir to drip lines or emitters.

iii) **Tubing**: Flexible tubing transports the nutrient solution from reservoir to plants. It can be made of various materials such as PVC, vinyl or polyethylene.

iv) **Drip emitters or drippers**: These devices attached to the tubing regulate the flow of nutrient solution to each plant. Drip emitters come in different types including adjustable drippers, pressure-compensating emitters and micro-sprayers.

v) **Growing medium**: Drip systems can utilize various growing media such as perlite, coco coir or rockwool to support plant roots and retain moisture.

vi) **Timer:** A timer controls the frequency and duration of watering cycles, ensuring that plants receive water and nutrients at regular intervals.

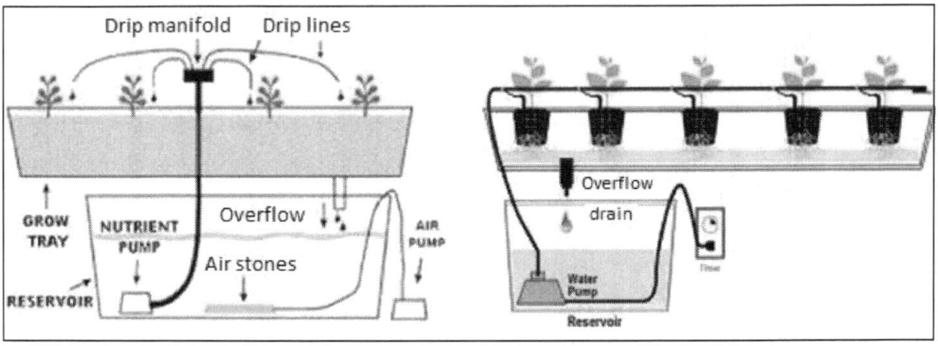

Designs for Drip Hydroponics System.

4.10.2 Operation of Drip Hydroponic System

In a drip hydroponic system the nutrient solution is pumped from the reservoir through the tubing to the drip emitters. The emitters release small controlled amounts of solution directly onto the base of each plant, ensuring precise delivery to the root zone. This targeted approach minimizes waste and allows for efficient nutrient uptake by plants. Excess solution not absorbed by roots is collected and returned to reservoir for reuse promoting water conservation and nutrient efficiency.

Drip hydroponics can be designed as either a **nutrient solution recovery system (closed loop)** or a **non-recovery system (open loop)**. In a **recovery system** excess solution is collected back in reservoir for reuse using a drain line. This setup typically uses a simple timer but requires periodic pH and nutrient strength adjustments due to nutrient depletion. In this setup, small feed hoses are often staked inside the pots or if hydroton is employed water drip rings are simply placed on top of growing medium. In a **non-recovery system**, a precise timer is used to control watering cycles, minimizing runoff. This system requires less maintenance as excess solution is not recycled, reducing variations in reservoir pH and nutrient strength. pH and nutrient adjustments are only needed when adding more solution to the reservoir.

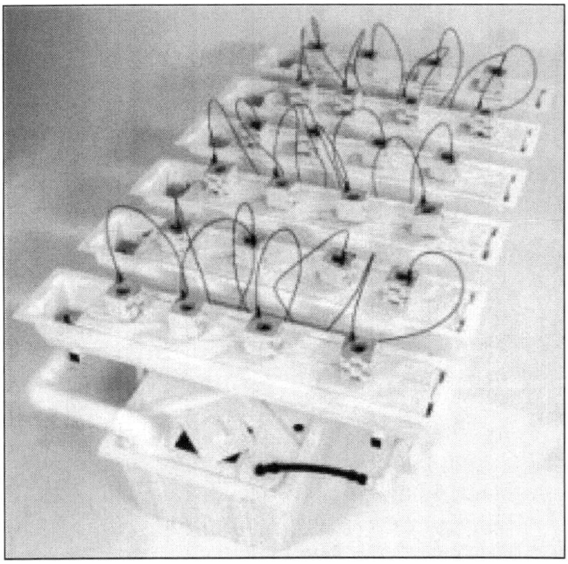

Working Model of Drip Hydroponic System.

Larger plants are typically positioned on a gravel base where supplied nutrient solution is allowed to drain into the ground without recirculation (non-recovery drip system). Conversely, smaller plants which can be spaced more closely together utilize a recirculating hydroponic system (recovery drip system).

4.10.3 Advantages of Drip Hydroponics

i) **Water efficiency**: Drip systems deliver water and nutrients directly to the root zone, reducing water usage compared to traditional soil-based agriculture.

ii) **Nutrient precision**: Precise control over nutrient delivery allows for optimized plant growth and higher yields.

iii) **Scalability**: Drip systems can be easily scaled to accommodate various crop sizes and growing environments from small-scale setups to large commercial operations.

iv) **Customization**: Growers can tailor nutrient solutions to meet specific needs of different plant varieties, optimizing growth and health.

v) **Automation potential**: Drip systems can be automated using timers and controllers, reducing labour requirements and ensuring consistent nutrient delivery.

4.10.4 Considerations and Maintenance

i) **Clogging:** Regular maintenance is essential to prevent clogging of emitters due to mineral buildup or debris in the nutrient solution.

ii) **pH and EC monitoring:** It is important to monitor and adjust the pH and EC of the nutrient solution to ensure optimal nutrient uptake by the plants.

iii) **Root health:** Proper aeration and drainage within the growing medium are crucial to prevent root rot and for maintaining healthy root systems.

iv) **System design:** Careful planning of tubing layout and emitter placement is necessary to ensure uniform water distribution and coverage across all plants.

4.10.5 Applications of Drip Hydroponics

Drip hydroponic systems are suitable for a wide range of crops including vegetables, herbs, flowers and fruits. These are commonly used in greenhouse cultivation, vertical farming and urban agriculture projects. Drip systems are favoured for their versatility, efficiency and ability to produce high-quality crops in controlled environments.

4.10.6 Conclusion

Drip hydroponic systems offer a precise, efficient and customizable method of growing plants hydroponically. By delivering water and nutrients directly to the root zone these systems promote healthy plant growth, maximize resource utilization and support sustainable agriculture practices. Whether used on a small scale for home gardening or in large commercial operations, drip hydroponics continues to be a popular choice for growers seeking reliable and productive cultivation methods.

4.11 Dutch Buckets or Bato Bucket System

It is a simple but highly efficient *cum* versatile method of soilless cultivation widely used in commercial agriculture and hobbyist setups alike. It is particularly popular in commercial greenhouse operations and is also widely used by hobbyists and home gardeners due to its ease of setup and operation. Originating in the Netherlands ~ a country renowned for its innovative agricultural practices, this system offers several advantages for growing a variety of crops in controlled environments. In this containerized drip hydroponic system the plants are grown in individual containers or buckets filled with a growing medium such as perlite, coconut coir or expanded clay pellets. These containers are typically connected to a central nutrient solution reservoir, allowing for efficient delivery of water and nutrients directly to the plant roots.

4.11.1 Components of the Dutch Bucket System

Dutch bucket system comprises several key components each playing a crucial role in its functionality:

i) **Bucket or container**: The primary component of the system is the bucket or container typically made of durable plastic. These buckets serve as individual growing units housing plants and the growing medium.

ii) **Growing medium**: Dutch bucket systems employ inert growing media such as perlite, coconut coir or expanded clay pellets. These materials provide support to the plant roots while allowing for proper drainage and aeration.

iii) **Nutrient solution reservoir**: A central reservoir holds the nutrient solution which is pumped or gravity-fed to each bucket through a network of tubing.

iv) **Submersible pump:** A submersible pump is used to circulate the nutrient solution from the reservoir to the buckets ensuring consistent delivery of water and nutrients to the plants.

v) **Delivery system**: Tubing or irrigation lines transport the nutrient solution from the reservoir to each bucket providing a controlled flow of water to the plants.

4.11.2 Operating Principles of the Dutch Bucket System

Dutch bucket system operates on the principle of recirculating nutrient solution, providing plants with a constant supply of water and essential nutrients:

i) **Continuous feeding**: Nutrient solution is continuously delivered to the buckets allowing plants to uptake water and nutrients as needed.

ii) **Drainage**: Excess nutrient solution drains freely from each bucket preventing waterlogging and ensuring proper aeration of the root zone.

iii) **Oxygenation**: Adequate oxygenation of the root zone is maintained through proper drainage and aeration thus promoting healthy root growth and nutrient uptake.

iv) **Customization**: The system offers flexibility in nutrient delivery so allowing growers to tailor nutrient formulations to meet specific needs of different plants and growth stages.

4.11.3 Advantages of the Dutch Bucket System

Dutch bucket system offers several advantages over traditional soil-based gardening and other hydroponic systems:

i) **Efficient use of space**: The compact design of the system makes it ideal for both indoor and outdoor cultivation, maximizing space utilization.

ii) **Water efficiency**: The recirculating nature of the system conserves water by minimizing runoff and evaporation, making it an environmentally sustainable option.

iii) **Increased yield**: The controlled environment provided by Dutch Bucket System promotes optimal plant growth thus resulting in higher yields compared to conventional gardening methods.

iv) **Versatility**: The system accommodates a wide range of plant species, making it suitable for commercial production as well as hobbyist gardening.

4.11.4 Challenges and Considerations

Despite its many benefits, the Dutch Bucket System also poses some challenges and considerations for growers:

i) **Nutrient management**: Proper nutrient management is essential to prevent nutrient deficiencies or toxicities thus requiring regular monitoring and adjustment of nutrient levels.

ii) **System maintenance**: Regular maintenance including cleaning and disinfection of components is necessary to prevent clogs, algal growth and disease outbreaks.

iii) **Initial investment**: Setting up a Dutch Bucket System requires an initial investment in equipment and materials though long-term savings on water and labour costs can offset these expenses.

4.11.5 Conclusion

Dutch Bucket System offers a versatile and efficient hydroponic growing solution suitable for a wide range of crops. With proper planning, maintenance and nutrient management growers can harness the benefits of this system to achieve high yields of healthy and vibrant plants. Whether used in commercial production or home gardening, the Dutch Bucket System continues to be a popular choice among hydroponic enthusiasts worldwide.

4.12 Run to Waste/Drain-to-Waste/Open Hydroponic System

A run-to-waste hydroponic system is a method where nutrient solution is delivered directly to plant roots and any excess solution is not recirculated but instead allowed to drain away. This active delivery distinguishes it from passive hydroponic systems which rely on natural forces like capillary action or wicking to deliver nutrients to the plants without any mechanical assistance. Therefore, run-to-waste hydroponics does not fall under the category of passive hydroponic systems.

4.12.1 Working

Here is how a run-to-waste hydroponic system typically works:

i) **Growing setup:** Plants are typically grown in containers filled with a growing medium such as perlite, coco coir, rockwool or soilless mix. Each container is placed in a tray or on a bench.

ii) **Nutrient delivery:** A nutrient solution is prepared according to the specific needs of the plants being grown. This solution is delivered directly to the root zone of each plant either manually or through an automated irrigation system.

iii) **Drainage:** In a run-to-waste system, any excess solution not absorbed by the roots is not recirculated back into the nutrient reservoir. Instead, it is disposed of or allowed to flow away, hence the term 'run-to-waste'.

iv) **Monitoring and maintenance:** Growers need to monitor pH and EC levels in the nutrient solution to ensure these remaining within the optimal range for plant growth. These also need to monitor the plants for signs of nutrient deficiencies or excesses and adjust the nutrient solution accordingly.

4.12.2 Advantages

Advantages of a run-to-waste hydroponic system include:

- ☆ Reduced risk of nutrient imbalances or buildup of salts in the root zone.
- ☆ Lower risk of diseases and pests spreading through recirculating nutrient solution.
- ☆ Less complex setup compared to recirculating systems, making it easier for beginners to manage.
- ☆ Ability to customize nutrient solutions for different plants or growth stages without affecting other plants in the system.

4.12.3 Disadvantages

However, there are also some disadvantages to consider:

- ☆ Higher water and nutrient usage compared to recirculating systems as excess solution is not reused.
- ☆ Potential environmental concerns if excess nutrient solution is not disposed of properly.
- ☆ Less efficient use of resources compared to recirculating systems.

4.13 Hanging Bag Technique

The hanging bag technique or 'verti-grow' technique utilizes 1-meter-long cylinder-shaped polythene bags treated with UV protection. These bags typically white with a black interior are filled with sterilized coconut fibre and sealed at the bottom. These are then suspended vertically from an overhead support using a small PVC pipe at the top, positioned above a channel for collecting the nutrient solution. Planting materials such as seedlings are placed in net pots and inserted into holes along the sides of the hanging bags.

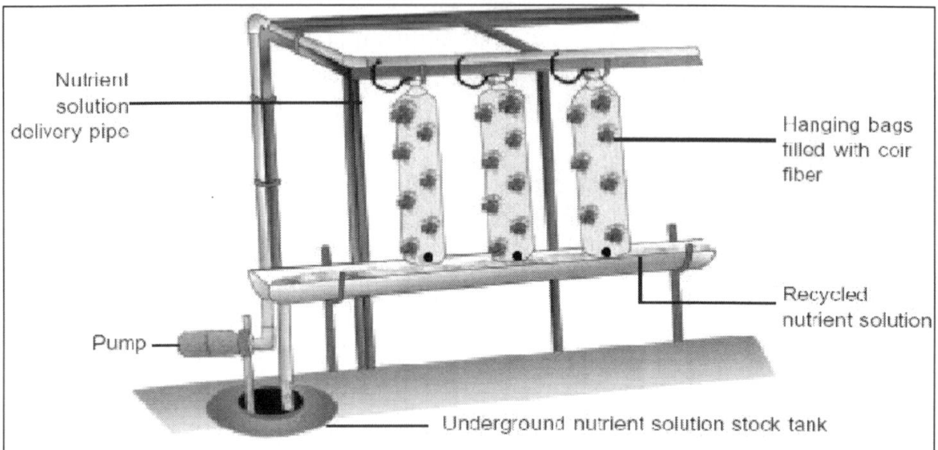

Hanging bag technique

A micro-sprinkler attached inside each bag at the top ensures even distribution of the nutrient solution throughout the bag. As the solution drips down it wets the coconut fibre and plant roots. Any excess solution is collected in the channel below through holes at the bottom of the bags where it flows back to the nutrient solution stock tank.

This system can be implemented in both open spaces and protected structures. In protected environments, proper spacing of the hanging bags ensures adequate sunlight reaching the inner rows. Despite being filled with coconut fibre, the bags are lightweight and durable, typically lasting for about two years. Number of plants per bag varies depending on the type of plant such as about 20 lettuce plants for each bag.

The hanging bag technique is well-suited for cultivating leafy vegetables, strawberries and small flower plants. To prevent mould growth, it is advisable to use black-coloured tubes for nutrient solution delivery.

4.14 Hybrid Hydroponic Systems

Hybrid hydroponic systems represent a fusion of different hydroponic techniques to leverage the advantages of each method while mitigating their individual limitations. These systems capitalize on the flexibility and efficiency of hydroponics by integrating various elements to create a customized growing environment tailored to specific crops and conditions.

- **Aquaponics**: One example of a hybrid hydroponic system is the aquaponics setup which combines hydroponics with aquaculture. In aquaponics, fish are raised in tanks and their waste-rich water is circulated through hydroponic growing beds. The plants utilize the nutrients from fish waste to grow and in turn these filter the water which is then recirculated back to the fish tanks ~ creating a symbiotic relationship between fish and plants.
- **Aeroponic-NFT hybrid system**: Another example is Aeroponic-NFT hybrid system which merges the benefits of aeroponics and NFT. In this setup, plants are suspended in a misted environment created by aeroponic nozzles, allowing for efficient nutrient absorption. The excess nutrient solution not absorbed by the roots is collected and recirculated through NFT channels, providing a continuous flow of nutrients to the plant roots.
- **NFT with drip irrigation**: In this hybrid system, plants are grown in channels as in NFT but instead of relying solely on thin film of nutrient solution a drip irrigation system is integrated to deliver additional water and nutrients directly to the root zone.
- **DWC with aeroponics**: This hybrid system combines the simplicity of DWC with efficient nutrient delivery of aeroponics. Plant roots are suspended in a nutrient solution reservoir but these are also periodically misted with nutrient-rich water to ensure optimal oxygenation and nutrient uptake.
- **Media-based flood and drain with capillary mat**: This system utilizes a flood and drain approach with a growing medium like perlite or coco coir but it also incorporates capillary mats to wick moisture from a reservoir to root zone providing additional hydration and nutrient access to the plants.
- **Ebb and flow with nutrient film technique**: Here, a traditional ebb and flow system is combined with NFT. Plants are periodically flooded with nutrient solution in grow trays but these also have access to a continuous thin film of nutrient solution flowing through channels, enhancing nutrient delivery and root oxygenation.

Some hybrid systems combine traditional soil-based gardening with hydroponic elements such as Kratky method which incorporates hydroponic principles by growing plants in containers filled with a nutrient solution but without need for aeration or recirculation systems. This method is particularly suitable for leafy greens and herbs thus offering a low-maintenance yet productive growing solution.

Moreover, vertical farming systems can be considered hybrid hydroponic setups as these often integrate multiple hydroponic techniques such as nutrient film technique, drip irrigation and vertical stacking to maximize space utilization and crop yield in indoor environments.

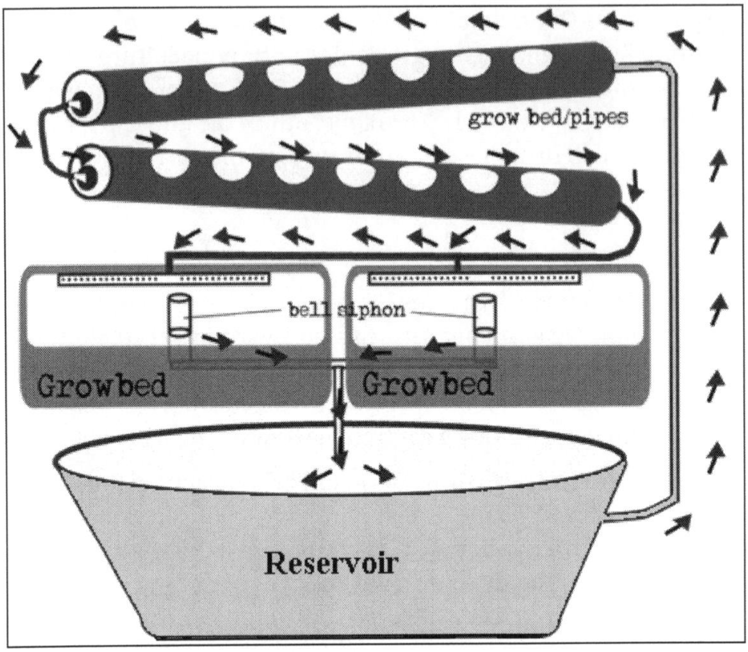

Hydroponics System with Two NFTs and One 'Ebb and Flow'.

These are just a few examples and there are many other ways to create hybrid hydroponic systems by combining different techniques to suit specific plant requirements and environmental conditions.

The figure at the right demonstrates the combination of **NFT** and **Flood and Drain** methods.

The solution is pumped from the nutrient reservoir to the topmost NFT from where it is drained back by gravity to another NFT below. Again from here the nutrient solution is drained by gravity to the two **Flood and Drain systems** that have bell siphons which are used to drain back the nutrient solution back to the nutrient solution reservoir.

4.15 Rotary Hydroponics Garden

A rotary hydroponic system (revolving/rotating hydroponic system) is a type of hydroponic setup where plants are grown on a rotating platform. This system is designed to maximize space utilization, optimize light exposure and provide efficient nutrient delivery to plants, representing a paradigm shift in plant cultivation and harnessing the power of innovation to redefine agricultural practices and usher in a new era of sustainable high-yield farming.

4.15.1 Working

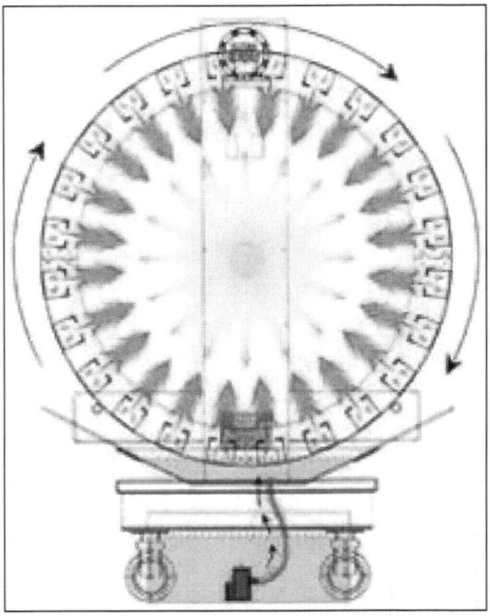

Rotary Hydroponics System

Unlike traditional methods, this innovative system operates within a circular or cylindrical platform that rotates slowly but continuously throughout the entire growth cycle of the plants. This platform is typically motorized and can rotate either horizontally or vertically. The rotation frequency varies based on its design with most systems completing a full rotation once per hour. This means that within a 24-hour period plants experience a staggering 24 rotations around the circle. Planting modules are attached to the rotating platform. These modules can vary in design but typically consist of trays, pots or channels where plants are grown. A nutrient solution is continuously circulated or intermittently supplied to the plants through a network of tubes, pipes, or channels. This ensures that the plants receive the necessary water and nutrients for optimal growth.

High intensity grow lights are often installed above the rotating platform to provide adequate light to the plants. The grow light and rotating frame is meticulously designed to simulate the natural rhythm of sunlight. The rotation of the platform ensures that all plants receive uniform light exposure, minimizing shading and maximizing photosynthetic efficiency. Enhancing this setup is a mechanized timer, ensuring optimal light exposure for each plant. Many rotary hydroponic systems are equipped with automation features such as timers, sensors and controllers to monitor and regulate factors like nutrient levels, pH, temperature and humidity.

4.15.2 Advantages

What truly sets this hydroponics system apart is its remarkable efficiency. Plants grown within the rotary setup exhibit accelerated growth rates compared to those cultivated in soil or conventional hydroponic systems. This accelerated growth can be attributed to the plant's continuous engagement in a battle against gravity, prompting robust development and bountiful yields.

i) **Space efficiency**: By utilizing vertical space and rotating the plant beds, rotary hydroponic systems can significantly increase the planting area compared to traditional hydroponic setups. It is ideal for environments where space is limited but high yields are desired such as urban agriculture settings or indoor farming facilities.

ii) **Uniform growth**: The rotation of the platform ensures that all plants receive uniform light exposure and nutrient distribution thus promoting consistent growth and reducing risk of uneven development.

iii) **Optimized resource use**: Typically this system uses water and nutrients more efficiently compared to soil-based agriculture as hydroponic setup allows for precise control over nutrient delivery and minimizing water wastage.

iv) **Reduced labour requirements**: Automation features in rotary hydroponic systems can help reduce the need for manual intervention, saving time and labour costs associated with traditional farming practices.

v) **Year-round cultivation**: With controlled environmental conditions and artificial lighting, rotary hydroponic systems enable year-round cultivation regardless of external weather conditions.

4.15.3 Considerations

i) **Initial investment**: Setting up a rotary hydroponic system requires a significant initial investment due to cost of equipment including rotating platform, planting modules, lighting and automation systems.

ii) **Maintenance**: While rotary hydroponic systems can reduce labour requirements, these still require regular maintenance to ensure proper functioning of components such as pumps, motors and nutrient delivery systems.

iii) **Technical expertise**: Operating and troubleshooting a rotary hydroponic system may require technical knowledge, particularly regarding nutrient management, environmental control and automation systems.

iv) **Energy consumption**: The use of artificial lighting and motorized components in rotary hydroponic systems can contribute to higher energy consumption compared to traditional farming methods.

Chapter 5

Environmental Conditions for Plants

Various plant species exhibit unique requirements for optimal growth and development, emphasizing the crucial interplay between environmental factors and plant productivity. Factors such as temperature, light intensity and duration, humidity levels, oxygen availability, carbon dioxide concentration, water supply, wind patterns, rainfall, nutrient availability and pH balance profoundly influence plant growth trajectories. Within hydroponic systems, growers must maintain two distinct environments: the air environment and the root environment. The root environment, particularly in hydroponic setups employing nutrient solutions and growing media, plays a pivotal role in supporting plant health and vigour as elaborated under Chapters 'Hydroponic Nutrient Supply', 'Factors affecting Nutrient Solution' and 'Substrates or Growing Medium'.

Giant Greenhouses in the Netherlands.

Hydroponic cultivation represents one of the most intensive methods of crop production within the modern agricultural landscape. Often integrated with greenhouse structures/growrooms and protective covers, hydroponic systems form the cornerstone of controlled agricultural environment. This integration enables growers to finely adjust air and root temperatures, manipulate light exposure, regulate water availability, optimize nutrient delivery and shield crops from adverse weather conditions. By harnessing these capabilities, growers can maximize productivity, minimize water and land usage and uphold environmental conservation efforts. Consequently, the amalgamation of hydroponics with controlled environment agriculture offers a potent avenue for sustainable and high-yield crop production while safeguarding environmental integrity.

Indoor cultivation offers flexibility for a wide range of plants, unlike outdoor growth which contends with fluctuating weather conditions such as heat and cold. Certain hydroponic plants demand high light exposure while others thrive in lower light environments.

5.1 Greenhouses

A greenhouse serves as a controlled environment where various growable plants can be cultivated year-round. This structure plays a pivotal role in providing necessary heat and light inputs especially during winter months when warm-weather crops are produced. Commercial greenhouses exemplify high-tech production facilities equipped with screening installations, heating, cooling, lighting systems and often managed *via* computer automation. Irrigation in most greenhouses is facilitated through sprinklers or drip lines.

The primary function of a greenhouse is to shield plants whether soil-based or hydroponic ~ from external elements such as heat, wind, cold and rain thereby enhancing overall crop productivity. Many large-scale hydroponic systems are situated within such controlled environments. The size and design of a greenhouse hinge upon various factors including capital investment, environmental conditions, market demands, labour requirements and personal preferences. Selecting an optimal building site with considerations like accessibility, drainage, available utilities and sun exposure is crucial.

A well-designed greenhouse must effectively protect crops from adverse weather conditions and pathogens. Particularly in cooler climates, maintaining an interior temperature between 4.5°C and 40°C is essential for plant health and growth. It is more efficient to regulate a building's temperature within this range rather than attempting to cool excessively high outdoor temperatures.

Furthermore, a greenhouse design should prioritize the following functionalities:

- ✰ Maximizing light transmission to the crop to facilitate photosynthesis.
- ✰ Transmitting the visible light spectrum, crucial for plant photosynthesis.
- ✰ Absorbing ultraviolet radiation and converting it into visible light beneficial for plants.
- ✰ Preventing overheating by reflecting or absorbing infrared radiation.

By fulfilling these criteria a well-designed greenhouse not only provides a conducive environment for plant growth but also optimizes resource utilization and enhances crop yields.

5.1.1 Covering Materials for a Greenhouse

Different types of greenhouse covering materials offer various advantages and considerations:

Glass Cover

- ✰ High emissivity for long-wave radiation leading to slightly higher nighttime air temperatures.
- ✰ Inflexible, heavy and expensive.

Greenhouse with Glass Cover

PE (Polyethylene) Sheet Film

- ✰ Easy to handle due to flexibility and lightweight nature.
- ✰ Relatively inexpensive compared to other options.
- ✰ Less durable than alternative materials.
- ✰ Widely used globally for greenhouse covering.

Air-inflated Blanket of PE

- ✰ Offers significant heat retention (30-40 per cent savings in winter) by utilizing air pressure between two layers of PE maintained by a small fan.
- ✰ Provides stability against high winds.

PVC (Polyvinyl chloride)

- ✰ High emissivity for long-wave radiation, similar to glass.
- ✰ Less flexible than PE.
- ✰ Narrow width compared to PE but can be heat welded to form larger sheets.
- ✰ Not photodegradable like PE.
- ✰ Not suitable for air-inflated roofs due to potential stretching and reduced structural strength.

Double-Skinned Panels of Polycarbonate and Acrylic

- ✰ Offer superior insulation properties but come at a higher cost.

Each type of greenhouse covering material has its own set of advantages and limitations whether choice depends on factors such as climate, budget and specific requirements of crops being grown.

5.2 Growroom

A growroom serves as a controlled environment where a diverse range of plants can be cultivated whether in soil or hydroponic systems. It essentially functions as a larger-scale version of a grow box thus providing optimal conditions for plant growth.

Utilization of a grow room for plant cultivation stems from various reasons. Some may employ it for cultivating illicit varieties while others may resort to indoor growing due to limited outdoor options. Primarily, the appeal lies in the ability to exert complete control over climatic variables such as temperature, humidity, light, CO_2 levels and nutrition while also effectively shielding plants from insect-pests and diseases.

However, when plants are subjected to artificial or natural light sources, the heat generated by high-power lamps can elevate the temperature within the growroom beyond optimal levels. In such instances, supplemental ventilation fans prove beneficial in maintaining suitable growing conditions.

Fully automated growroom system

The advantages of cultivating plants in a growroom are manifold:

- ✰ Off-season crops can be successfully grown, enabling year-round cultivation.

- ☆ Exotic plant species, typically suited to distinct climatic conditions elsewhere can thrive within a growroom environment.
- ☆ Enhanced crop yields are achievable through optimized growing conditions.
- ☆ Harvest times can precisely be scheduled minimizing losses and ensuring timely production.
- ☆ Reduced reliance on agrochemicals is feasible due to controlled environment.
- ☆ Airflow within and outside the growroom can be filtered and sterilized, rendering them ideal for cultivation purposes.

Most growrooms completely depend on artificial lighting. Different stages of plants require different levels of light. A lower light level for longer duration (usually 18 hours of light and 6 hours of darkness) is required during the vegetative stage of a crop whether a higher level of light for shorter time periods (usually 12 hours of light and 12 hours of darkness) is required during flowering stage of a crop. The lighting system in a growroom can be divided into 2 banks of lights which can be alternated. Only 1 bank of lights is used during vegetative stage and both banks are used during flowering stage to give more intensity. The adjustment of the height of the lights and the angles of reflector hoods can also increase or decrease light intensity.

Moving light which simulates the movement of the sun and provides light at a range of angles to the crop can be a good idea for a growroom. In this system, a rail long enough to fully cover the growing area of the crop is installed above the crop and lights are suspended from a silent moving motor which rides on the rail. The motor moves along the length of the rail, pauses at the end and then slowly travels back to other direction. This allows for more uniform growth of all the plants from top to bottom as these all receive the same amount of light with less shading of one plant over another.

5.3 Growbox

Growboxes (grow cabinets) are versatile enclosed systems designed for indoor plant cultivation which are ideal for year-round use in limited spaces. Serving as compact versions of growrooms, these vary in size and complexity to suit diverse needs. In certain contexts, particularly in the United States where cultivation laws may vary, growboxes are colloquially known as stealth cabinets, often associated with the cultivation of plants such as marijuana.

Available in hydroponic or soil-based configurations, growboxes offer protection against pests and diseases, making them popular choices for indoor gardening. These serve multiple purposes, catering to individuals facing constraints like limited outdoor space or those seeking to cultivate vegetables, herbs or flowers throughout colder months.

Sophisticated growboxes have comprehensive features including built-in grow lights, intake and exhaust fan systems for ventilation and hydroponic setups for nutrient delivery to plants. Advanced units may incorporate odour control filters and air conditioning systems to regulate internal temperatures effectively. CO_2 supplementation is also integrated to enhance plant growth rates. These advanced functionalities empower gardeners to maintain optimal conditions, including temperature, lighting, nutrient levels, and overall environmental parameters, tailored to the specific needs of their chosen plants.

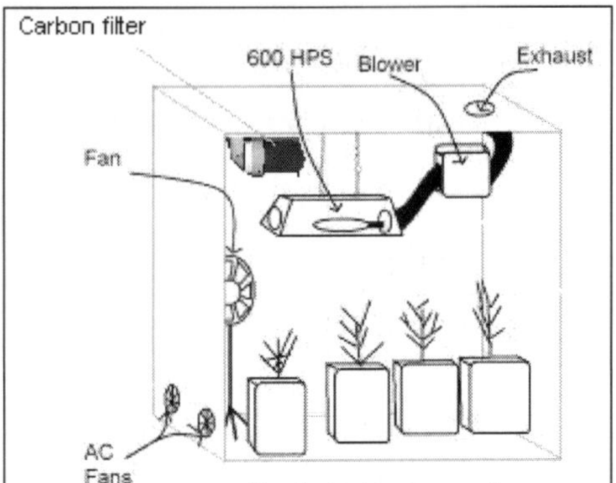

Simple design of grow box

5.3.1 Basic Equipments of Grow Cabinet

A grow cabinet may have different equipments required to improve plant growth and yields. These equipments, once all together, will allow its owner to control all conditions inside the cabinet to make them perfect for growth. Below is a table of different pieces of equipment on many commercially-built cabinets.

These units can come in a variety of sizes. Depending on size of plants to be grown, growers can choose the right size growbox. Basic equipment typically found in a grow-cabinet includes:

i) Odour Control Filter

- ☆ **Use/benefits:** Prevents odours from escaping the growbox ~ crucial for discreet operations or when cultivating aromatic plants.
- ☆ **Types:** Cylinder filters available in various sizes (heights and diameters).

ii) Lighting Systems

- ☆ **Use/benefits:** Essential for providing artificial light to plants, mimicking natural sunlight and controlling seasonal cycles. Facilitates increased energy uptake.
- ☆ **Types:** CFL, HID, LED, *etc.*

iii) Hydroponic Systems

- ☆ **Use/benefits:** Enables plants to grow in a nutrient solution, enhancing nutrient uptake and giving better growth rates compared to soil cultivation.
- ☆ **Types:** DWC, aeroponics, drip systems, ebb and flow systems, floater systems, wick systems. Nutrient delivery system includes reservoirs, pumps and tubing for nutrient delivery.

iv) Ventilation

- ☆ **Use/benefits:** Facilitates air circulation within the cabinet, controlling humidity and temperature conditions.
- ☆ **Includes:** Induct fan, rotating fan, light hood cooling fan.

v) CO_2 Systems

- ☆ **Use/benefits:** Controls CO_2 levels to enhance photosynthesis rates.
- ☆ **Includes:** Cans that slowly release CO_2 (disposable) or tank systems with controlled release from a tank.

vi) Water Filtering System

- ☆ **Use/benefits:** Filters nutrient solution to remove harmful particles.
- ☆ **Types:** Inline filters, filtering systems integrated into nutrient solution.

vii) Temperature Control

- ☆ **Use/benefits:** Regulates root and air temperature.
- ☆ **Includes:** Exhaust fan system, water heater and advanced units with air conditioning.

viii) Timers

- ☆ **Use/benefits:** Automates lighting schedules so ensuring the plants receiving appropriate light-dark cycles.
- ☆ **Types:** Timers for lighting systems.

ix) pH and EC Meter

- ☆ **Use/benefits:** Monitors and adjusts pH and nutrient levels in the water.

x) Support Structures

- ☆ **Use/benefits:** Supports plant growth and prevents collapse under their weight.
- ☆ **Types:** Trellises, stakes, *etc.*

These are the fundamental components of a grow-cabinet though more advanced setups may incorporate additional features for convenience and improved performance.

5.4 Optimal Growing Environments for Greenhouses and Growrooms

Greenhouse or growroom environmental conditions play a crucial role in determining plant growth, productivity and overall success of cultivation. Maintaining optimal environmental conditions within the greenhouse requires a comprehensive approach that addresses temperature regulation, humidity control, lighting management, CO_2 supplementation, air circulation and water management. By careful monitoring and adjusting these factors growers can create an ideal growing environment conducive to healthy plant growth, crop quality and high yields.

Here are some key factors and considerations regarding greenhouse environmental conditions:

5.4.1 Light

Light plays a fundamental role in plant biology serving as the primary energy source for photosynthesis ~ a vital process that converts light energy into chemical energy thus fuelling growth and development. Optimizing photosynthesis is paramount for robust plant growth, underscoring the importance of maximizing light transmission within the greenhouse environment. External obstructions like nearby buildings, trees or taller plants within the greenhouse can hinder light penetration so necessitating strategic orientation of structures and crops. Ideally, greenhouses and plant rows should align north to south ensuring even distribution of sunlight across each plant. Employing white floor coverings or paint can enhance light reflection within the greenhouse, effectively redirecting photons back towards the crops. While clean white paint offers superior reflectivity compared to metallic or foil alternatives, it is worth noting that foil may have the added benefit of confusing insects, potentially leading to decreased insect-pest damage, albeit marginally. Thus, careful consideration of light management strategies can significantly optimize plant growth and yield within greenhouse environments.

Though natural sunlight is the primary source of light in greenhouses, supplementary lighting becomes imperative during periods of low light intensity or shorter daylight durations. High-pressure sodium (HPS), metal halide and light-emitting diode (LED) lights serve as effective supplements, ensuring consistent illumination to optimize plant growth and flowering. Artificial lighting is typically not cost-effective for vegetable crops except for seed production. However, during winter months, particularly in northern latitudes, supplemental lighting may be necessary to ensure robust growth of seedlings.

Providing artificial lighting for 14-18h/day is generally recommended for most vegetable seedlings to simulate natural daylight and encourage healthy development. However, specific lighting requirements may vary slightly depending on the type of vegetable and its growth stage. Some vegetables may require slightly shorter or longer photoperiods so it is essential to find out the specific needs of the vegetables one is growing. Factors such as light intensity, spectrum and distance from the plants can also impact their growth and development.

5.4.1.1 Ideal Light Spectrum

Optimizing the light spectrum is paramount for promoting robust plant growth, with the ideal range falling between 400 and 700 nanometers (nm), encompassing the visible light spectrum from blue to red. Various artificial lighting options are available ranging from fluorescent to high-intensity discharge (HID) lighting each offering distinct advantages. In particular, HID lighting is favoured for larger growing areas due to its efficiency and coverage. It is crucial to note that plants perceive light differently from humans, with their response to light quantified in terms of PAR Watts, representing photo-synthetically active radiation conducive to photosynthesis. Peak photosynthesis occurs at wavelengths of 435nm (blue) and 675nm (red), highlighting the importance of optimizing light sources to accommodate these spectrums.

Contrary to human perception, where brightness or lumen intensity often dictates preference, the most effective lighting for plant growth may not be the brightest. While lumen output serves as a useful indicator of a lamp's efficiency it is essential to prioritize the spectral quality of light over sheer brightness. By understanding and tailoring artificial lighting to mimic natural sunlight's spectral composition growers can maximize photosynthetic efficiency and ultimately enhance plant growth and yield.

5.4.1.2 Light and Plant Growth

Optimal light intensity is crucial for promoting healthy plant growth and facilitating essential nutrient metabolism. Factors such as light quality, intensity and duration significantly influence overall plant development. Inadequate light levels can result in stunted growth, poor flowering and diminished yields while excessive light intensity may lead to leaf burn, photoinhibition and nutrient imbalances. By carefully optimizing light levels growers can enhance photosynthetic efficiency and foster robust plant growth in hydroponic environments. Furthermore, light exerts influence over various physiological processes in plants directly impacting nutrient uptake and utilization, further underscoring its importance in hydroponic cultivation.

5.4.1.3 Photosynthesis and Plant Growth

In hydroponic systems, the provision of adequate light intensity plays a pivotal role in fostering photosynthesis, chlorophyll synthesis and carbohydrate production within plants. Carbohydrates serve as the primary energy source utilized by plants for various metabolic processes including nutrient uptake and transportation. Optimal light exposure promotes efficient photosynthesis leading to heightened carbohydrate synthesis which subsequently facilitates enhanced nutrient absorption. For instance, in a hydroponic lettuce farm high-intensity LED grow lights are strategically employed to deliver the specific light spectrum necessary for photosynthetic activity. This ensures vigorous leafy growth and accelerated development resulting in amplified yield and superior quality of lettuce.

5.4.1.4 Light Spectrum and Plant Responses

Light spectrum plays a crucial role in influencing plant growth and development as different wavelengths of light trigger specific physiological responses in plants.

Photoreceptors within plants such as phytochromes and cryptochromes are sensitive to various parts of the light spectrum including red (R), blue (B) and far-red (FR) light. For instance, in controlled environments like hydroponic greenhouses, growers utilize supplemental LED lights tailored to specific growth stages. During fruiting stage of tomatoes, LED lights emitting a combination of red and blue wavelengths are strategically employed. Red light primarily stimulates flowering, initiating the reproductive phase while blue light promotes vegetative growth fostering robust foliage development and enhancing photosynthetic efficiency. This strategic use of light spectrum manipulation contributes to increased yields with improved crop quality thus exemplifying the importance of understanding plant responses to different light wavelengths in modern agriculture.

5.4.1.5 Light Intensity and Plant Productivity

Light intensity quantified by photosynthetic photon flux density (PPFD) plays a crucial role in determining plant productivity by directly impacting photosynthesis and growth rates. While higher light intensity generally correlates with enhanced productivity, it is important to strike a balance to avoid photoinhibition and potential damage to plant tissues. For instance, in vertical hydroponic farms cultivating herbs like basil, precision is key. LED fixtures equipped with adjustable intensity levels are strategically installed to optimize light distribution and intensity throughout the vertical setup. This meticulous approach ensures uniform growth and consistent yields of premium-quality basil across all levels of the farm thus maximizing efficiency and profitability.

5.4.1.6 Light Duration (Photoperiodism)

Light duration known as photoperiod plays a crucial role in governing various plant developmental processes including flowering and dormancy. Different plant species exhibit specific responses to changes in light exposure, influencing their growth patterns and reproductive cycles. In hydroponic cultivation as at a strawberry farm growers utilize techniques like blackout curtains or adjustable LED lighting systems to manipulate the photoperiod. By strategically adjusting light duration growers can induce flowering and fruiting during desired periods thereby synchronizing fruit production and maximizing yields of strawberries throughout the year. This precise control over photoperiodism allows for optimized crop management and consistent harvests in controlled environment agriculture.

5.4.1.7 Light Quality and Nutrient Uptake

Light quality plays a pivotal role in influencing nutrient uptake and utilization in plants. Different wavelengths of light have distinct effects on plant physiology including nutrient absorption, root growth and metabolic processes. Specific wavelengths within the light spectrum can stimulate the activity of chlorophyll and other pigments involved in photosynthesis so leading to increased nutrient assimilation. In commercial hydroponic operations, growers leverage advanced lighting technologies such as full-spectrum LED grow lights to provide plants with tailored light conditions that optimize nutrient uptake and overall growth. By mimicking natural sunlight, these LED lights ensure a balanced spectrum that promotes efficient photosynthesis and nutrient assimilation. This strategic approach results in robust plant development and enhanced yields as exemplified in the successful cultivation of cucumbers in hydroponic systems.

5.4.1.8 Stomatal Regulation

Stomatal regulation ~ a critical physiological process in plants is intricately linked to light availability. Stomata, the tiny pores dotting the leaf surface play a pivotal role in gas exchange particularly the intake of CO_2 essential for photosynthesis. When exposed to light, stomata open enabling the influx of CO_2 necessary for plant metabolism. However, this action extends beyond carbon dioxide uptake alone. Stomatal opening also governs the absorption of water and minerals including essential nutrients dissolved in the nutrient solution. Therefore, light-mediated stomatal regulation indirectly influences nutrient uptake, highlighting the interconnectedness of environmental factors in plant physiology and growth optimization.

5.4.1.9 Root Development

Root development in plants is intricately influenced by light exposure which plays a significant role in shaping root morphology and function. Within hydroponic systems where roots are directly exposed to nutrient solutions, the importance of adequate light exposure cannot be overstated. Optimal light conditions foster robust root growth characterized by increased surface area and enhanced absorptive capacity. These well-developed roots exhibit improved nutrient uptake efficiency, facilitating overall growth and productivity of plants. Proper lighting ensures that roots remain healthy and vibrant, contributing to the plant's resilience against environmental stressors and diseases. Therefore, maintaining appropriate light levels is essential for maximizing root development and optimizing the performance of hydroponic cultivation systems.

5.4.1.10 Metabolic Activity and Light Availability

The availability of light significantly impacts plant metabolism and nutrient requirements. In conditions where plants receive ample light their metabolic rates are elevated thus necessitating increased nutrient uptake to sustain growth and development. Consequently, the intensity and duration of light exposure can directly influence the rate at which plants absorb nutrients from their surroundings. By understanding this relationship between light availability and metabolic activity growers can adjust lighting conditions to optimize nutrient uptake and promote robust plant growth.

5.4.1.11 Management Considerations

Light management is crucial in both greenhouses and growrooms to ensure optimal plant growth and development. Here are some key aspects to consider for effective light management in both the environments:

i) **Light source selection**: The appropriate light source is choosen based on factors such as the type of plants being grown, budget, energy efficiency and available space. Common options include LED (light emitting diode), HID (high intensity discharge), fluorescent and natural sunlight.

ii) **Light intensity**: Light intensity, measured in foot-candles or lux, refers to amount of light reaching a given area within the greenhouse. Different crops have varying light intensity requirements based on their growth stage and light sensitivity. Greenhouse operators must carefully calibrate lighting systems to deliver appropriate light intensity levels for optimal plant growth and development. This can be achieved by adjusting the distance between the light source and the plants or by using dimmable lights.

iii) **Light quantity**: Providing plants with adequate light is fundamental for photosynthesis which is the process by which plants convert light energy into chemical energy to fuel growth. Greenhouse operators must ensure that plants receive the right amount of light for their specific needs considering the factors such as crop type, light intensity requirements and seasonal changes in sunlight availability.

iv) **Light quality**: The quality of light including its spectrum and wavelength significantly impacts plant growth and development. Different types of plants have varying light requirements with some species thriving under full-spectrum light while others preferring specific wavelengths such as red or blue light for optimal photosynthesis and flowering. Greenhouse lighting systems including high-pressure sodium (HPS), metal halide or light-emitting diode (LED) lights can be used to tailor light spectrum to meet specific needs of different crops.

v) **Light duration (photoperiod)**: Duration of light exposure influences plant growth, flowering and fruiting. Many plants require specific day-length conditions to initiate flowering or maintaining vegetative growth. Greenhouse operators can manipulate the photoperiod using techniques such as supplemental lighting or light deprivation systems to control flowering times and optimizing crop yields.

vi) **Light distribution**: Ensuring uniform light distribution throughout greenhouse is essential for maximizing plant growth and minimizing uneven growth patterns or shading effects. Proper placement and spacing of lighting fixtures as well as the use of reflective materials can help optimize light distribution and ensuring that all plants receive adequate illumination.

vii) **Light directionality**: Adjustment of the angle of light may be required to ensure uniform coverage and penetration into the canopy. This may involve using adjustable fixtures or hanging lights at different heights to provide more direct light to lower leaves and branches.

viii) **Light pollution and light leakage**: It is necessary to minimize light pollution and leakage to prevent disrupting natural cycles which affect neighbouring environments or attracting pests. Growrooms should properly be sealed and for shades light-blocking curtains should be used in greenhouses to contain light within the desired area.

ix) **Temperature management**: Considering the heat generated by the light source appropriate cooling systems such as ventilation or air conditioning should be used to maintain optimal temperatures for plant growth.

x) **Energy efficiency**: Energy-efficient lighting systems can help reduce operating costs and to minimize environmental impact. In particular, LED lighting offers several advantages including lower energy consumption, longer lifespan and customizable light spectra. By investing in energy-efficient lighting technologies and implementing smart lighting controls, greenhouse operators can maximize light efficiency and minimize energy consumption.

xi) **Regular monitoring and maintenance**: Light levels, spectrum and other environmental conditions should continuously be monitored to ensure these remaining within desired range. Regularly the light fixtures should be cleaned and as needed the bulbs replaced to maintain optimal performance.

By carefully managing light quantity, quality, duration, distribution, intensity and energy efficiency greenhouse operators can create an optimal growing environment that promotes healthy plant growth, enhances crop yields and ensures overall success of their operations.

5.4.1.12 Kelvin and CRI

Kelvin and CRI (colour rendering index) are crucial metrics for assessing the quality of artificial lighting in indoor cultivation environments. Kelvin serves as a measure of colour temperature, indicating the hue of light emitted by a source. Light output within the 380-500nm range known as the daylight spectrum closely resembles natural sunlight and is typically rated at 6500K. This colour temperature closely mimics the natural daylight thus making it ideal for promoting robust plant growth and development.

On the other hand, CRI quantifies ability of a lamp to accurately render colours compared to natural sunlight. It is expressed as a percentage with higher values indicating better colour rendition. For example, a lamp with a CRI of 65 is considered to be 65 per cent similar to natural sunlight in terms of colour rendering. Metal halide lamps often fall within this range ~ offering relatively good colour rendition suitable for plant growth.

In contrast, High-pressure sodium lamps (HPS lamps) typically have lower CRI values ranging from 25 to 28. This means that these render only colours at approximately 25 to 28 per cent of the fidelity of natural sunlight while HPS lamps are effective for promoting flowering and fruiting in plants due to their spectral output rich in red and orange wavelengths ~ they may not accurately represent colours in the same way as daylight-spectrum lighting sources.

Understanding the Kelvin and CRI values of lighting fixtures is essential for selecting the most appropriate lighting solutions for indoor cultivation. By choosing lighting systems with higher Kelvin temperatures and CRI values closer to natural sunlight, growers can ensure optimal conditions for plant growth, development and overall productivity.

5.4.1.13 Types of Grow Light

The selection of grow lights for a hydroponic system depends on several factors such as:

i) **Plant variety**: Different plant varieties have varying light requirements therefore it is essential to choose lights that meet the specific needs of the plants being grown.

ii) **Plant size**: Larger plants may require more intense light to cover their foliage adequately.

iii) **Plant stage**: Light requirements of plants may vary during different stages of growth such as flowering and fruiting.

iv) **Type and size of hydroponic system**: The size and layout of the hydroponic system will influence the placement and configuration of grow lights.

v) **Indoor or outdoor location**: Indoor environments may require artificial lighting to compensate for the lack of natural sunlight.

vi) **Seasonal variations**: Light availability can fluctuate depending on the season, with shorter days in winter necessitating additional artificial lighting.

Through careful considerations of these factors growers can select the most appropriate grow lights to meet the specific needs of their hydroponic plants to optimize their growth and productivity.

5.4.1.13.1 Fluorescent Lights

Fluorescent lighting whether traditional or compact offers only a fraction of the desired light spectrum compared to incandescent or HID lamps. While compact fluorescent bulbs are slightly less efficient than HID lamps in terms of light output per unit of power input (lumens), they typically do not provide enough concentrated light to serve as the primary light source in most indoor growing setups. However, fluorescent and compact fluorescent lamps are highly suitable for nurturing young plants such as seedlings. These serve as excellent supplementary light sources to enhance the illumination provided by primary HID lamps.

Fluorescent lights offer several advantages for indoor plant cultivation, especially in hydroponic systems. Their relatively cool operating temperatures allow them to be positioned close to the tops of plants, sometimes mere inches away without causing damage. This proximity maximizes the light intensity reaching the plants thus enhancing growth potential. While fluorescent grow lights may not emit the same high light output as metal halide or high-pressure sodium lights, these excel in providing a balanced spectrum of light wavelengths essential for plant development. Moreover, fluorescent lights are energy-efficient resulting in significant savings on electricity bills compared to other types of grow lights. T5 lamps and compact fluorescent lamps (CFLs) are common types of fluorescent lights used in indoor gardening. Though rated in fluorescent watts, their output is often converted into incandescent watts for easier comparison. Fluorescent lights have demonstrated success in cultivating lettuce and herbs, making them suitable for both seedlings and larger plants like tomatoes. Their versatility and cost-effectiveness make them a popular choice for growers seeking reliable lighting solutions for indoor cultivation.

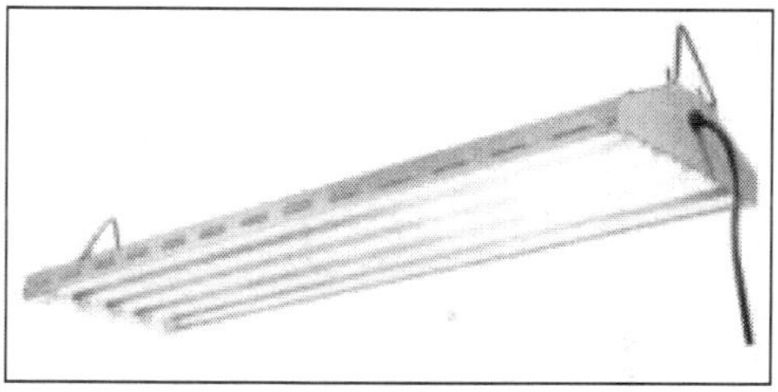

T5 Lamps

T5 technology represents a significant advancement in lighting efficiency for indoor horticulture ~ delivering impressive light output while consuming only a fraction of energy compared to traditional metal halide (MH) lamps. With a power consumption of approximately 216 watts, T5 fixtures can achieve the equivalent light output of a 400-watt MH lamp. These T5 lamps are available in two spectrums tailored for specific growth stages: a 6400K spectrum for vegetative growth and a 2700K spectrum for flowering crops. In addition to their energy efficiency, T5 lamps boast a longer lifespan and superior performance compared to standard fluorescent lamps ~ making them a preferred choice among growers seeking optimal lighting solutions.

While **T8** high output bulbs can also be utilized in indoor cultivation settings, they are better suited for smaller plants such as lettuce, herbs and alike. However, they fall short in terms of light output efficiency when compared to T5 lamps. The T5 technology's ability to deliver higher lumens makes it more suitable for promoting robust growth and development across a wider range of crops, especially those with higher light requirements. Therefore, for growers aiming to maximize productivity and quality in their indoor gardens, T5 fixtures stand out as a highly efficient and effective lighting solution.

Compact fluorescent lamps (CFLs) offer a versatile lighting solution with several advantages. Unlike traditional fluorescent tubes, CFLs are available in various wattages and lumen ratings, providing flexibility to meet diverse lighting needs. One notable feature of CFLs is their compatibility with standard 10-amp power points, eliminating the need for additional ballasts or specialized fixtures. Furthermore, CFL bulbs are available in different wavelengths, allowing growers to customize light spectra according to plant requirements. With an impressive lifespan of 8000 hours or more, CFLs offer long-term reliability and efficiency, making them a popular choice for indoor gardening applications.

5.4.1.13.2 HID (High intensity discharge) Systems

HID lighting systems typically employ high-wattage bulbs, often utilizing metal halide bulbs to emit blue light essential for plant growth or high-pressure sodium bulbs emitting red light crucial for flowering stages. In certain setups, both types of bulbs may be employed simultaneously to cater to various growth phases. Despite their efficacy in providing intense light that penetrates several metres down to plants, HID systems generate significant heat. Improper placement can lead to heat stress or even burning of plants. Originally designed for street and warehouse lighting, HID lamps have found adaptation in horticulture, particularly in environments with insufficient natural sunlight. Despite their heat output, HID systems remain popular due to their ability to provide high-intensity illumination conducive to robust plant development. However, their energy consumption and heat emission necessitate careful management to avoid adverse effects on plant health and overall greenhouse climate.

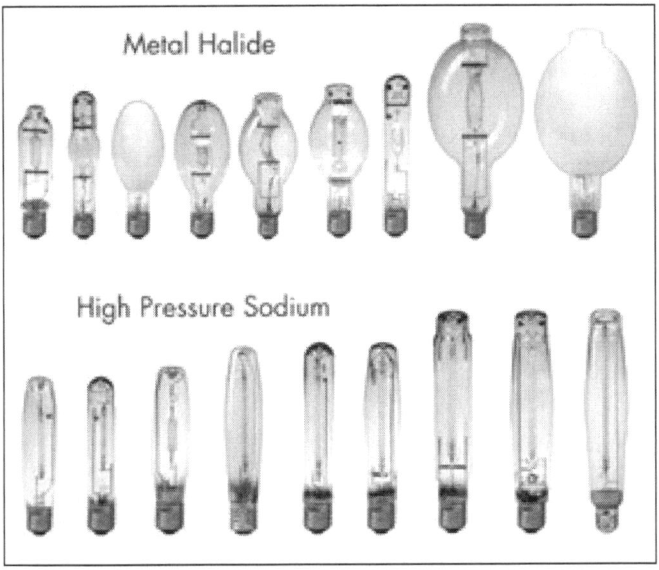

Metal halide and high-pressure sodium (HPS) grow lights are powerful options for indoor plant cultivation, suited particularly for larger hydroponic setups or areas requiring expansive coverage. However, due to their intense light output and heat production, careful placement is essential to prevent damage to plants. These growlights should be positioned several metres above the plants to avoid burning of the foliage as their higher heat output poses a risk of overheating. During colder months these can be utilized strategically to provide supplemental warmth and protect plants from extreme cold conditions. It is worth noting that while metal halide and HPS lights are effective for promoting flowering and supporting larger plants, these typically consume more electricity than fluorescent alternatives therefore resulting in higher energy bills. Growers should consider the specific needs of their plants, the size of their growing area and energy costs when selecting appropriate lighting solution for their hydroponic systems.

Metal halide lamps are favoured for their low colour temperature typically around 6500 Kelvin which results in a cooler operating temperature and a blue-white light that closely mimics natural sunlight. This similarity to sunlight is quantified by the Colour Rendering Index (CRI) which for MH lamps typically falls within the range of 60-70, indicating that they render colours about 60-70 per cent as accurately as natural sunlight. Despite this relatively high CRI, MH lamps have a lower lumen output compared to HPS lamps thus resulting in a shorter lifespan.

On the other hand, HPS lamps have a lower CRI typically in the range of 25-28 meaning they are only about 25-28 per cent as close to natural sunlight as MH lamps. However, these compensate for this by offering a higher lumen output and longer lifespan. HPS lamps emit light with a colour temperature of around 2700 Kelvin, producing a red-orange hue similar to the colour of the sky at sunset.

While both types of lamps have their advantages and limitations, growers often choose between them on the basis of their specific lighting needs and the plant growth stage. MH lamps due to their blue-white light are preferred during vegetative growth phase as this promotes compact and leafy growth though HPS lamps with their higher lumen output and red-orange light are typically used during flowering phase for robust flowering and fruiting. Some growers opt for a combination of both MH and HPS lamps known as dual-spectrum or conversion lamps to provide a broad spectrum of light throughout the entire growth cycle.

Sodium grow lights lacking in blue light predominantly emit red and yellow wavelengths which can lead to plants growing tall and leggy. Conversely, metal halide lights emit minimal red wavelengths. To counterbalance these deficiencies and for providing a more comprehensive light spectrum for optimal plant growth a combination of sodium and metal halide lights is often employed in hydroponic systems. This combination ensures that plants receive a balanced spectrum of light, promoting sturdy growth and robust development. It is noteworthy that sodium bulbs typically have a longer lifespan of up to 2 years while metal halide bulbs last usually around 1 year before requiring replacement. Regular maintenance and timely bulb replacement are essential to sustain optimal lighting conditions within the hydroponic setup and for supporting consistent plant growth.

5.4.1.13.3 LED Grow Lights (Light emitting diode)

LED grow lights or light emitting diode grow lights offer a tailored solution for optimizing plant growth with precise control over light spectrum wavelengths. Specifically designed to emit red and blue light at wavelengths ideal for plant growth, these LED lights leverage advancements in nanotechnology to provide targeted illumination. Unlike traditional HID bulbs, LED grow lights boast an impressive lifespan lasting 10-15 times longer while generating minimal heat and consuming significantly less power. This remarkable efficiency not only translates to substantial cost savings on electricity bills but also underscores their eco-friendly credentials. LED grow lights operate at an impressive 98 per cent efficiency, harnessing scientific principles to deliver optimal conditions for plant growth and development.

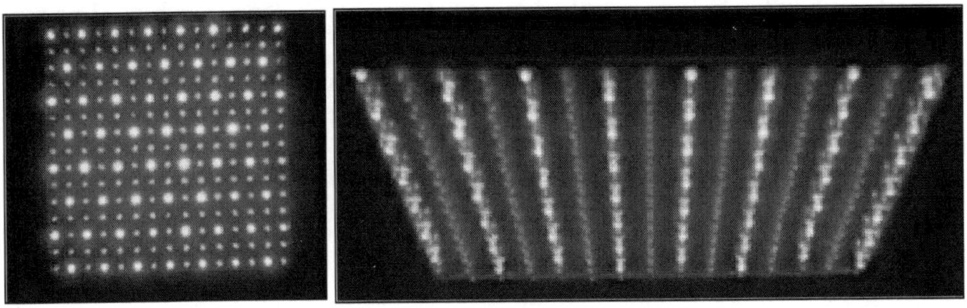

LED Grow Lights.

5.4.1.14 Utilizing Mylar Film or Foil

Irrespective of the specific growlight system implemented in a hydroponic setup, incorporating Mylar film or foil can significantly enhance light distribution and intensity for improved plant growth. By lining the walls surrounding the grow lights with Mylar, light rays are effectively reflected and redirected back into the plant area, ensuring maximum utilization of available light. This amplification of light not only optimizes photosynthesis but also helps mitigating any uneven light distribution within the growing space. Mylar film is readily available for purchase at reputable hydroponic stores, offering a cost-effective solution to boost light efficiency and promote healthier plant development.

5.4.2 HVAC (Heating, Ventilation and Air Conditioning) Systems

HVAC systems play a critical role in greenhouse operations providing climate control to create optimal growing conditions for plants. Here is an overview of HVAC systems in greenhouses:

5.4.2.1 Temperature Management ~ Heating and Cooling Systems

Greenhouse or growroom temperature management is paramount for ensuring optimal plant growth and development year-round. Temperature plays a crucial role in plant metabolism, enzyme activity and overall growth particularly in hydroponic systems where nutrient absorption and photosynthesis are directly impacted. Extremes in temperature can disrupt cellular processes, hinder nutrient uptake and increase susceptibility to stress and diseases. By maintaining temperatures within the ideal range growers can foster robust growth and vitality in their hydroponic crops. Interestingly, while overheating can occur, damage to plants is more commonly caused by cold weather and frosts. It is important to note that temperatures above the optimal range can adversely affect plant growth and even lead to plant death. Therefore, maintaining precise temperature control is essential for maximizing crop productivity and ensuring the success of greenhouse operations.

A well-designed greenhouse or growroom should effectively regulate both day and night temperatures to optimize plant growth. Adequate insulation is crucial to minimize external temperature fluctuations. Implementing separate day and night temperature settings with a gradual transition between the two mirrors the natural outdoor environment. Night temperatures should ideally be maintained approximately 5°C lower than daytime temperatures. Given that HID lighting contributes to heat buildup options such as air conditioning or air-cooled lighting systems can effectively manage internal temperatures. Inlet and exhaust fans play a vital role in temperature control by facilitating air circulation and heat dissipation within the growroom.

In colder climates, maintaining proper temperature levels is of utmost importance to prevent metabolic slowdown and ensure steady growth. While sunlight provides warmth during daytime nighttime temperatures may drop significantly, necessitating greenhouse coverings such as glass or polyethylene sheets to trap heat and create a stable microclimate.

Regular monitoring of greenhouse temperatures is essential for detecting fluctuations and ensuring that conditions remain within the optimal range for plant growth. Thermometers, temperature sensors and automated climate control systems are used to monitor and adjust temperature settings as needed providing precise temperature management. In small-scale setups multiple thermometers strategically placed throughout the greenhouse are used and cross-checked for accuracy. Conversely, large commercial operations often employ sophisticated computer-controlled systems that provide comprehensive management of temperature, humidity, irrigation, fertilization, carbon dioxide levels, light exposure and shading.

5.4.2.1.1 Heating Systems

To ensure consistent plant growth and productivity the heating systems play a crucial role in maintaining optimal temperatures within greenhouses especially during colder periods or at nights. Various heating methods including boilers, radiant heaters, geothermal systems, natural gas or electrical furnaces, *vis-a-vis* solar heating systems are used to provide supplemental warmth as needed.

i) **Insulation**: Adequate insulation helps minimizing the heat loss and to maintain consistent temperatures inside the greenhouse. Insulating materials such as double-layered polyethylene films, polycarbonate panels or fibreglass insulation are used to reduce heat transfer through walls, roofs and doors thus improving energy efficiency and temperature control.

ii) **Boilers**: Boilers are commonly used in larger greenhouse operations to provide central heating. These generate hot water or steam which is circulated through pipes or tubes installed throughout the greenhouse structure. This radiant heating method effectively warms the air and surfaces within the greenhouse creating a uniform temperature distribution.

iii) **Radiant heaters**: These are installed within the greenhouse to emit infrared radiation which directly heats objects and surfaces and provides warmth to the plants without significant heating the surrounding air. This method is efficient and provides localized heating where needed thus reducing energy consumption compared to heating entire greenhouse volume.

iv) **Geothermal heating**: Geothermal heating utilizes the natural heat stored beneath earth's surface to warm greenhouse structures. Geothermal systems typically involve the installation of heat exchange loops buried underground where temperature remains relatively stable throughout year. These loops transfer heat to a heat pump system which then distributes warmth throughout the greenhouse.

v) **Thermal mass**: Thermal mass materials such as water barrels, concrete floors or rock beds absorb and store heat during daytime and release it gradually at night thus helping to stabilize temperatures within the greenhouse. Proper placement and utilization of thermal mass can help regulate temperature fluctuations and improve overall energy efficiency.

vi) **Natural gas or electrical furnaces**: These are commonly used in smaller greenhouse operations or as supplemental heating sources. These generate heat through combustion or electrical resistance and distribute warm air throughout the greenhouse *via* fans or ductwork.

vii) **Solar energy**: Solar heating systems harness solar radiation to heat greenhouse structures. These systems typically consist of solar panels or collectors installed on the roof or walls of the greenhouse which absorb sunlight and convert it into thermal energy. This energy is then used to the warm air or water circulated within the greenhouse thus providing renewable and sustainable heating.

viii) **Waste heat recovery**: Waste heat from various sources such as industrial processes or livestock operations can be captured and utilized to heat greenhouses, *e.g.* placing a chicken coop inside a greenhouse allows the heat generated by the chickens to be recovered and circulated throughout the growing space thus reducing heating costs and energy consumption.

By employing these heating systems, greenhouse operators can maintain optimal temperatures year-round creating favourable growing conditions for a wide range of crops and maximizing productivity. Each heating method offers unique benefits and considerations allowing growers to choose the most suitable option based on their specific needs, budget and environmental conditions.

5.4.2.1.2 Cooling Techniques

Various techniques are employed to cool greenhouse environments and maintain optimal growing conditions for plants:

5.4.2.1.2.1 Air Conditioning

While air conditioning is less commonly utilized in greenhouses compared to heating and ventilation systems, it may be necessary in certain climates or during periods of extreme heat to maintain optimal growing conditions for plants. Here is an overview of air conditioning in greenhouses:

i) **Purpose**: The primary purpose of air conditioning in greenhouses is to control temperature and humidity levels particularly during hot summer months when ambient temperatures rise above the optimal range for plant growth. Air conditioning systems help prevent heat stress in plants thus maintaining uniform temperatures and creating a comfortable environment for greenhouse personnel.

ii) **Types of air conditioning systems**: Several types of air conditioning systems can be used in greenhouses such as:

- **Split system air conditioners**: These systems consist of indoor and outdoor units connected by refrigerant lines. The indoor unit is typically mounted on a wall or ceiling inside the greenhouse while outdoor unit is installed outside the structure.

- **Ductless mini-split systems**: Similar to split system air conditioners ductless mini-split systems do not require ductwork for installation. These are compact and versatile thus making these suitable for greenhouse applications where space is limited.

- **Evaporative or swamp cooling systems**: These systems use the evaporation of water to lower air temperatures. These systems are more energy-efficient than traditional air conditioners but may not be suitable for high-humidity environments.

iii) **Installation considerations**: When installing air conditioning systems in greenhouses several factors should be considered such as greenhouse size, layout, insulation and ventilation requirements. Proper sizing and placement of air conditioning units are essential to ensure uniform cooling and efficient operation.

iv) **Energy efficiency**: Energy-efficient air conditioning systems help minimize operating costs and reducing environmental impact. The systems should be with high energy efficiency ratio (EER) ratings and features such as variable-speed compressors, programmable thermostats and energy-saving modes.

v) **Humidity control**: In addition to cooling, some air conditioning systems offer humidity control features to maintain optimal humidity levels within the greenhouse. Controlling humidity is essential for preventing fungal diseases, promoting plant transpiration and ensuring overall plant health.

vi) **Maintenance and monitoring**: Regular maintenance of air conditioning systems is crucial to ensure their reliability and efficiency. This includes cleaning or replacing filters, checking refrigerant levels, inspecting ductwork for leaks and lubricating moving parts. Continuous monitoring of system performance allows growers to identify issues early and address them promptly.

vii) **Integration with environmental controls**: Air conditioning systems may be integrated with other environmental control systems in the greenhouse such as heating, ventilation and CO_2 supplementation. This integration enables automated control of multiple parameters to create an optimal growing environment for plants while minimizing manual intervention.

5.4.2.1.2.2 Other Less Expensive Methods for Cooling

- **Ventilation**: Adequate ventilation is crucial for regulating temperatures and humidity levels inside the greenhouse. Ventilation systems such as exhaust fans, side vents and roof vents allow hot air to escape and cooler air to enter thus helping to prevent overheating during sunny periods and stagnant air pockets.

- **Shading**: Shading systems such as shade cloths or retractable curtains are used to reduce solar radiation and to prevent excessive heating inside the greenhouse. These shading materials can be adjusted throughout the day to optimize light levels while minimizing temperature fluctuations.

- **Fan cooling**: Utilizing fans to circulate air within the greenhouse helps dissipate heat and reduce temperature extremes. This method enhances air movement thus promoting evaporation and creating a more uniform temperature throughout the growing space.

- **Evaporative pad cooling**: Evaporative pad cooling involves the use of porous pads made from cellulose material, aspen wood or multi-celled/honeycombed material. These pads are saturated with water and as air passes through them the water evaporates thus cooling the air in the process. This technique is often combined with fan systems to enhance cooling efficiency.

- **Fog or mist cooling**: Fog or mist cooling systems are effective in lowering temperatures particularly in environments with low humidity. Clean water free from soluble salts is essential to prevent the mist nozzles from clogging. These systems generate fine water droplets that evaporate quickly thus absorbing heat from the surrounding air and reducing the temperature in the greenhouse.

5.4.2.1.3 Thermostats

Thermostats play a vital role in greenhouse operations by regulating temperatures to create an optimal growing environment for plants. Here are several key considerations regarding the use of thermostats in greenhouses:

i) **Temperature control**: Thermostats are used to control heating and cooling systems within the greenhouse thus maintaining temperatures within a specified range. When temperatures fall below the set point the thermostat signals the heating system to activate while temperatures above the set point trigger the cooling system to turn on ensuring that temperatures remain within the desired range.

ii) **Precision and accuracy**: High-quality thermostats offer precise temperature control and accuracy thus allowing greenhouse operators to maintain optimal growing conditions for plants. Digital thermostats equipped with sensors provide real-time temperature readings and can be programmed with specific temperature set points to meet the needs of different crops.

iii) **Integration with heating and cooling systems**: Thermostats are integrated with heating and cooling systems such as boilers, radiant heaters, fans and ventilation systems to automatically adjust temperature settings based on environmental conditions. This automated control helps ensure consistent temperatures throughout the greenhouse thus minimizing temperature fluctuations and providing a stable growing environment for plants.

iv) **Remote monitoring and control**: Advanced thermostat systems offer remote monitoring and control capabilities thus allowing greenhouse operators to monitor temperature conditions and adjust settings from anywhere using a smartphone, tablet or computer. This remote access provides flexibility and convenience enabling growers to respond quickly to changing environmental conditions and optimize temperature management.

v) **Alarm systems**: Some thermostats are equipped with alarm systems that alert greenhouse operators to temperature deviations outside the desired range. These alarms can help prevent damage to plants due to extreme temperatures by providing timely notifications thus allowing growers to take corrective action and adjust heating or cooling systems as needed.

vi) **Energy efficiency**: Energy-efficient thermostats help reduce operating costs and minimize energy consumption by optimizing heating and cooling cycles based on temperature requirements. Programmable thermostats allow users to create temperature schedules tailored to plant growth stages and daily fluctuations in temperature thus ensuring efficient use of energy resources.

Overall, thermostats are indispensable tools in greenhouse management providing precise temperature control, remote monitoring capabilities and energy-efficient operation. By effectively utilizing thermostats, greenhouse operators can create an optimal growing environment that promotes healthy plant growth, maximizes crop yields and ensures overall success of their operations.

5.4.2.1.4 Temperature Requirements and Sensitivity to Frost of Various Vegetable Crops

Crop	*Temp. (°C)*		*Stability to Frost*	*Temperature Range for Normal Growth and Development (°C)*
	Minimal	*Optimal*		
Artichoke	—	—	Stable (perennial)	15.6 – 18.3
Asparagus	10	24	Stable (perennial)	15.6 – 18.3
Bean (common)	16	24	Sensitive	15.6 – 21.1
Beetroot	4	29	Partially stable	15.6 – 18.3
Broccoli	4	29	Stable	15.6 – 18.3
Brussells sprout	4	29	Stable	15.6 – 18.3
Cabbage	4	29	Stable	15.6 – 18.3
Carrot	4	29	Stable	15.6 – 18.3
Cauliflower	4	27	Partially stable	15.6 – 21.1
Celery	4	21	Partially stable	15.6 – 21.1
Chinese cabbage	4	29	Partially stable	15.6 – 21.1
Cucumber	16	24	Sensitive	18.3 – 23.8
Brinjal	16	29	Very sensitive	21.1 – 29.4
Endive	—	—	Partially stable	15.6 – 21.1
Garlic	2	27	Partially stable	12.8 – 23.8
Leek	2	27	Partially stable	12.8 – 23.9

Crop	Temp. (°C)		Stability to Frost	Temperature Range for Normal Growth and Development (°C)
	Minimal	Optimal		
Lettuce	2	24	Stable	15.6 - 18.3
Mustard	4	27	Stable	15.6 - 18.3
Onions	2	27	Partially stable	12.8 - 23.8
Parsley	4	27	Stable	15.6 - 18.3
Green peas	4	24	Stable	15.6 - 18.3
Pepper	16	29	Very sensitive	21.1 - 29.4
Sweet pepper	16	29	Sensitive	21.1 - 23.9
Potato	–	–	Partially stable	15.6 - 18.3
Sweet potato	–	–	Very sensitive	21.1 - 29.4
Pumpkin	16	35	Very sensitive	18.3 - 23.9
Radish	4	29	Stable	15.6 - 18.3
Spinach	1	21	Stable	15.6 - 18.3
Squash	16	35	Very sensitive	18.3 - 23.9
Sweet corn	10	29	Partially stable	15.6 - 23.9
Swiss chard	4	29	Partially stable	15.6 - 21.1
Tomato	10	24	Sensitive	21.1 - 23.9
Turnip	–	–	Stable	15.6 - 18.3
Watermelon	16	35	Sensitive	21.1 - 29.4

Source: From *Vegetable Production Guide for Eastern Ethiopia* (1969) by Haile Michael Kidane Mariam.

For best results, the ventilation and circulation system must work together with the heating, cooling and CO_2 systems. Thermostats can be used for controlling both cooling and heating devices in a grow-room or greenhouse.

5.4.2.2 Ventilation and Air-circulation

Adequate ventilation and air circulation are essential for success of a greenhouse or growroom as these play a pivotal role in regulating temperature to optimal levels. Proper ventilation ensures the movement of air thus preventing the buildup of plant pathogens like *Botrytis cinerea* together with facilitating the supply of fresh air crucial for photosynthesis and plant respiration. Further, it allows important pollinators to access greenhouse crops. Effective ventilation is necessary for cooling, heating, CO_2 replenishment and for removal of undesirable gases such as ethylene. A well-ventilated space typically yields higher production compared to poorly ventilated environments unless controlled by a complete climate control system. In such cases, where air-conditioners and CO_2 injection systems are employed, CO_2 is introduced through a designated injection system rather than airflow. It is worth noting that CO_2 injection is most effective in a fully controlled climate. Vents and recirculation fans (often automated) are commonly utilized for providing ventilation in greenhouses thus ensuring optimal growing conditions and maximizing crop yield.

Greenhouses in the Westland Region (Netherland).

5.4.2.2.1 Methods for Circulating Air

Several methods are employed to circulate air effectively within a greenhouse or growroom, ensuring optimal growing conditions and maximizing plant health and productivity:

i) **Natural ventilation**: Passive ventilation systems such as roof vents, side vents, louvres and windows utilize natural air currents and convection to circulate air within the greenhouse. These openings allow hot air to escape and fresh air to enter thus creating a continuous flow of air without the need for mechanical ventilation. Natural ventilation systems are particularly effective in moderate climates and can be supplemented with mechanical fans for enhanced airflow control.

ii) **Fans**: Mechanical fans are commonly used to circulate air throughout the greenhouse. These fans come in various sizes and configurations including ceiling-mounted fans, wall-mounted fans and portable floor fans. By moving air within the space fans help distributing heat, humidity and CO_2 evenly thus preventing stagnant air pockets and promoting uniform plant growth.

- **Inlet fans**: Inlet fans are installed to bring fresh air into the greenhouse or growroom. These help regulate temperature, humidity and CO_2 levels while providing plants with necessary air circulation. These fans are usually positioned strategically around the structure to ensure even distribution of fresh air. These may be installed at ground level or near the sidewalls to draw in air from outside. Inlet fans may be controlled manually or automatically depending on the specific setup. Automated systems often use sensors to monitor environmental conditions and adjust fan speed accordingly. The size and capacity of inlet fans depend on factors such as size of the greenhouse or growroom, the number of plants and desired air exchange rate. Proper sizing ensures adequate airflow without causing excessive drafts or temperature fluctuations.

- **Exhaust fans**: Exhaust fans are typically installed near the roof or at the highest point of the structure to facilitate removal of hot air that rises. These may also be placed strategically to ensure efficient airflow and circulation throughout the space. These fans help remove stale air, excess heat and humidity from the greenhouse or growroom thus maintaining a healthy growing environment. These help prevent the buildup of CO_2 and other harmful gases while maintaining proper ventilation. These are often paired with evaporative cooling pads to enhance cooling efficiency. Similar to inlet fans exhaust fans can be controlled manually or automatically. Automatic control systems may utilize thermostats, humidity sensors or CO_2 monitors to regulate fan operation based on preset parameters. The size and capacity of exhaust fans should be matched to the ventilation requirements of the greenhouse or growroom. Factors such as size of space, number of plants and desired air exchange rate will influence the selection of appropriate fan sizes.
- **Balancing inlet and exhaust**: It is important to maintain a balance between inlet and exhaust airflow to prevent negative pressure or air stagnation. Proper ventilation design considers the location, size and capacity of both inlet and exhaust fans to ensure effective air exchange without causing drafts or temperature inconsistencies.
- **Recirculation fans**: Recirculation fans are strategically placed within the greenhouse to recirculate air and maintain uniform temperature and humidity levels. These fans work by drawing air from the top of the greenhouse and redistributing it throughout the space thus ensuring consistent air movement and minimizing temperature differentials between different areas.
- **Horizontal airflow (HAF) fans or horizontal air circulation fans**: Such fans are specifically designed to create a horizontal airflow pattern within the greenhouse. These fans are typically mounted on sidewalls or overhead structures and are angled to direct air across the crop canopy. By promoting air movement at plant level, HAF fans help prevent microclimatic variations, in reducing humidity gradients and to enhance nutrient uptake and transpiration.
- **Vertical airflow system**: Vertical airflow systems use fan-jets to move air along the roof, downward at the end walls and along the floor through the crop. This method provides excellent mixing of air and brings warm air down into the plants ensuring optimal growing

conditions. Though highly efficient, this system is more complex and may require careful planning and maintenance.

iii) **Vent-tube system**: The vent-tube system is though widely used but is considered relatively inefficient. It operates by utilizing a fan-jet connected to a perforated plastic tube running the length of the greenhouse at ceiling height. The fan propels air through the tube, effectively displacing warm air from the roof space downwards to displace cooler air at the floor level.

iv) **Ducted air distribution systems**: Ducted air distribution systems consist of ductwork installed throughout the greenhouse with fans or blowers located at strategic points to circulate air effectively. These systems allow growers to control airflow direction and velocity ensuring uniform distribution of air and optimizing growing conditions for plants.

v) **Misting and fogging systems**: Misting and fogging systems can also contribute to air circulation by introducing fine water droplets into the air. As the water evaporates it creates a cooling effect and increases humidity levels thus promoting air movement and creating a more comfortable environment for plants.

By utilizing these methods for circulating air, greenhouse and growroom operators can create optimal growing conditions, minimizing the risk of disease and maximizing crop yields throughout the growing season.

5.4.2.3 Maintenance and Monitoring of HVAC Systems

Regular maintenance and monitoring of HVAC systems are essential to ensure their reliability and efficiency. This includes cleaning filters, lubricating moving parts, checking for leaks and calibrating thermostats and sensors. Continuous monitoring of system performance allows growers to identify issues early and implement corrective actions promptly.

5.4.2.4 Integration of HVAC Systems with Environmental Controls

HVAC systems are often integrated with other environmental control systems in the greenhouse such as irrigation, lighting and CO_2 supplementation. This integration enables automated control of multiple parameters to create an optimal growing environment for plants while minimizing manual intervention.

5.4.3 Air Filtration Systems

Air filtration systems are essential components of greenhouse operations contributing to optimal air quality and maintaining a healthy growing environment for plants. Here is an overview of the air filtration systems commonly used in greenhouses:

5.4.3.1 Particulate Filters

Particulate filters are integral components of greenhouse air filtration systems designed to capture and remove airborne particles to maintain optimal air quality for plant growth. Here's a closer look at particulate filters in greenhouses:

i) **Functionality**: Particulate filters work by trapping airborne particles such as dust, pollen, fungal spores and other contaminants as air passes through the filter media. These filters effectively remove particulate matter from the air thus preventing it from settling on plants and surfaces within the greenhouse.

ii) **Filter media**: Particulate filters typically consist of fibrous materials such as fibreglass, polyester or synthetic fibers arranged in a dense matrix. The filter media's structure allows air to pass through while capturing and retaining particles of various sizes ranging from large debris to fine particulate matter.

iii) **Efficiency**: The efficiency of particulate filters is determined by their ability to capture particles of different sizes. Filters are rated based on their 'minimum efficiency reporting value' (MERV) or 'high-efficiency particulate air' (HEPA) rating indicating their effectiveness at removing particles from the air. Higher MERV or HEPA-rated filters are capable of capturing smaller particles with greater efficiency.

iv) **Installation**: Particulate filters are typically installed within the ventilation system of the greenhouse either as standalone units or integrated into HVAC (heating, ventilation and air conditioning) systems. Filters are strategically placed to intercept incoming air ensuring that air entering the greenhouse is free from particulate contaminants.

v) **Maintenance**: Regular maintenance of particulate filters is essential to ensure their effectiveness and longevity. Filters should be inspected regularly and replaced or cleaned as needed to prevent clogging and to maintain optimal airflow. Dirty or clogged filters can restrict airflow, may reduce filtration efficiency and compromise air quality within the greenhouse.

vi) **Benefits**: Particulate filters play a crucial role in improving air quality within the greenhouse, reducing the risk of plant diseases, pest infestations and respiratory issues among workers. By removing airborne particles, these filters help create a clean and healthy growing environment conducive to plant growth and productivity.

Overall, particulate filters are essential components of greenhouse air filtration systems which provide effective filtration of airborne particles to maintain optimal air quality and support healthy plant growth.

5.4.3.2 High-Efficiency Particulate Air (HEPA) Filters

These are advanced filtration systems commonly used in greenhouses to ensure optimal air quality and to promote healthy plant growth. Here is an overview of HEPA filters in greenhouse applications:

i) **Filtration efficiency**: HEPA filters are designed to capture and remove airborne particles with exceptional efficiency. These filters can trap particles as small as 0.3 microns with a high degree of effectiveness, including dust, pollen, mould spores, bacteria and viruses. By effectively removing airborne contaminants, HEPA filters help maintaining a clean and healthy environment for plants.

ii) **Construction**: HEPA filters consist of a dense arrangement of fibreglass fibres or other synthetic materials arranged in a pleated configuration. The filter medium is enclosed within a sturdy frame typically made of aluminum or other durable materials to provide structural support and to ensure a securely fit within the ventilation system.

iii) **Installation**: HEPA filtres are installed within the greenhouse ventilation system to intercept incoming air and remove particulate contaminants before these enter the growing environment. These filters are often integrated into HVAC (heating, ventilation and air conditioning) systems or air handling units, ensuring that air circulating within the greenhouse is thoroughly filtered and cleaned.

iv) **Airborne pathogen control**: HEPA filters are highly effective at removing airborne pathogens including bacteria, viruses and fungal spores from air. By capturing and retaining these microscopic contaminants, HEPA filters help reduce the risk of disease outbreaks among plants thus minimizing crop damage and improving overall plant health.

v) **Maintenance**: Regular maintenance of HEPA filters is essential to ensure their continued effectiveness and longevity. Filters should be inspected periodically and replaced as needed to prevent clogging and to maintain optimal filtration efficiency. Proper maintenance practices help ensure that HEPA filters continue to provide high-quality air filtration in the greenhouse.

vi) **Benefits**: The use of HEPA filters offers several benefits to greenhouse operations including improved air quality, enhanced plant health and productivity, *vis-a-vis* reduced risk of disease transmission. By removing airborne particles and pathogens, HEPA filters create a clean and sanitary growing environment that supports optimal plant growth and maximizes crop yields.

Overall, HEPA filters are valuable components of greenhouse air filtration systems so providing high-efficiency filtration of airborne contaminants to create a healthy and conducive environment for plant growth.

5.4.3.3 Activated Carbon Filters

An indoor gardener may need to invest in some sort of air filtration system. The most common systems are carbon filters that offer the best air filtration system. Centrifugal fans are considered the best types of fans to use with carbon filters.

Carbon filters are usually of solid heavy construction. High humidity can affect the longevity of carbon filters as well as the overall yield of the crop so it is important to have good airflow.

Activated carbon filters are valuable components of greenhouse air filtration systems renowned for their ability to effectively remove odours, volatile organic compounds (VOCs) and other chemical contaminants from the air. Here is a closer look at activated carbon filters in the greenhouses:

i) **Odour control**: Greenhouse operations, such as composting, animal husbandry or certain plant species may generate unpleasant odours that can linger in the air. Activated carbon filters are highly efficient at adsorbing and neutralizing odour-causing compounds thus ensuring a fresh and odour-free environment within the greenhouse.

ii) **Volatile organic compounds (VOCs) removal**: Pesticides, fungicides and other chemical compounds used in greenhouse operations can release VOCs into the air posing health risks to plants and personnel. Activated carbon filters effectively adsorb VOC molecules, reducing their concentration in the air and minimizing potential harm to plants and workers.

iii) **Chemical filtration**: Activated carbon filters can also remove various chemical contaminants from the air including formaldehyde, benzene and ammonia which may be present in greenhouse environments due to agricultural activities or nearby industrial sources. These filters help maintain a clean and safe working environment for greenhouse personnel.

iv) **Filter construction**: Activated carbon filters consist of a porous carbon substrate treated with oxygen to create a highly porous structure with a large surface area. This structure enables the adsorption of a wide range of organic compounds and contaminants, effectively trapping them within the filter media.

v) **Integration**: Activated carbon filters are typically integrated into the ventilation system of the greenhouse either as standalone units or as part of HVAC (heating, ventilation and air conditioning) systems. Filters are strategically placed to intercept incoming air ensuring that air passing through the system is effectively purified before entering the greenhouse environment.

vi) **Maintenance**: Regular maintenance of activated carbon filters is essential to ensure their continued effectiveness. Filters should be inspected periodically and activated carbon media replaced or replenished as needed to prevent saturation and to maintain optimal adsorption capacity.

vii) **Benefits**: Activated carbon filters offer numerous benefits to greenhouse operations including improved air quality, odour control and chemical contaminant removal. By effectively removing airborne pollutants these filters help create a healthy and comfortable working environment for greenhouse personnel and to support optimal plant growth and productivity.

In a nutshell, activated carbon filters are invaluable tools for maintaining air quality and mitigating odours and chemical contaminants in greenhouse environments thus contributing to a safe, healthy and conducive growing environment for plants and workers alike.

5.4.3.4 Ultraviolet (UV) Air Sterilizers

Ultraviolet (UV) air sterilizers are advanced components of greenhouse air filtration systems offering efficient sterilization of airborne pathogens to maintain a healthy growing environment for plants. Here is an overview of UV air sterilizers in greenhouses:

i) **Functionality**: UV air sterilizers utilize ultraviolet light to disinfect and sterilize the air by destroying bacteria, viruses, fungi and other microbial contaminants. As air passes through the sterilizer unit it is exposed to UV-C light which disrupts the DNA and RNA of pathogens thus rendering them incapable of reproduction and causing their eventual destruction.

ii) **UV-C light spectrum**: UV air sterilizers primarily emit UV-C light which has a wavelength of 200 to 280 nanometers (nm). This range of UV-C light is highly effective at disinfecting and sterilizing the air thus making it an ideal solution for eliminating airborne pathogens in greenhouse environments.

iii) **Installation**: UV air sterilizers are typically installed within the ventilation system of the greenhouse either as standalone units or integrated into HVAC (heating, ventilation and air conditioning) systems. The sterilizer unit is positioned strategically to ensure that all air passing through the ventilation system is exposed to the UV-C light for maximum disinfection efficacy.

iv) **Effectiveness**: UV air sterilizers offer high levels of effectiveness in eliminating airborne pathogens including bacteria, viruses, mould spores and other fungi. By destroying these microbial contaminants, UV sterilization helps prevent the spread of diseases among plants thus reducing crop losses and improving overall plant health *cum* productivity.

v) **Maintenance**: Regular maintenance of UV air sterilizers is essential to ensure their continued effectiveness and longevity. The UV lamps should be inspected periodically and replaced according to the manufacturer's recommendations to maintain optimal disinfection performance. The sterilizer unit should be kept clean and free from debris to prevent obstruction of UV light.

vi) **Benefits**: UV air sterilizers provide numerous benefits to greenhouse operations including improved air quality, reduced incidence of plant diseases and enhanced crop yields. By effectively sterilizing the air these systems create a cleaner and safer growing environment for plants thus minimizing the need for chemical treatments and promoting sustainable cultivation practices.

Overall, UV air sterilizers are valuable tools for greenhouse growers offering efficient and environment-friendly solutions for controlling airborne pathogens and maintaining optimal plant health and productivity.

5.4.3.5 Electrostatic Precipitators

Electrostatic precipitators (ESP) are advanced air purification systems increasingly utilized in greenhouse operations to maintain clean and healthy air environments conducive to optimal plant growth. Here is an overview of electrostatic precipitators in greenhouse settings:

i) **Functionality**: Electrostatic precipitators work by utilizing electrostatic charges to remove airborne particles from the air. As air passes through the ESP unit it encounters an ionization chamber where particles are electrically charged. The charged particles then pass through a collector plate or surface with an opposite charge causing them to adhere to the surface and effectively removing them from the air stream.

ii) **Particle removal**: ESPs are highly efficient at removing a wide range of airborne particles including dust, pollen, fungal spores and other contaminants. The electrostatic attraction between charged particles and collector surfaces ensures effective particle capture and removal resulting in improved air quality within the greenhouse.

iii) **Customizable design**: Electrostatic precipitators can be customized to suit the specific needs and requirements of greenhouse operations. These are available in various sizes and configurations allowing growers to select the most suitable ESP unit based on size of greenhouse, airflow rate and particle removal efficiency desired.

iv) **Low maintenance**: One of the key advantages of electrostatic precipitators is their low maintenance requirements. Unlike traditional filters that require frequent replacement, ESP units can be cleaned periodically to remove accumulated particles from the collector surfaces. This maintenance process helps ensure continued performance and longevity of the ESP system.

v) **Environmental benefits**: Electrostatic precipitators offer environmental benefits by reducing the need for disposable filters and minimizing waste generation associated with air filtration. By providing efficient particle removal without need for disposable filter replacements, ESPs contribute to sustainable greenhouse practices and reduce overall environmental impact.

vi) **Health and safety**: Maintaining clean air within the greenhouse is essential for promoting plant health and productivity as well as ensuring health and safety of workers. Electrostatic precipitators help minimize the risk of airborne diseases, allergens and respiratory irritants thus creating a safer and more comfortable working environment for greenhouse personnel.

Overall, electrostatic precipitators are effective air purification solutions for greenhouse operations, offering efficient particle removal, low maintenance requirements and environmental sustainability benefits. By incorporating ESP units into greenhouse air filtration systems growers can maintain clean and healthy air environments thus supporting optimal plant growth and productivity.

5.4.3.6 Proper Maintenance and Cleaning

Proper maintenance and cleaning of air filtration systems are essential tasks in greenhouse operations to ensure their effectiveness and longevity. Here are important considerations for maintaining and cleaning air filtration systems:

i) **Regular inspection**: There should be routine inspection schedule of air filtration systems to assess their condition and performance. These should be checked for signs of damage, wear or clogging in filters, ductwork and other components. Identification of any areas requiring attention or replacement to maintain optimal filtration efficiency should be carried out.

ii) **Filter replacement or cleaning**: Manufacturer guidelines should be followed for filter replacement or cleaning intervals based on usage and environmental conditions. Replacement of disposable filters should be carried out according to recommended schedules or as specified the washable filters should be cleaned and reused. Dirty or clogged filters can restrict airflow and reduce filtration effectiveness thus compromising air quality within the greenhouse.

iii) **Cleaning procedures**: When cleaning filters, proper procedures should be followed to ensure thorough removal of accumulated debris, dust and contaminants. Use of appropriate cleaning agents such as mild detergent or specialized filter cleaners should be done along with rinsing of filters thoroughly to remove residues. Filters should be allowed to dry completely before reinstalling them to prevent mould or any bacterial growth.

iv) **Ductwork maintenance**: The ductwork and ventilation components should be inspected for debris buildup, obstructions, or leaks that may impair airflow or compromise filtration efficiency. Cleaning should be conducted as necessary to remove accumulated dust, dirt, or mould, and any damaged or deteriorated sections should be repaired to ensure proper airflow distribution.

v) **Equipment calibration**: Periodically the air filtration equipment including fans, blowers and sensors should be calibrated to ensure accurate operation and optimal performance. Fan speed, airflow rates and pressure differentials should also be verified to confirm that filtration systems are functioning as intended so that these provide adequate air exchange within the greenhouse.

vi) **Documentation and record-keeping**: Detailed records of maintenance activities including filter replacements, cleanings, repairs and equipment inspections should be maintained. Any observed issues, deviations from normal operation or corrective actions taken to address them should also be documented. Records should be kept in an organized way so that whenever needed these may become accessible for future reference and compliance purposes.

vii) **Training and education**: Greenhouse staff should also be properly educated and trained for proper maintenance procedures for air filtration systems. It should be ensured that personnel understand the importance of regular upkeep and cleaning to maintain air quality and promote plant health. Staff should be encouraged to report any maintenance concerns or issues promptly for timely resolution.

By implementing a proactive maintenance and cleaning regimen for air filtration systems, greenhouse operators can ensure optimal filtration efficiency, promote healthy air quality and support thriving plant growth within the greenhouse environment.

5.4.4 Humidity Levels

Optimal humidity levels are crucial for healthy plant growth particularly in hydroponic systems where plants rely on transpiration for nutrient uptake and photosynthesis. Adequate water availability coupled with moderate humidity and proper air circulation ensures optimal transpiration rates and nutrient distribution within the plant. Humidity plays a pivotal role in regulating calcium uptake and hormonal distribution by influencing transpiration, ion pumping and stomatal function. High humidity, especially in stagnant air conditions can impede transpiration cooling and potentially lead to heat stress in plants. Therefore, it is essential to monitor and control relative humidity levels around hydroponic crops to mitigate risks associated with excessive moisture.

5.4.4.1 Monitoring Humidity

Plants perceive humidity in terms of vapour pressure deficit (VPD) reflecting the difference between atmospheric and leaf vapour pressure. Both extremely low (high humidity) and high (low humidity) VPD values can contribute to nutrient deficiencies and hinder plant growth. High humidity levels inhibit stomatal opening thus reducing gas exchange and potentially causing nutrient imbalances.

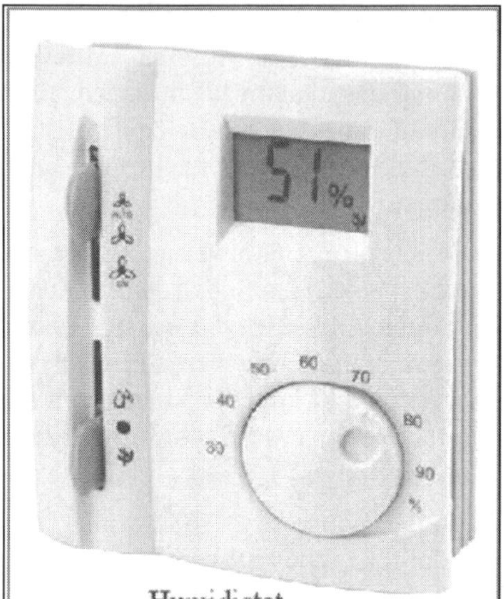

Humidistat

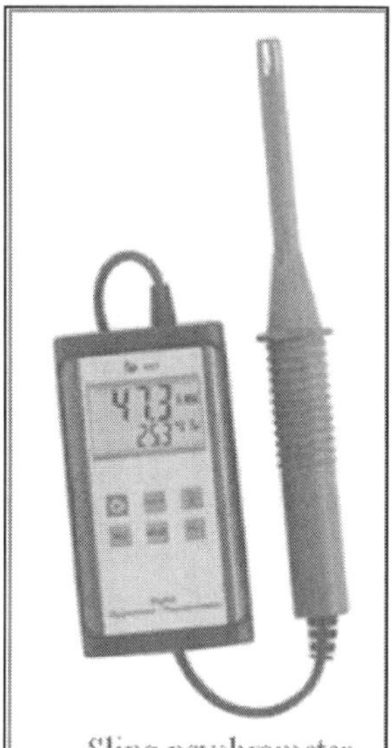
Sling psychrometer

To monitor humidity levels effectively, various tools such as sling psychrometers or humidistats are available from greenhouse supply companies. These instruments provide accurate measurements of relative humidity, enabling growers to adjust environmental conditions accordingly to optimize plant health and productivity. By maintaining optimal humidity levels, growers can create a favourable growing environment conducive to robust plant growth and abundant yields.

5.4.4.2 Impact of Humidity Level on Plant Biology

The humidity level within a greenhouse or growroom significantly impacts various aspects of plant biology including transpiration rates, water absorption, nutrient transport, cell turgor pressure and overall growth. Typically, a humidity range of 55 to 75 per cent is maintained or adjusted based on the specific plant species and growth stage. If air filtration systems are in use, it is advisable to keep humidity below 70 per cent to optimize efficiency. Nighttime humidity levels should generally be lower than daytime levels to reflect natural environmental fluctuations.

Excess humidity creates favourable conditions for development of mould, mildew and pests such as spider mites which compromise plant health and yield. High humidity reduces transpiration rates which limits circulation of essential nutrients such as calcium within the plant. Conversely, low humidity levels increase transpiration which can desiccate leaves and growing points therefore affecting pollination, *vis-à-vis* fruit development in crops like tomatoes.

5.4.4.3 Humidity Management

Misting and fogging systems offer effective means to augment humidity levels and lower temperatures within a greenhouse or growroom. However, improper usage of these systems can significantly elevate the risk of mildew and plant diseases while potentially causing corrosion to metals of greenhouse structures. In greenhouses/growrooms equipped with lighting and CO_2 injection systems, plants tend to exhibit vigorous growth and increased transpiration rates so leading to elevated humidity levels.

Proper ventilation facilitated by inlet and exhaust fans is instrumental in regulating humidity levels within the closed environment. Inlet fans introduce less humid air while outlet (exhaust) fans expel humid air with oscillating fans aiding in air circulation. Air conditioning units contribute to dehumidification, further optimizing the growroom environment. Oscillating fans further aid in maintaining air circulation so preventing stagnant pockets of humidity. Air conditioning units inherently possess a dehumidifying effect which contribute to humidity control within the greenhouse.

5.4.4.4 Optimum Air Humidity for Vegetable Crops

Crop	*Air Humidity (per cent)*	*Crop*	*Air Humidity (per cent)*
Cabbage	50 – 75	Savoy cabbage	60 – 70
Brussels sprouts	60 – 70	Cauliflower	70 – 80
Turnip	60 – 75	Rutabaga	60 – 75
Tomato	45 – 60	Brinjal	50 – 60
Pepper	50 – 60	Cucumber	70 – 80
Onion	70 – 80	Welsh onion	70 – 80
Garlic	50 – 70	Carrot	60 – 70
Beet	60 – 80	Parsley	60 – 80
Celery	60 – 80	Parsnip	60 – 80

Source: Gobeev, A. B. and K. V.Guber (1980). Optimum Growing Conditions for Vegetable Crops.

5.4.5 CO_2 Enrichment

Carbon dioxide is indispensable for plant growth, playing a pivotal role in the process of photosynthesis. During photosynthesis, plants utilize CO_2 from the air, water and light to produce sugars which serve as fundamental building blocks for various essential compounds such as carbohydrates, proteins, vitamins, enzymes and hormones. Oxygen is released as a byproduct of this process. Photosynthesis primarily occurs in the chloroplasts of plant leaves where mesophyll cells absorb CO_2 through stomata ~ small pore-like openings on the underside of leaves. While chloroplasts are light-responsive and typically require daylight for CO_2 absorption though certain plants like cacti have adapted to absorb CO_2 at night.

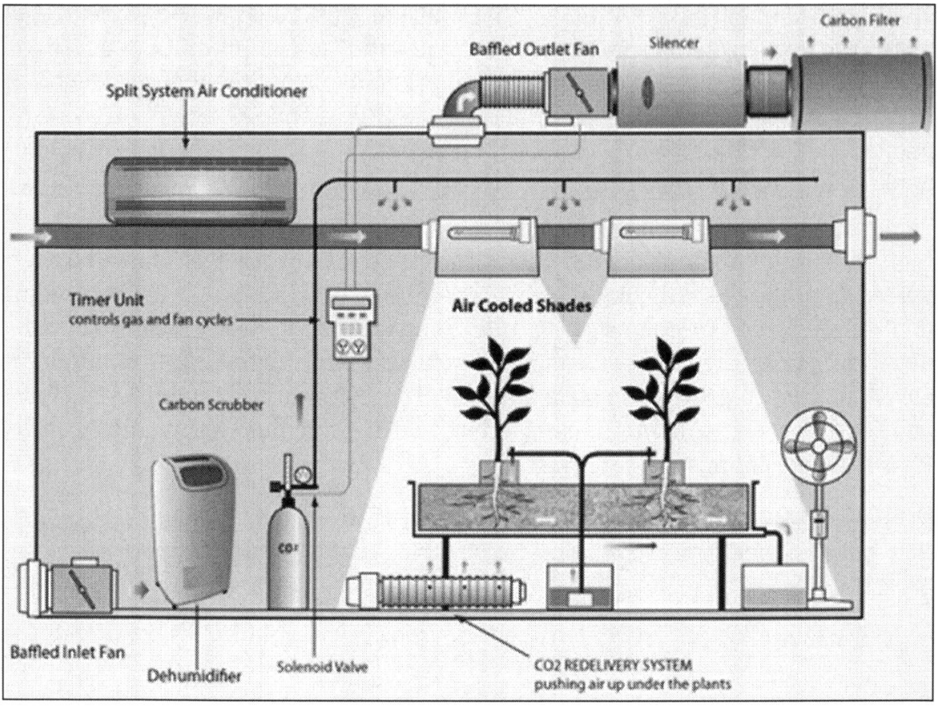

5.4.5.1 Impact of CO_2 Enrichment on Plant Growth

CO_2 enrichment significantly influences plant metabolism and growth rates. Elevated CO_2 levels in the growing environment can substantially enhance growth rates and ultimately improve yields. Employing CO_2 in greenhouses and growrooms not only fosters plant growth but also contributes to environmental sustainability.

5.4.5.2 CO_2 Enrichment Systems

Utilizing bottled CO_2 with a regulator and solenoid valve offers the most precise control over CO_2 injection. Specialized CO_2 controllers can monitor CO_2 levels within the growroom, ensuring these remaining within the desired range. Specialized equipment such as natural gas or propane burners connected to sensors or flue gases from hot water boilers burning natural gas can serve as CO_2 generators for large-scale operations.

Greenhouses situated in industrialized areas can utilize CO_2 emissions from local factories to enrich the growing environment which offers a solution to excess CO_2 production and potential mitigation of global warming.

5.4.5.3 Optimum CO_2 Level

The optimum CO_2 level for greenhouse or growroom cultivation typically hovers around 1100-1500ppm compared to surrounding atmospheric level of approximately 300ppm.

5.4.5.4 Effective Implementation of CO_2 Enrichment

Injecting CO_2 into the greenhouse during daylight hours when plants actively photosynthesize provide optimal results. For growrooms where grow lights are being used, operators should ensure that lights are switched on. During injection, it is essential to seal the greenhouse/growroom and deactivate inlet and outlet fans to prevent CO_2 loss by ventilation systems thus maximizing its absorption by plants. Air conditioning units can be employed to regulate temperature and maintain optimal growing conditions. Following a designated period to allow for CO_2 absorption, ventilation fans can be reactivated to resume airflow within the room.

5.4.6 Advanced HVAC Systems

Advanced HVAC systems in greenhouses incorporate innovative technologies and features to provide precise climate control, optimizing energy efficiency and enhancing crop productivity. Here is an overview of advanced HVAC systems in greenhouse operations:

i) **Climate control integration**: Advanced HVAC systems are often integrated with sophisticated climate control systems that monitor and regulate multiple environmental parameters including temperature, humidity, CO_2 levels and air circulation. This integration allows for automated adjustments to maintain optimal growing conditions for various crops throughout their growth cycles.

ii) **High-efficiency equipment**: Selecting HVAC equipment with high 'energy efficiency ratio' (EER) or 'seasonal energy efficiency ratio' (SEER) ratings is essential for maximizing energy efficiency. Growers should look for systems that meet or exceed industry standards for efficiency such as ENERGY STAR-certified equipment.

iii) **Variable speed technology**: Many advanced HVAC systems utilize variable speed technology which allows components such as fans, compressors and pumps to adjust their speed based on real-time demand. Variable speed operation improves energy efficiency by matching system output to current load requirements, reducing energy consumption and operating costs.

iv) **Energy recovery ventilation (ERV)**: ERV systems capture and transfer heat and humidity between incoming and outgoing air streams so maximizing energy efficiency and reducing heating and cooling loads. By recovering energy from exhaust air, ERV systems help maintain stable indoor temperatures while minimizing energy losses particularly in climates with significant temperature differentials between day and night.

v) **Direct expansion (DX) cooling**: DX cooling systems utilize refrigerant-based cooling technology to provide precise temperature control in greenhouses. These systems offer rapid cooling capabilities and precise temperature regulation, making them ideal for maintaining optimal growing conditions during periods of extreme heat.

vi) **Advanced filtration**: Advanced HVAC systems may incorporate high-efficiency filtration technologies such as HEPA filters or electrostatic precipitators to remove airborne particles, allergens and contaminants from the air. Clean air filtration improves air quality within the greenhouse so reducing the risk of plant diseases and promoting healthier plant growth.

vii) **Remote monitoring and control**: Many advanced HVAC systems feature remote monitoring and control capabilities, allowing growers to access real-time data and adjust system settings remotely *via* computer, smartphone or tablet. Remote monitoring enables proactive maintenance, troubleshooting and optimization of HVAC performance even when growers are off-site.

viii) **Smart controls and automation**: Implementing smart controls and automation features allows HVAC systems to operate more efficiently by optimizing settings based on real-time data and demand. Smart thermostats, sensors and programmable controllers enable automated adjustments and scheduling to minimize energy usage while maintaining optimal conditions.

ix) **Insulation and sealing**: Proper insulation and sealing of the greenhouse structure help minimize heat loss in cold weather and prevent heat gain in hot weather. Well-insulated walls, ceilings and glazing materials reduce the workload on HVAC systems by maintaining stable indoor temperatures with less energy input.

x) **Renewable energy integration**: Some advanced HVAC systems integrate renewable energy sources such as solar panels or geothermal heat pumps to supplement conventional energy sources and reduce greenhouse gas emissions. Renewable energy integration further enhances the sustainability and environmental performance of greenhouse operations while reducing reliance on fossil fuels.

Overall, advanced HVAC systems represent a significant advancement in greenhouse technology, offering growers unprecedented control over environmental conditions, energy efficiency and crop productivity. By leveraging cutting-edge technologies and features, growers can optimize growing conditions, minimize operational costs and maximize the success of their greenhouse operations.

5.5 Alarm Functions

In a professional greenhouse or growroom, the implementation of an integrated detection and alarm system stands as a cornerstone for ensuring optimal conditions and safeguarding plant health. This sophisticated system serves as a vigilant guardian, promptly identifying and alerting growers to any deviation from predetermined parameters such as temperature, humidity or lighting intensity.

With meticulous precision the detection component of the system continuously monitors key environmental variables critical for plant growth. Any anomalies or equipment failures are swiftly identified, triggering immediate alerts to the growers. Whether it is a sudden spike in temperature, a drop in humidity levels or a fluctuation in lighting intensity the alarm system acts as an invaluable early warning mechanism.

By empowering growers with real-time insights and notifications, this comprehensive alarm system enables proactive intervention, preventing potential damage to crops and ensuring maintenance of optimal growing conditions. Ultimately, it serves as an indispensable tool in the quest for consistent high-quality yields in greenhouse and growroom environments.

Chapter 6

Hydroponic Nutrient Supply

Hydroponic nutrient solutions play a fundamental role in hydroponic agriculture by providing plants with essential nutrients in a controlled and efficient manner. These solutions are precisely formulated mixtures of essential mineral nutrients dissolved in water, designed to provide plants with all the elements necessary for growth and development. These mineral nutrients are mostly in **inorganic** and **ionic** form though some organic compounds such as iron chelates may also be used. However, organic hydroponics utilizes only organic fertilizers in the nutrient solution which also contains microorganisms to degrade organic fertilizers into inorganic nutrients. Main dissolved **cations** (positively charged ions) in the nutrient solution are **calcium, magnesium** and **potassium**. Major nutrient **anions** in nutrient solution are **nitrate, sulphate** and **dihydrogen phosphate.**

Many variations of the concentrated nutrient solutions can be purchased from commercial nutrient manufacturers. Although their nutrient solution for a plant may have different chemical combinations, the final compositions would be nearly the same. Commonly used chemicals for macronutrients are potassium nitrate, calcium nitrate, potassium phosphate and magnesium sulphate. The micronutrients that are added to nutrient solution include iron, manganese, copper, boron, zinc, chlorine and molybdenum. Manufacturers sometimes may add chelating agents (to keep iron soluble) and humic acids (to increase nutrient uptake) to the nutrient solution.

Growers equipped with knowledge of specific nutrient requirements whether macronutrients or micronutrients essential for their target plant species can prepare nutrient solutions for hydroponic systems at home. Typically, these solutions are created by dissolving fertilizer salts in water. The selection of salts hinges on several factors including solubility in water, cost, availability and the desired parts per million (ppm) concentration of each element.

Nutrients appropriate for the plant are mixed in the proper proportion in the pure irrigation water (water is usually passed through a RO-filter). This system of combining the supply of water and nutrients is known as fertigation. This saves both water and fertilizers and simultaneously enhances yield and quality of the crops. The fertigation technology is being adapted both under protected and open conditions in Israel, USA, Australia, France, Greece and in many of other developed countries. A fertigation system is used to distribute the nutrient solution to the plants in any one of many ways such as drip and micro-sprinkler methods. The distribution of the nutrient solution is done according to a schedule known as the fertigation schedule. Fertigation schedule is of critical importance to the health and well-being of the seedlings or plants.

6.1 Essential Elements

In a hydroponic system usually the plants require 17 elements for healthy growth and development and these elements are carbon, hydrogen, oxygen, nitrogen, phosphorus, potassium, calcium, magnesium, sulphur, iron, copper, zinc, manganese, molybdenum, boron, chlorine and nickel. Nutrient composition of solution influences its electrical conductivity and osmotic potential. Below is a compilation of essential minerals in recommended ppm form for most plants including their respective ionic forms absorbed by plants.

6.1.1 Carbon, Hydrogen and Oxygen

These are absorbed from the air and water in presence of macronutrients and micronutrients and are involved in enzymatic process and oxidation-reduction process. The deficiency of these elements occur only in extreme conditions of water deficit and waterlogging due to which there is severely restricted uptake of hydrogen and oxygen by the plants. This in turn also causes nitrogen deficiency which consequently results in lower leaf yellowing followed by wilting due to water deficit.

- ☆ **Carbon:** Primarily the carbon in ionic CO_2 form is acquired from atmospheric carbon dioxide through stomatal pores of the leaves. It occurs in the cell walls, in chlorophyll and in sugars manufactured by chlorophyll. It forms up to 50 per cent of the dry weight of a plant and is a constituent of all organic compounds that are found in plants.
- ☆ **Hydrogen:** Usually this is absorbed by plants in H_2O ionic form from water and helps in nutrient uptake from roots. It is also essential for formation of sugar and starch. It comprises 6 per cent of dry weight of the plant and is a part of all organic compounds of which carbon is a constituent. It is critical in the cation exchange in plant-soil/medium relationship and is a major constituent of plant structure when combined with oxygen to create water.

- ☆ **Oxygen:** It is essential for respiration and formation of sugar, starch and cellulose. Oxygen is about 88 per cent of the composition of water and plays a critical role in growth of plants. Plants acquire oxygen in O_2 ionic form through stomata on the leaves, through the roots *via* water and through the process of photosynthesis. It is involved in anion exchange between the roots and surrounding medium.

6.1.2 Mineral Nutrients

Plants require a range of mineral nutrients from growing media or nutrient solution to support their growth and development:

- ☆ Primary macronutrients: Nitrogen, phosphorus and potassium.
- ☆ Secondary macronutrients: Calcium, sulphur and magnesium.
- ☆ Micronutrients/trace minerals: Chlorine, manganese, iron, zinc, copper, boron, molybdenum and nickel.

6.1.2.1 Macronutrients

Macronutrients are required by plants in relatively large quantities (g/kg dry matter) compared to micronutrients. These include nitrogen, phosphorus, potassium, calcium, magnesium and sulphur and these are usually supplied in the form of salts such as calcium nitrate, potassium phosphate and magnesium sulphate. These nutrients are essential for basic physiological functions and are major components of plant structures and metabolic processes. Macronutrients are often involved in processes such as photosynthesis, respiration, protein synthesis and cell structure.

i) Nitrogen (N) ~ essential for vegetative growth, leaf development and protein synthesis

Nitrogen is indispensable for synthesis of amino acids, co-enzymes and chlorophyll. It plays a crucial role in fostering stem and leaf development and is a fundamental component in the construction of plant cells. Moreover, nitrogen undergoes conversion into proteins essential for proliferation of new cells ~ thus a vital element for comprehensive growth of plants. Nitrogen is crucial for maintaining optimal oxygen levels within plants. Oxygen derived from nitrate (NO_3) nitrogen facilitates breakdown of carbohydrates produced during photosynthesis. It plays a pivotal role in protein synthesis with its highest demand observed at critical growth points such as young leaves, root tips, flowers and fruits. Nitrogen being a mobile element readily translocates to young buds, shoots and leaves albeit at a slower rate to older foliage. Nitrogen deficiency primarily manifests in older leaves which turn pale yellow and may eventually perish. Consequently, new growth appears weak and spindly. Conversely, an excess of nitrogen leads to soft *cum* feeble growth and may even impede flower and fruit production if allowed to accumulate unchecked. Most plants exhibit a preference for obtaining nitrogen in nitrate form. Urea and ammonium sulphate are deemed unsuitable for hydroponic systems as these require conversion from ammonia-type to nitrate-type by soil bacteria before plants can effectively utilize them. In sunnier climates typically plants require higher nitrogen levels compared to those in cooler climates. Leafy greens such as lettuce, cabbage and silverbeet often have a higher nitrogen requirement than less foliar-intensive plants. Among the preferred nitrate-type fertilizers in hydroponics are potassium nitrate and calcium nitrate. Nitrogen is typically available to plants in two primary forms: nitrate (NO_3^-) and ammonium (NH_4^+). While nitrate is generally preferred by most plants, excessive use of ammonium can interfere with calcium (Ca^{2+}) uptake and may even become toxic to plants. To balance pH during nitrogen absorption a recommended practice is to maintain a 3:1 ratio by weight of nitrate-nitrogen (NO_3^--N) to ammonium-nitrogen (NH_4^+-N) (Benton, 2004). Plants exhibit different responses depending on the nitrogen form they receive. Ammonium being positively charged prompts the plant to expel a proton (H^+) for every NH_4^+ taken up leading to a reduction in rhizosphere pH. Conversely, when supplied with nitrate the plant may release bicarbonate (HCO_3^-) thereby increasing rhizosphere pH. These alterations in pH can significantly impact the availability of other essential plant nutrients such as Zn, Ca and Mg (McNear, 2013). Its requirement in plants is usually 10—250ppm ~ minimum concentration during growth phase being 160ppm though typical minimum concentration during blooming phase is 130ppm.

ii) Phosphorus (P) ~ important for root development, flowering and fruiting

Phosphorus plays a crucial role in supporting growth and development of fruit-producing vegetables such as peas and beans. Acting as a catalyst for energy transfer within the plant, phosphorus is indispensable for photosynthesis. It fosters robust root growth and is essential for production of flowers, fruits and seeds. Its demard during seed germination and flowering stages is notably high. Concentrated in seeds, fruits and meristematic tissues phosphate (PO_4) is vital for various physiological processes. As a mobile element its deficiency symptoms manifest initially in older leaves. Insufficient phosphorus can lead to an overall reduction in plant growth as characterized by deep green leaves with occasional brown or purple spots. Its ionic forms absorbed by plants are $H_2PO_4^-$, PO_4^{3-} and HPO_4^{2-}. Excess of NO_3^- tends to inhibit PO_4^{3-} absorption. The ratio of iron to PO_4^{3-} can affect co-precipitation reactions (Sholto Douglas, 1985). Its requirement in plants is usually 30-60ppm, typical minimum during growth phase being 30ppm while typical minimum during blooming phase being 60ppm.

iii) Potassium (K) ~ aids in overall plant health, disease resistance and fruit quality

Potassium plays a crucial role throughout all the stages of plant growth with particular significance during fruit development. It serves to stimulate robust root development and is indispensable for overall plant vitality. Acting as a catalyst, potassium facilitates various biochemical processes within the plant. Integral to energy assimilation from photosynthesis, potassium aids in the production of sugars and starches. It fosters robust root growth, enhances water uptake and activates enzymes vital for disease resistance. Potassium is instrumental in protein synthesis, increases chlorophyll content in foliage and regulates stomatal openings thus optimizing light and air utilization. Plants in cooler climates have an increased demand for potassium compared to those in warmer and sunnier regions. Deficiency in potassium manifests as taller yet unhealthy-looking plants with older leaves displaying yellowing between veins that progresses to dark yellow and eventual necrosis. Flower and fruit drop are also common symptoms of potassium deficiency. As a mobile element, potassium is distributed throughout the plant but is primarily concentrated in regions of heightened physiological activity. Its ionic form absorbed by plants is K^{+}. Its high concentrations interfere with the function of Fe, Mn and Zn. Zinc deficiencies often are the most apparent (Benton, 2004). Usually its requirement in the plants is 100-300ppm, typical minimum concentration in plants during growth phase being 230ppm while during blooming phase is 300ppm.

iv) ***Magnesium (Mg) ~ necessary for chlorophyll production, photosynthesis and enzyme activation***

Magnesium serves essential functions in plant biology including enzyme synthesis, chlorophyll formation and absorption of light energy. It plays a pivotal role in distributing phosphorus throughout the plant and aids in nutrient utilization while neutralizing harmful acids and toxins produced by plant itself. Deficiency in magnesium manifests as yellowing of older leaves starting from the centre and spreading outward. The tips and edges of leaves may exhibit discolouration, turning lime-green in growing tips and curling upwards. As a mobile element within the plant, magnesium deficiency symptoms typically appear first in older leaves. Its ionic form absorbed by plants is Mg^{2+}. Due to competitive uptake, the Ca^{2+} concentration should not exceed in the plants (Benton, 2004). Its requirement in the plants is usually 30-70ppm.

v) ***Calcium (Ca) ~ supports cell wall structure, enzyme activity and nutrient uptake***

Calcium serves as the primary element essential for cell wall formation in plants. It plays a pivotal role in fostering robust root growth and facilitates potassium absorption. Calcium mitigates the toxic effects of other mineral salts and contributes to protein synthesis. As a fundamental component, calcium forms the structural basis of plant cell architecture. Its immobility within the plant means that deficiency symptoms typically manifest first in young growth. Signs of deficiency include deficient leaf tips and edges with new growth exhibiting browning and potential necrosis. Conversely, an abundance of calcium in the early stages of plant development can lead to stunted growth. Its ionic form absorbed by plants is Ca^{2+}. Its requirement in the plants is usually 80-140ppm. Excess of Ca^{2+} inhibits Mg^{2+} uptake (Benton, 2004).

vi) Sulphur (S) ~ required for protein synthesis, enzyme function and plant metabolism

Sulphur plays a vital role in various aspects of plant health and development. It contributes significantly to root growth and towards chlorophyll ~ the green pigment essential for photosynthesis. Sulphur serves as a fundamental building block for plant proteins, amino acids, coenzyme A and vitamins such as thiamine and biotin. Furthermore, sulphur participates actively in the generation of plant energy and enhances the efficiency of phosphorus utilization. It aids in crucial processes like protein synthesis, water uptake and formation of fruits and seedlings. Its presence in substantial quantities within leaves influences the flavour and aroma of many plants. Sulphur is particularly crucial for maintaining health of root system and plays a role in nitrogen assimilation. As an immobile element within the plant, deficiency symptoms primarily manifest in younger leaves. Signs of sulphur deficiency include the paling of young leaves, slowed growth and development of narrower-than-normal leaves which may become brittle over time. Its ionic form absorbed by plants is SO_4^{2-}. Usually its requirement in the plants is 50-120ppm. Unlike most nutrients, plants can tolerate a high concentration of SO_4^{2-} thus selectively absorbing the nutrient as needed (Gericke, 1940; Sholto Douglas, 1985; Benton, 2004).

6.1.2.2 Micronutrients (trace elements)

These include iron (Fe), manganese (Mn), zinc (Zn), copper (Cu), boron (B), molybdenum (Mo) and chlorine (Cl) that are required in smaller quantities but are nonetheless essential for plant growth. Micro-nutrient additives are often provided in chelated form to ensure these remaining available to plants in hydroponic solutions.

i) Iron (Fe) ~ essential for chlorophyll synthesis and enzyme activity

Iron is essential for chlorophyll formation and plays a crucial role in photosynthesis. It acts as both an oxygen carrier and an enzyme catalyst. Additionally, iron helps regulate manganese levels, thereby preventing manganese toxicity. Being an immobile element, iron deficiency symptoms initially manifest in younger leaves, which turn pale yellow or white while the veins remain green. However, plants absorb iron slowly, making it somewhat challenging for them to access. To address this, chelated iron, which is immediately available to plants, is recommended in nutrient mixes. In hydroponic systems, it's crucial to consider the solubility of iron. Inorganic forms of iron, such as iron sulphate, can precipitate out completely in alkaline solutions with high pH levels, leaving no iron available for plants. To ensure optimal iron availability, it's advisable to use iron-EDTA, which is the iron salt of ethylene di-amine tetra acetic acid. Maintaining the correct pH is essential to prevent iron from combining with phosphates, carbonates, and hydroxyl ions, which could reduce its availability to plants. Its ionic forms absorbed by plants are Fe^{2+} and Fe^{3+}. It is required in the plants usually at 1-3ppm where typical minimum concentration during growth and blooming phase remains 2ppm.

pH values above 6.5 greatly decreases iron solubility. Chelating agents (DTPA, citric acid or EDTA) are often added to increase iron solubility over a greater pH range (Benton, 2004).

ii) Chlorine (Cl) ~ plays a role in photosynthesis, osmosis and ionic balance

Chlorine is indispensable in the composition of cytochromes and plays roles in photosynthesis, osmosis and ion balance within plant cells. It acts as a catalyst in the enzymatic process of liberating oxygen from water. Chlorine serves as a key enzyme activator, particularly in facilitating the release of oxygen molecules during photosynthesis. Moreover, its presence significantly contributes to enhancing the drought resistance of plants by regulating tissue water content. Its ionic form absorbed by plants is Cl^-. In the plants it is required >75ppm. This micronutrient can interfere with NO_3^- uptake in some plants though in some plants as in asparagus at 5ppm it may be beneficial though usually remains absent in conifers, ferns and most bryophytes (Sholto Douglas, 1985). It is needed in small quantities (<50–100µM in nutrient media) for healthy growth of plants. High concentrations of chloride ions can lead to toxicity symptoms such as leaf burn, wilting and reduced growth.

iii) Manganese (Mn) ~ involved in photosynthesis, enzyme activation and nitrogen metabolism

Manganese serves as a vital catalyst in growth processes of plants. It aids plant enzymes in reducing nitrates ~ a crucial step before protein synthesis. Manganese plays a pivotal role in photosynthesis by contributing to the formation of oxygen. Its involvement in carbohydrate metabolism and chlorophyll formation further underscores its significance in plant development. Working in tandem with nitrogen, manganese accelerates plant growth. Typically concentrated in leaves, it acts as an immobile element with deficiency symptoms manifesting first in younger foliage. These symptoms are characterized by appearance of mottled yellow or brown patches on young leaves. Its ionic forms absorbed by plants are Mn^{2+} and Mn^{4+}. In nutrient media it is needed at 0.5-1ppm. Its uptake is enhanced by high PO^{3-}_4 concentrations (Benton, 2004).

iv) Zinc (Zn) ~ necessary for enzyme function, hormone regulation and carbohydrate metabolism

Zinc plays crucial roles in various physiological processes within plants. It is integral to chlorophyll formation, facilitating respiration and aiding in nitrogen metabolism. As a vital component in the energy transference process, zinc is essential for optimal plant growth albeit required only in trace amounts. The uptake of zinc by plants is closely linked to light availability ~ higher light exposure enhances zinc absorption leading to increased metabolic activity. Zinc serves as an enzyme activator and contributes to the synthesis of IAA ~ a plant growth hormone. Being a mobile element, zinc deficiency manifests first in older leaves resulting in stunted growth, yellowing and curling of foliage. However, excess zinc can be detrimental potentially leading to toxicity and even plant death. Its ionic form absorbed by the plants is Zn^{2+}. Its requirement in the plant is 0.2-0.6ppm. Excess of zinc is highly toxic to plants though is essential at low concentrations. The zinc content of commercially available plant-based food ranges from 3-10μg/g fresh weight (Waldner and Gunther, 1996).

v) Copper (Cu) ~ aids in photosynthesis, respiration and lignin formation

Copper plays vital roles in plant physiology primarily in chlorophyll production and enzyme activation ~ essential for photosynthesis and respiration. It enhances disease resistance in plants. Due to its immobility within plants, copper deficiency initially manifests in younger leaves resulting in wilting and irregular growth. Conversely, excessive copper levels can lead to plant death. Its ionic form absorbed by plants is Cu^{2+} It is required in plants at 0.08-0.2ppm, minimum being 0.02ppm during growth as well as blooming phases. Plant sensitivity to copper is highly variable as to some plants even at 0.1ppm can be toxic (Benton, 2004) while up to 0.5ppm for certain plants is often considered ideal (Sholto Douglas, 1985).

vi) Boron (B) ~ required for cell division, carbohydrate metabolism and pollination

Boron is indispensable as a trace element in the cultivation of vegetables, acting to rectify various undesirable traits such as brittle stems, deteriorating growing tips and development of hollow-stemmed cauliflowers. Collaborating with calcium, boron contributes significantly to the formation of cell walls ~ an essential process for plant structure and strength. Moreover, boron is pivotal for cell division and protein synthesis. Furthermore, boron assumes an active role in the reproductive phase of plants ~ aiding in pollination and seed production. Its influence extends to the regulation of the ratio of anions to cations absorbed by plants thus facilitating a balanced uptake of essential nutrients. Boron enhances assimilation of cations while restricting absorption of anions ~ optimizing nutrient utilization within plants. Moreover, boron significantly impacts carbohydrate and nitrogen metabolism, essential processes for energy production and nitrogen utilization within plants. Facilitating the utilization of calcium, boron ensures its effective incorporation into plant tissues. Boron is an immobile element within plants so deficiency symptoms typically manifest first in younger leaves. This underscores the critical role of boron in ensuring the overall health and vitality of plants. Plants can absorb boron in various ionic forms, primarily as borate ions. The main forms of boron that plants can absorb are borate ions ($B(OH)_4$-), tetraborate ions ($B_4O_7^{2-}$), boric acid (H_3BO_3) and hydrogen borate ions (H_2BO_3-). Borate ions are readily available form of boron which are highly soluble in water and are absorbed by plants mainly in this form. Tetraborate ions in the solution are formed under certain conditions but are less common and less readily absorbed. Boric acid though exists in the solution as undissociated molecules but can ionize to form borate ions [$B(OH)_{4-}$] in presence of water. In this case plants can absorb boron from boric acid through passive diffusion. Hydrogen borate ions are other forms of boron found in soil or solution. Though less common but plants can still absorb boron in this form albeit less efficiently compared to borate ions. In plants its requirement is 0.2-0.5ppm. However, in certain plants as in citrus it shows its toxicity symptoms at 0.5ppm (Sholto Douglas, 1985) though is an essential element.

vii) Molybdenum (Mo) ~ facilitates nitrogen fixation, enzyme activity and sulphur metabolism

Molybdenum plays a vital role in protein synthesis, nitrogen metabolism and fixation. It facilitates conversion of nitrogen gas from air into soluble nitrogen compounds through the action of nitrogen-fixing microorganisms. As an immobile element within plants, its deficiency symptoms manifest primarily in younger leaves. Signs of deficiency include the paling of leaves with scorched fringes, *vis-a-vis* occurrence of irregular leaf growth. Its ionic form absorbed by plants is MoO_4^{2-} (molybdate). Its requirement in the plant is 0.04-0.08ppm, usually 0.05ppm being minimum during both the phases, *i.e.* growth and blooming. It is a component of enzyme nitrate reductase which is required by rhizobia for nitrogen fixation (Benton, 2004).

viii) Nickel (Ni) ~ primarily involved in various enzymatic reactions particularly those related to nitrogen metabolism

Nickel is a cofactor for certain enzymes such as urease and hydrogenase which play crucial roles in nitrogen assimilation and its fixation processes. In higher plants, nickel is absorbed primarily as Ni^{2+} ions. It plays a crucial role in activating urease ~ an enzyme vital for nitrogen metabolism and urea processing. Without adequate nickel, urea accumulates to toxic levels leading to formation of necrotic lesions. In lower plants, nickel acts as a catalyst for various enzymes involved in diverse physiological processes. Nickel can substitute for zinc and iron as a cofactor in certain enzymes, further highlighting its importance across different biological systems. It is essential to many plants, especially legumes and many grain crops (Benton, 2004). It is also used in enzyme urease.

Certain plant species for their good growth may require silicon (Si), aluminum (Al), cobalt (Co), vanadium (V), sodium (Na), iodine (I) and selenium (Se). These are considered beneficial because some of them can stimulate growth or can compensate the toxic effects of other elements, *vis-a-vis* may replace essential nutrients in a less specific role (Trejo-Téllez *et al.,* 2007).

Note: Typical minimum concentrations of above elements are given for high quality general purpose hydroponic formulations during the growth and bloom phase.

In hydroponic agriculture, it is crucial to tailor nutrient solutions according to Liebig's law of the minimum, ensuring that each specific plant variety receives the essential elements it needs for optimal growth (Sholto Douglas, 1985). While nutrient requirements vary between plant species, there are generally accepted concentration ranges that provide a framework for nutrient solution formulation (Steiner, 1984). Typically, nutrient solutions are formulated to fall within the range of 1,000 to 2,500ppm (Gericke, 1940). This total ppm figure encompasses concentrations of individual nutrient ions which are critical for plant health. It is important to note that concentrations below these ranges often result in nutrient deficiencies while exceeding these may lead to nutrient toxicity. Achieving optimal nutrition concentrations for specific plant varieties involves empirical methods based on experience or through plant tissue tests (Sholto Douglas, 1985). This approach ensures that plants receive the appropriate balance of essential nutrients for healthy growth and development.

ix) Silicon (Si) and Hydroponics

Field-grown plants typically contain abundant silicon ~ a component often absent from hydroponic solutions. Silicon plays a crucial role in reinforcing cell walls thereby enhancing plant vigour, resilience and productivity. Its deposition in plant cell walls has been associated with improved tolerance to heat and drought as well as heightened resistance to insect-pests and fungal pathogens. Furthermore, silicon aids in mitigating the adverse effects of excessive levels of manganese, iron, phosphorus and aluminum while addressing zinc deficiency issues.

While silicon is not universally recognized as an essential plant nutrient due to the ability of most plant species to complete their life cycles without it but it is acknowledged as a beneficial element. Certain plant species can accumulate silicon at concentrations exceeding those of many essential macronutrients. Despite its advantageous properties, silica is typically excluded from hydroponic nutrient formulations due to its alkaline nature and limited solubility in concentrated nutrient solutions. Consequently, silicon must be introduced separately into the nutrient reservoir as an additive.

Notable benefits have been observed through experimentation involving increased silicon concentrations in the nutrient solutions of cucumber, roses and courgettes (zucchinis). Enhanced silicon levels in the root environment lead to overall increased yields and suppression of powdery mildew ~ underscoring the potential advantages of optimizing silicon availability for hydroponically cultivated plants.

Silicon enhances structural cell strength and bolstering plants' defense mechanisms against insect-pests and diseases. By boosting resistance to various insects/pests, airborne fungal pathogens (such as *Botrytis*) and waterborne pathogens, it fortifies plants against potential attacks. Silicon mitigates the toxic effects of metals and enhances stress tolerance. It effectively addresses toxicity from manganese, iron, phosphorus and aluminum, *vis-a-vis* counters zinc deficiency. Furthermore, silicon aids in increasing drought resistance and salt tolerance, further enhancing the plant's resilience to adverse environmental conditions.

Silicon (Si) plays a crucial role in suppressing fungal pathogens like *Botrytis*. Its addition could significantly benefit crops such as cucumbers especially when cultivated in regions where local water supply lacks sufficient silicon, particularly in recirculating solutions or Si-deficient media like peat.

The treatment with silicon may effectively mitigate the development of powdery mildew by triggering host defense mechanisms in plants. It is widely believed that the deposition of silicon at sites of fungal penetration serves as a fundamental aspect of the plant's defense response across various plant families.

Moreover, silicon is deposited within root cell walls, acting as a protective barrier against invasion by parasites and pathogens. Potassium silicate has shown efficacy in preventing *Pythium ultimum* infections. Soluble silicon polymerizes rapidly and disease suppression occurs only when silicon is present in a soluble form, hence to curb disease development effectively, a continuous supply of soluble silicon is imperative in hydroponic systems.

6.2 Nutrient Sources

Although ready-made nutrient formulations for various crops can be bought and mixed in fresh water to prepare nutrient solution but the growers of large commercial hydroponics systems prepare their own solutions through commonly used salts. The mineral nutrients usually are available in powder, liquid, resin or tablet form which are added to water to make nutrient solution or to the aggregate or reservoir.

Here is a list of fertilizer salts or other sources of macro and micro nutrients commonly used in hydroponics, along with their molecular weights, the nutrients they supply and the ionic forms required by plants:

Nutrient Sources	*Molecular Weight*	*Nutrients Supplied*
KNO_3 (Potassium nitrate)	101.10 g/mol	Potassium (K+), nitrate (NO3-)
KOH (Potassium hydroxide)	56.11 g/mol	Potassium (K+), hydroxide (OH-)
NH_4NO_3 (Ammonium nitrate)	80.04 g/mol	Ammonium (NH4+), nitrate (NO3-)
$Ca(NO_3)_2$ (Calcium nitrate)	164.09 g/mol	Calcium (Ca^2+), nitrate (NO3-)
HNO_3 (Nitric acid)	63.01 g/mol	Nitrate (NO_3-)
$(NH_4)_2SO_4$ (Ammonium sulphate)	132.14 g/mol	Ammonium (NH4+), sulphate (SO4^2-)
$(NH_4)_2HPO_4$ (Ammonium hydrogen phosphate)	132.06 g/mol	Ammonium (NH_4+) and dihydrogen phosphate (HPO4^2-)
$NH_4H_2PO_4$ (Ammonium dihydrogen phosphate)	115.03 g/mol	Ammonium ions (NH4+), dihydrogen phosphate ions (H_2PO_4-)
K_2SO_4 (Potassium sulphate)	174.26 g/mol	Potassium (K+), sulphate (SO4^2-)
KH2PO4 (Monopotassium phosphate)	136.09 g/mol	Potassium (K+), phosphate (H2PO4-)

Nutrient Sources	*Molecular Weight*	*Nutrients Supplied*
$MgSO_4 \cdot 7H_2O$ (Magnesium sulphate heptahydrate, commonly known as epsom salt)	246.47 g/mol	Magnesium (Mg^2+), sulphate (SO4^2-) (Magnesium content=9.86 per cent, sulphur content=13 per cent)
K_2HPO_4 (Potassium phosphate dibasic)	174.18 g/mol	Potassium (K+), phosphate (HPO4^2-)
$Ca(H_2PO_4)2$ (Calcium dihydrogen phosphate)	234.05 g/mol	Calcium (Ca^2+), phosphate (H2PO4-)
Fe-EDTA (Iron ethylenediaminetetraacetate), a chelated form of iron commonly used as a micronutrient source in hydroponic systems. The chelating agent, EDTA, helps to keep the iron soluble and stable in the nutrient solution, preventing it from precipitating out and becoming unavailable to plants.	Variable (depending on chelating agent) calculation based on the sum of the atomic weights of iron and EDTA is ~ Fe: 55.85 g/mol + EDTA: 292.24 g/mol ≈ 348.09 g/mol	Iron (Fe^2+ or Fe^3+) Some sources report the molecular weight of FeEDTA as approximately 367.20 g/mol. while others report it as approximately 372.13 g/mol. Some sources might consider the hydrated form of FeEDTA, which could lead to a molecular weight around 382.1 g/mol.
FeEDTA (iron chelate ~ Iron (III) diethylenetriaminepentaacetate), a chelated form of iron used as a micronutrient source in hydroponic systems.	The molecular weight of FeDTPA can vary depending on the specific form or salt used.	Iron (Fe^3+)FeDTPA with sodium (Na+) as the counter ion: molecular weight is approximately 393.1 g/molFeDTPA with ammonium (NH4+) as the counter ion: molecular weight is approximately 435.2 g/molwhile some sources might consider the hydrated form of FeDTPA, which could lead to a molecular weight around 468.20 g/mol
$CuSO_4 \cdot 5H_2O$ (Copper sulphate pentahydrate)	249.68 g/mol	Copper (Cu^2+), sulphate (SO4^2-)
$ZnSO_4 \cdot 7H_2O$ (Zinc sulphate heptahydrate)	287.56 g/mol	Zinc (Zn^2+), sulphate (SO4^2-)
ZnEDTA (zinc ethylenediaminetetraacetic acid), a chelated form of zinc commonly used in hydroponic systems	Variable approximately 431.6 g/mol	Zinc (Zn^2+)
$MnSO_4 \cdot H_2O$ (Manganese sulphate monohydrate)	169.02 g/mol	Manganese (Mn^2+), sulphate (SO4^2-)
$MnSO_4 \bullet 4H_2O$ (Manganese Sulphate Tetrahydrate)	223.07 g/mol	Manganese (Mn^2+), sulphate (SO4^2-), Water (H2O)

Nutrient Sources	*Molecular Weight*	*Nutrients Supplied*
MnEDTA (manganese ethylenediaminetetraacetic acid), a chelated form of Mn commonly used in hydroponic systems.	381.2can vary depending on the specific form and counterion.	Mn^{++} The two common variants are: MnEDTA with sodium (Na+) as the counterion (approximately 374.1 g/mol) MnEDTA with NH4+ as the counterion (approximately 381.2 g/mol)
$FeSO_4 \cdot 7H_2O$ (Iron sulphate heptahydrate)	278.01 g/mol	Iron (Fe^2+), sulphate (SO4^2-)
KCl (Potassium chloride)	74.55 g/mol	Potassium (K+), chloride (Cl-)
K2CO3 (Potassium carbonate)	138.21 g/mol	Potassium (K+), carbonate (CO3^2-)
H_3BO_3 and H_3BO_4 (Boric acid)	61.83 g/mol	Boron (B(OH)4-)
$Na_2B_4O_7 \cdot 10H_2O$ (Sodium tetraborate decahydrate) or borax	381.37 g/mol	Boron ($B4O_7$^2-) or ($B(OH)_4$-)
$Na_2B_4O_7$ (sodium tetraborate)	201.22 g/mol	Boron ($B(OH)_4$-)
$(NH_4)_6Mo_7O_{24} \cdot 4H_2O$ (Ammonium molybdate tetrahydrate)	1235.85 g/mol	Molybdenum (MoO_4^2-)
Na_2MoO_4 (Sodium molybdate)	205.92 g/mol	Molybdenum (MoO4^2-)
$CaSO_4$ (Calcium sulphate)	136.14 g/mol	Calcium (Ca^2+), sulphate (SO4^2-)
$CaCl_2$ (Calcium chloride)	110.98 g/mol	Calcium (Ca^2+), chloride (Cl-)
$MgSO_4$ (Magnesium sulphate)	120.37 g/mol	Magnesium (Mg^2+), sulphate (SO4^2-)
$MgCl_2$ (Magnesium chloride)	95.21 g/mol	Magnesium (Mg^2+), chloride (Cl-)
K_2SiO_3 (Potassium silicate)	154.28 g/mol	Potassium (K+), silicate (SiO3^2-)
Na_2SiO_3 (Sodium metasilicate)	122.06 g/mol	Sodium (Na+), silicate (SiO3^2-)
H_2SiO_3 (Orthosilicic acid)	Variable	Silicon (H4SiO4)
H_4TiO_4 (Titanium dioxide)	284.81 g/mol	Titanium (TiO4^2-)
$CoSO_4$ (Cobalt sulphate)	155.00 g/mol	Cobalt (Co^2+), sulphate (SO4^2-)
$NiSO_4$ (Nickel sulphate)	154.76 g/mol	Nickel (Ni^2+), sulphate (SO4^2-)
$NiCO_3$ (Nickel carbonate)	118.70 g/mol	Nickel (Ni^2+), carbonate (CO3^2-)
Nickel chloride ($NiCl_2$)	129.6 g/mol	Nickel (Ni^2+), chloride (Cl-)
Nickel nitrate (Ni(NO3)2)	182.7 g/mol	Nickel (Ni^2+), nitrate (NO3-)
NaHCO3 (Sodium bicarbonate)	84.01 g/mol	Sodium (Na+), bicarbonate (HCO3-)

Nutrient Sources	Molecular Weight	Nutrients Supplied
NaOH (Sodium hydroxide)	40.00 g/mol	Sodium (Na+), hydroxide (OH-)
$VOSO_4$ (Vanadium sulphate)	182.00 g/mol	Vanadium (VO^2+), sulphate (SO4^2-)
Li_2SO_4 (Lithium sulphate)	109.94 g/mol	Lithium (Li+), sulphate (SO4^2-)
LiCl (Lithium chloride)	42.39 g/mol	Lithium (Li+), chloride (Cl-)
LiOH (Lithium hydroxide)	23.95 g/mol	Lithium (Li+), hydroxide (OH-)
NaCl (Sodium chloride)	58.44 g/mol	Sodium (Na+), chloride (Cl-)
$FeSO_4$ (Iron(II) sulphate)	151.91 g/mol	Iron (Fe^2+), sulphate (SO4^2-)
$FeCl_3$ (Iron(III) chloride)	162.20 g/mol	Iron (Fe^3+), chloride (Cl-)
H_3PO_4 (Phosphoric acid)	98.00 g/mol	An acid form of phosphorus and is typically used for pH adjustment in nutrient solutions

The list provided includes a mix of fertilizer salts, compounds and chemicals. While many of the entries are indeed fertilizer salts such as potassium nitrate (KNO3), calcium nitrate ($Ca(NO_3)_2$) and magnesium sulphate ($MgSO_4$), others are compounds or chemicals that may serve as nutrient sources or supplements in hydroponic systems.

For instance, compounds like potassium hydroxide (KOH) and phosphoric acid (H_3PO_4) can be used to adjust pH levels in hydroponic nutrient solutions while also providing potassium and phosphorus, respectively. The salts like lithium sulphate (Li_2SO_4) and vanadium sulphate ($VOSO_4$) may not typically be considered traditional fertilizer salts for hydroponics but these contain elements that can potentially benefit plant growth under specific circumstances.

6.3 Required Amounts of Common Salts to make Nutrient Solution

The table below adapted from Jensen and Malter (1995), outlines commonly used fertilizer salts and the corresponding amounts required to make 1000 liters of a 1ppm solution for each specified nutrient. To determine the grammes of fertilizer needed per 1000 liters, simply multiply the desired ppm value by the amount listed for each salt.

Fertilizer Salts	Element Supplied	Grammes of Fertilizer Needed per 1000 litres of Water to Provide 1 mg/l (ppm) of the Nutrient Specified
Boric acid [H_3BO_3]	B	5.64
Calcium nitrate [$Ca(NO_3)_2 \cdot 4H_2O$]	N	6.45
	Ca	4.70
Cupric chloride [$CuCl_2 \cdot 2H_2O$]	Cu	2.68
Copper sulphate [$Cu(SO_4) \cdot 5H_2O$]	Cu	3.91
Chelated iron (9 per cent)	Fe	11.10

Fertilizer Salts	*Element Supplied*	*Grammes of Fertilizer Needed per 1000 litres of Water to Provide 1 mg/l (ppm) of the Nutrient Specified*
Ferrous sulphate [$FeSO_4$]	Fe	5.54
Magnesium sulphate [$MgSO_4 \cdot 7H_2O$] (epsom salts)	Mg	10.75
Manganese chloride [$MnCl_2 \cdot 4H_2O$]	Mn	3.60
Manganese sulphate [$MnSO_4 \cdot 4H_2O$]	Mn	4.05
Molybdenum trioxide [MoO_3]	Mo	1.50
Monopotassium phosphate [KH_2PO_4]	K	3.53
	P	4.45
Potassium chloride [KCl]	K	2.05
Potassium nitrate [KNO_3]	N	7.30
	K	2.70
Potassium sulphate [K_2SO_4]	K	2.50
Zinc sulphate [$ZnSO_4 \cdot 7H_2O$]	Zn	4.42

Adapted from Jensen and Malter, 1995.

If we want to prepare a hydroponic nutrient solution with a concentration of 20ppm of potassium (K), referring to the table we see that potassium nitrate [KNO_3] provides both nitrogen (N) and potassium (K) and for potassium alone 2.7 grammes of fertilizer is required per 1000 liters to achieve 1ppm. To reach 20 ppm of potassium we should multiply 2.7 grammes by 20 to achieve 54 grammes of potassium nitrate needed for every 1000 liters of the solution.

There exist various formulations of nutrient solutions, often developed empirically. Presented below in the table are some of the examples:

Nutrient	*Hoagland and Arnon (1950)*	*Hewitt (1966)*	*Cooper (1988)*	*Steiner (1984)*
N	210	168	200-236	168
P	31	41	60	31
K	234	156	300	273
Ca	160	160	170-185	180
Mg	34	36	50	48
S	64	48	68	336
Fe	2.5	2.8	12	2-4
Cu	0.02	0.064	0.1	0.02
Zn	0.05	0.065	0.1	0.11
Mn	0.5	0.54	2.0	0.62

Nutrient	Hoagland and Arnon (1950)	Hewitt (1966)	Cooper (1988)	Steiner (1984)
B	0.5	0.54	0.3	0.44
Mo	0.01	0.04	0.2	Not considered

Concentration ranges of essential mineral elements according to various authors.

Adapted from Cooper, 1988; Steiner, 1984; Windsor and Schwarz, 1990.

6.4 Selecting Best Fertilizer

Several factors should be considered when selecting the best fertilizer salt for use in a hydroponic nutrient solution:

i) **Nutrient requirements of the plant:** The nutrient requirement, particularly nitrogen, phosphorus and potassium (N-P-K) for different plants may be different which depends on the species and variety of plants, growth stage of plants, weather and season of the year as tomatoes require higher levels of phosphorus, calcium and potassium whereas leafy vegetables such as lettuce need higher levels of nitrogen.

ii) **Ionic composition:** Ionic composition of fertilizer salts should be examined and the balance between cations (positively charged ions) and anions (negatively charged ions) provided by each fertilizer salt should be considered. The choice between potassium nitrate (KNO_3) and calcium nitrate [$Ca(NO_3)_2$] highlights the importance of considering the balance between cations and anions in the nutrient solution. While both salts provide nitrate ions (NO_3^-), calcium nitrate yields one calcium ion (Ca^{++}) and two nitrate ions per molecule ($2NO_3^-$) whereas potassium nitrate yields one potassium ion (K^+) and one nitrate ion per molecule (NO_3^-). Therefore, if the goal is to supply ample nitrate anions while minimizing the introduction of cations into the solution, calcium nitrate may be the preferred option.

iii) **Solubility:** Solubility of fertilizer salt in water should be evaluated. It should be ensured that chosen salt dissolves readily in the hydroponic solution to provide nutrients to the plants efficiently.

iv) **pH considerations:** Impact of the chosen fertilizer salt on the pH of the nutrient solution should be considered. Some salts may have acidic or alkaline properties thus affecting the overall pH balance of the solution.

v) **Cost and availability:** Cost and availability account of the fertilizer salts should be considered. Options that are cost-effective and readily accessible for the said hydroponic setup should be chosen

It is crucial to maintain a balanced nutrient profile within the solution. In addition to nutrient content, water quality such as salinity and the presence of potentially harmful elements such as boron, sodium and chlorides must be taken into account. Moreover, pH levels require attention as these tend to fluctuate as plants absorb nutrients, particularly during later stages of growth. The presence of ammonium/nitrate significantly influences pH.

6.5 NPK and PPM

NPK values represent the relative proportions of these three primary nutrients in a fertilizer, typically expressed as percentages by weight. NPK values are crucial for selecting the right fertilizer formulations to meet the nutrient requirements of plants at different growth stages. Different plant species and growth stages have varying needs for nitrogen, phosphorus and potassium so it is essential to choose fertilizers with appropriate NPK ratios to provide necessary nutrients.

Nutrients are available in many different formulations expressed as percentages. By selecting different NPK formulations the growth, development and flowering of the plants can be controlled to a certain extent. The product label for the fertilizer shows the list of ingredients (percentage of each ingredient) available in the fertilizer where N-P-K is always listed first, *i.e.* if it is 24-8-16 it is understood that it contains 24 per cent nitrogen, 8 per cent phosphorus and 16 per cent potassium ~ all in soluble form, however, minor elements may be listed as 0.02 per cent boron, 0.07 per cent copper, 0.15 per cent iron, 0.05 per cent manganese, 0.0005 per cent molybdenum and 0.06 per cent zinc. Nutrient solutions in hydroponics have both ~ the macro (NPK) and micro both in ppm.

6.6 Nutrient Solution Requirements of some Plants

Given below is the list of some plants with their nutrient solution requirements:

Crops	*PH*	*CF*	*EC*	*ppm*
	Vegetables			
Artichoke	6.5-7.5	8-18	0.8-1.8	560-1260
Asparagus	6.0-6.8	14-18	1.4-1.8	980-1260
Bean (common)	6.0	20-40	2-4	1400-2800
Beetroot	6.0-6.5	8-50	0.8-5	1260-3500
Broad bean	6.0-6.5	18-22	1.8-2.2	1260-1540
Broccoli	6.0-6.8	28-35	2.8-3.5	1960-2450
Brussells sprouts	6.5-7	25-30	2.5-3.0	1750-2100
Cabbage	6.5-7.0	25-30	2.5-3.0	1750-2100
Capsicum	6.0.6.5	18-22	1.8-2.2	1260-1540
Carrots	6.3	16-20	1.6-2.0	1120-1400
Cauliflower	6.0-7.0	5-20	0.5-2.0	1050-1400
Celery	6.5	18-24	1.8- 2.4	1260-1680
Cucumber	5.5-6.0	17-25	1.7-2.5	1190-1750

Crops	*PH*	*CF*	*EC*	*ppm*
Eggplant	5.5-6.5	25-35	2.5-3.5	1750-2450
Endive	5.5	20-24	2.0-2.4	1400-1680
Garlic	6.0	14-18	1.4-1.8	980-1260
Leek	6.5-7.0	14-18	1.4-1.8	980-1260
Lettuce	5.5-6.5	8-12	0.8-1.2	560-840
Marrow	6.0	18-24	1.8-2.4	1260-1680
Okra	6.5	20-24	2.0-2.4	1400-1680
Onions	6.0-6.7	14-18	1.4-1.8	980-1260
Pak choi	7.0	15-20	1.5-2.0	1050-1400
Parsnip	6.0	14-18	1.4-1.8	980-1260
Peas	6.0-7.0	8-18	0.8-1.8	980-1260
Pepino	6.0-6.5	20-50	2.0-5.0	1400-3500
Potato	5.0-6.0	20-25	2.0-2.5	1400-1750
Pumpkin	5.5-7.5	18-24	1.8-2.4	1260-1680
Radish	6.0-7.0	16-22	1.6-2.2	840-1540
Spinach	5.5-6.6	18-23	1.8-2.3	1260-1610
Silverbeet	6.0-7.0	18-23	1.8-2.3	1260-1610
Sweet corn	6.0	16-24	1.6-2.4	840-1680
Sweet potato	5.5-6.0	20-25	2.0-2.5	1400-1750
Taro	5.0-5.5	25-30	2.5-3.0	1750-2100
Tomato	5.0-6.5	20-50	2.0-5.0	1400-3500
Turnips	6.0-6.5	18-24	1.8-2.4	1260-1680
Zucchini	6.0	18-24	1.8-2.4	1260-1680
Herbs				
Basil	5.5-6.5	10-16	1.0-1.6	700-1120
Chicory	5.5-6.0	20-24	2.0-2.4	1400-1600
Chives	6.0-6.5	18-22	1.8-2.2	1260-1540
Fennel	6.4-6.8	10-14	1.0-1.4	700-980
Lavender	6.4-6.8	10-14	1.0-1.4	700-980
Lemon balm	5.5-6.5	10-16	1.0-1.6	700-1120
Marjoram	6.0	16-20	1.6-2.0	1120-1400
Mint	5.5-6.0	20-24	2.0-2.4	1400-1680
Mustard cress	6.0-6.5	12-24	1.2-2.4	840-1680
Parsley	5.5-6.0	8-18	0.8-1.8	560-1260
Rosemary	5.5-6.0	10-16	1.0-1.6	700-1120
Sage	5.5-6.5	10-16	1.0-1.6	700-1120

Crops	*PH*	*CF*	*EC*	*ppm*
Thyme	5.5-7.0	8-16	0.8-1.6	560-1120
Watercress	6.5-6.8	4-18	0.4-1.8	280-1260
Flowers				
African Violet	6.0 - 7.0	12 - 15	1.2-1.5	840 - 1050
Anthuurium	5.0 - 6.0	16 - 20	1.6-2.0	1120- 1400
Antirrhinim	6.5	16 - 20	1.6-2.0	1120-1400
Aphelandra	5.0 - 6.0	18 - 24	1.8-2.4	1260-1680
Aster	6.0 - 6.5	18 - 24	1.8-2.4	1260-1680
Begonia	6.5	14 - 18	1.4-1.8	980 - 1260
Bromeliads	5.0 - 7.5	8 - 12	0.8-1.2	560 - 840
Caladium	6.0 - 7.5	16 - 20	1.6-2.0	1120-1400
Canna	6.0	18 - 24	1.8-2.4	1260-1680
Carnation	6.0	20 - 35	2.0-3.5	1260-2450
Chrysanthemum	6.0 - 6.2	18 - 25	1.8-2.5	1400-1750
Cymbidium	5.5	6 - 10	0.6-2.5	420 - 560
Dahlia	6.0 - 7.0	15 - 20	1.5-2.0	1050-1400
Dieffenbachia	5.0 - 6.0	18 - 24	1.8-2.4	1400-1680
Dracaena	5.0 - 6.0	18 - 24	1.8-2.4	1400-1680
Fern	6.0	16 - 20	1.6-2.0	1120-1400
Ficus	5.5 - 6.0	16 - 24	1.6-2.4	1120-1680
Freesia	6.5	10 - 20	1.0-2.0	700 – 1400
Impatiens	5.5 - 6.5	18 - 20	1.8-2.0	1260-1400
Gerbera	5.0 - 6.5	20 - 25	2.0-2.5	1400-1750
Gladiolus	5.5 - 6.5	20 - 24	2.0-2.4	1400-1680
Monstera	5.0 - 6.0	18 - 24	1.8-2.4	1400-1680
Palm	6.0 - 7.5	16 - 20	1.6-2.0	1120-1400
Rose	5.5 - 7.0	15 - 25	1.5-2.5	1050-1750
Stock	6.0 - 7.0	16 - 20	1.6-2.0	1120-1400
Fruits				
Banana	5.5 - 6.5	18 - 22	1.8 - 2.2	1260-1540
Black Currant	6.0	14 - 18	1.4 - 1.8	980-1260
Blueberry	4.0 - 5.0	18 – 20	1.8 – 2.0	1260-1400
Melon	5.5 – 6.0	20-25	2.0-2.5	1400-1750
Passion fruit	6.5	16-24	1.6-2.4	840-1680
Paw-Paw	6.5	20-24	2.0-2.4	1400-1680
Pineapple	5.5-6.0	20-24	2.0-2.4	1400-1680

Crops	PH	CF	EC	ppm
Red Currant	6.0	14-18	1.4-1.8	980-1260
Rhubarb	5.0-6.0	16-20	1.6-2.0	840-1400
Strawberries	5.5-6.5	18-22	1.8-2.2	1260-1540
Watermelon	5.8	15-24	1.5-2.4	1260-1680

It is important to note that these ranges are provided as general guidelines and may vary depending on factors such as regional climate, seasonal variations and different growth stages of the plants. Adjustments in nutrient solutions, pH and other factors should be made accordingly. Regular monitoring for nutrient deficiencies, toxicities, pests and diseases is of paramount importance. Environmental factors like water quality, temperature, humidity and light intensity should be closely monitored and managed. As such, nutrient solution formulae need to be modified based on specific growing conditions and systems.

6.7 Nutrient Uptake

Understanding how plants absorb nutrients is essential for optimizing nutrient uptake in hydroponic systems. Here is an overview of the process and how to adjust solution composition accordingly:

Plants exhibit varying nutrient requirements based on their developmental stage and metabolic activities. While carbon is primarily assimilated from atmospheric carbon dioxide through stomatal pores in leaves, all other essential nutrients are absorbed through the root system. Nutrients in the solution move from areas of higher concentration in the solution to areas of lower concentration in the root zone through a process called diffusion. Plants actively transport ions such as nitrate (NO_3-), potassium (K+), and calcium (Ca2+) across their root membranes using specialized transport proteins.

In hydroponic systems, particularly those employing medium culture, water and nutrient status can fluctuate rapidly. It is therefore crucial to monitor and adjust parameters routinely such as electrical conductivity (EC), pH and nutrient levels using simple measurement tools and techniques. Although tracking pH and EC can signal changes in the nutrient solution, it may not pinpoint alterations in the preferential uptake of specific ions. Hence, in large commercial closed hydroponic setups, comprehensive nutrient solution analysis becomes imperative. If such analysis is not feasible, consider replacing the nutrient solution entirely at least every two weeks.

To gauge medium changes particularly post-fertigation, regular monitoring of fertigation input lines and drainage points is advisable. Rough estimations of nutrient consumption can be inferred by observing fluctuations in nitrate (NO3-) levels within the nutrient solution. These fluctuations in relation to a reference point such as the drip-line position serve as indicators of changes in plant uptake rates. Furthermore, keeping an eye on nitrite (NO2-) concentrations offers valuable insights into the overall conditions of the growth medium. Over-irrigation can lead to water accumulation so causing waterlogging and reduced oxygen availability. Consequently, ammonium may undergo chemical conversion to nitrate leading to nitrite accumulation ~ an adverse condition for plant roots.

6.8 Nutrient Lockout

Nutrient lockout in hydroponics refers to a situation where plants are unable to uptake essential nutrients from nutrient solution despite the nutrients being present in the solution. This can occur due to various factors including imbalances in pH levels, excess salts or interactions between different nutrients.

6.8.1 Symptoms of Nutrient Lockout

Leaf chlorosis (yellowing), curling *cum* yellowing and foliage shedding, stunted *cum* slow plant growth and flowering *cum* fruiting may be adversely affected in case when nutrients are locked in the solution.

6.8.2 Causes and Prevention

i) **pH imbalance:** Incorrect pH levels can hinder nutrient uptake by plants. Nutrient availability is influenced by pH as certain nutrients may become unavailable to plants at too high or too low pH. Regular monitoring and adjustment of pH levels of nutrient solution falling within the pH range is essential for nutrient availability (typically 5.5 to 6.5 for most plants).

ii) **Excess salts:** Accumulation of salts in the growing medium or nutrient solution can lead to nutrient imbalances thus interfering with nutrient uptake by plants. Measurement and control of proper EC levels of nutrient solution can prevent excessive salt buildup.

iii) **Imbalance of nutrients:** Imbalanced ratios or discrepancies among various nutrients can create competition for absorption by plant roots, potentially resulting in excess or deficiencies of specific nutrients. Growers should provide plants with a balanced nutrient solution containing appropriate ratios of essential macro and micronutrients.

iv) **Poor water quality:** Contaminated water can significantly impact nutrient availability and ultimate plant health. It is crucial to utilize only clean and pure water devoid of any contaminant when preparing the nutrient solution.

v) **Interactions and precipitation**: Some nutrients may interact with each other or form insoluble compounds under certain conditions, leading to precipitation and nutrient lockout, *e.g.* calcium may precipitate with phosphate or sulphate ions so reducing their availability to plants. Balanced nutrient formulations should be used to minimize the likelihood of interactions. It is advised to monitor nutrient solution parameters regularly and maintain proper ratios of essential elements. It is also recommended to consider chelated micronutrients which are more stable and less prone to forming insoluble compounds.

vi) **Temperature**: Root uptake rates are influenced by water temperature. Warmer temperatures generally increase metabolic activity and nutrient uptake while colder temperatures may slow it down. Therefore optimum range of temperatures, *i.e.* 18-24°C should be maintained to promote efficient nutrient absorption. It Is advisable to install a thermometer in the nutrient reservoir and check the temperature regularly. The nutrient solution should remain within the optimal temperature range during the hottest and coolest parts of the day. Nutrient uptake may significantly decrease if temperature of solution drops below 16°C.

vii) **Oxygen levels**: Adequate oxygen levels in the root zone are crucial for nutrient uptake. Oxygen facilitates nutrient absorption by promoting root respiration and maintaining root health. Therefore growers must ensure proper aeration of the nutrient solution.

viii) **Root zone issues**: Poor aeration in the solution causes lockout of the nutrients, *vis-a-vis* root rot so plants are unable to draw even available nutrients. Growers should maintain proper root health through adequate aeration, proper drainage and regular monitoring for signs of root diseases or deficiencies. Sufficient oxygen levels should be ensured in the root zone by using appropriate aeration techniques such as air stones or oxygen pumps.

ix) **Regular maintenance:** Regular maintenance of hydroponic systems should be conducted through periodic flushing of the system to remove excess salts and through replenishing the nutrients.

By implementing these preventive measures and closely monitoring plant health and nutrient levels, growers can minimize the risk of nutrient lockout and maintain healthy and productive crops in hydroponic systems. Timely corrective actions should be taken if symptoms of nutrient lockout are observed to mitigate any adverse effects on plant growth and yield.

6.9 Nutrition in Closed Systems and Open System

While root zone solutions remain largely consistent between closed and open hydroponic systems but feed solutions often differ due to variations in nutrient uptake rates.

6.9.1 Nutrition in Closed Systems

A thriving plant will absorb necessary nutrients as long as solution in the root zone maintains optimal levels. In a fully enclosed hydroponic system the nutrient intake solely depends on what is available in the root zone. To uphold equilibrium in nutrient solution surrounding the roots the nutrients supplied must precisely match uptake by plants. Any deviation from this balance will cause the solution to become imbalanced ~ leading to further disruptions if left unaddressed. In such cases the solution must be replenished or replaced to restore equilibrium.

Maximum nutrient uptakle in plants is essential to promote accelerated growth rates. However, the uptake process for different nutrients varies in efficiency and speed. Enhancing the concentration of the root zone solution for these nutrients amplifies the driving force thereby augmenting their uptake rate.

6.9.2 Nutrition in Open Systems

In open hydroponic systems, nutrient delivery involves both plant uptake and runoff. As the nutrient solution is supplied to the roots, it accumulates in the growing medium until it reaches the point of runoff. This gradual buildup ensures that nutrient solution achieves necessary balance within the medium before being expelled.

Maintaining adequate runoff is crucial as it helps flushing out any imbalance present in the nutrient solution. Unlike closed systems where precise nutrient composition is critical due to lack of runoff, open systems offer more flexibility in nutrient management. As a result, nutrient management is generally simpler and more straightforward in open systems compared to closed ones.

The inherent advantages of open systems including easier nutrient management and the ability to flush out imbalances make these a preferred choice for many commercial hydroponic operations.

6.10 Commercial vs. Homemade Solutions

Choosing between commercial pre-mixed nutrient solutions and homemade formulations in hydroponic systems involves weighing several factors. Here are the pros and cons of each option:

6.10.1 Commercial Pre-mixed Nutrient Solutions

i) Pros

- ☆ **Convenience**: Commercial nutrient solutions are pre-formulated so save time and effort in measuring and mixing individual components.
- ☆ **Consistency**: Commercial products are designed to provide a balanced blend of nutrients, ensuring consistent plant nutrition and minimizing the risk of deficiencies or toxicities.
- ☆ **Quality assurance**: Reputable brands often conduct rigorous testing to ensure product quality and effectiveness so growers are relaxed with peace of mind.
- ☆ **Ease of use**: Commercial solutions typically come with instructions for proper usage, making them suitable for beginners or growers without extensive knowledge of nutrient formulation.

ii) **Cons**

- ☆ **Cost**: Commercial nutrient solutions may commensurately be quite expensive compared to homemade formulations especially over the long term for large-scale operations.
- ☆ **Limited customization**: While some commercial products offer different formulations for specific plant types or growth stages, these may not provide same level of customization as homemade solutions.
- ☆ **Dependency**: Relying solely on commercial products can limit grower's understanding of nutrient management and their ability to adjust formulations based on specific crop needs or environmental conditions.

6.10.2 Homemade Formulations

i) **Pros**

- ☆ **Cost-effectiveness**: Mixing own nutrient solutions using individual components can be more cost-effective, especially for larger-scale operations or recurrent use.
- ☆ **Flexibility**: Homemade formulations offer greater flexibility and customization, allowing growers to adjust nutrient concentrations and ratios based on plant requirements, growth stages and environmental factors.
- ☆ **Learning opportunity**: Creating homemade solutions provides an opportunity for growers to deepen their understanding of plant nutrition and hydroponic system management.
- ☆ **Control over ingredients**: Growers have full control over quality and source of ingredients used in homemade formulations, potentially reducing the risk of contaminants or impurities.

ii) **Cons**

- ☆ **Complexity**: Formulating nutrient solutions from scratch requires knowledge of plant nutrition, chemistry and nutrient interactions which may be challenging for inexperienced growers.
- ☆ **Time and effort**: Mixing homemade solutions can be time-consuming and labour-intensive, requiring precise measurements and careful attention to detail.
- ☆ **Risk of imbalance**: Without proper understanding and expertise, homemade formulations may result in nutrient imbalances or deficiencies, potentially impacting plant health and yield.
- ☆ **Quality assurance**: Homemade solutions lack the quality control and testing procedures of commercial products, increasing the risk of variability in nutrient composition and effectiveness.

Ultimately, the choice between commercial pre-mixed nutrient solutions and homemade formulations depends on factors such as budget, expertise, desired level of customization and preference for convenience. Some growers may opt for commercial products for simplicity and consistency while others may prefer the flexibility and cost-effectiveness of creating their own nutrient blends. Whichever option is chosen, careful monitoring and adjustment of nutrient solutions are essential for maximizing plant growth and productivity in hydroponic systems.

6.11 Recipe Development

Formulating nutrient solutions tailored to specific plant species and growth stages is essential for optimizing plant health, growth, and yield in hydroponic systems. Here are some guidelines for developing nutrient solution recipes:

i) **Understanding plant requirements:** Specific nutrient requirements of the plant species should be worked out exclusively for seedlings, vegetative growth, flowering and fruiting. Different plants have varying needs for macro-, micro- and other essential elements.

ii) **Selection of base nutrient solution:** A high-quality base nutrient solution that provides a balanced blend of essential nutrients for hydroponic plants should be selected specifically formulated for the type of plants to be grown along with their growth stages.

iii) **Adjustment of nutrient ratios:** For optimum harvest, nutrient solution composition should be modified by adjusting the ratios of individual macro- and micro-nutrients to match specific requirements of plants to be grown.

iv) **Consideration of water quality and pH:** For preparing nutrient solution, one should start with a suitable pH and clean *cum* filtered water and never with contaminated one. Most plants thrive in slightly acidic conditions with a pH range of 5.5 to 6.5.

v) **Monitoring and adjusting:** Regularly the pH and nutrient levels, *i.e.* EC or TDS of the nutrient solution should be monitored through testing kits. One should be in constant touch with the system looking into plant response, nutrient uptake rates and environmental conditions. Moreover, required changes in the nutrient solution at different stages of growth should be met with.

vi) **Considering supplemental nutrients:** Depending on the plant species and growing conditions, consider addition of supplemental nutrients or supplements such as calcium, magnesium or beneficial microbes to enhance plant health and productivity. Any specific nutrient requirements or deficiencies commonly associated with the plant species should be taken into account.

vii) **Documentation and evaluation:** Detailed records of the nutrient solution recipes used including nutrient concentrations, pH levels and any adjustments made should properly be maintained. The performance of nutrient solution recipes over time should be recorded taking note of plant growth, yield and overall health. Based on observations and results recipes as needed should also be adjusted.

By following these guidelines for recipe development, growers can create nutrient solutions tailored to the specific requirements of their chosen plant species and growth stages leading to healthier and more productive plants in hydroponic systems. Regular monitoring and adjustment are key to maintaining optimal nutrient levels and ensuring successful crop cultivation.

6.12 Methods to Monitor Changes in Nutrient Levels

Several methods can be employed to monitor changes in nutrient levels in a hydroponic solution:

i) **Water testing kits**: These kits are readily available and provide a quick and easy way to test nutrient levels in the solution. These typically include test strips or liquid reagents for various nutrients including nitrate.

ii) **EC meters**: These measure the electrical conductivity of the solution which correlates with concentration of dissolved ions including nutrients like nitrate. Changes in EC can indicate fluctuations in nutrient levels.

iii) **pH meters**: While pH meters primarily measure the acidity or alkalinity of the solution, these can also indirectly indicate nutrient levels. Certain nutrients including nitrate can influence pH levels so monitoring pH can provide insights into nutrient availability.

iv) **Laboratory analysis**: For more precise and comprehensive nutrient analysis, the samples of hydroponic solution can be sent to a laboratory for testing. This method provides accurate quantitative data on nutrient concentrations.

v) **Visual inspection**: While not as precise as the other methods, observing the condition of the plants regularly can offer clues about nutrient deficiencies or imbalances. Yellowing of leaves or stunted growth may indicate nutrient issues.

vi) **Plant tissue analysis**: Analyzing plant tissue samples can provide insights into nutrient uptake and deficiencies. By testing the leaves or other plant parts, nutrient levels can be assessed and the hydroponic solution can be adjusted accordingly.

By employing a combination of these methods, growers can effectively monitor and manage nutrient levels in their hydroponic solution ~ ensuring optimal conditions for plant growth.

6.13 Management Techniques for Nutrient Solution

Here are some refined management techniques for nutrient solution in hydroponic systems:

i) **Regular sampling and analysis**: Growers should take frequent samples of the nutrient solution focusing on areas around the plant roots or runoff in open systems. They must analyze each sample for pH and electrical conductivity (EC) to ensure optimal nutrient uptake and plant health.

ii) **Comprehensive chemical analysis**: Periodically a comprehensive chemical analysis of the nutrient solution should be conducted. This analysis can be tied to leaf analysis for a more thorough understanding of plant nutrient status.

iii) **Maintaining detailed records**: Keep A detailed diary documenting all sample results including pH, EC or any finding towards chemical analysis should be recorded. Any additional information such as weather conditions, crop performance, symptoms of nutrient deficiencies or toxicities and observation of pests or diseases should also have records.

iv) **Responsive management**: Nutrient solution should be managed by observing trends over time and making gentle adjustments to correct any imbalances. Avoid drastic actions that could disrupt plant health or nutrient balance.

v) **Appropriate nutrient strength**: Growers must monitor plant health and growth regularly to identify any sign of nutrient deficiencies or imbalances. These should consider the specific nutrient requirements of the plant species being grown and so solution composition is adjusted accordingly as different plants may have different nutrient preferences and uptake rates. Nutrient concentrations in the solution should be adjusted by adding or reducing specific nutrient components as needed. Common nutrients to adjust include N, P, K, Ca, Mg and micronutrients such as Fe, Zn and Mn. It is advised to ensure that Ca, Mg and S levels in the nutrient solution are kept higher ~ potentially up to double the strength of the feed as these elements have slower uptake rates. Nitrogen levels can be maintaineed at a consistent strength while potassium levels can be kept relatively low due to their faster uptake rate.

vi) **Gradual building of nutrient balance**: Building up of intended nutrient balance in the root zone should be made gradually over time. Using special starter solutions with elevated levels of Ca, Mg and S to establish desired nutrient balance more quickly should also be considered.

By understanding the process of nutrient uptake and adjusting solution composition accordingly, growers can optimize nutrient availability and promote healthy growth and development in hydroponic plants. Regular monitoring and adjustment of nutrient levels are essential components of successful nutrient management in hydroponic systems.

6.13.1 Adjusting Nutrient Levels

Techniques for modifying nutrient concentrations based on plant response and environmental factors should be adopted. Adjusting nutrient levels in hydroponic systems is crucial for maintaining optimal plant health and growth. Here are techniques for modifying nutrient concentrations based on plant response and environmental factors:

i) **Monitoring plant health:** Plant inspection for signs of nutrient deficiencies or toxicities such as yellowing or browning of leaves, stunted growth or leaf curling should be done regularly. Plant tissue or leaf sap analysis to determine nutrient levels within the plant should regularly be done so that specific nutrient imbalances are made known for correction.

ii) **Monitoring nutrient solution parameters:** pH and EC or total dissolved solids (TDS) of nutrient solution should regularly be measured using appropriate testing kits or meters.

- ❖ **pH**: pH of the nutrient solution should be adjusted to the appropriate range (5.5-6.5) for the plant species being grown.
- ❖ **EC/TDS**: Nutrient levels/concentrations in the solution should be adjusted as needed akin to every crop.

iii) **Adjusting nutrient solution strength:** Strength of nutrient solution through increase or decrease in the strength of individual nutrients should be adjusted. Plants showing signs of nutrient deficiency due to a particular nutrient, its concentration should be increased, however, in case of toxicity (higher EC//TDS) of a nutrient the solution is required to be diluted through addition of fresh water.

iv) **Consideration of environmental factors:** Environmental conditions such as temperature, humidity, light intensity and CO_2 levels can affect nutrient uptake and plant growth so in such cases concentration adjustment is must because during high humidity or temperature nutrient requirement is more due to increased metabolic activities. Moreover, it would be better if greenhouse temperature is maintained at optimum levels, *i.e.* 18-22°C.

v) **Changes in documentation and evaluation:** Detailed records of nutrient solution adjustments including changes in nutrient concentrations, pH levels and plant responses should be maintained.

By implementing these techniques for adjusting nutrient levels in hydroponic systems, growers can ensure that plants receive the optimal balance of nutrients for healthy growth and maximum yield. Regular monitoring and adjustment are essential to maintain nutrient levels within the desired range and responding effectively to changing plant and environmental conditions.

6.13.2 Replenishment and Recycling of Nutrient Solution

Replenishment of depleted solution along with recycling of the same solution will be cost-effective as the waste would be minimized and almost people apply the same practice everywhere::

i) **Continuous dosing:** Implement a system for continuous nutrient dosing, where fresh nutrient solution is continuously added to the system to replace depleted nutrients. Use of automated dosing systems equipped with sensors to monitor nutrient levels in real-time adjustment is a wise step. This method ensures a steady supply of nutrients to plants while minimizing fluctuations in nutrient levels.

ii) **Periodic replenishment:** Nutrient levels in the solution and replenishment of depleted nutrients on a scheduled basis should be monitored regularly. Nutrient solution analyses should be conducted regularly to determine nutrient concentrations and to adjust the replenishment accordingly.

iii) **Recirculating systems:** Recirculating hydroponic systems should be implemented where nutrient solution is continuously circulated through the growing medium and root zone. Reservoirs should be set up to collect and recirculate excess nutrient solution reducing waste and allowing for reuse. Filtration systems should be used to remove debris and prevent clogging in recirculating systems.

iv) **pH and EC adjustment:** Regularly pH and EC levels in the solution should be monitored and adjusted as needed to maintain optimal nutrient availability. pH buffers and pH adjustment solutions should be used to maintain pH within the desired range. Nutrient concentrations should be adjusted to ensure that EC levels remain within the optimal range for plant growth.

v) **Nutrient solution reuse:** After plants have absorbed nutrients from the solution the nutrient solution should be collected and filtered to remove any organic matter or debris. Filtered nutrient solution for subsequent crop cycles or for replenishing should be reused. Periodic analyses should be conducted to assess nutrient levels and adjust solution composition as needed before reuse.

vi) **Complementary fertilization:** Hydroponic nutrient solutions are supplemented with organic or supplemental fertilizers to provide additional nutrients and enhance nutrient cycling. Organic additives such as compost tea or seaweed extract should be used to supplement nutrient solutions to improve plant health. Beneficial microorganisms into nutrient solutions should be incorporated to promote nutrient cycling and enhance nutrient availability to plants.

vii) **Nutrient solution disposal:** Spent or excess nutrient solutions should properly be disposed off according to local regulations and environmental guidelines. Nutrient solutions should be diluted with water before disposal to reduce salt concentration and minimize environmental impact.

By implementing these methods for replenishment and recycling of nutrient solutions, growers can maintain optimal nutrient levels, minimize waste and reduce costs in hydroponic systems. Regular monitoring, adjustment, and responsible nutrient management practices are key to maximizing efficiency and sustainability in hydroponic cultivation.

Chapter 7

Factors affecting Nutrient Solution

Several factors influence nutrient uptake by plant roots and overall root function. Below are key considerations to ensure the effective preparation and maintenance of nutrient solutions:

7.1 Water Quality

Water quality is a crucial factor in hydroponic systems as this directly impacts plant health and growth. Poor-quality water used in preparing nutrient solutions can lead to a range of issues such as nutrient deficiencies, pH fluctuations, toxicity problems, clogging of irrigation systems and can manifest in various plant growth issues such as stunted growth and disease contamination. Moreover, impurities or contaminants like high levels of salts especially excessive sodium content, heavy metals or other pollutants in water sources used for hydroponic systems can interfere with nutrient uptake by plants and may lead to lockout. Water with high dissolved solids may contain components that affect nutrient availability to plants. While soft water may require specific nutrient combinations, hard water or water contaminated with elements like sodium, sulphide or heavy metals necessitates filtration methods like reverse osmosis (RO) or deionization (DI) filtration systems.

Commonly found salts such as calcium and magnesium carbonates in water can benefit plant growth within reasonable levels. However, excessive levels can lead to nutrient lockout rendering essential elements unavailable to plants as excess of calcium may bind with phosphorous forming insoluble compounds such as calcium phosphate. Therefore, it is crucial to begin with pure water and tailor nutrient solutions accordingly to ensure optimal plant nutrition. Sterilizing water before use can eliminate harmful microorganisms like *Pythium*. Water can also be treated through methods such as UV radiation, heat treatment, membrane filtration or chemical agents like ozone or hydrogen peroxide. Hydrogen peroxide also acts as a root disease preventative but loses effectiveness once root diseases are present.

The selection and concentration of fertilizers in nutrient solution depend on raw water quality as it may contain minerals such as Ca, Mg, S, *vis-a-vis* trace elements such as B, Mn, Fe and Zn. High concentration of unwanted minerals sodium, sulphide, chloride or fluoride in raw water can render it unsuitable for hydroponics. Therefore, testing raw water before formulating fertilizer solutions is imperative to ensure optimal plant nutrition and growth in hydroponic systems.

7.1.1 Common Ions in Water Influencing Plant Nutrition

The composition of water supplies can significantly impact plant nutrition, with several common ions playing key roles:

- **Sodium (Na^+) and chloride (Cl^-):** These ions are often present together in water supplies. Though essential in traces but most plants do not actively uptake sodium and chloride which in water are usually abundant beyond plant needs. However, excesses are detrimental.
- **Iron (Fe):** Despite being a micronutrient, iron in its oxidized form readily precipitates as rust so rendering it unavailable for plant uptake. This can lead to issues such as clogging of irrigation systems. Aeration of water and subsequent settling or filtration can expedite the precipitation process.
- **Calcium (Ca^{++}) and magnesium (Mg^{++}):** These ions are commonly found in hard water and serve as essential macronutrients in nutrient solutions. Their concentrations should be considered when formulating nutrient solutions.
- **Bicarbonate (HCO_3):** Present in hard water, bicarbonate is not a nutrient but can influence pH levels due to its alkalinity. Acidification typically with phosphoric or nitric acid may be necessary to neutralize bicarbonate. The amount of acid added should be factored into formulation calculations.
- **Boron (B):** While boron is a micronutrient, it has a narrow optimal range. If present in water, it may be unnecessary to include it in formulations. However, concentrations exceeding 1ppm or lower for sensitive crops can lead to issues. Fortunately, such high concentrations are rare in most water supplies.

Understanding the presence and influence of these ions in water supplies is crucial for formulating nutrient solutions and optimizing plant nutrition in hydroponic systems. Regular monitoring and appropriate adjustments can help mitigate any potential issues arising from these common ions.

7.1.2 High Salt Levels in Water Supplies

The tolerance level of salts in water varies based on their composition and the specific crop being cultivated. Even low concentrations, *e.g.* 50ppm of sodium can prove detrimental to sensitive plants such as lettuce, strawberries and roses. Conversely, certain crops like tomatoes can withstand even higher levels exceeding 200ppm. However, as dissolved solids accumulate in water supplies, effective management becomes increasingly crucial.

Beyond a certain threshold which varies depending on the crop type the management of recirculating hydroponic systems can become impractical due to accumulation of salts. Non-recirculating systems may still be viable but may require higher levels of runoff to mitigate salt build-up. Ultimately, if water quality deteriorates to an extent where it is unusable, only viable solution is to remove majority of salts from water, typically through RO processes.

7.1.3 Water Treatment

When managed effectively, RO systems can significantly reduce dissolved salts to minimal levels. Primary and often costliest component of an RO system is its membrane. The performance and longevity of this membrane heavily rely on the quality of water passing through it. In many cases, pretreatment steps such as iron removal are necessary to optimize equipment efficiency and extend membrane lifespan. Another option is the utilization of rainwater. If adequate quantities are available, rainwater can be used directly or blended with less favourable water sources to enhance their quality.

Prior to use, it is crucial to eliminate suspended particles from water. Employing robust filtration methods such as sand filtration can help prevent common issues like dripper blockages encountered by some growers. Furthermore, water must be devoid of pathogens to ensure plant health. While municipal water supplies are typically safe, other water sources may require sterilization, often achieved through chlorination. Water should be screened for any phytotoxin to avoid adverse effects on plant growth and development.

7.2 EC, TDS and CF

In hydroponics, EC, TDS and CF are essential parameters used to measure concentration of dissolved salts or nutrients in the nutrient solution. Here is a brief overview of each:

i) **Electrical conductivity (EC)**: EC is a fundamental parameter for assessing nutrient strength and managing nutrient levels in hydroponic systems. It measures ability of a solution to conduct electricity due to presence of dissolved ions including nutrients. It is typically expressed in units of millisiemens per centimeter (mS/cm) or microsiemens per centimeter (μS/cm). Higher EC values indicate a greater concentration of dissolved salts or nutrients in the solution.

ii) **Total dissolved solids (TDS)**: TDS represents the total concentration of dissolved substances such as inorganic salts and minerals, suspended solids and organic compounds in the solution. It is often measured in parts per million (ppm) or milligrams per litre (mg/l). TDS meters measure conductivity and then convert the reading into ppm or mg/l based on a conversion factor. TDS measurement provides an estimate of overall nutrient concentration in the solution and is closely related to EC. Growers either use TDS or EC as the standard of measurement. In a laboratory, TDS is found by allowing the liquid to evaporate and weighing the left over particles. In hydroponics, TDS is estimated using a conversion from the measure of EC. Different manufacturers use different conversion rates. Different TDS meters may show a discrepancy of as much as 600ppm when reading the same solution. When calibrating TDS meters, the correct calibrating solution from the correct manufacturer should be used.

☆ **Conversion of EC to TDS and *vice versa***: There is a positive correlation between TDS and EC. while electrical conductivity (EC) refers to the ability of ions in water to conduct an electrical current. Converting readings between TDS (Total Dissolved Solids) and EC (Electrical Conductivity) involves using a conversion factor specific to the measurement units being used. Here's how to convert readings from TDS to EC and vice versa:

From TDS to EC: To convert TDS readings in ppm to EC readings in µS/cm a conversion factor can be used. The most commonly used conversion factor is 0.5, and thus EC (µS/cm) is = TDS (ppm) × 0.5, *e.g.* if TDS reading is 700 ppm the corresponding EC would be: EC = 700 ppm × 0.5 = 350 µS/cm

From EC to TDS: To convert EC readings in µS/cm to TDS readings in ppm, the EC value can be divided by the conversion factor (0.5). In case EC reading is 600 µS/cm the corresponding TDS would be 600 µS/cm + 0.5 = 1200ppm.

iii) **Conductivity factor (CF)**: CF is a parameter used in hydroponics to measure EC of a nutrient solution. It represents the ratio between the actual conductivity of the solution and a reference standard typically expressed as a multiplication factor. CF is especially useful in hydroponic systems where growers may prefer a simpler measurement scale compared to EC or TDS. Here is an example to illustrate CF in hydroponics: In case if EC reading for a hydroponic nutrient solution is 2.0 mS/cm, so while converting it to CF we will have to consider the reference standard or conversion factor. In case if it has a CF reference standard of 1.0, it means that an EC of 1.0 mS/cm corresponds to a CF of 1.0. In case measured EC is 2.0 mS/cm the CF would be 2.0 (2.0 mS/cm ÷ 1.0 = 2.0 CF). Similarly, in case of a CF reference standard of 0.5 it means that an EC of 1.0 mS/cm corresponds to a CF of 0.5. In this case, when measured EC is 2.0 mS/cm, the CF would be 4.0 (2.0 mS/cm ÷ 0.5 = 4.0 CF). Therefore, CF provides a convenient way for hydroponic growers to gauge the nutrient concentration of their solution based on a chosen reference standard ~ helping them maintain optimal growing conditions for their plants.

It's important to note that these conversion factors are commonly used but may vary slightly depending on factors such as the composition of the nutrient solution and the calibration of the measuring instruments. Always refer to the manufacturer's guidelines or calibration standards for accurate conversions. In hydroponic systems monitoring EC, TDS and CF allows growers to maintain optimal nutrient levels so preventing nutrient imbalances or deficiencies and ensuring healthy plant growth. Regular measurement and adjustment of these parameters are essential for optimizing nutrient delivery and maximizing crop yields.

7.2.1 Monitoring and Maintaining EC

Maintaining optimal EC levels in hydroponic nutrient solutions is crucial for robust plant growth. This can be achieved by formulating specialized nutrient blends tailored to different plant types and growth stages. Regular monitoring of EC in the nutrient solution is imperative to ensure proper nutrient concentrations. Purified water will exhibit minimal conductivity whereas nutrient-rich solutions will register higher EC values. EC or ppm meters are used to measure EC of the nutrient solution providing insights into its mineral content. Low EC readings indicate low nutrient concentrations which lead to nutritional deficiencies and stunted growth. Conversely, high EC levels signify ample nutrients but excessively high readings can harm plants. Maintaining appropriate nutrient levels is essential to ensure plants receiving sufficient nutrients without risking toxicity.

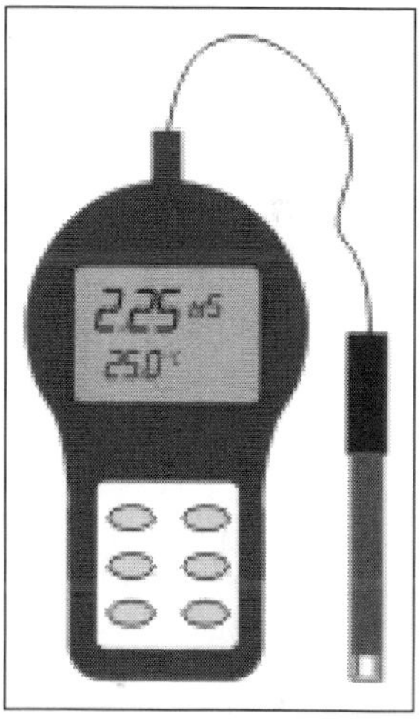

EC Meter

Nutrient concentrations as indicated by EC or TDS influence nutrient uptake. Imbalances or incorrect ratios between nutrients can lead to competition for uptake by plant roots resulting in deficiencies. Diluted nutrient solutions may fail to supply adequate nutrients while excessive fertilization can lead to nutrient buildup and imbalances. Regular monitoring of nutrient strength using an EC/ ppm meter helps adjusting nutrient concentrations based on plant requirements and growth stages. Growers should adhere to recommended dosages provided by nutrient manufacturers and fine-tune nutrient strength as necessary to optimize plant health and productivity.

Growers should measure EC when initially preparing the nutrient solution and afterwards at regular intervals depending on the plant type and growth stage. If EC levels become too high, dilution with water can be employed to return it to the original value. Conversely, if EC drops below 70 per cent of the original value simply adding nutrients may not suffice as the solution's composition has likely been altered significantly by plant uptake. In such cases, it is advisable to discard the solution and start afresh as the exact ionic composition cannot be determined solely from EC readings.

In closed hydroponic systems where nutrient solution is recirculated, certain elements and ions not absorbed by plants such as sodium, chloride and fluoride accumulate over time. Since plants absorb nutrients selectively based on their needs, EC monitoring alone cannot pinpoint changes in specific ion uptake. Therefore, in commercial closed systems, comprehensive nutrient analysis is imperative to gauge the concentration of each nutrient accurately. If conducting regular nutrient analysis is impractical, complete solution replacement every one or two weeks is recommended to maintain optimal nutrient balance and plant health.

In a commercial hydroponics setup, plant tissue analysis emerges as a vital tool for growers to diagnose potential nutrient absorption issues within their crops. These challenges could stem from various factors such as pH fluctuations, elevated cation exchange capacity in the growing media, excessive humidity or presence of diseases and nematodes. By conducting plant tissue analysis growers can pinpoint specific issues hindering nutrient uptake enabling them to implement targeted solutions and optimize plant health and productivity.

7.2.2 Symptoms Due to Excessive EC

When the total salt content in water surpasses acceptable levels it is imperative to prevent it from coming into contact with foliage. Excessive salt content as indicated by elevated EC can lead to following symptoms:

- Initial wilting of leaves and stems accompanied by stunted growth.
- Leaf margins may exhibit scorching, followed by wilting or necrosis.
- Premature leaf drop and dieback of shoots may occur.

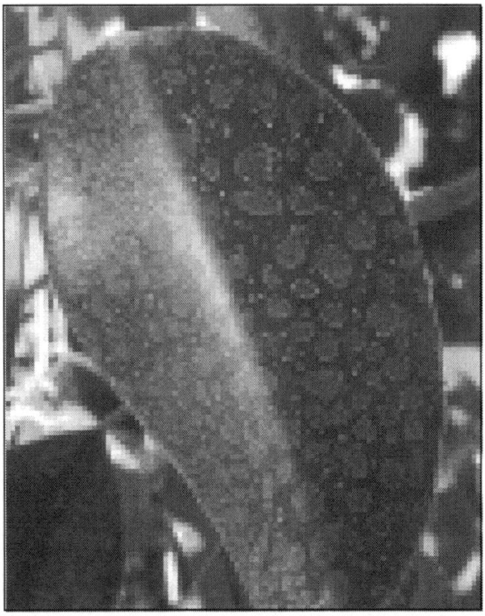

Leaf Burn Due to Excessive EC Nutrient Solution Touching the Foliage.

While these symptoms are often associated with high salt content, these can also signal other underlying issues such as water deficiency, disease, nutrient imbalance or excessive light and heat exposure. Therefore, careful observation and diagnosis are essential for effective management and mitigation of plant stressors in hydroponic systems.

7.2.3 Impact of pH Fluctuations on EC and TDS Measurements

Monitoring of Electrical Conductivity (EC) in hydroponic systems must be conducted under consistent pH conditions. Fluctuations in pH can significantly alter EC readings since pH-affecting ions exert a substantial influence on EC values. Moreover, it's crucial to calibrate the conductometer using a solution of known conductivity to ensure accurate measurements. Without proper calibration, comparisons between readings can be misleading.

7.2.4 Impact of Temperature on EC, and TDS Measurements

Temperature has a significant impact on EC and TDS measurements in hydroponics. As the temperature of a solution changes so does the movement of ions within it leading to fluctuations in conductance. To ensure accurate measurements it is crucial to compensate for temperature variations. Typically, a temperature coefficient of 2 per cent per degree Celsius is employed with a reference point set at 25°C. This means that for every degree Celsius change, the meter adjusts the absolute reading by 2 per cent before displaying the corrected value. For instance, a meter with automatic temperature compensation (ATC) will display a consistent reading of 1000 µS/cm at both 25°C and 26°C as it adjusts for the temperature-induced variation in ion movement.

Conversely, a meter lacking ATC or temperature compensation coefficient will exhibit different readings at varying temperatures. For instance, it may display 1000 µS/cm at 25°C and 1020 µS/cm at 26°C. Conductivity meters are available with either adjustable or fixed temperature compensation coefficients. Some meters feature an exposed temperature probe which facilitates faster response times by swiftly evaluating temperature changes for compensation.

7.3 The pH (Potential hydrogen)

pH quantifies the concentration of hydrogen ions (H+) present in the solution. It is measured on a scale from 0 to 14 where pH 7 represents neutrality. Values below 7 indicate acidity while values above 7 indicate alkalinity. In the context of hydroponics, pH plays a crucial role as it directly affects nutrient availability and uptake by plants.

It is crucial to utilize a well-buffered nutrient solution that ensures pH stability throughout the growth cycle. Recommended pH range typically falls between 5.5 and 6.5 with slight variations for different growth stages as for vegetative growth it should be 6.0 to 6.3 while during flowering and fruiting phases it should be in the range of 5.7 and 5.9.

In a hydroponic system, plants receive all their essential nutrients from a nutrient solution rather than from soil. The pH level of this solution is critical because it influences the availability of essential nutrients to plants. Here is a detailed look at pH and its role in a nutrient solution:

i) Nutrient Availability

pH affects the solubility and availability of nutrients in the solution. Different nutrients have different levels of solubility at different pH levels as at lower pH levels micronutrients such as Fe, Mn and Zn are more readily available to plants. However, excessively low pH levels can lead to nutrient toxicity so hindering the absorption of essential elements like Ca and Mg. Conversely, at higher pH levels, certain nutrients may become less available resulting in deficiencies despite their presence in the solution. Maintaining pH within optimal range ensures that all the essential nutrients are readily available to plants for uptake. The pH of a growing medium or nutrient solution significantly impacts availability of both micro and macro nutrients to plants in hydroponic systems. Here is how pH affects availability of these nutrients:

Macronutrients

- ☆ **Nitrogen:** Nitrogen availability is relatively unaffected by pH within the typical range (5.5-6.5) for hydroponics. However, extremely high or low pH levels can affect nitrogen uptake. At very high pH levels, nitrogen can become less available due to volatilization while at a very low pH levels nitrogen may be converted into ammonia gas thus reducing its availability.
- ☆ **Phosphorus:** Phosphorus availability decreases as pH levels increase above 7.0. At higher pH levels, phosphorus tends to form insoluble compounds with calcium thus reducing its availability to plants. Maintaining pH within the recommended range ensures optimal phosphorus uptake.
- ☆ **Potassium:** Potassium availability is generally unaffected by pH within the typical range, however, extreme pH levels can affect potassium uptake indirectly by influencing root health and nutrient transport.
- ☆ **Calcium:** Calcium availability decreases as pH levels decrease below 5.0 or increase above 7.0. At lower pH levels, calcium can become less available due to competition with hydrogen ions for binding sites on soil colloids or other nutrients. Conversely, at higher pH levels calcium may precipitate out of solution so reducing its availability to plants.
- ☆ **Magnesium:** Magnesium availability decreases as pH levels increase above 7.0. At higher pH levels, Mg tends to form insoluble compounds with hydroxide ions so reducing its availability to plants. Maintaining pH within recommended range ensures optimal magnesium uptake.

Micro nutrients

- ☆ **Iron:** Iron availability decreases as pH levels increase above 6.5. At higher pH levels iron tends to form insoluble compounds such as iron hydroxides so reducing its availability to plants. Iron deficiency is common in alkaline growing media or nutrient solutions.

- **Manganese:** Manganese availability decreases as pH levels increase above 6.5. At higher pH levels manganese tends to form insoluble compounds such as manganese oxides so reducing its availability to plants.
- **Zinc:** Zinc availability decreases as pH levels increase above 6.5. At higher pH levels, zinc tends to form insoluble compounds so reducing its availability to plants.
- **Copper:** Copper availability decreases as pH levels increase above 6.5. At higher pH levels copper tends to form insoluble compounds so reducing its availability to plants.
- **Boron:** Boron availability decreases as pH levels increase above 7.0. At higher pH levels boron tends to form borates so reducing its availability to plants.
- **Molybdenum:** Molybdenum availability decreases as pH levels decrease below 5.5. At lower pH levels, molybdenum tends to form insoluble compounds so reducing its availability to plants.
- **Nutrient uptake:** pH influences the uptake of nutrients by plant roots. This is because the pH of the nutrient solution affects the charge of ions in the solution, which in turn affects the ability to move through the root system of plants. Inadequate pH levels can lead to nutrient deficiencies or toxicities even if the nutrients are present in the solution.

ii) Plant Health and Growth

Proper pH levels are essential for healthy plant growth and development. When pH levels are outside the optimal range, plants may exhibit symptoms of nutrient deficiencies or toxicities such as stunted growth and leaf yellowing or necrosis. Maintaining the pH within the recommended range promotes optimal nutrient uptake, vigorous plant growth and overall plant health.

iii) Microbial Activity

pH also influences the activity of beneficial microorganisms in the nutrient solution. Many beneficial microorganisms such as bacteria and fungi play crucial roles in nutrient cycling and plant health. Too low or too high pH levels can inhibit growth and activity of these beneficial microbes thus affecting nutrient availability and plant health.

7.3.1 Ideal pH Range

Ideal pH range for a nutrient solution in hydroponics typically falls between 5.5 and 6.5. Within this range nutrients are most readily available to plants promoting optimal growth and development. Deviations from this range can lead to nutrient deficiencies or toxicities ultimately affecting plant health and productivity. A well-buffered nutrient solution should be used to promote pH stability.

- Ideal pH range is considered to be between 5.8 and 6.3.

- ☆ pH fluctuations within the range of 5.5 to 7.0 can be tolerated.
- ☆ For vegetative growth, the optimal pH range is 6.0 to 6.3.
- ☆ For flowering and fruiting stages, the optimal pH range is 5.7 to 5.9.

7.3.2 Measuring pH

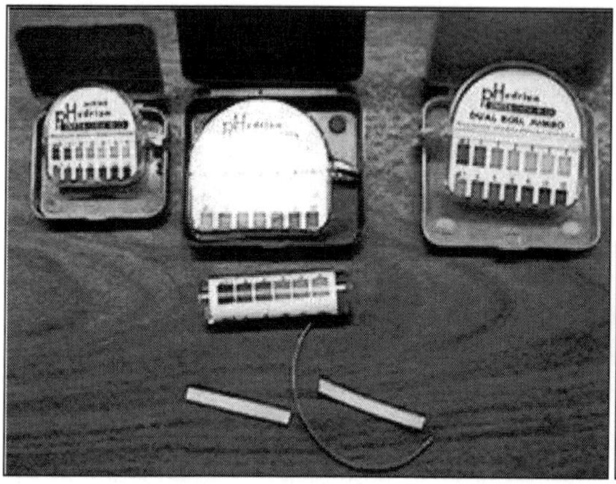

pH Test Strips

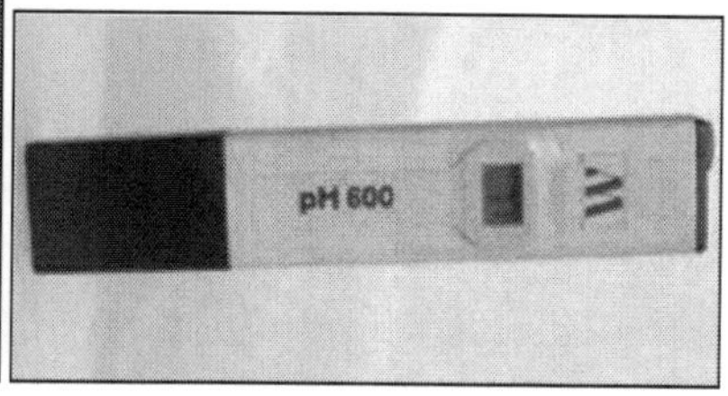

pH Meter

Measuring pH accurately is essential for maintaining optimal nutrient solutions in hydroponics. Various techniques can be employed for pH measurement, including:

i) pH Meters

pH meters are essential tools for accurately measuring the acidity or alkalinity of a solution including nutrient solutions in hydroponic systems. pH meters typically consist of a probe that is inserted into the nutrient solution along with a digital display that shows pH value.

- ☆ **Principle of operation:** pH meters work based on principle of detecting concentration of hydrogen ions (H+) in a solution. These typically consist of a glass electrode and a reference electrode immersed in the solution. The glass electrode generates a voltage proportional to pH of the solution which is measured and displayed on a digital screen.
- ☆ **Accuracy and precision:** pH meters offer high accuracy and precision compared to other pH testing methods such as pH test strips or liquid pH indicators. These can provide pH measurements with a resolution of 0.01 pH units or better allowing for precise adjustments of nutrient solutions in hydroponic systems.
- ☆ **Calibration:** Calibration is crucial for ensuring the accuracy of pH measurements. pH meters need to be calibrated using buffer solutions with known pH values. Typically, two or three buffer solutions are used

to calibrate the pH meter across the desired pH range. Calibration should be performed regularly preferably before each use to maintain accuracy.

- **Maintenance:** Proper maintenance is essential to prolong the lifespan and accuracy of pH meters. This includes rinsing the electrodes with distilled water before and after each use, storing the pH meter in a protective case with a storage solution (usually a potassium chloride solution) and following manufacturer instructions for cleaning and maintenance.
- **Temperature compensation:** pH measurements are temperature-dependent and pH meters often feature automatic temperature compensation (ATC) to correct for temperature variations. ATC ensures accurate pH readings regardless of temperature changes which is particularly important in hydroponic systems where nutrient solutions may fluctuate in temperature.
- **Types of pH meters:** Various types of pH meters are available ranging from basic handheld models to more advanced laboratory-grade meters. Handheld meters are portable and suitable for field measurements while benchtop meters offer higher accuracy and additional features such as data logging and connectivity options.
- **Applications:** pH meters are widely used in various industries and applications including agriculture (for soil and nutrient solution analysis), environmental monitoring, water quality testing, food and beverage production, pharmaceuticals and in scientific research.

ii) pH Test Strips

pH test strips are simple and convenient tools used to measure the acidity or alkalinity of a solution. These consist of narrow strips of paper or plastic impregnated with a pH-sensitive indicator dye. Due to their use and affordability, these strips are widely used in various fields including hydroponics, aquarium keeping, swimming pool maintenance and laboratory experiments.

- **Principle of operation:** pH test strips work based on the principle of colorimetric analysis. The indicator dye on the strip changes colour in response to the pH of the solution being tested. Different indicator dyes are used to cover different pH ranges, allowing for versatility in testing various solutions.
- **pH range:** pH test strips are available in different pH ranges, typically covering the entire pH scale from 0 to 14. However, some strips may have a narrower pH range depending on their intended application. It is essential to select test strips with a pH range suitable for the solution being tested.
- **Usage:** To use pH test strips, simply the strips are dipped for a few seconds into solution being tested. The indicator dye on the strip will

change colour which can be compared to a colour chart provided by the manufacturer. The colour chart indicates the corresponding pH value of the solution based on the colour observed on the test strip.

- ☆ **Accuracy:** While pH test strips provide a quick and convenient method for pH measurement, these may not offer the same level of accuracy as electronic pH meters or laboratory-grade equipment. The accuracy of pH test strips can be influenced by factors such as ambient light, humidity and the age of the strips. However, these are generally accurate enough for most applications where precise pH measurements are not critical.
- ☆ **Storage and handling:** pH test strips should be stored in a cool and dry place away from direct sunlight and moisture to prevent degradation of the indicator dye. It is essential to follow manufacturer's instructions regarding storage and handling to ensure accurate results.
- ☆ **Applications:** pH test strips find applications in various industries and fields including: hydroponics, aquarium keeping, swimming pool maintenance and laboratory experiments.

iii) pH Indicator Solutions

pH indicator solutions are substances used to determine the pH of a solution by changing colour in response to the acidity or alkalinity of the solution. By comparing the resulting colour to a pH colour chart, the pH of the solution can be determined. For quick and easy assessment of solution pH, these are commonly employed in various fields including chemistry, biology, environmental science and agriculture.

- ☆ **Principle of operation:** pH indicator solutions work based on the principle of acid-base indicators. These indicators are usually weak acids or bases that undergo a colour change as the pH of the solution changes. The colour change occurs due to a shift in the equilibrium between the protonated and deprotonated forms of the indicator molecule.
- ☆ **Types of pH indicators:** There are several types of pH indicators each with its own pH range over which it changes colour. Common types include:
    - ❖ **Universal indicators:** These are mixtures of several different indicators that cover a wide pH range, typically from pH 1 to pH 14.
    - ❖ **Single-range indicators:** These indicators change colour over a specific pH range and are used for more targeted pH measurements.
    - ❖ **Natural indicators:** Some natural substances such as red cabbage juice or turmeric can also act as pH indicators and change colour in response to changes in pH.
- ☆ **Colour changes:** pH indicators exhibit different colours at different pH levels such as from red to yellow in acidic solutions, from green to blue in neutral solutions and from blue to purple in alkaline solutions. The

specific colour changes depend on the chemical structure of the indicator molecule and its behaviour in solution.

- ☆ **Applications:** pH indicator solutions find widespread use in various applications, including:
  - ❖ **Laboratory experiments:** pH indicators are commonly used in chemistry and biology laboratories for qualitative pH measurements.
  - ❖ **Water quality testing:** They are used to assess the acidity or alkalinity of water in environmental monitoring and water treatment processes.
  - ❖ **Agriculture:** pH indicators can be used to monitor soil pH in agriculture to ensure optimal conditions for plant growth.
  - ❖ **Aquarium maintenance:** pH indicators help aquarium enthusiasts monitoring and maintaining pH of aquarium water which is crucial for health of aquatic organisms.
- ☆ **Limitations:** While pH indicators are convenient for quick pH measurements but also have limitations. These provide qualitative rather than quantitative results and the colour changes can be subjective depending on factors such as lighting conditions and individual perception. For precise pH measurements, more accurate methods such as pH meters or titration are recommended.

7.3.3 Correcting pH ~ pH adjusters

pH adjusters such as acids and bases are used to regulate acidity or alkalinity of the nutrient solution. While maintaining, the proper pH level is essential for nutrient availability and uptake. pH adjusters themselves are not nutrients. These serve to optimize nutrient uptake by ensuring that nutrient solution pH is within the appropriate range for plant absorption.

i) pH Up and pH Down Adjusters

These commercial products are specifically formulated to adjust pH levels in hydroponic solutions. Both, pH up and pH down adjusters are essential tools for maintaining pH of nutrient solutions within optimal range for plant growth. It is crucial to use them carefully and follow manufacturer instructions to avoid overcorrection or drastic pH changes which could harm plant health. Regular monitoring of pH levels is also recommended to ensure that adjustments are made as needed to support healthy plant growth in hydroponic systems.

- ☆ **Up pH adjusters:** Up pH adjusters are solutions or substances used to raise the pH of a solution that is too acidic. These usually contain alkaline compounds such as potassium hydroxide (KOH), potassium carbonate (K_2CO_3) or sodium hydroxide (NaOH). KOH and K_2CO_3 provide potassium nutrients to plants together with increasing pH levels to create a more alkaline environment. NaOH is highly alkaline and should be used with caution to avoid overdosing. Adding up pH adjusters to the solution increases the concentration of hydroxide ions (OH-) relative to hydrogen ions (H+), thereby raising the pH level. pH up adjusters are commonly used when pH of the nutrient solution falls below desired range.

- **Down pH adjusters:** Down pH adjusters are solutions or substances used to lower pH of a solution that is too alkaline. These typically contain acidic compounds such as phosphoric acid (H_3PO_4), citric acid ($C_6H_8O_7$) or sulphuric acid (H_2SO_4). Phosphoric acid is effective in acidifying the solution while providing essential phosphorus nutrients to plants. H_2SO_4 provides sulphur nutrients to plants while decreasing pH levels. $C_6H_8O_7$ is less concentrated than phosphoric acid but can still adjust pH downwards effectively. Adding pH down adjusters to solution increases concentration of hydrogen ions (H+) relative to hydroxide ions (OH-) thereby lowering pH level. pH down adjusters are commonly used when pH of nutrient solution rises above desired range.

ii) Natural pH Adjusters

Natural pH adjusters can be useful in hydroponic systems for maintaining or correcting pH levels. However, these methods may not provide precise control over pH and can potentially introduce other substances into nutrient solution. Effectiveness of natural pH adjusters may vary depending on factors such as water quality, nutrient composition and plant species. Here are some common natural pH adjusters:

- **Citric acid:** Citric acid found naturally in citrus fruits such as lemons and oranges can be used to lower pH levels in hydroponic solutions. It is effective at lowering pH levels gradually without causing abrupt changes.
- **Vinegar:** Common household distilled white vinegar can also be used to lower pH levels in hydroponic nutrient solutions. It contains acetic acid which helps in reducing pH levels when added in small quantities.
- **Ascorbic acid (vitamin C):** Ascorbic acid can be used as a mild pH adjuster in hydroponic systems. It is found naturally in fruits such as strawberries and kiwis and can help lower pH levels when dissolved in nutrient solution.
- **Sulphur:** Elemental sulphur in the form of powdered sulphur or sulphur-containing compounds can gradually be used to lower pH levels in hydroponic solutions. Sulphur reacts with water to form sulphuric acid which helps reducing pH levels over time.
- **Peat moss:** Peat moss is derived from decomposed plant material which has natural acidic properties and can be used to lower pH levels in hydroponic systems. It can be mixed into the growing medium or put in a mesh bag and placed in the nutrient solution to release acids gradually and thus lowering pH levels.

iii) Buffering Agents

Buffering agents play a crucial role in maintaining pH stability in hydroponic nutrient solutions. These agents help resist changes in pH by absorbing excess hydrogen ions (H+) or hydroxide ions (OH-) when these are added to or removed from the solution. Buffering agents ensure that pH remains within the desired range so providing a stable environment for plant roots and optimizing nutrient availability. When selecting buffering agents for hydroponic nutrient solutions it is essential to consider factors such as desired pH range, nutrient composition and compatibility with other components in the solution. Properly buffered nutrient solutions can help ensure optimal plant growth and productivity in hydroponic cultivation. Common buffering agents used in hydroponics include:

- **Phosphate buffers:** Phosphate compounds such as monopotassium phosphate (KH_2PO_4) or dibasic potassium phosphate (K_2HPO_4) are commonly used as buffering agents in hydroponic nutrient solutions. These compounds can help maintain pH stability within the desired range by reacting with excess acids or bases in the solution.
- **Carbonate buffers:** Carbonate compounds such as potassium carbonate (K_2CO_3) or calcium carbonate ($CaCO_3$) can also act as buffering agents in hydroponic systems. These compounds react with acids or bases in the solution to maintain pH stability.
- **Organic acids:** Organic acids such as citric acid or acetic acid can serve as buffering agents in hydroponic nutrient solutions. These acids can help stabilize pH by donating or accepting hydrogen ions to resist changes in acidity or alkalinity.
- **Biological buffers:** Some microbial organisms in the nutrient solution can produce organic acids that act as buffering agents thus helping to maintain pH stability. These biological buffers can contribute to a more stable pH environment in hydroponic systems.

It is important to note that choice of pH adjuster depends on several factors including initial pH of the water source, nutrient composition of the solution and the specific requirements of the plants being grown. Some pH adjusters may contain other nutrients or compounds that can affect plant growth so it is essential to use them judiciously and monitor the nutrient solution regularly.

When adjusting pH in hydroponic systems, it is recommended to make gradual changes and monitor the pH closely to avoid sudden swings that could stress the plants. pH meters and test kits are commonly used to measure pH of nutrient solution accurately. By maintaining the pH within the appropriate range, growers can ensure optimal nutrient uptake and healthy plant growth in hydroponic systems.

7.4 Media Quality and pH

The choice of growing media can have a significant impact on nutrient solution pH in hydroponic systems. Growers should consider the buffering capacity, cation exchange capacity (CEC), microbial activity and chemical composition of the media when selecting a growing medium to ensure optimal pH stability and nutrient availability for plant growth. Regular monitoring and adjustment of pH may be necessary to maintain optimal growing conditions in hydroponic systems.

7.4.1 Choice of Growing Media

The choice of growing media in hydroponic systems can influence pH of nutrient solution in several ways:

i) **Growing media pH buffering:** Certain growing media have inherent pH buffering capacities, meaning that these can resist changes in pH levels such as media like cococoir and peatmoss which tend to be slightly acidic and can help maintain a stable pH in the nutrient solution. However, other media like rockwool and gravel may have limited buffering capacity thus making them less effective at stabilizing pH.

ii) **Cation exchange capacity (CEC):** Growing media with a high CEC such as clay pellets or expanded clay balls have the ability to absorb and retain cations (positively charged ions) from nutrient solution. This can affect pH by removing cations that contribute to acidity from the solution thus leading to an increase in pH over time potentially.

iii) **pH drift:** Certain growing media ~ particularly those with a low buffering capacity or low CEC may contribute to pH drift in the nutrient solution such as inert media like perlite or vermiculite have little to no buffering capacity and may allow the pH of the nutrient solution to drift over time as acids or bases are added or removed.

iv) **Microbial activity:** Some growing media such as soilless mixes or peatmoss contain organic matter that can support microbial activity. Microbial metabolism can produce acids or bases that may influence pH of nutrient solution. Presence of beneficial microbes may help stabilize pH by aiding in nutrient cycling and maintaining a healthy root environment.

v) **Interaction with nutrient solution:** The composition of growing media can interact with nutrient solution affecting pH, *e.g.* media containing limestone or other alkaline materials may increase the pH of the nutrient solution over time though media with acidic properties may decrease the pH.

7.4.2 Poor Quality Growing Media

Poor quality growing media, particularly industrial-grade rockwool and gravel can contribute to pH instability in hydroponic systems. These media may contain impurities or residues that leach into nutrient solution thus affecting its pH. Industrial-grade rockwool may release alkaline compounds into the solution thus causing rise of pH levels constantly. In cases where poor quality growing media are used, the pH of the nutrient solution may steadily increase to dangerously high levels. This rise in pH can occur due to the leaching of alkaline substances from the media or inadequate pH buffering capacity. High pH levels can lead to nutrient imbalances, nutrient lockout and reduced nutrient availability to plants thus ultimately impacting plant health and growth.

7.4.3 Evaluating pH Stability in Hydroponic Growing Media

Testing a new growing medium is simple and crucial for maintaining pH stability in hydroponic systems. It should begin by placing a sample of medium (such as rockwool, gravel or sand) into a clean container such as a cup or jar and immerse it in distilled or deionized water. After allowing it to sit for a period pH of the water should be tested and recorded. This test should be repeated periodically for about a week until the pH stabilizes. It is important to note that pH may initially rise to levels as high as 8.0, 9.0 or even 10.0 which can be detrimental to plant health. pH of growing medium influences plant growth significantly which is why water culture hydroponic methods are gaining popularity over media-based hydroponics. Using fewer media in hydroponic systems reduces pH instability and salt accumulation. Solution culture systems require less water and nutrients due to their higher efficiency and reduced evaporation.

7.4.4 Monitoring and Adjustment

To mitigate the effects of poor quality growing media on pH stability it is crucial for hydroponic growers to regularly monitor the pH of nutrient solution and take corrective actions as needed. This may involve using pH-adjusting solutions or additives to bring pH levels back within the optimal range. Selecting high-quality growing media with better pH buffering capacity can help maintain more stable pH levels in hydroponic systems.

7.5 Optimal Nutrient Solution Composition

Growers should always try to specialize in one or two particular plant varieties so that they can adjust their nutrient mix to suit that particular crop variety. It will be better if growers find an all round mix which works for just about everything by just adjusting the strength. Too much nutrients appear to be toxic and are often worse as compared to not enough nutrients. Fewer nutrients will not only cost less but the growers don't really have to panic if they are going away for a weekend. Even when their heavy feeder plants are coming into full fruit, fewer nutrients fortnightly will just slow them down and it generally won't kill them.

7.6 Root Zone Moisture

Adjusting moisture levels in the root zone is a critical aspect of managing plant growth and development in hydroponic systems. Commercially available hydroponic setups often provide growers with the flexibility to control moisture levels to suit specific needs of their crops.

Maintaining a drier root zone can have beneficial effects particularly for aromatic crops such as basil and mint. In these plants a drier environment can stimulate production of essential oils, enhancing their aroma and flavour profiles. Through careful management of moisture levels growers can optimize quality and yield of aromatic herbs thus ensuring desirable characteristics for culinary or medicinal purposes.

On the other hand, increasing moisture levels in the root zone can be advantageous for promoting vegetative growth especially in large fan-leaved plants. A wetter root zone enhances transpiration and photosynthetic efficiency providing plants with the hydration and resources needed to support robust vegetative growth. By adjusting moisture levels accordingly, growers can encourage vigorous foliage development and overall plant vitality.

7.7 Temperature and Dissolved Oxygen

Effective temperature management is vital in hydroponic cultivation as different plants thrive within specific temperature ranges, such as cucumbers excel in warmer environments. Temperature levels and oxygen concentrations in the root zone profoundly impact root respiration rates. Maintaining the nutrient solution within the optimal temperature range is crucial for promoting healthy root development. Extremes in temperature can stress the roots and inhibit nutrient absorption leading to nutrient deficiencies, *vis-à-vis* stunted growth. Most plants thrive within a root zone temperature range of 18 to 24°C with adjustments needed for winter and tropical crops. Some plants may have specific temperature preferences outside the general range. Local climate conditions and system design factors can influence temperature management in hydroponic systems. It is essential to find out the temperature requirements of the crops being cultivated so that the same is adjusted accordingly for optimal growth. Regular monitoring and adjustment of temperature are key to ensuring optimal plant health and growth. The air temperature in the room should be around 20°C.

Nutrient solution temperatures exceeding 24°C can lead to low dissolved oxygen levels dangerously thus necessitating proper aeration measures. Conversely, temperatures below 16°C can cause plants to slow their metabolic processes. Some growers strategically lower nutrient temperatures towards the end of the flowering stage to facilitate ripening. By managing temperatures carefully within the optimal range and addressing fluctuations promptly, hydroponic growers can create an environment conducive to healthy plant growth and optimal yields.

Extreme temperatures can hinder plant growth and vitality. Cold temperatures can impede seed germination, root development and overall plant growth while excessively high temperatures can lead to oxygen deficiency or thermal stress in the root zone thus resulting in plant death. With diminishiong of oxygen level in root zone, roots may experience hypoxia thus hindering respiration rates due to limited oxygen availability. This phenomenon can affect various root functions critical for crop production. Adequate aeration and circulation are essential to mitigate this risk and for ensuring optimal oxygen levels for root health.

Maintaining optimum nutrient solution temperature is crucial for promoting robust plant growth and high-quality yields. Utilizing a high-quality aquarium heater during winter months can raise nutrient solution temperatures while cooling fans or chillers can mitigate overheating issues in summer months. Employing aquarium heaters and shade cloths can prevent abrupt temperature fluctuations for smaller hydroponic setups while larger systems benefit from greenhouse structures for temperature regulation. Larger water volumes such as those exceeding 1,000 litres provide stability and buffer against sudden temperature shifts in the system. It is essential to match the temperature of added water to avoid temperature shocks with that of the reservoir. Rapid temperature fluctuations especially in the root zone are unfavourable for plant health.

The demand for O_2 rises in tandem with increasing temperatures which underscores the importance of adequate aeration in the root zone. CO_2 levels without sufficient aeration can elevate which can affect root health adversely. Oxygen concentration within nutrient solution is also contingent upon crop demand, peaking during heightened photosynthetic activity. Inadequate dissolved oxygen levels may impede root growth thus manifesting a brown discolouration ~ a telltale sign of oxygen deficiency.

Temperature plays a pivotal role in water and nutrient uptake across different plant species, *e.g.* in roses where colder nutrient solutions tend to boost nitrate (NO_3-) uptake while yielding thin white roots; however, water uptake diminishes under such conditions. Each plant species exhibits distinct thresholds for growth under minimum, optimum and maximum temperature ranges. Consequently, the implementation of heating or cooling systems becomes imperative to maintain an optimal nutrient solution temperature conducive to plant health and vitality.

7.8 Management of Dissolved Oxygen (DO) Levels

High temperatures above 24°C can lead to a decline in dissolved oxygen levels which may impact plant health. Adequate aeration is essential to prevent DO levels from dropping too low, especially in warmer environments. Conversely, temperatures below 16°C can slow plant metabolism so affecting growth rates. Some growers strategically lower nutrient temperatures during flowering stage to aid in ripening. Here are some key strategies for managing dissolved oxygen in hydroponic systems:

i) **Aeration**: Proper aeration of nutrient solution is essential to maintain adequate levels of dissolved oxygen. Aeration systems, *e.g.* air pumps and air stones are commonly used to introduce oxygen into the nutrient solution. Air stones create fine bubbles that increase the surface area of the water in contact with air so facilitating oxygen exchange.

ii) **Water movement**: Circulating the nutrient solution helps to prevent stagnation and stratification thus ensuring uniform distribution of oxygen throughout the system. Submersible pumps or recirculating systems can be employed to maintain water movement and promote oxygenation.

iii) **Temperature control**: Lower water temperatures hold more dissolved oxygen than warmer water. Monitoring and controlling the temperature of the nutrient solution can help maintain optimal oxygen levels. Using coolers or chiller units may be necessary especially in warmer climates or during hot weather.

iv) **Oxygen-enriched atmosphere**: Increasing the concentration of oxygen in the growing environment can enhance oxygen levels in nutrient solution. CO_2 enrichment systems should be properly regulated to prevent excessive buildup as high CO_2 levels can displace oxygen.

v) **Avoiding overcrowding**: Plant overcrowding in the grow tray can restrict oxygen availability to roots. Proper spacing and pruning of plants help ensure adequate access to oxygen thus preventing root suffocation.

vi) **Regular monitoring**: Monitoring DO levels using a dissolved oxygen meter is essential for ensuring optimal conditions. DO levels should typically be maintained above 5ppm for healthy plant growth.

vii) **Maintaining cleanliness**: Keeping the hydroponic system clean and free from debris, algae and other contaminants helps prevent oxygen depletion. Regular maintenance including cleaning of reservoirs, tubing and equipment is essential for preserving water quality and oxygen levels.

viii) **Hydrogen peroxide (H_2O_2):** Hydrogen peroxide can be added to nutrient solution to increase oxygen levels and to prevent growth of algae, bacteria and other pathogens. On breaking down of hydrogen peroxide it releases oxygen molecules which help oxygenate the root zone so promoting healthy root development. Hydrogen peroxide has mild antiseptic properties which can help sterilize the nutrient solution and prevent the buildup of harmful microorganisms.

7.9 Bicarbonate (HCO_3^-)

Bicarbonate (HCO_3-) is an essential component in nutrient solutions used in hydroponic systems. It is commonly found in underground water sources such as bore water. Due to its alkaline nature, bicarbonate can elevate pH levels of solutions. When high alkalinity makeup water is added to nutrient solutions it can lead to an undesirable increase in pH. Here is a comprehensive overview of its role and impact:

i) **Buffering agent:** Bicarbonate serves as a vital buffering agent in nutrient solutions, playing a pivotal role in stabilizing pH levels. Its buffering capacity allows it to absorb surplus hydrogen ions (H+) when pH decreases, and release them when pH rises. This capability is essential for maintaining optimal pH in hydroponic systems, as pH fluctuations can adversely affect nutrient availability and overall plant health.

ii) **pH increase:** Within hydroponic setups, bicarbonate ions can induce pH elevation, particularly in alkaline environments. This phenomenon arises from the reaction of bicarbonate ions with water, yielding carbonic acid (H_2CO_3). Subsequently, carbonic acid dissociates into bicarbonate (HCO_3^-) and hydrogen (H^+) ions. The release of hydrogen ions augments solution acidity, thereby causing a surge in pH levels.

iii) **Carbon source:** Bicarbonate serves as a source of carbon for plants in hydroponic systems. Plants utilize carbon dioxide (CO_2) during photosynthesis to produce carbohydrates and other organic compounds essential for growth and development. Bicarbonate in the nutrient solution can be converted into CO_2 by plants, providing them with the carbon they need for photosynthesis.

iv) **Alkalinity regulation:** Bicarbonate helps regulate the alkalinity of the nutrient solution; alkalinity refers to the solution's capacity to resist changes in pH when an acid or base is added. It can neutralize excess acids in the solution, preventing a rapid decrease in pH levels. This is particularly important in systems where pH fluctuations are common due to factors such as plant uptake, nutrient imbalances or microbial activity. However, higher concentrations of bicarbonate ions result in higher alkalinity, which can make it more challenging to adjust pH levels in the nutrient solution. Hence, it's essential to regularly monitor pH levels when using high alkalinity makeup water. To mitigate the effects of alkalinity, pH adjustment with acid is necessary to restore balance in the nutrient solution.

v) **Nutrient uptake:** Bicarbonate can influence the uptake of other essential nutrients by plants. It can interact with cations (positively charged ions) and anions (negatively charged ions) in the solution, affecting their availability to plants. Additionally, bicarbonate ions can compete with other ions for binding sites on root surfaces, potentially impacting nutrient uptake efficiency.

vi) **Ca and Mg precipitation:** Bicarbonate ions can react with calcium (Ca^{2+}) and magnesium (Mg^{2+}) ions in the nutrient solution forming insoluble precipitates. This can lead to precipitation of $CaCO_3$ and $MgCO_3$ so reducing the availability of these essential nutrients to plants. To prevent this it is essential to maintain proper pH levels and avoid excessive bicarbonate concentrations in the nutrient solution.

7.9.1 Buffering Capacity of Bicarbonate Ions

Bicarbonate ions are indeed weak alkaline and their ability to raise the pH of a solution is limited. Bicarbonate ions in most cases are unable to increase the pH significantly above 8.3. While bicarbonate ions do act as a buffer in solutions their buffering capacity is considered relatively weak compared to other buffering systems such as phosphate or carbonate buffers. Bicarbonate ions can resist pH changes to some extent when acid is added but their effectiveness as a buffer may vary depending on factors such as concentration and presence of other buffering agents.

7.9.2 Managing pH Fluctuations: Understanding Groundwater Dynamics

When utilizing groundwater, it is common for pH levels to initially rise shortly after being lowered. This phenomenon occurs due to addition of acid to a nutrient solution containing bicarbonates. When acid is introduced, the by-products such as CO_2 and H_2O are generated. The presence of free CO_2 tends to lower pH by forming carbonic acid (H_2CO_3) when it reacts with water. However, concentrations of CO_2 exceeding approximately 0.5mg/l in water become unstable when exposed to atmospheric pressure at sea level. In such cases, excess CO_2 gradually escapes into the atmosphere which results in a loss of acidity and subsequent rise in pH. To mitigate these effects, it is advisable to aerate water with high alkalinity or CO_2 levels before use. Initially it is recommended to adjust pH of the nutrient solution to around 5.0 upon preparation or after the addition of top-up water.

7.10 Change-Interval of Nutrient Solution

Regular changing of nutrient solution is essential to prevent salt accumulation, nutrient deficiencies and buildup of plant waste. With transpiration of the plants, moisture and nutrient levels in the solution decrease which lead to an increase in EC or nutrient strength potentially to hazardous levels. The frequency of nutrient solution changes depending on various factors such as plant species, quantity *cum* size of plants, growth stage, water quality, reservoir capacity, nutrient quality, environmental conditions (temperature and humidity) and the type of hydroponic system employed. Monitoring and record-keeping are essential components of this process.

Initiating a hydroponic system with fresh nutrient solution is paramount. Record-keeping should include dates, pH levels and EC or ppm measurements. As the system operates, the solution level naturally decreases. Regular recording and adjusting of EC/ppm levels by adding fresh water and nutrients as needed is crucial. The amount of water added to top-up the reservoir should be recorded carefully. Frequent testing of the nutrient solution helps determine the appropriate timing for either replacing the solution entirely or diluting it with fresh water. This procedure with each reservoir top-up should be followed consistently.

When total volume of water added equals the capacity of the reservoir, it is time to drain the existing nutrient solution and replace it entirely with fresh solution. This systematic approach ensures continuous provision of balanced nutrients and optimal growing conditions for hydroponic plants.

7.11 Cleaning and Sterilization of Nutrient Solution Reservoir

Maintaining cleanliness in the nutrient solution reservoir is essential for promoting healthy plant growth and preventing the proliferation of pathogens and pests. Here is a systematic approach to clean and sterilize the reservoir effectively::

i) Remove entire plant matter and debris from the water to ensure a clean surface.

ii) Drain the reservoir completely to prepare for cleaning.

iii) Using a paper towel, wipe out entire residue and buildup as possible from reservoir interior.

iv) Refill the reservoir with fresh water and add a new batch of nutrient solution to maintain optimal nutrient levels.

Reservoir Sterilization

i) It is advisable to sterilize the reservoir every fortnight or whenever changing the nutrient solution to prevent accumulation of harmful pathogens and pests.

ii) Choose an appropriate sterilizing agent such as a 10 per cent H_2O_2 or 10 per cent bleach solution.

iii) In case of quite small reservoir to handle it can be immersed in a sink or large basin for sterilization though larger reservoirs are sterilized by filling it with the sterilizing solution and allowing it to sit for several hours.

iv) Hydrogen peroxide is advantageous as it naturally breaks down over time and does not require rinsing. However, in case of a bleach solution, thorough rinsing till the end is a must.

v) When handling concentrated peroxide or bleach one should wear gloves to protect hands from potential irritation or burns.

By following these steps diligently the hydroponic growers can maintain a clean and sterile reservoir environment which will promote optimal plant health and productivity.

7.12 Salt Accumulation in Hydroponic Media

Salt accumulation poses a significant challenge in hydroponic cultivation impacting plant growth and overall health. Salt build-up occurs due to transpiration and evaporation that leads to concentration of undesirable salts within the growing medium. Some salts are absorbed by plants but an excessive accumulation can hinder growth and nutrient uptake. Here is a comprehensive approach to manage and mitigate salt build-up effectively:

7.12.1 Effective Management Practices

i) Adequate drainage is crucial to prevent salt build-up. Supplying extra water during irrigation cycles promotes sufficient leaching of excess salts.

ii) Irrigation water should permeate the entire medium volume to facilitate proper dilution and leaching of accumulated salts.

iii) Implement a 30-50 per cent increase in water volume during daily irrigation cycles to enhance salt dilution and removal.

iv) Utilize regular chemical treatments to prevent excess salt build-up and maintain optimal nutrient levels.

7.12.2 Considerations for Recirculating Systems

i) **Frequent nutrient solution replacement:** In recirculating systems where nutrient solutions are reused, frequent complete replacements are necessary to address salt accumulation.

ii) **Monitoring salinity levels:** Regular monitoring of salt content and its contribution to total EC to gauge the extent of build-up accurately.

7.12.3 Chemical Treatment to Manage Salt Build-ups

Implementing regular chemical treatments is essential for maintaining optimal nutrient levels and preventing salt accumulation. Here is a detailed guide on effective chemical treatment methods:

i) **Identification of issue**: Regular monitoring of EC levels and visual inspection of plant health can help identify salt build-up issues promptly.

ii) **Choosing of suitable chemicals**: Appropriate chemicals based on type and severity of salt accumulation should be selected. Here common options include:

- ❖ **Acidic solutions**: Acidic compounds such as phosphoric acid or citric acid can help lower pH levels and dissolve salt deposits.
- ❖ **Chelating agents**: Chelating agents such as EDTA (ethylenedia minetetraacetic acid) or DTPA (diethylenetriaminepentaacetic acid) can bind to excess ions thus preventing them from accumulating in the growing medium.
- ❖ **Enzymes**: Enzyme-based treatments can break down organic matter and prevent biofilm formation.
- ❖ **Surfactants**: These aid in breaking down salt crusts and facilitating their removal during irrigation.

iii) **Application methods**:

- ❖ **Direct application**: Dilution of the chosen chemical treatment should be done according to manufacturer instructions and this should be applied directly to the affected areas or into the nutrient solution. Direct application to the root zone can prevent salt build-up in the medium thus enhancing nutrient absorption by plants.

- ❖ **System flushing**: Flushing of entire hydroponic system with chemical solution should be done to remove salt build-ups from the growing medium, pipes and reservoirs.

iv) **Treatment frequency**: Establishing a routine schedule for chemical treatments based on severity of salt accumulation and specific needs of plants should be done. Manufacturer guidelines should be followed for dosage and application frequency to prevent overuse and potential damage to plants. Treatment frequency should be adjusted based on severity of salt build-up and the specific needs of the hydroponic system.

v) **Monitoring and adjustments**: Regular monitoring of EC levels and plant responses to treatment should be done to gauge effectiveness. Treatment methods or frequencies should be adjusted based on observed outcomes and changing environmental conditions.

vi) **Record-keeping**: Detailed records of chemical treatments, the type of chemical used, concentration, application method and treatment schedule should be recorded.

vii) **Safety precautions**: Appropriate personal protective equipment (PPE) such as gloves and eye protection when handling concentrated chemical solutions should be worn. Concentrated chemicals should be diluted as per instructions. Adequate ventilation should also be ensured to minimize exposure to fumes. Chemicals should be stored securely by following proper handling and disposal protocols to minimize environmental impact.

7.13 Hardness of Water

Water hardness ~ primarily attributed to the combined concentrations of calcium and magnesium, poses challenges in re-circulating systems. With successive additions of top-up water the calcium level may surge which may lead to formation of a white precipitation of calcium sulphate within the system thus obstructing plumbing components like pipes and drippers. This process also diminishes sulphate availability. A water sample containing over 150mg/l of bicarbonates (HCO_3) is typically classified as hard as characterized by elevated levels of mineral salts and often exhibiting a high pH. To mitigate this, hydroponic growers commonly introduce phosphoric acid (H_3PO_4) to lower the pH which results in increased phosphate levels over time. Gradual buildup of phosphate can detrimentally affect the uptake of crucial nutrients thus resulting in substantial imbalances within the system potentially.

7.13.1 Impact on Plant Growth and System Health

Plugging of Pipe Due to Ca Precipitation which Cements and Chokes the Internal Body.

The presence of water hardness in hydroponic nutrient solutions exerts a significant impact on both plant growth and system health. Elevated levels of calcium and magnesium inherent in hard water impede the vital absorption of nutrients by plants, resulting in potential deficiencies and imbalances. Moreover, the heightened pH characteristic of hard water may disrupt the delicate pH balance of the nutrient solution, necessitating adjustments with acidic solutions. These adjustments introduce additional elements into the mix, potentially causing stress to the plants. Furthermore, hard water's propensity for mineral buildup poses a risk to the hydroponic system, particularly in irrigation lines, pipes and reservoirs. This accumulation can lead to clogging and reduced water flow, ultimately impeding the efficient distribution of nutrients to plants and inducing stress within the system.

7.13.2 Hard Water Detecting

Detecting hard water can be done through several methods:

i) Hard water often leaves white or cloudy spots on dishes and glassware after washing. It can also cause soap scum to form in sinks and bathtubs.

ii) Hard water tends to produce less lather with soap compared to soft water. If it requires more soap to produce suds while washing hands or clothes, it indicates presence of hard water.

iii) Water testing kits are available for home use and can provide a more precise measurement of water hardness. Typically these measure concentration of Ca and Mg ions in the water.

iv) Water hardness can also be assessed by sending a water sample to a laboratory for analysis. This method provides accurate and detailed information about mineral content of water.

7.13.3 Dealing with Hard Water

Growers can employ various strategies to manage water hardness effectively. Before introducing the water into the hydroponic system, water is firstly treated through filtration or reverse osmosis to reduce water hardness. Secondly, alternative water sources with lower hardness levels can be explored such as rainwater collection or purchasing distilled water as feasible alternatives to mitigate the impact of hard water on hydroponic systems. Thirdly, regular monitoring and adjustment of nutrient solution pH are crucial to maintain optimal conditions for nutrient uptake. Nutrient formulations can also be adjusted to account for presence of hard water, however, chelated micronutrients can be used to ensure adequate nutrient availability in high pH environments. Lastly, routine system maintenance such as flushing with clean water and cleaning of irrigation lines and reservoirs helps prevent mineral buildup and ensures smooth operation of hydroponic system.

Through these measures growers can mitigate adverse effects of water hardness and support healthy plant growth in their hydroponic setups.

7.14 Iron in Water

Iron is a vital trace element crucial for plant growth in its natural chemical state. However, when exposed to the atmosphere iron fails to form a stable solution in water with pH values above 3. Iron is frequently present in bore- or ground-water. While water containing iron levels exceeding ~0.3mg/l may initially appear clear and colourless upon extraction but exposure to air triggers conversion of iron to iron oxide resulting in a brown and cloudy appearance. Upon standing, the iron oxide often settles, forming a brown precipitate. This phenomenon commonly leads to blocked drippers and unsightly brown stains on surfaces that come into contact with the water.

To prevent blockages, it is essential to remove iron from makeup water before usage. One method involves iron oxidation removal through aeration followed by 48 hours of gravity settling. In cases where aeration proves ineffective, a small amount of chlorine can be added followed by 5 minutes of gentle agitation, then gravity settling.

7.15 Corrosion

Corrosion in Iron Pipe.

The propensity for corrosion of metal components escalates with heightened total salt content and acidity which is attributed to CO_2 presence. CO_2 serves as the primary catalyst for corrosion in natural groundwater environments. In instances where water exhibits corrosive properties, employing pumps equipped with stainless steel, ceramic or plastic components becomes imperative. Furthermore, reservoirs and plumbing infrastructure should also be constructed from durable plastic materials to withstand the corrosive effects of such water compositions. This proactive approach ensures the longevity and reliability of hydroponic systems ~ safeguarding against the detrimental impacts of corrosion on equipment and overall system integrity.

7.16 Harmful Elements

Chloride deficiency in plants is uncommon due to its minimal requirement though found typically in most water sources. However, excessive chloride levels can be detrimental leading to leaf chlorosis and hampering plant growth or yield.

Potassium chloride and ammonium chloride are potential sources of chloride toxicity. Sodium accumulation poses a risk in recirculating hydroponic systems as it gradually builds up in the solution. The optimal concentration of sodium and chloride for hydroponically grown plants is generally below 75ppm.

Ammonium toxicity can arise from an overuse of ammonium nitrate where excessive ammonium levels can damage roots and stem bases particularly in young plants resulting in stunted growth and reduced yields. It is important to note that plants typically do not uptake nitrogen in the ammonium (NH_4+) form though in nitrate (NO_3-) form through oxidation before absorption. Therefore, maintaining a balanced nitrogen source is crucial with ammonium concentration ideally constituting no more than approximately 20 per cent of the total required nitrogen concentration.

7.17 Pathogens

Contaminated water sources or sick plants can introduce pathogens into the nutrient solution so spreading diseases among entire crop. This issue is often influenced by regional and seasonal factors. Dirty growing conditions, warm water temperatures and high nutrient strengths can create favourable conditions for root diseases such as *Pythium* and *Fusarium*. Implementing proper sanitation practices, monitoring water temperature and adjusting nutrient strength can help prevent the onset of root diseases and ensure healthy plant growth. To prevent such problems, it is crucial to maintain cleanliness by working in a hygienic environment, *vis-a-vis* planting only disease-free healthy plants with regular monitoring crop health.

In the event of a disease outbreak, prompt action is essential. Diseased plants must be removed immediately, the nutrient solution should be drained from the reservoir and reservoir should thoroughly be cleaned and replenished with fresh nutrient solution. Flushing the system with fresh water without nutrients for a day can further help cleanse the system. It will be better to consider performing this flushing process between every three or four nutrient changes to maintain cleanliness in the root zone and entire hydroponic system. For gravel systems, periodic flushing is particularly beneficial in removing salt accumulation from the medium.

Ensuring proper growing conditions including adequate light and air ventilation will minimize the risk of pests or diseases infiltrating the system. Diligent attention to these factors is essential for maintaining a healthy hydroponic environment and promoting successful crop growth.

7.18 Algal Growth

Algal growth poses a significant concern in hydroponic systems as it competes with plants for nutrients, reduces oxygen levels and can potentially clog irrigation systems. Ensuring that the container or reservoir holding the nutrient solution remains light-proof is essential to prevent proliferation of algae in solution. By maintaining a light-excluded environment within the reservoir, growers can effectively mitigate the risk of algal infestation and safeguard the integrity of the nutrient solution. This precautionary measure not only preserves the nutrient balance critical for plant health but also contributes to the overall stability and efficiency of the hydroponic system.

7.19 Impact of Light on Nutrient Solution

Light intensity indirectly affects nutrient solution management by influencing plant growth rates and nutrient uptake. Higher light intensity generally stimulates faster growth and nutrient uptake. The container (reservoir) of nutrient solution should exclude light to prevent algae growth in the nutrient solution. Here's how light affects the nutrient solution and its uptake:

i) **Plant growth:** Adequate light intensity promotes healthy plant growth and nutrient metabolism. Light quality, intensity and duration affect overall plant growth and development. Rapidly growing plants have higher nutrient requirements to support their metabolic processes. Therefore, light conditions that promote vigorous plant growth can indirectly influence nutrient uptake rates in hydroponic systems.

ii) **Stomatal regulation:** Light regulates the opening and closing of stomata ~ the microscopic pores on the leaf surface responsible for gas exchange. In presence of light stomata open to facilitate uptake of CO_2 for photosynthesis. This process indirectly influences nutrient uptake because stomatal opening also affects the uptake of water and minerals including nutrients dissolved in the nutrient solution.

iii) **Root development:** Light exposure influences root development and morphology in plants. In hydroponic systems where roots are directly exposed to nutrient solution adequate light exposure promotes healthy root growth. Well-developed roots have increased surface area and absorptive capacity, allowing for efficient nutrient uptake from the solution.

iv) **Metabolic activity:** Light availability influences plant metabolism and nutrient demand. Plants grown under sufficient light exhibit higher metabolic rates and increased nutrient demand to support growth and development. As a result, nutrient uptake rates may be influenced by light intensity and duration.

7.20 Impact of CO_2 on Nutrient Solution

Impact of CO_2 on nutrient solution in hydroponic systems is indirect but significant as it influences the overall growth and health of plants which in turn affects nutrient uptake and utilization. Here is how CO_2 can impact nutrient solution in hydroponics:

i) **Enhanced photosynthesis**: CO_2 is a critical component of photosynthesis, the process by which plants convert light energy into chemical energy thus producing glucose and oxygen. Increased levels of CO_2 can enhance photosynthetic rates leading to greater carbohydrate production and biomass accumulation in plants. This enhanced photosynthetic activity can stimulate nutrient uptake from the nutrient solution as plants require various nutrients for synthesizing carbohydrates and supporting growth.

ii) **Stimulation of plant growth**: Elevated CO_2 levels can stimulate plant growth and development by promoting cell division, elongation and expansion. As plants grow more vigorously so these may exhibit increased demand for nutrients leading to higher uptake rates from nutrient solution. This heightened metabolic activity can result in improved nutrient absorption and utilization efficiency, potentially impacting the nutrient composition and concentration in solution.

iii) **pH regulation**: CO_2 dissolved in water forms carbonic acid (H_2CO_3) which can lower pH of nutrient solution. In hydroponic systems, maintaining optimal pH levels is crucial for nutrient availability and uptake by plants. Elevated CO_2 levels can contribute to a decrease in pH which may affect nutrient solubility and plant nutrient uptake. Growers may need to monitor and adjust pH levels accordingly to ensure that nutrient concentrations remain within the desired range for optimal plant growth.

iv) **Indirect impact on nutrient solution**: While CO_2 primarily affects plant physiology and growth, its indirect impact on the nutrient solution lies in its influence on plant nutrient demand and uptake. As CO_2 levels increase and stimulate plant growth, plants may require higher concentrations of essential nutrients from the solution to support their increased metabolic activity. This can lead to changes in nutrient uptake rates *cum* nutrient depletion rates in the solution and the need for adjustments in nutrient replenishment strategies to maintain optimal nutrient levels for plant growth.

7.21 Impact of Humidity on Nutrient Solution

Humidity levels influence the rate of evaporation from the nutrient solution reservoir. Higher humidity can lead to increased evaporation rates and potential fluctuations in nutrient concentrations. Humidity affects transpiration rates in plants so influencing water and nutrient uptake through the roots. Low humidity can lead to increased transpiration and water stress while high humidity may reduce transpiration and nutrient uptake.

Chapter 8

Additives

Additives in hydroponics are substances added to the nutrient solution to provide essential elements and promote plant growth. In both organic and conventional hydroponic systems, the addition of specific compounds can significantly enhance plant nutrition acquisition and uptake. For instance, chelating agents and humic acid have demonstrated effectiveness in boosting nutrient uptake (Adania *et al.*, 1998; Benton, 2004). Furthermore, plant growth-promoting rhizobacteria (PGPR) extensively employed in field and greenhouse agriculture exhibit notable benefits for hydroponic plant growth and nutrient absorption (Lee and Lee, 2015). Certain PGPR strains have been identified to enhance nitrogen fixation ~ particularly a valuable trait even in hydroponic systems with well-maintained fertilizer regimens, *e.g.* the genera like *Azospirillum* and *Azotobacter* help sustaining mobilized nitrogen forms in environments with heightened microbial activity in the rhizosphere.

In traditional hydroponic setups nitrogen accumulation within plant tissues at harvest is a common issue associated with conventional fertilizer methods. However, *Rhodopseudomonas palustris* has emerged as a promising solution significantly enhancing nitrogen use efficiency, elevating yields and notably reducing nitrate concentrations by up to 88 per cent at harvest compared to conventional hydroponic practices particularly in leafy greens (ShuHua *et al.,* 2015). Numerous bacterial species such as *Bacillus* spp., *Pseudomonas* spp. and *Streptomyces* spp. play crucial roles in converting otherwise unavailable phosphorus forms into soluble anions by altering soil pH, liberating chelated phosphorus for plant uptake across a broader pH spectrum and mineralizing organic phosphorus compounds.

Further, recent studies have highlighted the capacity of *Bacillus* inoculants to mitigate high salt stress in hydroponic leaf lettuce thereby bolstering growth under conditions that would otherwise impede it (Moncada *et al.,* 2020). This discovery holds particular promise for regions grappling with elevated EC or salt content in their water sources, potentially circumventing the need for costly reverse osmosis filtration systems while sustaining robust crop yields.

Additives are also required for supplying those substances which have a dose rate that varies throughout the growth cycle at a different rate than the base nutrient. This is often the case with PGR's or hormone additives.

8.1 Breakdown of Common Additives

The additives can include various types of nutrients, supplements, pH adjusters and growth enhancers. Here is a breakdown of common additives used in hydroponics:

i) **Macronutrients**: These are primary nutrients (N, P, K, Ca, Mg and S) required by plants in relatively large quantities. Macronutrient additives are typically supplied in the form of water-soluble salts such as potassium nitrate, calcium nitrate and magnesium sulphate.

ii) **Micronutrients (trace elements)**: Though required in smaller quantities, micronutrients (Fe, Mn, Zn, Cu, B, Mo and Cl) are essential for plant growth. Micronutrient additives are often provided in chelated forms to ensure their availability in hydroponic solutions.

iii) **pH adjusters**: Maintaining the correct pH level is crucial for nutrient uptake by plants. pH adjusters are used to raise or lower the pH of the nutrient solution as needed. Common pH adjusters include pH up (potassium hydroxide or potassium carbonate) and pH down (phosphoric acid or citric acid).

iv) **Beneficial microorganisms**: Some hydroponic systems incorporate beneficial bacteria and fungi to promote plant health and nutrient uptake. These microorganisms (mycorrhizal fungi and various strains of beneficial bacteria) can help break down organic matter, fix nitrogen and protect plants from pathogens.

v) **Plant growth regulators**: These additives (auxins, cytokinins, gibberellins and abscisic acid) influence plant growth and development. These can stimulate root growth, increase flowering or enhance nutrient absorption.

vi) **Enzymes**: Enzyme additives can help breaking down organic matter and improve nutrient availability in hydroponic systems. They can also aid in prevention of root rot and other diseases by breaking down dead root material.

vii) **Organic supplements**: While hydroponic systems are typically based on inorganic nutrients, some growers may choose to supplement their nutrient solutions with organic additives to enhance plant health and flavour. These may include seaweed extracts, fish emulsion, compost teas and humic acids.

viii) **Rooting hormones**: These additives stimulate root growth and development so helping the plants establishing a healthy root system. Rooting hormones are commonly used when propagating plants from cuttings in hydroponic systems.

ix) **Biostimulants**: These are substances that promote plant growth and development, improve stress tolerance and enhance nutrient uptake. These can include amino acids, vitamins and other organic compounds that stimulate plant metabolism and physiological processes.

When using additives in hydroponics, it is essential to follow manufacturer recommendations carefully and monitor the nutrient solution regularly to ensure that plants receive the appropriate balance of nutrients for optimal growth. Maintaining proper sanitation practices and water quality is crucial to prevent nutrient imbalances and disease outbreaks in hydroponic systems.

8.2 Nutrient Additives

Nutrient additives (micronutrient supplements, organic supplements and mineral supplements such as silica) are substances added to the nutrient solution to supplement essential elements that may not be present in sufficient quantities or may not be stable in concentrated nutrient formulations. These additives provide essential nutrients or beneficial compounds that support plant growth, stimulate root development, enhance flowering and improve disease resistance. Nutrient additives are primarily focused on providing essential nutrients and supporting overall plant health and development.

Note: While enzymes, PGRs, pH adjusters and beneficial microorganisms play crucial roles in hydroponic systems, these are not typically classified as nutrient additives because these directly do not supply essential nutrients to plants. Instead, these contribute to nutrient availability, plant growth regulation, pH management and overall health of the system through various mechanisms.

8.2.1 Phosphate Potassium (PK) Additives

Phosphate potassium additives are supplements used in hydroponic systems to provide additional P and K to plants during specific growth stages or to address deficiencies in the nutrient solution. P and K are essential macronutrients required by plants for various physiological processes including energy transfer, photosynthesis, root development and fruit formation. Plants with robust fruiting stages such as tomatoes and peppers exhibit heightened requirements for phosphorus and potassium. While bloom nutrients typically encompass these essential elements, supplementing with PK additives becomes crucial to meet the elevated demands during intense fruit production. Growers commonly integrate PK additives into their nutrient regimen starting from the budding stage until flowering or fruits reaching full maturity. This strategic supplementation ensures optimal nutrient availability thus preventing deficiencies and supporting vigorous fruit development.

Here is how PK additives can benefit hydroponic cultivation:

i) **Promoting flowering and fruit formation**: Phosphorus and potassium play critical roles in flower initiation, pollination and fruit development in plants. Supplementing hydroponic nutrient solutions with PK additives during flowering and fruiting stages can enhance flower and fruit quality, increase fruit set and improve overall crop yield.

ii) **Enhancing root growth and nutrient uptake**: Phosphorus is essential for root development and elongation while potassium regulates water uptake, osmotic balance and nutrient transport within the plant. PK additives can stimulate root growth, increase root branching and improve nutrient uptake efficiency in hydroponically grown plants thus leading to healthier root systems and improved nutrient absorption.

iii) **Improving stress tolerance**: Potassium plays a crucial role in plant stress tolerance, helping plants withstanding environmental stressors such as drought, salinity and temperature extremes. Supplementing nutrient solutions with PK additives can enhance plant resilience to stress and minimize the negative impacts of adverse growing conditions on crop productivity and quality.

iv) **Balancing nutrient ratios**: Maintaining optimal nutrient ratios including phosphorus to potassium ratio is essential for maximizing plant growth and yield in hydroponic systems. PK additives can help adjust nutrient ratios in nutrient solution to meet specific requirements of different plant species, growth stages and environmental conditions, ensuring balanced nutrition and optimal plant performance.

v) **Addressing nutrient deficiencies**: Phosphorus and potassium deficiencies can occur in hydroponic systems due to insufficient nutrient levels in the nutrient solution or impaired nutrient uptake by plants. PK additives can be used to correct nutrient deficiencies thus providing plants with additional P and K they need for healthy growth and development.

PK additives are available commercially in various formulations including liquid concentrates, powders and granules designed to be easily dissolved and mixed into hydroponic nutrient solutions. When using PK additives in hydroponic cultivation, it is essential to follow manufacturer recommendations regarding application rates, timing and compatibility with other nutrients to ensure optimal results and to prevent nutrient imbalances. It is crucial for growers to choose PK additives that are pH neutral as alkaline additives can raise the pH of nutrient solution significantly. Failure to adjust the pH promptly after dosing can lead to destabilization of the nutrient solution, potentially affecting plant health and growth.

8.2.2 Silica Additives

Silica additives in the form orf silicon oxide (SiO_2) are commonly used in hydroponic systems to enhance plant growth and resilience. It is one of the most abundant elements on the earth's crust which is naturally present in many soils. However, it is often deficient in hydroponic systems where plants grow in a soilless medium. It improves structural integrity, increases resistance to stressors and thus enhances nutrient uptake.

8.2.2.1 Application Methods

Silica though beneficial to the plants at vegetative as well as flowering phases but is usually left out of hydroponic nutrient mixes due to its alkalinity and low solubility. Electron microscopy and X-ray analysis have demonstrated that existing silica within the plant is immobile and incapable of benefiting new growth. To ensure optimal growth, silica should always be present in nutrient solution which should be added separately in the reservoir. Some of the silica supplements are available in liquid form while others in powder or granule forms. Silica additives can be applied in hydroponic systems through various methods to ensure plants receiving adequate silicon supplementation. Here are some common application methods:

i) **Dissolved in nutrient solution**: Silica additives can be dissolved directly into the reservoir of the nutrient solution for uniform distribution and with each watering cycle. To avoid adverse effects due to over-application, it is essential to follow the manufacturer's recommended dosage.

ii) **Foliar spray**: Diluted silica solution can also be misted on foliage for direct and rapid absorption. This method is particularly useful for providing immediate support during periods of stress such as high temperatures or pest infestations. However, it should be noted that foliar spraying may not provide long-term silicon supplementation as effectively as dissolved silica in the nutrient solution.

iii) **Hydroponic substrate amendment**: Some growers choose to incorporate silica into the hydroponic substrate or growing medium itself. This can be achieved by mixing silica granules or powder into the substrate before planting or by using pre-fertilized media containing silica. By amending the substrate with silica the plants can access silicon as their roots grow and develop, providing continuous supplementation throughout the growth cycle.

iv) **Hydroponic supplements**: Commercial hydroponic supplements specifically formulated to enhance silicon uptake and utilization are available. These supplements often contain silicon in a readily available form, *i.e.* potassium silicate. These can be added to the nutrient solution according to the manufacturer's instructions.

v) **Continuous dosing systems**: Some advanced hydroponic setups utilize automated dosing systems to maintain optimal nutrient levels including silica. These systems can be programmed to deliver precise amounts of silica additive to the nutrient solution at regular intervals thus ensuring consistent silicon supplementation to the plants.

8.2.2.2 Sources

There are several sources of silica additives available for hydroponic use. Some common sources are:

i) **Potassium silicate**: This is a widely used source of soluble silica in hydroponics, providing both potassium and silica which help strengthening cell walls and enhancing plant resilience. Potassium silicate supplements are available in liquid form and can be easily added to the nutrient solution.

ii) **Silica powders and granules**: Silica powders and granules derived from natural sources such as diatomaceous earth or synthetic sources are also commonly used in hydroponics. These products contain finely ground silica particles that can be mixed into the growing medium or dissolved in the nutrient solution to provide plants with silicon supplementation.

iii) **Silica gel**: Silica gel products often marketed as silica supplements for plants are available in various forms such as granules or sachets. These products are designed to release silica slowly into growing medium or nutrient solution over time thus providing a continuous source of silicon for plant uptake.

iv) **Liquid silica supplements**: Some manufacturers offer liquid silica supplements specifically formulated for hydroponic use. These products typically contain highly soluble forms of silica that can be easily absorbed by plants. Liquid silica supplements are convenient to use and can directly be added to nutrient solution.

v) **Sources**: Silica supplements for hydroponics are available commercially under various brand names. These products may contain potassium silicate or other forms of soluble silica.

vi) **Organic options/natural sources**: For growers seeking organic hydroponic solutions, there are silica additives derived from natural sources such as seaweed extracts, diatomaceous earth or potassium silicate derived from natural minerals. While these sources may provide additional benefits beyond silica supplementation, their silica content may vary so it is essential to choose products with a guaranteed silica content.

8.2.2.3 Optimum ppm of Si in Solution

Utilizing liquid silicon denoted as Si (as opposed to SiO_2 which contains 46.743 per cent Si and 53.25 per cent O_2) within nutrient solution offers significant benefits to plants within the range of 25-100ppm. Typically, it is not integrated into nutrient concentrates at these levels requiring growers to add it separately. However, caution is advised when using potassium silicate products due to their alkalinity. Prior to incorporation it is advisable to pre-dilute any Si product in approximately 5-8 liters of water and adjust the pH to 5.8-6.0 before introducing it into the nutrient tank or reservoir.

For coco substrates, it is recommended to apply Si at the lower end of the scale, *i.e.* 25-35ppm. In inert media and water-based systems, Si can be utilized at higher rates, *i.e.* 50-75ppm.

8.2.2.4 Benefits

Here are some key points about silica additives in hydroponics:

i) **Plant benefits**: Silica plays a pivotal role in fortifying plant cell walls, enhancing cell structure and promoting plant rigidity. This fosters the growth of healthier and more robust plants which exhibit heightened resistance to various stressors including diseases, pests and environmental fluctuations. Silica enhances plant resilience to heat stress and wilting while promoting superior leaf orientation, increased photosynthesis and overall growth rates. Formation of a rigid and quartz-like matrix within plant cell walls upon silica absorption aids in efficient healing of pruning wounds particularly beneficial for crops like tomatoes and cucumbers prone to disease risks associated with regular pruning.

ii) **Chemical tolerance**: Silica contributes to ability of plants to tolerate harmful chemicals such as sodium and chloride.

iii) **Disease and pest resistance**: Silica augments innate defense mechanisms of plants and bolstering resistance to fungal diseases such as mildews and botrytis as well as pests such as aphids and mites. It accumulates around sites of fungal attack hindering ingress and potentially reducing reliance on chemical interventions for pest and disease management.

iv) **Increased yield**: By strengthening cell walls silica can help plants supporting heavier fruits and flowers thus leading to increased yields. This is particularly beneficial for crops like tomatoes, cucumbers and peppers.

v) **Improved nutrient uptake**: Silica enhances ability of plants to absorb and utilize nutrients particularly P and K. This can lead to overall improved nutrient uptake efficiency thus resulting in healthier plants.

vi) **pH buffering**: Silica can act as a pH buffer helping to stabilize pH of nutrient solution in hydroponic systems. This can be useful in systems where pH fluctuations are common.

8.2.2.5 Considerations

Though silica additives are generally compatible with most hydroponic nutrient solutions, it is essential to check compatibility and follow manufacturer recommendations to prevent any adverse reactions. Growers should also be aware that some commercial silica solutions may have limited shelf life as becomes evident when concentrated solution becomes cloudy. Regular monitoring and replacement may be necessary to maintain efficacy. When selecting a silica additive for hydroponic use, it is important to consider factors such as solubility, purity and ease of application. Dosage and frequency of silica application can vary depending on the specific product and the plants being grown. Hence, manufacturer recommendations must be followed for dosage and application to ensure optimal results without causing nutrient imbalances or adverse effects on plant health.

8.2.3 Calcium, Magnesium and Iron Additives

A blend of Ca, Mg and Fe additives is commonly employed to address or prevent deficiencies of these essential elements. Deficiency in any of these nutrients can lead to various symptoms including blossom-end-rot in fruits, upward leaf curling, poor flower *cum* fruit development, interveinal chlorosis and dieback of terminal buds *cum* root tips. High pH levels can induce insolubility of both calcium and iron so impeding their uptake by plants. Excessive humidity can hinder nutrient translocation within plant so contributing to calcium deficiency. It is advisable to avoid using this additive with excessively hard water as it may lead to accumulation of white scale in pipes causing potential blockages. Moreover, excessive use of this additive might reduce the availability of sulphate ~ a vital plant nutrient.

8.2.4 Organic Supplements

Organic supplements can be beneficial additions to hydroponic systems even though hydroponics is traditionally associated with inorganic nutrients. Use of organic additives such as kelp and fish products ~ both serving as nutrient stimulants as well as fertilizers. Here are some of the examples of organic supplements that can be used in hydroponic systems:

8.2.4.1 Seaweed Extracts

Seaweed extracts are rich in micronutrients, trace elements, polysaccharides, amino acids, proteins and growth-promoting plant hormones such as cytokinins, auxins and gibberellins that can promote plant growth, enhance nutrient uptake and improve stress tolerance. These are usually derived from various types of seaweed such as kelp and are available in liquid or powdered form. When applied to hydroponic systems, seaweed extracts provide organic matter and nutrients that support microbial growth and enzyme synthesis. These enzymes can contribute to nutrient cycling, root health and overall plant vigour.

8.2.4.2 Fish Emulsion

Fish emulsion is made from decomposed fish remains which is a good source of N, P and K as well as micronutrients. When used in hydroponic systems, fish emulsion provides organic matter and nutrients that stimulate microbial activity and enzyme production. These enzymes can help break down organic matter, release nutrients and improve root health.

8.2.4.3 Compost Tea

Compost tea is a liquid extract made from compost or vermicompost and is rich in organic matter, beneficial microorganisms and microbial metabolites. These teas in the soil can enhance microbial activity, improve nutrient cycling and suppress plant diseases when used as a supplement in hydroponic systems. When applied to hydroponic systems, compost tea provides a source of organic matter and microbial diversity so stimulating microbial activity and enzyme production. These enzymes can then help break down organic matter, release nutrients and promote health of overall system.

8.2.4.4 Molasses

Molass is a byproduct of sugar making which is rich in carbohydrates, amino acids, vitamins and minerals. Molasses when added to hydroponic systems serve as a food source for beneficial microorganisms thus stimulating microbial growth and enzyme synthesis. These enzymes can then contribute to organic matter decomposition, nutrient cycling and plant health promotion. It can serve as a food source for beneficial microorganisms in the hydroponic system, promoting microbial activity and improving nutrient availability for plants.

8.2.4.5 Alfalfa Meal

Alfalfa meal is made from dried alfalfa plants which contains proteins, carbohydrates, vitamins, nitrogen, phosphorus, potassium and various others. It can be used as a natural fertilizer and soil conditioner in hydroponic systems thus providing slow-release nutrients and promoting beneficial soil microorganisms. When incorporated into hydroponic systems, alfalfa meal provides organic matter and nutrients that support microbial growth and enzyme production.

8.2.4.6 Guano

Guano as well as bat dropping is a rich source of N, P and K and trace elements. It can be used as a natural fertilizer in hydroponic systems to promote plant growth and improve nutrient availability.

8.2.4.7 Humic and Fulvic Acids

Additives containing humic acid and fulvic acid are renowned for their ability to boost the cation exchange capacity of growing media so serving as effective chelators to facilitate nutrient uptake. Moreover, these play pivotal roles in fostering root development, enhancing soil water retention capabilities and fostering the proliferation of beneficial bacteria.

8.2.4.7.1 Sources

Humic and fulvic acids are naturally occurring organic compounds derived from the decomposition of organic matter such as plant material, peat, coal and humus-rich soils. These can be found in various natural sources and are often extracted or derived from these materials for use as supplements in agriculture including hydroponic systems. Here are some common sources of humic and fulvic acids:

i) **Leonardite**: It is a naturally occurring material resembling coal that is rich in humic substances such as humic and fulvic acids. It is formed through decomposition of ancient plant material and is commonly mined for extraction of humic substances. This decomposition process often contributes to colouration of surface water. Fulvic acid represents the soluble fraction of humic acid, remaining soluble across a wide pH spectrum. Fulvic acid is recognized for its superior efficacy compared to humic acid.

ii) **Peat**: Peat moss is another natural source of humic and fulvic acids. It forms over thousands of years as organic matter decomposes in waterlogged acidic environments such as bogs and wetlands. Peat moss is often used in gardening and horticulture and can be a source of humic substances.

iii) **Compost**: Compost is produced through decomposition of organic wastes such as food scraps, yard waste and manure. It contains humic and fulvic acids. These substances are formed during composting process as organic matter breaks down and undergoes microbial decomposition.

iv) **Lignite (brown coal)**: Lignite is a type of coal formed from compression of peat over millions of years. It contains humic and fulvic acids and can be used as a source of these compounds.

v) **Seaweed**: Some types of seaweed such as kelp and other marine algae contain humic and fulvic acids. These substances are formed through decomposition of organic matter in marine environments and can be extracted from seaweed for use in agriculture.

vi) **Soil**: Humic and fulvic acids are naturally present in soils rich in organic matter. These are formed through decomposition of plant and animal residues by soil microorganisms and play essential roles in soil fertility and plant nutrition.

These sources can be utilized to extract or derive humic and fulvic acids for use as supplements in hydroponic systems and other agricultural applications. Commercial products containing humic and fulvic acids are often derived from these natural sources and are available in liquid, powder or granular form for use in soilless growing systems.

8.2.4.7.2 Benefits in Hydroponic Systems

Here is how humic and fulvic acids can benefit hydroponic cultivation:

i) **Nutrient chelation**: Humic and fulvic acids have chelating properties as these can bind essential nutrients in the nutrient solution so making these more available to plants. By chelating nutrients such as iron, manganese, zinc and copper, humic and fulvic acids help prevent nutrient precipitation and improve nutrient uptake efficiency in hydroponic systems.

ii) **pH buffering**: Humic and fulvic acids can act as natural pH buffers so helping to stabilize pH of nutrient solution and maintaining it within optimal range for plant growth. This buffering capacity reduces risk of pH fluctuations and helps ensure consistent nutrient availability to plants.

iii) **Soil conditioning**: In soilless growing systems, humic and fulvic acids contribute to soil conditioning by improving soil structure, water retention and aeration. These organic acids help create a favourable environment for root growth and microbial activity so leading to healthier root systems and improved nutrient absorption by plants.

iv) **Stimulating microbial activity**: Humic and fulvic acids serve as food sources for beneficial microorganisms in the rhizosphere so promoting microbial growth and activity. Increased microbial activity enhances nutrient cycling, breaking down organic matter and suppressing harmful pathogens therefore contributing to overall plant health and disease resistance in hydroponic systems.

v) **Plant growth promotion**: Humic and fulvic acids contain bioactive compounds that stimulate plant growth, root development and stress tolerance. These organic acids act as natural biostimulants which enhance plant vigour, resilience to environmental stressors and overall crop yield in hydroponic cultivation.

Humic and fulvic acids in hydroponic systems are typically added to the nutrient solution at recommended concentrations or applied as soil drenches in media-based systems. These are available commercially as concentrated liquid formulations or powdered extracts derived from natural sources. When used appropriately, humic and fulvic acids can significantly enhance nutrient availability, soil health and plant growth in hydroponic cultures.

It is important to exercise caution when introducing organic matter in water-based systems like NFT and aeroponics as excessive organic content can lead to unwanted microbial proliferation thus depriving the root zone of oxygen potentially. When using organic supplements in hydroponic systems, it is essential to ensure that these are compatible with specific system and nutrient solution being used. Organic supplements may introduce organic matter into the system so regular monitoring of nutrient levels, pH and water quality is necessary to prevent imbalances and ensuring optimal plant growth.

8.3 PGR Additives (Plant Hormones)

While both nutrient additives and PGRs can contribute to plant growth and development in hydroponic systems though these serve different purposes. Nutrient additives primarily provide essential nutrients and support overall plant health while PGRs regulate plant physiological processes directly to achieve specific growth objectives or overcome growth limitations.

PGRs are compounds that affect root growth, flowering, stem elongation or shortening. PGRs are usually applied in precise concentrations to manipulate specific growth processes rather than to provide essential nutrients. These can either be synthetic compounds that mimic the role of natural hormones or can be hormones extracted from plant tissues. These are applied either as a foliar spray or a liquid drench to the root zone.

There are 5 groups of PGR's:

i) Auxins

These produce several growth responses in plants including the formation of adventitious roots, promotion of apical dominance, growth, flower formation and fruit-set. Common examples of auxins include:

a) **Indole-3-butyric acid (IBA)**: IBA is a synthetic auxin hormone that promotes root initiation and development. It is the most commonly used rooting hormone for hydroponic propagation. IBA is available in various formulations such as powders, gels and solutions for dipping or applying to the cut ends of plant cuttings.

b) **Naphthaleneacetic acid (NAA)**: NAA is another synthetic auxin hormone used to stimulate root formation in plant cuttings. It works similarly to IBA but may have different effectiveness depending on the plant species and cutting type. NAA is available in liquid form for application to cuttings or as part of rooting hormone formulations.

c) **Indole-3-acetic acid (IAA)**: IAA is a natural auxin hormone produced by plants which is involved in various growth and developmental processes including root initiation. While less commonly used in rooting hormone formulations, IAA can still be effective for stimulating root growth in some plant species.

d) **Auxin-based formulations**: Rooting hormone products often contain a combination of synthetic or natural auxins along with other ingredients such as vitamins, nutrients and fungicides to enhance root development and protect cuttings from diseases. These formulations may be available as liquids, gels, powders or concentrates for different application methods.

When using rooting hormones in hydroponic propagation, it is essential to follow manufacturer instructions regarding application rates, methods and timing to ensure optimal results. Rooting hormone treatments are typically applied to the cut ends of cuttings before insertion into the rooting medium. Proper environmental conditions such as humidity, temperature and lighting should also be maintained to support root development and minimize stress on the cuttings. With right rooting hormone treatment and care, hydroponic growers can achieve high success rates in propagating plants from cuttings and establishing healthy root systems for future growth.

ii) Gibberellins

Gibberellins are a group of plant hormones known to regulate various aspects of plant growth and development including stem elongation, leaf expansion, seed germination and flowering. While gibberellins are not commonly used directly in hydroponic systems but their effects on plant growth can be indirectly influenced through certain practices and environmental conditions in hydroponics. Growers may adjust nutrient formulations, light regimes and other parameters to optimize plant growth and flowering so indirectly affecting gibberellin levels and responses in hydroponically grown plants. Gibberellins break seed dormancy and induce flowering. Gibberellic acid can be used to influence the timing of flowering, flower gender and flower size. Here is how gibberellins may play a role in hydroponic cultivation:

a) **Seed germination**: Gibberellins are often used to promote seed germination in traditional agriculture. While hydroponic systems typically use vegetative propagation methods such as cuttings or tissue culture rather than seeds, gibberellins may still indirectly influence seedling growth and development if seeds are germinated before transplanting into hydroponic systems.

b) **Stem elongation**: Controlling stem elongation is important in certain hydroponic crops for managing plant height and optimizing space utilization. While gibberellins can promote stem elongation in some plants, their application in hydroponic systems may not be common due to the risk of excessive elongation and lodging. However, specific cultivars or growth stages of certain crops may benefit from gibberellin treatments to encourage controlled elongation.

c) **Flowering induction**: Gibberellins play a role in flowering induction and flower development in many plant species. In hydroponic cultivation, manipulating photoperiod, light intensity, temperature and other environmental factors can influence flowering and fruiting processes. While gibberellins themselves may not be directly applied in hydroponics growers may employ strategies to optimize flowering through environmental manipulation which could indirectly affect gibberellin levels and signalling pathways in plants.

d) **Tissue culture**: Gibberellins are sometimes used in tissue culture techniques to induce shoot elongation and proliferation *in vitro*. While tissue culture is not a primary method of plant propagation in most hydroponic systems, it may be used for certain specialized applications or to establish disease-free planting material.

iii) Cytokinins

Cytokinins are a class of plant hormones known to regulate various aspects of plant growth and development. These can be useful for stimulating cell division, inducing shoots and delaying aging *cum* death (senescence). While cytokinins are not commonly used directly in hydroponic systems, their effects on plant growth and development can be indirectly influenced through certain practices and environmental conditions. **Benzyl-amino-purine (6-BA)** is a common cytokinin used for inducing side branching. This is useful for development of short and bushy plants. Here is how cytokinins play a role in hydroponic cultivation:

a) **Shoot and root growth**: Cytokinins promote cell division and shoot growth in plants. In hydroponic systems, managing nutrient levels particularly nitrogen can influence cytokinin production and distribution within the plant. Providing balanced nutrient solutions with adequate nitrogen content can support healthy shoot growth and development in hydroponically grown plants.

b) **Root development**: Cytokinins also play a role in regulating root growth and development. In hydroponics, maintaining optimal oxygen levels in the root zone, proper nutrient availability and appropriate root-zone temperatures can indirectly influence cytokinin signalling and root development. Well-aerated nutrient solutions and adequate oxygenation of the root zone are crucial for promoting healthy root growth in hydroponically grown plants.

c) **Tissue culture**: Cytokinins are commonly used in tissue culture techniques to induce shoot proliferation and to promote development of shoots from explants. While tissue culture may not be a primary method of plant propagation in most hydroponic systems, it may be used for certain specialized applications or to establish disease-free planting material.

d) **Flowering and fruit development**: Cytokinins interact with other hormones such as auxins and gibberellins to regulate flowering and fruit development in plants. In hydroponics, managing environmental factors such as photoperiod, light intensity and temperature can indirectly influence cytokinin signalling pathways and flowering responses in plants. Optimal environmental conditions can promote flowering and fruiting in hydroponically grown crops.

iv) Ethylene

Ethylene is a gaseous plant hormone that plays a significant role in regulating various physiological processes including fruit ripening, leaf abscission, flower senescence and response to stress. Ethylene is commonly used to ripen fruits in preparation for sale. While ethylene is not commonly used directly in hydroponic systems, its effects on plant growth and development can still be important considerations for hydroponic growers. Here is how ethylene may play a role in hydroponic cultivation:

a) **Fruit ripening**: Ethylene is known to stimulate fruit ripening in many plant species. In hydroponic systems where fruit-bearing crops are grown, managing ethylene levels is important to control the timing and uniformity of fruit ripening. Fruits such as tomatoes and bananas are particularly sensitive to ethylene and may require special attention to prevent premature ripening during cultivation.

b) **Flower senescence**: Ethylene also promotes flower senescence leading to wilting and eventual death of flowers. In hydroponic systems where flowering crops are grown, controlling ethylene levels can help prolong flower lifespan and maintaining flower quality. This may involve optimizing environmental conditions such as temperature, humidity and air circulation to minimize ethylene accumulation around flowers.

c) **Stress response**: Plants produce ethylene in response to various environmental stresses such as drought, flooding and infection of pathogens. In hydroponic systems where plants may be exposed to suboptimal growing conditions or disease pressure, monitoring ethylene levels can provide insights into plant stress responses and help growers identify and address potential issues affecting plant health and productivity.

d) **Growth regulation**: Ethylene can also influence plant growth and development, including stem elongation, leaf expansion and root growth. In hydroponic systems, managing ethylene levels may indirectly affect these growth processes and contribute to overall plant vigour and productivity. However, excessive ethylene exposure can lead to abnormal growth patterns and reduced yields in hydroponically grown crops.

v) Abscisic Acid (ABA)

This is a plant growth inhibitor. It causes abscission of leaves, fruits and flowers and causes stomata to close. During periods of drought when stomata are closed, high concentrations of ABA are found in leaf guard cells. While ABA is not commonly used directly in hydroponic systems, its effects on plant growth and development can be indirectly influenced through certain practices and environmental conditions. Here is how ABA may play a role in hydroponic cultivation:

a) **Drought tolerance**: ABA is known as the 'stress hormone' as here plants are produced in response to water stress. In hydroponic systems where water availability can be precisely controlled, managing ABA levels may indirectly affect plant responses to drought stress. Optimizing irrigation practices by maintaining proper moisture levels in the root zone and avoiding overwatering can help minimize ABA accumulation and improve drought tolerance in hydroponically grown plants.

b) **Stress responses**: In addition to drought stress, plants produce ABA in response to other environmental stresses such as high salinity, extreme temperatures and pathogenic infection. In hydroponic systems where plants are exposed to suboptimal growing conditions or disease pressure, managing ABA levels can help modulate stress responses and improve plant health and resilience. Providing optimal nutrient solutions, maintaining proper pH and EC levels and implementing disease prevention measures can help minimize stress-induced ABA production in hydroponically grown crops.

c) **Seed germination**: ABA also plays a role in regulating seed dormancy and germination. While hydroponic systems typically use vegetative propagation methods rather than seeds, managing ABA levels may indirectly influence seedling growth and development if seeds are germinated before transplanting into hydroponic systems. Providing optimal germination conditions such as proper moisture and temperature levels can help promote uniform and vigorous seedling establishment in hydroponics.

d) **Flowering and fruit development**: ABA interacts with other hormones such as gibberellins to regulate flowering and fruit development in plants. In hydroponic cultivation, manipulating environmental factors such as photoperiod, light intensity and temperature can indirectly influence ABA signalling pathways and flowering responses. Optimizing these environmental conditions can promote flowering and fruiting in hydroponically grown crops.

8.4 Polysaccharides

Polysaccharides play several important roles in hydroponics contributing to plant health, root development, nutrient retention and overall system resilience. Such polysaccharides (traditional PGRs) may not regulate plant growth directly, *viz.* auxins, gibberellins, cytokinins and abscisic acid but these can play important roles in creating favourable conditions for plant growth and development. As such these are often used as supplements or components of biostimulants in agriculture and horticulture to support overall plant health and productivity. Here is how polysaccharides can be utilized in hydroponic systems:

i) **Root health and development**: Polysaccharides serve as a source of energy for beneficial microorganisms in the rhizosphere ~ promoting their growth and activity. In turn, these microorganisms help enhance root health and development by improving nutrient availability, producing growth-promoting substances and suppressing soil-borne pathogens. Polysaccharides can stimulate root growth directly by providing a substrate for root elongation and branching.

ii) **Water retention and soil structure**: Polysaccharides have water-absorbing properties that can help retain moisture in the root zone so reducing the risk of water stress and promoting steady plant growth. Furthermore, polysaccharides produced by soil microorganisms contribute to soil aggregation and structure, improving water infiltration, aeration and drainage in hydroponic systems.

iii) **Nutrient cycling and availability**: Certain polysaccharides such as those derived from microbial sources or organic supplements serve as substrates for beneficial microorganisms involved in nutrient cycling. These microorganisms break down complex organic compounds into simpler forms so releasing nutrients that are readily available for plant uptake. Polysaccharides can also chelate or bind nutrients making them more accessible to plants and reducing nutrient leaching in hydroponic systems.

iv) **Biostimulant properties**: Polysaccharides derived from natural sources such as seaweed extracts or microbial polysaccharides may exhibit biostimulant properties that enhance plant growth, stress tolerance and overall health in the system. These polysaccharides contain growth-promoting substances, vitamins and other bioactive compounds that stimulate various physiological processes in plants leading to improved yield and quality.

v) **Microbial activity and disease suppression**: Polysaccharides support the growth and activity of beneficial microorganisms in hydroponic systems which can help suppressing soil-borne pathogens and promoting disease resistance in plants. Certain polysaccharides act as prebiotics ~ providing a food source for beneficial microbes and creating a competitive environment that inhibits pathogen proliferation.

In hydroponic systems, polysaccharides may be introduced through organic supplements, microbial inoculants or natural sources such as compost teas or seaweed extracts. By incorporating polysaccharides into hydroponic cultivation practices, growers can enhance root health, nutrient cycling and overall system resilience ~ leading to healthier plants and improved crop yields.

8.5 Enzymes

Enzymes are biological catalysts that facilitate chemical reactions including nutrient breakdown and organic matter decomposition. These aid in nutrient availability and plant health indirectly though as such do not provide any nutrient, this is why the enzymes are not classified as nutrient additives. Here is how enzymes are used in hydroponic systems:

i) **Nutrient availability**: Enzymes help break down complex organic compounds into simpler forms that are readily available for plant uptake. Enzymes such as phosphatases can convert organic phosphorus into inorganic forms that plants can absorb so enhancing phosphorus availability in the nutrient solution.

ii) **Organic matter decomposition**: In hydroponic systems where organic amendments or supplements are used enzymes aid in the decomposition of organic matter. Enzymes produced by microorganisms break down organic compounds into soluble nutrients so releasing nitrogen, phosphorus, potassium and other essential elements for plant uptake.

iii) **Root health**: Enzymes can promote root health by facilitating root growth, improving nutrient uptake efficiency and enhancing root system development. Certain enzymes such as cellulases and hemicellulases help in the break down of cell walls allowing roots to penetrate the growing medium more easily to access nutrients.

iv) **Disease prevention**: Enzymes produced by beneficial microorganisms in hydroponic systems can help suppress soil-borne pathogens and reduce the risk of root diseases. Enzymes with antimicrobial properties can inhibit growth of harmful bacteria and fungi so protecting plant roots from infection.

v) **Oxygenation**: Enzymes involved in aerobic respiration such as oxidases and catalases help oxygenating nutrient solution by facilitating the breakdown of oxygen-containing compounds. Proper oxygenation is crucial for root health and overall plant vigour in hydroponic systems.

vi) **Detoxification**: Peroxidases and cytochrome P450 enzymes can help detoxify harmful compounds that may accumulate in the nutrient solution or growing medium. These enzymes break down toxins and convert them into less harmful forms so reducing stress on plants.

vii) **pH regulation**: Certain enzymes such as urease and phosphatase can influence the pH of the nutrient solution by releasing acidic or alkaline byproducts during enzymatic reactions. This can help buffer changes in pH and maintain a stable growing environment for plants.

Enzymes can be introduced into hydroponic systems through various means including microbial inoculants, organic supplements and enzymatic products specifically designed for hydroponic cultivation. Regular monitoring of nutrient levels, pH and overall plant health is essential to optimize effectiveness of enzymes in hydroponic systems and ensuring healthy plant growth.

Enzymatic products in hydroponics are specialized formulations containing enzymes designed to enhancing nutrient availability, improving root health, promoting plant growth and optimizing overall system performance. These products are often derived from natural sources or produced through biotechnological processes and are specifically formulated for use in hydroponic systems. Here are some common types of enzymatic products used in hydroponics:

i) **Enzyme complexes**: These products contain a blend of enzymes targeting various functions such as nutrient breakdown, organic matter decomposition and root health promotion. These among others may include enzymes such as cellulases, hemicellulases, proteases, lipases, phosphatases and amylases.

ii) **Root health enhancers**: Enzymatic products formulated to support root health typically contain enzymes that facilitate root growth, improve nutrient uptake efficiency and enhance root system development. These products may include enzymes such as chitinases, glucanases and pectinases which break down root cell wall components and promote root branching.

iii) **Organic matter decomposition**: Enzymatic products designed to break down organic matter in hydroponic systems contain enzymes that facilitate decomposition of organic compounds releasing nutrients for plant uptake. These products may include cellulase, ligninase and xylanase enzymes which break down complex carbohydrates, lignin and hemicellulose found in organic matter.

iv) **pH regulation**: Some enzymatic products contain enzymes that influence pH of nutrient solution by catalyzing reactions that release acidic or alkaline byproducts. These enzymes help buffer changes in pH and maintain a stable growing environment for plants such as urease and phosphatase enzymes affecting pH through enzymatic reactions.

v) **Biostimulants**: Enzymatic products may also function as biostimulants so promoting plant growth, stress tolerance and overall system resilience. These products may contain enzymes that stimulate plant metabolism, enhance nutrient uptake and improve plant vigour. Enzymatic biostimulants can include a wide range of enzymes depending on their specific mode of action and intended benefits.

When using enzymatic products in hydroponic systems it is essential to follow manufacturer recommendations regarding application rates, frequency and compatibility with other inputs. Regular monitoring of nutrient levels, pH and plant health is also crucial to optimize effectiveness of enzymatic products so ensuring healthy plant growth in hydroponic cultures.

8.6 Hydrogen Peroxide (H_2O_2) ~ Multifaceted Additive

Hydrogen peroxide (H_2O_2) is a chemical compound consisting of two hydrogen atoms and two oxygen atoms. It is a clear colourless liquid with a slightly acidic taste which gives a characteristic 'fizzing' reaction when exposed to certain substances. It is a versatile compound that serves various purposes such as disinfectant, oxidizer and bleaching agent. When it breaks down a single atom of oxygen is released along with a single water molecule. It leaves no residue or waste behind. The single atom of oxygen is what makes it so useful.

Hydrogen peroxide is relatively unstable and can decompose over time particularly in presence of light, heat or certain catalysts. Its half-life in water can range from several hours to a few days depending on factors such as concentration and environmental conditions.

8.6.1 Benefits

Hydrogen peroxide serves as a versatile and indispensable component of hydroponic cultivation, contributing to plant health as well as system maintenance. Its multifaceted benefits make it a valuable asset for hydroponic growers seeking to create optimal growing conditions and maximizing plant yield. Here is some information about hydrogen peroxide and its use in hydroponics:

i) **Disinfectant properties**: Hydrogen peroxide is a powerful oxidizing agent that is effective at killing a wide range of microorganisms including bacteria, viruses, fungi and algae. It acts as a powerful sterilizing agent effectively disinfecting water sources, reservoirs and irrigation systems. This capability is invaluable for controlling microbial growth, preventing root rot and ensuring a clean and sterile environment conducive to healthy plant growth. It works by releasing oxygen radicals that react with and neutralize microbial cells ~ thereby disinfecting water and surfaces. It is important to note that hydrogen peroxide acts only as root disease preventatives though once root disease is present in the crop this is ineffective so in this case other means for controlling the disease should be sorted out.

ii) **Oxidation potential**: Hydrogen peroxide having a high oxidation potential make it an effective disinfectant and oxidizer. It can break down organic matter and remove contaminants so helping to improve water quality and preventing the buildup of harmful pathogens in hydroponic systems. It plays crucial role in oxygenating nutrient solutions thus facilitating the delivery of vital oxygen to plant roots. This oxygenation enhances root health and promotes robust growth.

iii) **Chlorine neutralizer**: Hydrogen peroxide can serve as a valuable tool for neutralizing chlorine commonly found in tap water which can be harmful to plants in hydroponic systems. It can neutralize chlorine by oxidizing it into chloride ions which are less harmful to plants. A systematic approach is necessary to utilize hydrogen peroxide for chlorine removal as given below:

- ❖ Chlorine level in the water should be measured using a chlorine test kit.
- ❖ Appropriate amount of hydrogen peroxide needed based on the chlorine concentration in the water and the volume of water in your hydroponic system should be calculated.
- ❖ Calculated amount of hydrogen peroxide to water reservoir or nutrient solution should be added.
- ❖ Water should be mixed thoroughly to ensure that hydrogen peroxide is evenly distributed.
- ❖ Water should be allowed to sit for a short period to ensure complete chlorine neutralization.
- ❖ Water should again be tested with a chlorine test kit to confirm that the chlorine level has been reduced to a safe level for plants.

8.6.2 Dosage

Hydrogen peroxide is added to water or nutrient solutions at specific concentrations to achieve desired level of disinfection. The dosage and application method may vary depending on factors such as water quality, system size and the level of microbial contamination. It is essential to follow recommended guidelines and avoid overdosing to prevent potential harm to plants.

It is crucial to exercise caution when using hydrogen peroxide in hydroponic systems as excessive application can pose risks to plant roots and disrupt the balance of the system. Therefore, it is essential to use hydrogen peroxide judiciously adhering to recommended dosage guidelines and closely monitoring its effects on water quality and plant health throughout the process. If a burning sign or leaf yellowing is observed in the plants the strength of H_2O_2 used should be reduced.

Further growers must be vigilant in selecting the appropriate type of hydrogen peroxide for hydroponic use. Specifically, only food-grade hydrogen peroxide intended for horticultural or hydroponic applications should be employed. This ensures purity and safety of hydrogen peroxide solution, *vis-a-vis* minimizing the risk of adverse effects on plants and the hydroponic environment.

By exercising diligence and employing high-quality hydrogen peroxide products designed for hydroponic use growers can effectively harness the benefits of this versatile solution while safeguarding the health and vitality of their plants.

8.6.3 Safety Considerations

Hydrogen peroxide is considered generally safe when used properly though at high concentrations it can be corrosive and may cause irritation or damage to skin, eyes and mucous membranes. Direct contact with H_2O_2 should be avoided so when using this appropriate protective measures should be adopted. Hydrogen peroxide should be mixed in a separate container. Hydrogen peroxide being an oxidant is not suitable for use where organic media (coco substrate, *etc.*) or organic additives are used as this can break down organic matters there.

8.7 Nutrient Disinfectants (Sterilizing Agents)

In hydroponic systems, plants lack protective root cover found in soil-grown plants so exposing them to potential fungal pathogens such as *Pythium*, *Phytophthora* and *Fusarium*. These pathogens can quickly spread in presence of excess water so necessitating regular disinfection of nutrient solution and growing medium as a preventive measure. Various nutrient disinfectants or sterilizing agents are employed in hydroponics to combat these pathogens effectively.

8.7.1 Half-life Duration and Oxidizing Potential

Among most commonly used disinfectants in hydroponic systems are oxidizing agents (chlorine dioxide, sodium hypochlorite, monochloramine, hydrogen peroxide, UV light and ozone) which do not permeate plant tissues, ensuring that disinfection treatments can safely be continued throughout the fruiting period without posing harm to consumers. Their suitability for hydroponics largely depends on factors such as their half-life duration and oxidation potential.

Half-life of a disinfectant refers to the time it takes for half of the chemical to degrade or become inactive. Longer half-lives ensure that disinfectant remains chemically active in the nutrient solution for an extended period, effectively treating the entire system with each nutrient cycle. Monochloramine boasts the longest half-life among oxidizing agents whereas chlorine dioxide and sodium hypochlorite have relatively shorter half-lives. On the other hand, ozone and UV light have short half-lives and disinfect the nutrient solution only at the point of application thus rendering these less effective for treating entire system.

Oxidation potential measures an oxidizing agent's ability to eliminate fungi, algae and other organic organisms. Sufficient oxidation potential is crucial for eradicating food sources that attract pests such as fungus gnats. While monochloramine, chlorine dioxide and sodium hypochlorite possess adequate oxidation potential, only monochloramine is suitable for use in hydroponics. Chlorine dioxide and sodium hypochlorite can potentially harm roots, organic media and growth promoters.

By understanding the characteristics of different disinfectants and their suitability for hydroponic systems, growers can effectively manage disease risks and maintain a clean and healthy environment for their plants, ultimately maximizing yields and quality.

8.7.2 Examples of some Nutrient Disinfectants used in Hydroponics

Apart from H_2O_2 here are some common nutrient disinfectants used in hydroponics along with their half-life duration and oxidation potential:

8.7.2.1 Monochloramine

Monochloramine is a chemical compound commonly used as a disinfectant in water treatment, *vis-à-vis* in hydroponic systems to control microbial growth and to ensure water quality. Here is some information about monochloramine including its properties and use in hydroponics:

i) **Chemical composition (NH_2Cl)**: Monochloramine is formed by reaction of ammonia (NH_3) with chlorine (Cl_2) or sodium hypochlorite (NaOCl) in water.

ii) **Disinfectant properties**: It is an effective disinfectant with broad-spectrum antimicrobial activity. It is capable of killing bacteria, viruses, algae and other microorganisms present in water thus helping to maintain a clean and sterile environment in hydroponic systems.

iii) **Stability and half-life**: Relatively it is stable compared to free chlorine (hypochlorous acid and hypochlorite ions) and has a longer-lasting residual effect. Its half-life in water can range from several days to weeks depending on factors such as pH, temperature and the presence of organic matter.

iv) **Oxidation potential**: Monochloramine is a weaker oxidizing agent compared to free chlorine. While it is effective at disinfecting water and controlling microbial growth but it typically has a lower oxidation potential and is less reactive with organic and inorganic compounds in water.

v) **Use in hydroponics**: In hydroponic systems, it can be used as a disinfectant to treat water sources, reservoirs and irrigation systems. It helps preventing growth of harmful pathogens such as bacteria, fungi and algae thereby reducing the risk of disease outbreaks and ensuring the health and productivity of plants.

vi) **Application and dosage**: Monochloramine is typically added to water at specific concentrations to achieve desired level of disinfection. Dosage and application method may vary depending on factors such as water quality, system size and level of microbial contamination.

vii) **Safety considerations**: While monochloramine is effective as a disinfectant, it is essential to use it properly and follow recommended guidelines to avoid potential health risks and adverse effects. Care should be taken to ensure that residual monochloramine levels in water are within safe limits for plants and human consumption.

8.7.2.2 Calcium Hypochlorite

Calcium hypochlorite is a chemical compound commonly used as a disinfectant and sanitizer in various applications including water treatment and sterilization in hydroponic systems. Here is some information about calcium hypochlorite and its use in hydroponics:

i) **Chemical composition**: Calcium hypochlorite $Ca(ClO)_2$ is a solid compound composed of calcium ions (Ca2+) and hypochlorite ions (ClO-). It is available typically in granular or tablet form and is highly soluble in water.

ii) **Disinfectant properties**: Calcium hypochlorite is a potent disinfectant with strong oxidizing properties. It releases chlorine gas when dissolved in water and hypochlorous acid which are highly effective at killing bacteria, viruses, fungi and algae. It works by oxidizing and destroying microbial cells, disrupting their cellular structures and metabolic processes.

iii) **Stability and half-life**: Calcium hypochlorite has a relatively long shelf life when stored properly in a cool and dry place away from moisture and sunlight. Calcium hypochlorite once dissolved in water releases chlorine which has a relatively short half-life and can dissipate quickly, especially when exposed to light, air, and organic matter.

iv) **Oxidation potential**: Calcium hypochlorite has a high oxidation potential so making it a potent disinfectant and oxidizing agent. It is effective at breaking down organic contaminants, removing stains and controlling microbial growth in water.

v) **Use in hydroponics**: Calcium hypochlorite in hydroponic systems is used to sterilize water sources, reservoirs, irrigation systems and equipment. It helps controlling microbial contamination, preventing spread of diseases and maintaining water quality for plant health. Calcium hypochlorite can be added to water at specific concentrations to achieve desired level of disinfection. It is essential to follow recommended guidelines for dosing and application to avoid potential harm to plants and aquatic life. Residual chlorine levels should be monitored regularly to ensure they are within safe limits for plants. Excessive chlorine concentrations can be harmful to plants and may cause leaf burn or stunted growth.

8.7.2.3 Sodium Hypochlorite

Sodium hypochlorite is a chemical compound commonly used as a disinfectant and sanitizer in various applications including water treatment and sterilization in hydroponic systems. Here is some information about sodium hypochlorite and its use in hydroponics:

i) **Chemical composition**: Sodium hypochlorite (NaClO) is a liquid compound composed of sodium ions (Na+) and hypochlorite ions (ClO-). It is typically available as a pale greenish-yellow solution with a characteristic chlorine odour.

ii) **Disinfectant properties**: Sodium hypochlorite is a potent disinfectant with strong oxidizing properties. It is effective at killing bacteria, viruses, fungi and algae commonly found in water and hydroponic systems. When dissolved in water it releases chlorine gas and hypochlorous acid which work together to oxidize and destroy microbial cells and to disrupt their cellular structures and metabolic processes.

iii) **Stability and half-life**: Sodium hypochlorite solutions have a relatively short shelf life and degrading over time especially when exposed to light, heat and air. The half-life of sodium hypochlorite in water can vary depending on factors such as concentration, pH, temperature and presence of organic matter. Generally, higher concentrations and warmer temperatures can accelerate degradation.

iv) **Oxidation potential**: Sodium hypochlorite has a high oxidation potential making it a potent disinfectant and oxidizing agent. It is effective at breaking down organic contaminants, removing stains and controlling microbial growth in water and hydroponic systems.

v) **Use in hydroponics**: In hydroponic systems, sodium hypochlorite is used to sterilize water sources, reservoirs, irrigation systems and equipment. It helps controlling microbial contamination, preventing spread of diseases and maintaining water quality for plant health. Sodium hypochlorite can be added to water at specific concentrations to achieve the desired level of disinfection. It is essential to follow recommended guidelines for dosing and application to avoid potential harm to plants and aquatic life. Residual chlorine levels should be monitored regularly to ensure they are within safe limits for plants. Excessive chlorine concentrations can be harmful to plants and may cause leaf burn or stunted growth.

8.7.2.4 Ozone (O_3)

Ozone is a naturally occurring gas composed of three oxygen atoms. It is a powerful oxidizing agent and disinfectant that is commonly used in various applications including water treatment and sterilization in hydroponic systems. Here is some information about ozone and its use in hydroponics:

i) **Chemical composition**: Ozone is a molecule composed of three oxygen atoms bonded together. It is formed when oxygen molecules (O_2) are exposed to high-energy sources such as UV light or electrical discharges which cause them to split apart and recombine as ozone.

ii) **Disinfectant properties**: Ozone is a potent oxidizing agent with strong disinfectant properties. It is highly effective at killing bacteria, viruses, fungi and algae by oxidizing and destroying their cellular structures and metabolic processes. Unlike chlorine-based disinfectants, ozone does not leave behind harmful disinfection byproducts and does not contribute to the formation of chloramines or trihalomethanes.

iii) **Stability and half-life**: Ozone is relatively unstable and has a short half-life in water and air. It readily decomposes back into oxygen molecules (O_2) through a process called ozone decomposition. The half-life of ozone depends on factors such as temperature, pH and the presence of organic matter. Generally, warmer temperatures and higher pH levels can accelerate ozone decomposition.

iv) **Oxidation potential**: Ozone has a high oxidation potential so making it a powerful oxidizing agent capable of breaking down organic contaminants, removing stains and controlling microbial growth in water and hydroponic systems. It works by releasing highly reactive oxygen molecules that react and neutralize microbial cells so effectively sterilizing the water and surfaces in hydroponic systems.

v) **Use in hydroponics**: Ozone in hydroponic systems is used to sterilize water sources, reservoirs, irrigation systems and equipment. It helps controlling microbial contamination, preventing spread of diseases and maintaining water quality for plant health. Ozone can be generated on-site using ozone generators or ozone injection systems. It is typically injected into water as a gas or dissolved in water as ozonated water. Proper dosing and monitoring are essential when using ozone in hydroponics to ensure that residual ozone levels are within safe limits for plants and aquatic life. Ozone can also be used to oxygenate nutrient solutions and provide supplemental oxygen to plant roots which can improve nutrient uptake and overall plant health.

8.7.2.5 Quaternary Ammonium Compounds (Quats)

Quats are a group of chemicals commonly used as disinfectants and sanitizers in various applications including in hydroponic systems. Here is some information about Quats and their use in hydroponics:

i) **Chemical composition**: Quaternary ammonium compounds are organic salts derived from ammonia (NH_3) in which four organic groups are attached to a central nitrogen atom. The most common type of Quats used as disinfectants is benzalkonium chloride (BAC).

ii) **Disinfectant properties**: Quats are effective disinfectants with broad-spectrum antimicrobial activity. These work by disrupting the cell membranes of microorganisms leading to cell leakage and eventual death. Quats are particularly effective against bacteria, fungi and some viruses commonly found in water and hydroponic systems.

iii) **Stability and half-life**: Quats are relatively stable compounds with long-lasting effects compared to other disinfectants. They can persist on surfaces for extended periods providing ongoing antimicrobial protection. However, the effectiveness of Quats can be reduced in the presence of organic matter or hard water minerals so requiring higher concentrations or longer contact times for adequate disinfection.

iv) **Oxidation potential**: Quats are not oxidizing agents like chlorine or ozone but instead disrupt microbial membranes and inhibit enzymatic processes. These are effective against a broad spectrum of microorganisms including bacteria, fungi and some viruses.

v) **Use in hydroponics**: Quats in hydroponic systems are used to sterilize water sources, reservoirs, irrigation systems and equipment. These help control microbial contamination, preventing spread of diseases and maintaining water quality for plant health. Quats are available in various formulations including liquid concentrates and wipes so making these easy to apply in hydroponic settings. It is essential to follow recommended guidelines for dosing and application when using Quats in hydroponics to ensure effective disinfection while minimizing the risk of phytotoxicity or other adverse effects on plants. Residual Quat levels should be monitored regularly to ensure these are within safe limits for plants and aquatic life. Excessive concentrations can be harmful to plants and may cause damage to roots or inhibit growth.

8.7.2.6 UV (Ultraviolet) Light

Ultraviolet light is a form of electromagnetic radiation with wavelengths shorter than those of visible light but longer than X-rays. It is categorized into three main types based on wavelength: UV-A, UV-B and UV-C. Here is some information about UV light and its use in hydroponics:

i) **UV spectrum**

- UV-A (315-400nm): This is the longest wavelength of UV light which is closest to visible light. UV-A light is less energetic and is often used for inducing photomorphogenesis in plants so influencing processes like flowering and pigmentation.
- UV-B (280-315nm): UV-B light has intermediate wavelengths and is more energetic than UV-A. It plays a role in plant stress responses including defense mechanisms against pathogens and production of protective compounds.
- UV-C (100-280nm): UV-C light has the shortest wavelengths and is the most energetic. It is highly effective at killing microorganisms including bacteria, viruses and fungi by damaging their DNA and disrupting cellular functions.

ii) **Disinfection and sterilization**: UV-C light is commonly used for disinfection and sterilization purposes in hydroponic systems. It is effective at killing or inactivating harmful microorganisms thereby reducing the risk of disease outbreaks and maintaining water quality. UV-C light can be used to treat water sources, reservoirs, irrigation systems and equipment in hydroponic setups. It provides a chemical-free and environment friendly method of controlling microbial contamination.

iii) **UV dosage**: The effectiveness of UV light for disinfection depends on factors such as UV dosage (intensity and duration of exposure), water quality and presence of organic matter or other substances that may interfere with UV transmission. Proper UV dosage is essential to ensure adequate disinfection without causing harm to plants or aquatic life. UV dosimeters or sensors can be used to measure UV intensity and optimizing disinfection protocols.

iv) **UV-C sterilization systems**: UV-C sterilization systems consist of UV lamps or LEDs that emit UV-C light at specific wavelengths. These systems are designed to expose water or surfaces to UV-C light so effectively killing or inactivating microorganisms present. UV-C sterilization systems are available in various configurations including in-line units for water treatment and surface-mounted fixtures for equipment sterilization.

v) **Safety considerations**: UV-C light can be harmful to humans and plants if exposure levels are too high. Proper safety precautions such as wearing protective eyewear and avoiding direct exposure to UV-C light should be followed when using UV-C sterilization systems. UV-C light can also degrade certain materials over time so it is essential to use UV-resistant components in hydroponic systems and equipment.

8.8 Microbial Inoculants Containing Beneficial Microorganisms

To harness the benefits of beneficial microorganisms in hydroponic systems growers can incorporate microbial inoculants or supplements containing beneficial microorganisms such as bacteria, fungi and other microbes into the nutrient solution and/or growing medium to enhance nutrient availability, promoting plant growth and improving health of overall system. The integration of these beneficial microorganisms into the nutrient solution offers significant advantages for root health and overall plant growth. These microorganisms play a vital role in maintaining clean and robust root zones by breaking down dead roots and enhancing nutrient absorption efficiency.

To support the survival and proliferation of beneficial bacteria it is crucial to provide optimal environmental conditions within the hydroponic setup. Since hydroponic nutrients typically lack organic carbon sources, ensuring an adequate supply of food and oxygen is essential for microbial thriving. Molasses stand out as a particularly effective food source for these bacteria.

Bacteria tend to colonize inorganic substrates more efficiently whereas fungi thrive better in organic substrates. Cocofibre harbours a significant population of *Trichoderma harzianum* while rockwool systems host a higher concentration of fluorescent pseudomonads bacteria. Organic media rich in lignocellulose such as straw, wood bark and coconut fibre create an optimal environment for *Trichoderma* colonization that make cocofibre particularly conducive to its growth.

Beneficial microorganisms play a crucial role in nutrient cycling, disease suppression and overall plant vitality. While they don't function as traditional nutrient additives, they are indispensable soil amendments that enhance plant productivity while reducing reliance on chemical inputs. Incorporating these microorganisms into hydroponic systems offers a sustainable approach to maximizing plant growth and health.

8.8.1 Some Common Beneficial Microorganisms

Here are some common beneficial microorganisms used in hydroponics:

i) **Mycorrhizal fungi**: Mycorrhizal fungi form symbiotic relationships with plant roots enhance nutrient uptake with special reference to phosphorus and micronutrients. These can improve plant resilience to environmental stressors and diseases.

ii) **Trichoderma**: *Trichoderma* species (*T. harzianum, T. viride, T. virens, T. koningii, T. hamatum* and certain other species) are beneficial fungi known for their ability to suppress plant pathogens through competition, antibiosis (production of antifungal compounds) and induction of systemic resistance in plants. These can protect plant roots from diseases such as root rot and damping-off. *Trichoderma* species may also enhance biomass production promoting root development and dissolving insoluble phosphate containing minerals. Some beneficial fungi are explained in the table below:

iii) **Rhizobacteria**: Certain strains of rhizobacteria such as those belonging to genera like *Bacillus, Pseudomonas* and *Azospirillum* can promote plant growth by producing growth-promoting substances, fixing atmospheric nitrogen and suppressing pathogenic microorganisms.

iv) **Beneficial protozoa and nematodes**: Certain species of protozoa and nematodes can contribute to nutrient cycling in hydroponic systems by breaking down organic matter and releasing nutrients in forms accessible to plants. These also prey on pathogenic microorganisms thus helping in controlling diseases.

v) **Actinomycetes**: Actinomycetes are a group of filamentous bacteria known for their ability to decompose organic matter and produce antibiotics. These can help suppress soil-borne diseases and promote nutrient cycling in hydroponic systems.

vi) **Algae**: Certain types of algae (Chlorella and Spirulina) can contribute to nutrient cycling and providing additional nutrients to plants. These can also help oxygenate the nutrient solution and reduce the risk of pathogen growth.

vii) **Composting microorganisms**: Microorganisms involved in the composting process such as bacteria, fungi and actinomycetes can help breaking down organic matter added to hydroponic systems thus releasing nutrients and improving soil structure.

viii) **Denitrifying bacteria**: Denitrifying bacteria can convert nitrate (NO_3-) to nitrogen gas (N_2) reducing the risk of nitrate accumulation and toxicity in hydroponic systems.

When incorporating beneficial microorganisms into hydroponic systems, it is essential to select strains that are compatible with the system's conditions and the specific needs of the plants being grown. Microbial inoculants can be added to the nutrient solution or applied to the growing medium to establish beneficial microbial populations. Regular monitoring of microbial populations and nutrient levels can help optimize the effectiveness of beneficial microorganisms in hydroponic systems.

Employing a blend of beneficial bacteria *cum* fungi and enzymes cultivates a vigorous and expanded root zone thus facilitating faster and more efficient nutrient absorption by plants. Maintaining optimal environmental conditions such as proper oxygenation and pH levels can support the growth and activity of beneficial microorganisms in nutrient solution. Regular monitoring and adjustment of nutrient levels and system parameters can help ensure a healthy microbial community and maximize the benefits for hydroponic plant growth.

8.8.2 Key Roles

Some of the key roles of beneficial microorganisms in nutrient solutions include:

8.8.2.1 Enzyme Production

Though microbial inoculants may not directly contain enzymes themselves but these can introduce microbial populations that produce enzymes capable of breaking down organic matter, improving nutrient cycling and enhancing root health in hydroponic systems. Many beneficial microorganisms naturally produce enzymes as part of their metabolic processes. These enzymes can include cellulases, hemicellulases, proteases, lipases, phosphatases and certain others involved in breaking down organic matter, releasing nutrients and promoting root health. By introducing microbial inoculants containing enzyme-producing microbes into hydroponic systems growers can enhance enzyme activity and nutrient availability for plants.

Similarly while organic supplements can indirectly promote enzyme production in hydroponic systems, it is essential to use them judiciously and monitor nutrient levels, pH and overall health of the system regularly. Excessive use of organic supplements can lead to nutrient imbalances, microbial overgrowth and other issues that may negatively impact plant growth.

8.8.2.2 Disease Suppression ~ Antagonistic Organisms

Certain beneficial microorganisms have antagonistic effects against plant pathogens thus helping to suppress/inhibit growth and activity of plant pathogens ~ fungi (*Trichoderma* spp.), bacteria (*Bacillus subtilis, Pseudomonas fluorescens*) and nematodes in the nutrient solution and root zone. These organisms offer sustainable benefits and not like those of sterilization methods such as UV, ozone, monochloramine, chlorine and hydrogen peroxide. Some beneficial fungi and bacteria are explained below:

8.8.2.2.1 Bacteria ~ Bacillus subtilis and Pseudomonas spp.

Both these bacteria have been observed to control *Fusarium oxysporum* in hydroponic settings. These bacteria are applied to a wide range of agricultural crops to suppress pathogens. These produce hydrolytic enzymes and antibiotics. *Bacillus subtilis* can produce great number of antibiotics such as lipopeptide antibiotics which exhibit wide spectrum antifungal activity. pH neutral environment is best for bacteria species. Optimum pH for *Bacillus subtillis* is 6.5 – 7.0 and optimum temperature should be 40–47°C, however, the temperature range of 20-22°C and pH 5.5-6.0 are the best. However, bacterioa thrive well in a high carbon environment such as humates (fulvic and humic acids), kelp and fish emulsions, as well as molasses though quite costly. Always, it would be better to maintain optimum temperatures and pH for optimal plant growth in hydroponics systems.

8.8.2.2.2Fungi ~ Trichoderma species (T. harzianum, T. viride, T. virens, T. koningii, T. hamatum, and some other species)

These are biological control agents against a wide range of pathogenic fungi such as *Rhizoctonia* spp., *Pythium* spp., *Botrytis cinerea* and *Fusarium* spp. These are also effective against *Phytophthora palmivora*, *P. parasitica* and various other species. *Trichoderma harzianum* is reported to be most widely used as an effective bio-inoculant. *T. harzianum* controls or eliminates all types of pathogens in both inorganic and organic media. *Trichoderma harzianum* is noted to increase the uptake and concentration of a variety of nutrients (copper, phosphorus, iron, manganese and sodium) in hydroponic culture.

Cellulases (enzymes) produced by *Trichoderma* spp. are the most efficient enzyme system for the complete hydrolysis of cellulosic matter (*i.e.* decaying root matter) into glucose. *Trichoderma viride, T. reesei, T. harzianum* and *T. asperellum* produce high levels of cellulase enzymes. *Trichoderma* spp. possesses an inherent resistance to most agricultural chemicals including fungicides though strains may differ in their resistance.

The optimum temperature for rapid colonization and bioactivity of most of the commonly applied *Trichoderma* spp. is 25-30°C. Their colonization will be slow if temperature is either too cold or too warm and then these will gradually disappear. Optimum pH for *Trichoderma* fungi may vary between species, however, fungi thrive in semi-acidic conditions at pH 5.5 – 5.8.

For entomopathogenic fungi, bacteria and nematodes see 'Pests and Diseases' under 'Troubleshooting Common Issues'.

8.8.2.3 Organic Matter Decomposition and Nutrient Cycling

Beneficial bacteria and fungi break down organic residues and dead plant material in the nutrient solution thus preventing the accumulation of organic matter that could lead to nutrient imbalances or the proliferation of harmful microorganisms. These participate in nutrient cycling, converting organic compounds into forms readily absorbed by plants.

Microbial inoculants containing decomposer organisms, such as bacteria and fungi can accelerate breakdown of organic matter in hydroponic systems. These microbes produce enzymes that degrade complex organic compounds such as cellulose, lignin and hemicellulose into simpler forms that plants can readily absorb. This process releases nutrients for plant uptake and improves overall nutrient cycling in the system. Some examples of decomposer organisms including bacteria and fungi that can accelerate breakdown of organic matter in hydroponic systems are:

- **Bacteria**: *Pseudomonas* spp., *Bacillus* spp., *Clostridium* spp., *Streptomyces* spp. and *Escherichia coli* (although primarily associated with pathogens, certain strains can contribute to organic matter decomposition).
- **Fungi**: *Aspergillus* spp., *Penicillium* spp., *Trichoderma* spp., *Mucor* spp. and *Fusarium* spp.

8.8.2.4 Root Health

Some microbial inoculants contain beneficial bacteria and mycorrhizal fungi that colonize plant roots and promote root health. These beneficial microorganisms form symbiotic relationships with plant roots promoting root growth and enhancing nutrient uptake. These can also help protect roots from pathogens and environmental stressors thus improving overall plant health and vigour.

These microbes produce enzymes that help break down organic matter in the rhizosphere, release nutrients and stimulate root growth and development. By enhancing root health, microbial inoculants indirectly contribute to improved enzyme activity and nutrient uptake in hydroponic systems. Some examples of beneficial bacteria and mycorrhizal fungi that colonize plant roots and promote root health in hydroponic systems are:

- **Beneficial bacteria**: *Azospirillum brasilense, Bacillus subtilis, Pseudomonas fluorescens, Rhizobium* spp. and *Streptomyces* spp.
- **Mycorrhizal fungi**: *Funneliformis mosseae* (formerly known as *Glomus mosseae*), *Glomus* spp., *Rhizoctonia* spp. (can be both beneficial and pathogenic depending on the strain), *Rhizophagus irregularis* (formerly known as *Glomus intraradices*) and *Trichoderma* spp. (primarily it was

considered as beneficial fungi but its some species can act as mycorrhizal helpers).

8.8.2.5 pH Regulation

Certain microbial inoculants contain bacteria capable of influencing pH through metabolic processes. By introducing pH-regulating microbes into hydroponic systems growers can indirectly influence enzyme activity and nutrient availability by modulating pH levels. Bacteria play a vital role in regulating pH in hydroponic systems through various metabolic processes. While not all bacteria directly regulate pH there are some that can indirectly influence pH levels by producing acids or alkalis during their metabolic activities. Here are some examples of bacteria known for their ability to influence pH in hydroponics:

i) *Acidifying Bacteria*

These bacteria produce acids as metabolic byproducts which can lower pH of the nutrient solution, such as:

- ***Nitrosomonas* spp.**: *Nitrosomonas* bacteria are involved in the first stage of nitrification ~ converting ammonia (NH_3) into nitrite (NO_2-). This process releases hydrogen ions (H+) which can lower pH of hydroponic solution. *Nitrobacter* bacteria further oxidize nitrite (NO_2-) to nitrate (NO_3-) as part of the nitrification process. These play a crucial role in converting nitrogen compounds into forms usable by plants.
- ***Thiobacillus* spp.**: These are sulphur-oxidizing bacteria that oxidize elemental sulphur or various sulphur compounds to produce sulphuric acid (H_2SO_4). This acidification process can decrease pH of the hydroponic nutrient solution.
- ***Acidithiobacillus ferrooxidans***: This bacterium oxidizes ferrous iron (Fe_2+) to ferric iron (Fe_3+), releasing protons (H+) in the process which can acidify hydroponic solution.
- ***Lactobacillus* spp.:** Certain strains of *Lactobacillus* bacteria produce lactic acid as a metabolic byproduct during fermentation. This lactic acid production can contribute to pH reduction in the hydroponic system.
- ***Acetobacter* spp.:** *Acetobacter* bacteria produce acetic acid (vinegar) as a metabolic byproduct during oxidation of ethanol or other alcohols. The production of acetic acid can lower pH of the hydroponic solution.

ii) *Alkalizing Bacteria*

Some bacteria can contribute to raising pH levels by metabolizing organic compounds and releasing alkaline byproducts as mentioned below:

- ***Pseudomonas***: Several species of *Pseudomonas* bacteria are capable of denitrification ~ converting nitrate (NO_3-) to nitrogen gas (N_2) under anaerobic conditions. This process consumes hydrogen ions (H+) which can raise pH of the surrounding environment.

- ***Bacillus* spp.:** Certain species of *Bacillus* can produce alkaline compounds such as ammonia, through metabolic processes leading to an increase in pH.
- ***Cyanobacteria* (*Spirulina* spp.):** These photosynthetic bacteria can raise pH levels by consuming carbon dioxide during photosynthesis thus resulting in release of oxygen and increasing the alkalinity of solution.

8.8.2.6 Nutrient Fixation

Some beneficial microorganisms are capable of fixing atmospheric nitrogen ~ converting it into forms that plants can use for growth. This reduces the need for external nitrogen inputs and enhances nutrient availability in the nutrient solution.

When selecting microbial inoculants for hydroponic systems, it is essential to choose products containing strains compatible with specific conditions and nutrient solutions used in hydroponics. Regular monitoring of microbial populations, nutrient levels, pH and plant health is crucial to optimize effectiveness of microbial inoculants and ensuring healthy plant growth in hydroponic cultures.

8.9 Negative Side-Effects of Additives

While additives can enhance yields, these may also lead to undesirable outcomes, including:

- **Chemical compatibility**: Some additives may be unstable when mixed with nutrient solutions or other additives. It is crucial for growers to adhere to the recommended feed schedule provided by the nutrient manufacturer.
- **pH fluctuations**: Additives can alter pH of nutrient solution. For instance, silica and many phosphorus *cum* potassium flowering additives require alkaline conditions for stabilization. Addition of these additives can result in significant pH changes thus necessitating immediate monitoring and adjustment.
- **Algae and slime growth**: Carbohydrate-based additives often promote the growth of algae and slime in nutrient solutions. This not only attracts pests but also leads to plumbing blockages.

Chapter 9

Substrates (Growing Media)

Hydroponic growing media serve as vital components in plant cultivation ~ offering more than just structural support. These play a crucial role in facilitating oxygen and carbon dioxide exchange apart from serving as reservoirs for nutrient absorption. An ideal hydroponic medium exhibits eco-friendliness, pH stability, effective nutrient retention, moisture regulation and efficient drainage. Matching the right medium with specific hydroponic technique is paramount for optimal plant growth. Moreover, proper utilization of growing media must be integrated seamlessly with other elements of hydroponic system to ensure its effectiveness so maximizing overall productivity.

Note: The use of a biostimulator containing humic acid, lactic acid and *Bacillus subtilis* has shown potential to enhance yield across various substrates.

9.1 Hydroton/Expanded Clay Aggregate/Grow Rocks

Hydroton (expanded clay aggregate or grow rocks) represents a cornerstone material in hydroponic cultivation systems. These clay pellets undergo a process of high-temperature baking ~ often exceeding 1200°C thus resulting in a highly porous structure with exceptional capillary action. Primarily utilized in hydroponic setups where nutrient solutions are meticulously controlled, hydrotons offer a range of beneficial attributes.

Characterized by inertness, lightness and pH neutrality, hydrotons lack intrinsic nutrient content, making them ideal for environments where nutrient levels must be precisely regulated. Furthermore, their porous nature facilitates robust oxygen exchange within root zone thus promoting healthy plant growth. Hydrotons exhibit eco-friendly qualities and can be reused after thorough cleaning, sterilization and rinsing with substances such as white vinegar, chlorine bleach or hydrogen peroxide.

Key features of hydroton include its versatility across various hydroponic systems, lightweight composition and ability to resist compaction. When combined with cocopeat, hydrotons enhance moisture retention thus prolonging intervals between watering sessions. However, despite their numerous advantages, hydrotons present certain challenges.

Expanded Clay Aggregate

Their rapid drainage and drying properties necessitate frequent watering intervals ~ typically every 2 to 4 hours to prevent root desiccation. Sourcing hydrotons may prove challenging in some regions. Moreover, while hydrotons foster healthy root development, there is a risk of root entanglement within the pellets though this issue is easily remedied through regular maintenance and cleaning procedures. While hydrotons offer a versatile and effective medium for hydroponic cultivation, growers must be mindful of their unique characteristics and requirements to optimize plant health and productivity.

9.2 Cocofibre (Cocopeat or Coir)

Cocofibre derived from coconut husks stands out as a premier organic medium for growth and flowering within hydroponic systems. Its composition is rich in millions of capillary micro-sponges thus facilitating rapid nutrient absorption and gradual release according to plant needs. Its porous structure fosters ample oxygenation and aeration in the root zone ~ vital for optimal plant health.

Rose Plants in Cocopeat

In hydroponic applications, cocofibre is often used in the form of coir bricks or blocks and can be utilized either as a standalone medium or blended with other growing media to create a customized substrate thus offering versatility to growers. Commonly it is used in systems such as drip irrigation, flood and drain (ebb and flow) and passive hydroponics.

Prior to use, proper washing preferably with hot water is recommended to eliminate dust and stabilize pH levels. In the rose growing industry, cocopeat has emerged as a favoured substrate due to its exceptional drainage capabilities coupled with moisture retention. Its fine consistency fosters the development of delicate root hairs crucial for efficient nutrient and water uptake by plants. Remarkably, cocopeat remains stable and impervious to fertilizer effects thus maintains a neutral pH and EC.

One notable feature of coco-fibre is its ability to retain nutrients and moisture over extended periods thus promoting robust root development and facilitating gas exchange including carbon dioxide. It not only retains moisture but also provides good drainage. This balance helps preventing waterlogging thus ensuring that plant roots have access to oxygen for promoting healthy root development. Coco fibre is pH neutral and typically free from pests and diseases thus providing a clean and stable growing environment for plants. Moreover, its innate pH range of 6.5 to 7.0 along with anti-fungal properties further enhances its suitability for plant growth. As a hormone-rich medium, coco-fibre supports accelerated plant growth while being environmentally friendly and biodegradable at the end of its life cycle.

Despite its numerous advantages, salt buildup can pose challenges over time. Nevertheless, the benefits of coco-fibre far outweigh its drawbacks making it a preferred choice for hydroponic cultivation. Its compatibility with various hydroponic systems, ability to hold water up to eight times its dry weight and cost-saving compactibility underscore its indispensable role in modern horticulture. Regular watering, typically every 3-5 hours ensures optimal hydration for thriving plant growth.

9.3 Coco Chips

The coco fibre particle size is about the same as potting soil while coco chips are larger pieces of coconut husk or shells that provide excellent aeration and drainage in hydroponic systems. Similar to coco fibre, coco chips are pH neutral and offer good water retention properties so making these suitable for a variety of plants.

Coco Chips

One of the primary advantages of coco chips is their durability and longevity. These break down more slowly compared to finer coco fibre thus allowing for multiple growing cycles without need for frequent replacement. Coco chips also provide ample space for root growth and are less prone to compaction so can restrict root development and nutrient uptake.

In hydroponic applications, coco chips are often used in systems such as deep water culture (DWC) and nutrient film technique (NFT) where these provide support for plant roots while allowing for efficient nutrient delivery. These can be used alone or mixed with other growing media to create a balanced substrate that meets the specific needs of different plant species.

9.4 Grow Stones

Growstones

Derived from recycled glass waste, growstones represent a sustainable and innovative growing medium with distinct advantages over traditional alternatives such as perlite, peat, and parboiled rice hulls. Growstones boast a unique composition that affords them superior air and water holding capacities compared to their counterparts. Their lightweight, porous and irregularly shaped structure enables efficient air and water circulation within the root zone thus fostering optimal conditions for plant growth.

The production process involves crushing waste glass bottles into a fine powder which is then melted and combined with calcium carbonate. Upon cooling, the mixture is broken into small rock-like lava pieces thus forming the characteristic 'growstones'. This manufacturing method not only repurposes discarded glass but also contributes to the lightweight nature and sustainability of the product.

Growstones exhibit remarkable wicking ability ~ capable of drawing water up to 10cm above the water line thus ensuring consistent hydration for plants. Moreover, their reusable nature aligns with eco-conscious practices thus minimizing waste and reducing environmental impact.

However, despite their numerous advantages growstones are not without limitations. One notable drawback is their challenging cleaning process as at post-harvest their roots tend to adhere to the porous surface which causes these requiring thorough cleaning to prepare for subsequent use. Strong grip of plant roots to growstones can make transplanting difficult as this causes potential root damage during relocation.

9.5 Parboiled Rice Hulls

Rice hulls ~ the protective shells surrounding rice grains agricultural byproducts that decompose slowly over time. Renowned for their porous nature, these facilitate effective drainage albeit possessing lower water retention capabilities compared to alternatives like growstones. Parboiled rice hulls undergo a specialized process involving steaming and drying after the rice has been milled. This sterilizes the material effectively by eliminating spores, bacteria and microorganisms.

One significant advantage of parboiled rice hulls lies in their pH stability typically ranging from 5.7 to 6.5 thus making them slightly acidic. However, it

Rice Hulls

is essential to note that phosphorus and potassium leach down from rice hulls to some extent. To optimize their performance, rice hulls may require maturation or mixing with substrates like 'kuntan' or 'perlite' which offer superior water retention capabilities.

Fresh or composted rice hulls may contain elevated levels of manganese thus posing a potential risk of toxicity. However, maintaining a pH above 5 can mitigate this concern though it is important to recognize that this pH level is generally below the normal range for hydroponic systems.

Rice hulls are categorized as fresh, aged, composted, parboiled or carbonized. Fresh rice hulls are typically unsuitable for hydroponic use due to the heightened risk of contaminants such as rice grains, fungal spores, bacteria, decaying organisms and weed seeds.

9.6 Kuntan Culture

Kuntan culture ~ a method employed in hydroponic cultivation utilizes carbonized rice hulls as its primary medium. Renowned for its lightweight composition and exceptional water retention capacity, kuntan is predominantly favoured in hydroponic systems employing the drip irrigation method. Its porous structure facilitates efficient water distribution and nutrient uptake by plant roots, contributing to robust growth and development.

However, despite its advantageous qualities, kuntan culture is not without its drawbacks. One notable disadvantage lies in its high pH values which can adversely affect nutrient availability and uptake by plants ~ potentially leading to deficiencies and impaired growth. Kuntan is known to contain elevated levels of water-soluble potassium which may require careful monitoring and management to prevent imbalances in nutrient levels within the hydroponic system.

9.7 Rockwool

Rockwool (mineral wool) stands as a prevalent inert substrate utilized in hydroponic cultivation. It originates from molten rock, basalt or slag. Rockwool undergoes a process wherein it is spun into filament fibres and bonded to create a medium capable of capillary action. It is available in various forms such as blocks, sheets, cubes, slabs or flocking, Rockwool is prized for its sterility, porosity and non-degradable properties thus making it a favoured choice among hydroponic growers.

Characterized by its capacity to readily absorb water, rockwool is protected from common microbiological degradation thus ensuring a stable growing

Rockwool Cube

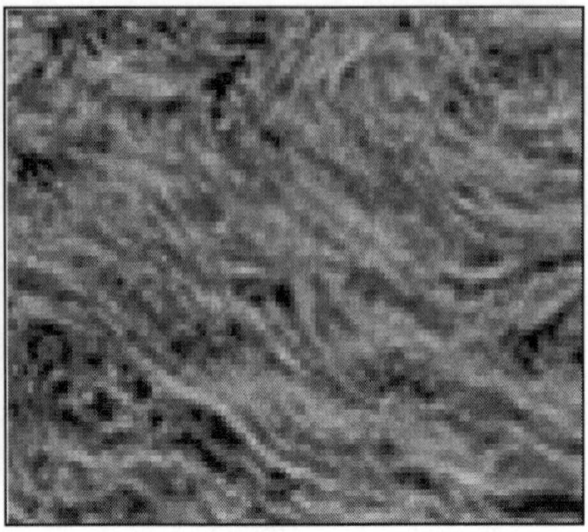

Rockwool Sheets

environment. However, it is essential to balance its pH before use by soaking it in pH-balanced water, as its nature may vary depending on the raw materials utilized. Despite its widespread use and several advantageous features, rockwool does present certain considerations for growers. Rockwool excels in seed starting applications and being excellent in water-holding capabilities can retain water and nutrients for extended periods. Furthermore, its high oxygen levels ~ approximately 18 per cent contribute to favourable root oxygenation thus fostering healthy plant growth. Nevertheless, the frequency of watering required varies from 1 to 5 times a day depending on the specific crops being cultivated.

However, alongside its benefits rockwool exhibits several disadvantages as it raises environmental concerns due to its non-biodegradable nature. Rockwool tends to have a high and sometimes unstable pH so necessitating careful monitoring and adjustment. Its cation exchange capacity (CEC) is notably low thus impacting nutrient retention and availability for plants. Furthermore, the fibres and dust present in rockwool can pose health risks so irritating the lungs and skin which highlights its importance of proper handling and safety precautions. Disposal of rockwool also presents challenges as its thin fibres of melted rock persist indefinitely thus adding to waste management concerns.

In a nutshell, rockwool offers advantages such as suitability for seed starting, long-term water *cum* nutrient retention and high oxygen levels so growers must weigh these benefits against environmental considerations and health risks associated with its use. Proper management and handling practices are essential to maximize the benefits of rockwool while minimizing its drawbacks in hydroponic systems.

9.8 Oasis Cubes

Oasis cubes represent an inert growing medium akin to rockwool cubes yet these offer distinct advantages over their counterparts. Unlike rockwool, oasis cubes exhibit a reduced tendency to become waterlogged so providing a more balanced moist environment conducive to healthy plant growth. These cubes are particularly favoured for their affordability and suitability for germination and for early seedling development stages.

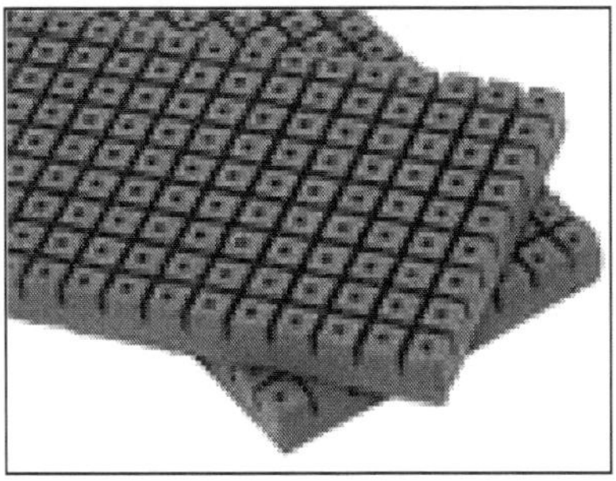

Oasis Cubes **Lettuce in Oasis Cubes**

It is important to note that 'oasis cubes' bear resemblance to rigid green or white floral foam often utilized in floral arrangements. This similarity underscores their open-cell structure allowing for both water and air absorption. The porous nature of oasis cubes facilitates wicking of moisture throughout the material thus promoting optimal hydration for developing roots. Furthermore, the open-cell configuration offers ample space for root expansion thus enabling robust growth.

While oasis cubes offer distinct advantages though these have certain limitations. These are not considered organic which may be a concern for growers prioritizing organic cultivation methods. Their utility is primarily restricted to germination and early seedling stages rather than serving as a comprehensive growing medium for mature plants.

In a nutshell, oasis cubes provide an economical and efficient option for germination and early seedling growth due to balanced moisture retention and ample root development space. While they may not align with organic farming principles and have limited utility beyond initial growth stages, their affordability and effectiveness make these a viable choice for hydroponic and soilless cultivation endeavours.

9.9 Perlite

Perlite, a product of natural volcanic processes involving the fusion of granite, obsidian, pumice and basalt at high temperatures holds a significant place in horticulture as a versatile growing medium. Its formation imbues it with unique properties that contribute to its widespread use in soil mixes and hydroponic systems alike.

Renowned for its role in enhancing soil aeration, perlite serves as a soil-free medium so making it particularly valuable in hydroponic setups where precise control over growing conditions is paramount. Its neutral pH, combined with exceptional wicking action *cum* high porosity fosters optimal root development by ensuring adequate oxygenation and moisture distribution throughout the root zone.

Perlite

Despite its lightweight nature which causes it to float over water the porous structure of perlite allows it to hold moisture while still providing sufficient aeration to plant roots. To further capitalize on its benefits, perlite is often blended with other media like coco coir or vermiculite to enhance overall oxygen retention levels and water-holding capacity.

In hydroponic systems such as 'top drip' or NFT, perlite finds its niche ~ providing a supportive environment for plant growth when paired with appropriately designed containers. However, due to its tendency to fall through the holes of standard net pots, perlite may not be suitable for all the hydroponic setups particularly those employing 'ebb and flow' systems.

Versatility of perlite extends to its application methods allowing it to be used either loosely or within plastic sleeves immersed in water, depending on the specific requirements of the growing system and crop being cultivated. While it demonstrates environmental friendliness as a growing medium, its lower water-holding capacity necessitates careful monitoring and adjustment of watering schedules, typically ranging from 3 to 5 times a day to meet plant requirement.

Perlite stands as a reliable and adaptable component in modern horticulture offering growers a valuable tool for optimizing plant health and productivity across a range of growing environments from traditional soil mixes to advanced hydroponic systems.

9.10 Vermiculite

Vermiculture

Vermiculite, a versatile hydroponic medium akin to perlite, undergoes a transformative process originating from volcanic rock. Subjected to intense heat, this rock is expanded into lightweight pebbles imparting unique properties ideal for hydroponic cultivation. Unlike perlite, vermiculite boasts exceptional water retention capabilities thus making it an invaluable component for maintaining optimal moisture levels within hydroponic systems. Its porous structure not only absorbs water but also possesses a natural wicking ability so ensuring efficient distribution of moisture throughout the root zone thus fostering robust plant hydration and growth.

Furthermore, vermiculite showcases a commendable cation-exchange capacity, enabling it to effectively absorb and retain essential nutrients within its porous matrix. This feature enhances nutrient availability to plants, supporting their nutritional needs over prolonged periods. However, the lightweight nature of vermiculite poses challenges in certain hydroponic configurations where it may float atop nutrient solutions or pass through net pot holes so limiting its suitability for use in such setups.

To mitigate these challenges and capitalize on its unique properties, vermiculite is often combined with perlite in hydroponic media formulations. This synergistic blend not only enhances oxygenation in the root zone but also optimizes moisture retention thus serves as a balanced growing medium tailored to specific requirements of diverse crops. The versatility of vermiculite thus facilitates creation of highly customized hydroponic systems capable of supporting a wide array of plants from delicate herbs to robust vegetables with precision and efficacy. In essence, vermiculite stands as a cornerstone of hydroponic cultivation, offering growers a dynamic tool to optimize plant health and productivity in controlled growing environments.

9.11 Pumice

Pumice

Akin to perlite, pumice emerges as a noteworthy component within the realm of hydroponics. It is distinguished by its origins as a lightweight volcanic rock. This mineral is meticulously mined and subsequently crushed to find application in select hydroponic systems due to its advantageous characteristics. Pumice boasts a notable capacity for retaining oxygen ~ a critical factor in facilitating healthy root development and overall plant vigour within hydroponic environments. Its porous structure contributes to efficient aeration within the growing medium thus ensuring that roots receive ample oxygen for optimal growth.

Furthermore, pumice exhibits exceptional drainage properties, preventing waterlogging and reducing the risk of root rot ~ a common concern in hydroponic setups. Its lightweight nature renders it easy to handle and to incorporate into various hydroponic systems, adding to its appeal among growers seeking efficient and manageable growing media.

In addition to its utility in hydroponics, pumice finds applications across diverse industries including horticulture, construction and cosmetics. Its versatility extends beyond its role in hydroponic cultivation, underscoring its value as a multifaceted resource.

While pumice shares similarities with perlite in its lightweight composition and utility in hydroponic systems, each possesses distinct characteristics that cater to specific growing requirements. High oxygen retention level of pumice sets it apart making it a preferred choice for growers aiming to optimize root health and overall plant performance in hydroponic environments. As with any growing medium, understanding its properties and compatibility with different crops is key to harnessing its full potential within hydroponic systems.

9.12 Sand

Sand serves as an easily accessible and cost-effective growing medium, often available at minimal expense or even free of cost. However, its utilization in hydroponic systems necessitates thorough sterilization before each use to eliminate potential pathogens and contaminants. Despite its affordability and widespread availability, sand possesses certain characteristics that warrant consideration in hydroponic cultivation.

Compared to alternative growing media, sand is notably heavy and may pose challenges in handling and transportation. Furthermore, its limited water retention capacity renders it susceptible to waterlogging so potentially hindering optimal root health and growth. Though sand shares similarities with rocks in composition but its smaller particle size allows for slower moisture drainage,*vis-a-vis* mitigating rapid drying but still necessitating caution to prevent over-saturation.

To address the inherent limitations of sand, it is often combined with additives such as vermiculite, perlite or coco coir. These amendments serve dual purposes ~ enhancing moisture retention, *vis-a-vis* promoting aeration within the growing medium thus fostering conducive conditions for robust root development.

Despite its drawbacks, sand offers several advantages in hydroponic cultivation. Its affordability and wide availability make it an attractive option for growers seeking cost-effective solutions. Moreover, the ease of sourcing sand contributes to its accessibility particularly in regions where specialized growing media may be scarce or prohibitively expensive.

However, the heavy nature of sand coupled with its relatively poor water retention capabilities, *vis-a-vis* small particle size may limit its suitability for certain hydroponic systems. Growers must weigh these considerations carefully when determining the most appropriate growing medium for their specific needs and environmental conditions. While sand presents both advantages and disadvantages in hydroponic cultivation, its integration into a balanced growing system can contribute to successful plant growth when managed effectively.

9.13 Gravel

Gravel, often recognized as the same medium utilized in aquarium setups, offers a robust option for hydroponic cultivation. Characterized by its weightiness, gravel presents an economical choice that boasts efficient drainage, ease of maintenance and resistance to waterlogging. Prior to use, it is imperative to thoroughly clean gravel to ensure optimal growing conditions. However, it is important to note that in systems lacking continuous water flow, there exists a risk of plant roots drying out which necessitates a consistent water supply.

Among its advantages, gravel stands out for its affordability, widespread availability and ability to facilitate proper drainage within hydroponic setups. Its ease of cleaning and reusability further contribute to its appeal, offering growers a sustainable option for multiple growing cycles. Gravel maintains its structural integrity over time, resists saturation and prevents waterlogging which can be detrimental to plant health in other media.

However, gravel does come with its share of drawbacks. Its weight can pose logistical challenges, particularly in larger-scale hydroponic systems. Moreover, the risk of root desiccation persists if water delivery is inconsistent, highlighting the importance of maintaining a reliable water supply. While gravel may be suitable for many hydroponic setups, its pH stability can fluctuate, requiring vigilant monitoring and potential adjustments to maintain optimal growing conditions.

Gravel represents a practical choice for hydroponic cultivation, offering a balance of affordability, durability and drainage efficiency. By understanding its characteristics and addressing its limitations, growers can effectively incorporate gravel into their hydroponic systems to foster healthy plant growth and maximize crop yields.

9.14 Wood Fibre

Wood fibre derived from steam friction of wood stands as a resilient and eco-conscious medium tailored for hydroponic cultivation. Through the intricate

process of steam friction, wood fibres are meticulously crafted, resulting in a durable and enduring structure that can withstand the rigours of hydroponic environments over extended periods. This inherent longevity ensures that the substrate remains intact, providing a stable foundation for sustained plant growth and development.

In the realm of hydroponics, wood fibre emerges as a standout choice offering a plethora of advantages. Its organic composition not only aligns with sustainability principles but also contributes to the creation of a wholesome growing environment. Functioning as a pure organic medium, wood fibre facilitates optimal moisture retention so fostering an environment conducive to robust root proliferation and nutrient uptake. Moreover, its porous nature facilitates ample aeration thus promoting delivery of oxygen to plant roots ~ a crucial factor in ensuring their vitality.

However, it is imperative to acknowledge potential considerations associated with wood fibre employment in hydroponic systems. While its biodegradable nature underscores its environmental friendliness, it may also introduce complexities related to sterility. Untreated wood fibre runs the risk of harbouring pests and pathogens, potentially compromising plant health and productivity. Consequently, stringent measures for pest management and sterilization are paramount to mitigate these risks effectively.

Notwithstanding these challenges, the allure of wood fibre remains undiminished particularly for growers committed to sustainable practices. Its biodegradability not only minimizes environmental impact but also offers a renewable alternative to traditional substrates. By adopting proper sterilization protocols and integrating comprehensive pest management strategies growers can harness the myriad benefits of wood fibre, cultivating thriving and resilient plants within hydroponic systems while championing environmental stewardship.

9.15 Brick Shards

Brick shards derived from crushed bricks represent a budget-friendly option for horticultural use though their suitability varies depending on the application. These shards akin to gravel in texture, possess advantageous attributes alongside notable considerations.

Similar to gravel, brick shards are valued for their affordability and ease of cleaning albeit with some caveats. Before reuse, thorough cleaning is imperative to eliminate residual brick dust, ensuring optimal growing conditions and for preventing potential issues associated with debris accumulation. One of the primary advantages of brick shards lies in their efficient drainage properties, facilitating proper moisture management within the root zone. Their permeable structure enables excess water to drain away, mitigating the risk of waterlogged conditions detrimental to plant health.

However, it is important to acknowledge the potential drawbacks associated with brick shards. Their use may inadvertently alter pH of growing medium so impacting nutrient availability and potentially compromising plant health. Moreover, due to their propensity to retain residual dust and debris, brick shards necessitate more rigorous cleaning procedures compared to other growing media. The weight of brick shards can pose logistical challenges particularly in larger-scale cultivation operations where handling and transportation are key considerations. Furthermore, the porous nature of brick shards may contribute to accelerated moisture evaporation, potentially leading to excessive dryness in the root zone and subsequent root desiccation.

In summary, while brick shards offer cost-effective drainage solutions and ease of cleaning, growers must exercise caution regarding their potential pH-altering effects so it should be ensured that thorough cleaning procedures are adhered to. By carefully considering these factors, growers can effectively harness the benefits of brick shards while mitigating associated challenges and thus optimizing plant health and overall cultivation success.

9.16 Sheep Wool

Though often overlooked but sheep wool is being increasingly recognized as a valuable component in horticulture. It offers unique properties that benefit plant growth. This natural material possesses a remarkable air capacity, allowing for efficient aeration within the root zone which is crucial for fostering healthy root development and facilitating nutrient uptake. However, it is important to note that air capacity of sheep wool may diminish over time with repeated use as fibres compact and settle. However, the water capacity of sheep wool tends to increase with use thus making it a suitable medium for retaining moisture and promoting consistent hydration levels for plants.

In addition to its capacity for air and water retention, sheep wool presents several other advantageous qualities for gardening applications. This renewable resource is biodegradable, contributing to sustainable gardening practices and minimizing environmental impact. Furthermore, sheep wool is naturally resistant to pests and diseases thus reducing the need for chemical interventions and promoting a healthier growing environment. Its fibrous texture also aids in the maintenance of soil structure, preventing compaction and improving overall soil health. Moreover, sheep wool is an excellent insulator so providing protection to plant roots against extreme temperatures and fluctuations in weather conditions. This insulation property helps creating a more stable and favourable environment for plant growth particularly in regions with variable climates or harsh weather patterns.

Sheep wool stands as a versatile and eco-friendly option for enhancing soil quality and supporting plant growth in gardening and other horticultural endeavours. While it may require periodic replenishment due to changes in air and water capacities over time, its numerous beneficial attributes make it a valuable addition to the toolkit of conscientious gardeners seeking sustainable and effective solutions for hydroponic culture.

9.17 Polystyrene Packing Peanuts

Polystyrene packing peanuts ~ a common material in packaging has found a niche application in certain hydroponic systems. These packing peanuts are characterized by their low cost, lightweight nature, and superior drainage properties that make them suitable particularly for closed-tube hydroponic setups. However, it is important to note that while biodegradable alternatives may seem environmentally friendly, these should be avoided in hydroponic systems due to risk of decomposition which may lead to undesirable sludge formation.

Primarily utilized in NFT systems, polystyrene packing peanuts offer efficient nutrient delivery to plant roots while maintaining optimal aeration. Despite their advantages, growers should exercise caution due to potential risks associated with polystyrene. There is a possibility that plants may absorb styrene from these packing peanuts, posing a contamination risk. Therefore, it's crucial to monitor plant health closely and consider alternative growing media if contamination concerns arise.

To elaborate in a nutshell, polystyrene packing peanuts offer certain benefits such as affordability and excellent drainage so growers must weigh these advantages against potential risks including styrene absorption by plants. Careful consideration of alternative options and vigilant monitoring of plant health are essential to mitigate any adverse effects associated with use of polystyrene in hydroponic systems.

9.18 Sawdust

Sawdust culture represents a unique approach to horticulture wherein sawdust serves as the primary substrate for cultivating crops. Typically, sawdust is arranged within U-shaped or V-shaped grooves or packed into polyethylene bags providing a structured environment for plant growth. It is worth noting that characteristics of sawdust can vary considerably depending on the types of trees from which it is derived thus influencing factors such as porosity, moisture retention and nutrient content.

This method finds particular prominence in regions where large-scale lumber industries exist as it offers a sustainable solution for utilizing leftover sawdust from wood processing operations. By repurposing this organic material, growers not only mitigate waste but also capitalize on its potential as a growing medium.

Sawdust has a widespread use with advantage. This provides a rich source of organic matter and nutrients so fostering robust plant growth when properly managed. Its porous structure promotes aeration and drainage creating an optimal environment for root development.

However, the sawdust culture may be influenced by factors such as its composition, pH levels and presence of contaminants. Some types of sawdust may contain high levels of lignin which can inhibit microbial activity and nutrient availability so necessitating supplementation with fertilizers or soil amendments.

Furthermore, sawdust culture requires careful monitoring of moisture levels to prevent waterlogging or desiccation as sawdust has a tendency to retain moisture unevenly. Proper irrigation management and periodic testing of nutrient levels are essential to ensure optimal crop performance and to mitigate the risk of nutrient imbalances or deficiencies.

In a nutshell, sawdust culture offers a sustainable and resource-efficient approach to horticulture particularly in regions with abundant wood processing industries. By leveraging the unique properties of sawdust and implementing appropriate management practices growers can harness its potential as a versatile growing medium while contributing to environmental stewardship and agricultural sustainability.

9.19 Water-Absorbing Polymer Crystals

Though water-absorbing polymer crystals are widely utilized in various industries such as childcare products and sport's equipments but are yet to gain significant traction as a hydroponic growing medium. These crystals have remarkable ability to expand to several times their original size as these absorb water ~ with just one pound capable of holding up to 50 gallons. Available in a range of sizes from fine powder to marble or even golf ball dimensions, their absorption process can take upwards of an hour or two to reach full capacity. Once dried out, these crystals can be stored and reused repeatedly so offering a cost-effective solution for growers.

While primarily found in large nurseries as soil amendments, water-absorbing polymer crystals find application in various settings. Florists employ them in vases to prolong freshness of cut flowers with coloured variants enhancing decorative displays. It is not a conventional choice for hydroponic systems but its affordability and reusability are making this increasingly popular. Larger crystals are particularly well-suited for hydroponic use as their size facilitates the retention of air pockets between particles so it is crucial for root health. Moreover, combining these crystals with river rock or similar media further enhances aeration within the growing medium.

Despite its lesser-known status in hydroponics, water-absorbing polymer crystals present a promising option for growers seeking an economical and sustainable growing medium. As their usage continues to grow, their potential benefits in hydroponic cultivation are likely to be further explored and appreciated.

9.20 Polyurethane Foam Insulation

Polyurethane foam insulation often abbreviated as polyfoam presents an unconventional yet promising option in hydroponic cultivation particularly as a substitute for traditional starter cubes like rockwool or oasis cubes. While not commonly utilized in hydroponics, polyfoam has garnered attention for its affordability and accessibility. Unlike specialized hydroponic substrates, polyfoam can readily be sourced from hobby stores or fabric suppliers so making it a convenient choice for growers seeking an alternative medium for germinating seeds or propagating cuttings.

Originally designed for insulation purposes, polyfoam is commonly employed in furniture manufacturing and is sometimes marketed as 'foam batting'. It is available in various forms including sheets and rolls with a range of sizes and thicknesses to accommodate diverse growing needs. Its lightweight and flexible nature make it easy to handle and customize for different applications within hydroponic systems.

In hydroponic setups, polyfoam demonstrates notable versatility, offering a stable and supportive environment for seedlings and cuttings. Its porous structure facilitates air circulation and moisture retention so promoting optimal conditions for root development. Moreover, polyfoam's inert composition minimizes the risk of introducing unwanted contaminants or altering the pH of nutrient solution thus contributing to a more controlled growing environment.

While polyfoam may not be as widely recognized as other hydroponic substrates but its affordability, accessibility and performance merit consideration among growers exploring innovative alternatives. By harnessing the advantages of polyfoam in hydroponic cultivation, growers can experiment with new methods and potentially achieve favourable results in early growth stages as well as in plant propagation.

9.21 Composted and Aged Pine Bark

Pine bark ~ once regarded as a mere byproduct has emerged as a versatile material with multiple applications such as ground mulch and a substrate for hydroponically cultivated crops. Unlike some other forms of tree bark, pine bark exhibits superior resistance to decomposition and contains lower levels of organic acids that may otherwise leach into the nutrient solution. This inherent durability and chemical composition render pine bark particularly suitable for hydroponic systems where stability and longevity are of paramount considerations.

Within the realm of pine bark utilization, distinctions are made between fresh, composted and aged varieties. In particular, composted and aged pine barks offer enhanced benefits for hydroponic cultivation. Through composting and aging processes pine bark undergoes transformation resulting in improved structural integrity, reduced acidity and increased nutrient availability. These characteristics contribute to a more stable and balanced substrate conducive to optimal plant growth and development.

A notable advantage of composted and aged pine bark lies in its availability and accessibility. This versatile material can be readily sourced from suppliers specializing in ground mulch or horticultural products. Its widespread availability ensures convenience for growers seeking reliable substrates for hydroponic systems.

Furthermore, the utilization of pine bark in hydroponics aligns with principles of sustainability and resourcefulness. By repurposing what was once considered waste into valuable resources, growers contribute to environmental conservation efforts while simultaneously reaping the benefits of a cost-effective and efficient growing medium.

In a nutbshell, the integration of composted and aged pine bark into hydroponic cultivation practices represents a symbiotic relationship between innovation and tradition. Through careful selection and utilization of this versatile material, growers can foster healthier plants, minimize environmental impact and can optimize the productivity of their hydroponic endeavours.

9.22 Pine Shavings

Pine shavings represent a cost-effective and widely utilized hydroponic growing medium, favoured particularly by commercial growers operating on a large scale often in hydroponic drip irrigation systems. It is essential to differentiate pine shavings from sawdust as the latter tends to compact and become waterlogged quickly so rendering it unsuitable for hydroponic applications. Prior to usage, growers must ensure that the pine shavings are derived from kiln-dried wood and are devoid of any chemical fungicides so ensuring a safe and conducive environment for plant growth.

Pet supply stores serve as convenient sources for acquiring pine shavings commonly used as bedding material for small animals such as hamsters and rabbits. However, it is imperative to scrutinize the packaging to confirm the absence of chemical additives like fungicides or odour inhibitors which could potentially harm plants or interfere with growth. Pine shavings find utility as bedding material in horse stalls offering comfort and absorbency for equine occupants.

The effectiveness of pine shavings as a hydroponic medium hinges on the size of air pockets between the shavings which directly impacts root aeration. Given their composition as a wood product, pine shavings readily absorb water so posing the risk of waterlogging if not managed carefully. Therefore, diligent monitoring of moisture levels and drainage is crucial to prevent adverse effects on plant health and performance.

In a nutshell, while pine shavings offer a cost-efficient option for hydroponic cultivation growers must exercise caution to ensure proper sourcing and management to optimize their efficacy and mitigate potential risks associated with waterlogging and chemical additives.

9.23 River Rock

River rock readily available from home improvement stores or pet supply outlets catering to aquarium enthusiasts, presents an affordable and versatile option for hydroponic growing media. Characterized by its rounded shape and smooth edges, river rock originates from natural erosion processes as stones tumble down riverbeds. While manufactured river rock undergoes a similar tumbling process using large mechanical tumblers, it achieves the same smooth texture. However, before use river rocks require thorough cleaning and sanitation. Before use, it is advisable to wash these meticulously followed by soaking overnight in a solution of bleach water to eliminate any potential contaminant. Following this treatment, the rocks should be rinsed thoroughly before incorporation into hydroponic systems.

Despite its ease of use and cost-effectiveness, river rock does have limitations. It lacks porosity since it can not retain moisture well within the root zone of hydroponic plants. Consequently, growers must carefully manage watering schedules to prevent root dehydration. The uneven surface of river rock results in numerous air pockets between the stones so facilitating oxygenation of the root system while allowing water to drain freely. To address the moisture retention issue, growers often mix river rocks with moisture-retaining growing media like coco chips. This combination not only improves moisture retention but also enhances drainage so preventing water saturation that could otherwise lead to root rot or other moisture-related issues.

In conclusion, while river rock offers benefits such as affordability, good drainage and ample oxygenation for plant roots, it is essential for growers to supplement its moisture-retaining capabilities with other media. By combining river rock with materials like coco chips, hydroponic enthusiasts can strike a balance between effective drainage and adequate moisture retention thus ensuring optimal conditions for healthy plant growth.

9.24 Floral Foam

Floral foam, although primarily associated with floral arrangements, finds utility as a growing medium in hydroponics, offering a viable alternative to traditional substrates. Comparable to oasis cubes in function, floral foam distinguishes

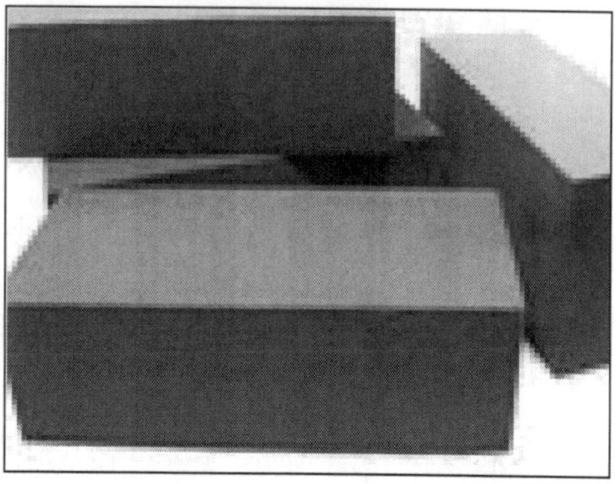

itself with larger cell sizes, providing ample space for root development and nutrient absorption. However, despite its potential, several drawbacks also merit consideration. One significant limitation is its tendency to crumble easily, potentially leaving particles in the water and compromising water quality. Floral foam's propensity to become waterlogged poses challenges as excessive moisture can impede oxygenation and foster conditions conducive to root rot. Nevertheless, its inherent ability to absorb water efficiently remains advantageous for maintaining adequate hydration levels within the root zone.

It is worth noting that while floral foam may serve as a convenient option for certain hydroponic applications its limitations necessitate careful consideration and management to optimize plant health and system performance. Growers may need to implement strategies to mitigate issues such as crumbling and waterlogging, potentially supplementing floral foam with additional support structures or incorporating other growing media to enhance stability and drainage. By understanding the strengths and limitations of floral foam in hydroponics, growers can make informed decisions to achieve successful and sustainable cultivation outcomes.

Chapter 10

Troubleshooting Common Issues

Common issues in hydroponics can arise due to various factors such as nutrient imbalances, environmental conditions, pests, diseases and equipment malfunctions. Here are some of the most prevalent issues encountered in hydroponic systems:

- ☆ **Nutrient imbalances:** Incorrect nutrient levels or imbalances can lead to deficiencies or toxicities thus affecting plant growth and health ~ ultimately making them more susceptible to pests and diseases.
- ☆ **pH fluctuations:** Fluctuations in pH levels can impact nutrient availability to plants so leading to nutrient deficiencies or toxicities.
- ☆ **Root rot:** Root rot caused by pathogens like *Pythium* and *Fusarium* can occur due to poor water quality, inadequate aeration or overwatering thus impacting plant health.
- ☆ **Algae growth:** Algae can thrive in hydroponic systems with excess light and nutrients thus competing with plants for resources and potentially clogging irrigation systems.
- ☆ **Pest infestations:** Pests such as aphids, thrips, spider mites and whiteflies can infest hydroponic crops causing damage to foliage and reducing yields.
- ☆ **Disease outbreaks:** Diseases such as powdery mildew, damping-off and leaf spot can spread rapidly in hydroponic environments due to high humidity and poor airflow.
- ☆ **Equipment failure:** Malfunctioning equipment such as pumps, timers and pH meters can disrupt nutrient delivery and environmental control so affecting plant health.

- **Water quality issues:** Poor water quality including high levels of salts, heavy metals or pathogens can negatively impact plant growth and health.
- **Environmental stress:** Inadequate lighting, improper humidity levels and poor air circulation can stress plants and make them more susceptible to diseases and pests.
    - **Temperature fluctuations and extremes:** Extreme temperatures can stress plants and affect nutrient uptake, transpiration rates and overall growth. Plants become weak and thus more susceptible to pests and diseases.
    - **Humidity imbalance**: High humidity levels can promote fungal diseases such as powdery mildew and botrytis while low humidity can lead to dehydration and stress.
- **Nutrient lockout:** Excessive levels of certain ions in the nutrient solution can lead to locking of nutrients so preventing plants from absorbing essential nutrients.
- **Overcrowding:** Overcrowding of plants can lead to competition for light, nutrients and space thus resulting in reduced yields and increased susceptibility to diseases.

To mitigate these issues, hydroponic growers should regularly monitor and maintain optimal nutrient levels, pH, temperature, humidity and airflow. Implementing proper sanitation practices, integrated pest management strategies and preventive measures can also help prevent problems before they arise. Prompt addressing any issues that arise can minimize their impact on crop health and productivity.

This section will delve into nutritional disorders alongside preventative and remedial strategies for algae and pathogens given that other concerns have been previously discussed.

10.1 Nutritional Disorders

Nutritional disorders encompass deficiencies, toxicities and imbalances of nutrients and other elements alongside fluctuations in acidity levels. Optimizing plant growth requires a crop-specific diagnosis of nutrient deficiencies which is subjectively assessed through careful observation. These disorders are influenced not only by nutrient levels but also by factors such as temperature, humidity, day length and disease ~ culminating in complex syndromes that may not align with any single disorder.

Nutrient deficiencies in plants are categorized as acute or chronic. Acute deficiencies occur when a nutrient is abruptly unavailable while chronic deficiencies arise from inadequate nutrient supply to the plants. Hydroponic studies often manifest acute deficiencies due to the immediate availability of nutrients leading to sudden deficits upon depletion.

Symptoms such as tip burns, chlorosis or necrosis characteristic of certain nutrient deficiencies can also stem from other stresses such as disease, chemical exposure or salinity. Identifying the location of symptom expression on the plant aids in diagnosing nutrient disorders.

Before delving into specifics, it is beneficial to examine external leaf structure and common leaf abnormalities resulting from nutrient deficiencies.

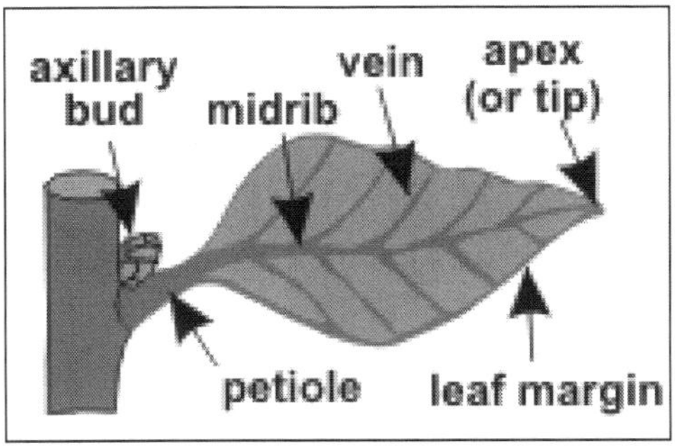

External Leaf Structure

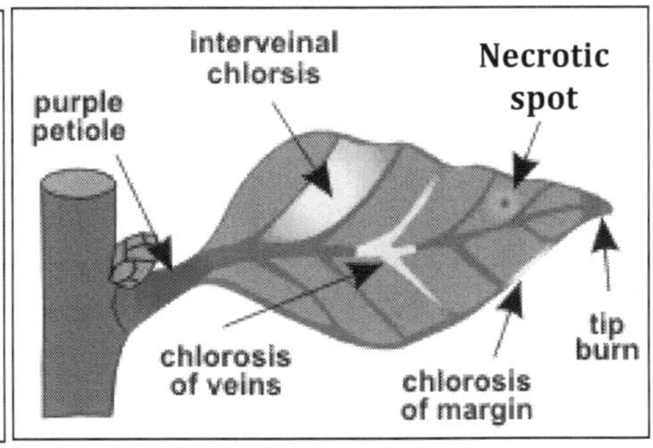

Abnormalities Resulting from Nutrient Deficiencies

Typically the nutrient deficiencies manifest uniformly across the entire crop whereas impact of plant diseases and chemical sprays may vary among affected plants. Understanding the history of chemical applications can aid in distinguishing between symptoms of nutrient deficiencies and those caused by spray burns of chemicals. If symptoms emerge shortly after a chemical application these may be linked to the effect of sprayed chemical. Damage from salinity such as tip burns could be mistaken for nutrient deficiencies. Marginal leaf burns might be misinterpreted as potassium deficiency.

10.1.1 Recognizing Plant Deficiencies

Timely identification of nutritional deficiencies is crucial for addressing issues before plants undergo significant stress. Deficiency signs and symptoms vary based on growth stage, plant species, climate and growing conditions. Gardeners must familiarize themselves with various deficiencies and their impacts on each crop. Photographs are invaluable for identifying deficiencies as symptoms may differ from those shown in textbooks. Even seasoned professionals may struggle with diagnosis due to multiple deficiencies, similar symptoms and potential confusion with pest or disease damage, genetic factors or improper watering. Early detection is key as deficiency symptoms are often distinctive and easier to distinguish at this stage. Early assessment allows for checking other factors such as EC, pH, humidity, *etc.* which can contribute to symptom development.

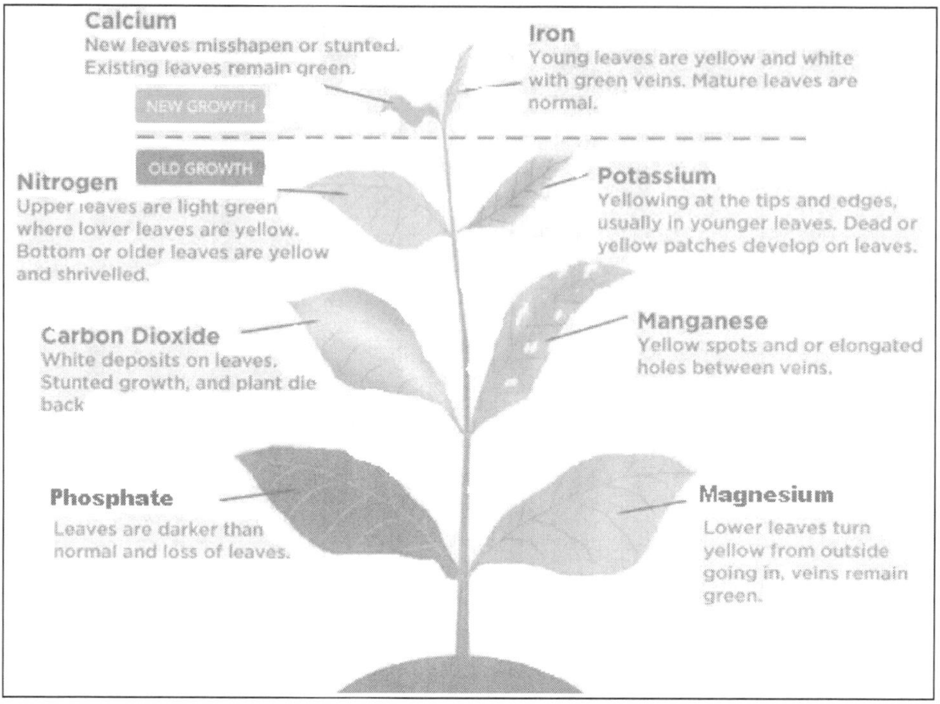

10.1.2 Understanding Nutrient Mobility and Deficiency Symptoms

A critical aspect of identifying nutrient deficiencies lies in comprehending the role of nutrient mobility in symptom manifestation. Mobile nutrients such as N, P and K can translocate from older leaves to younger ones. Consequently, deficiency symptoms of mobile nutrients typically emerge first on old mature leaves before progressing to actively growing parts such as younger leaves. N, P and K exhibit high mobility while Mg, S, Fe and Mn display moderate mobility. Mobility of sulphur can vary among different plant species. Conversely, deficiency symptoms of immobile nutrients such as Ca and B tend to appear initially in newer growth as these nutrients cannot be redistributed from older leaves to new growth.

Common deficiency symptoms of nutrients include chlorosis (yellowing); purpling; local necrosis (tissue death); stunted growth; poor-quality buds, flowers or fruit; inadequate root development and distorted leaves such as cupping or twisting.

10.1.2.1 Generalized Chlorosis Characterized by Overall Yellowing

- ✰ Chlorosis results from deficiencies in essential elements required for photosynthesis or chlorophyll production. Leaves or entire plants may exhibit a light green or yellow hue which can also manifest in localized areas such as yellowing between veins (interveinal chlorosis).

- ☆ Nitrogen or sulphur deficiencies particularly noticeable on older leaves often trigger chlorosis.
- ☆ Nitrogen deficiency-induced chlorosis is commonly observed in soil-grown plants.
- ☆ Chlorosis may precede the demise of older leaves when subjected to excessive levels of sodium, potassium, chloride or phosphate.
- ☆ Partial chlorosis on mature leaves may indicate magnesium deficiency.
- ☆ Factors unrelated to nutrition such as aging, inadequate root aeration or insufficient light can also contribute to generalized chlorosis.

10.1.2.2 Interveinal Chlorosis (Yellowing)

Interveinal chlorosis characterized by yellowing between leaf veins where veins retain their green colour is indicative of deficiencies in essential micronutrients such as Fe, Mn, Zn or Cu. Particularly iron deficiency can be exacerbated by alkalinity in the nutrient solution which reduces solubility of these metals. Apart from nutritional factors, interveinal chlorosis can also be caused by virus diseases. It is worth noting that certain ornamental plants naturally exhibit yellowing between the veins as part of their normal growth patterns.

10.1.2.3 Purpling

- ☆ Purple or red discolouration typically appears on stems or along leaf petioles, veins or margins stemming from abnormal levels of anthocyanin accumulation induced by plant stress. This phenomenon can also be triggered by physical stressors such as cold temperatures, drought or diseases.
- ☆ It manifests as an overall dark green colouration with a purple, red or blue hue often signalling phosphate deficiency.
- ☆ Some plant species and varieties, however, exhibit yellowing rather than purpling in response to phosphate deficiency.
- ☆ Certain healthy ornamentals naturally display purpling.
- ☆ Moreover, purpling may arise from drought conditions and aluminum toxicity ~ common issues in soil-based cultivation but less prevalent in hydroponic systems.

10.1.2.4 Necrosis (Tissue death, papery appearance, dry patches, scorching)

Necrosis is characterized by death of tissues usually presenting as a papery appearance with dry patches *cum* scorching on leaves.

- ☆ Leaf burn manifests as patches or spots of death on leaves which turn brown with ultimate death. Such symptoms usually occur in later stages of deficiency when plant parts are severely stressed.

- Deficiencies in potassium, magnesium and occasionally molybdenum may contribute to necrosis.
- Non-nutritional factors such as frost, prolonged drought (or low humidity) and sun scorch can also induce necrosis.
- The appearance, location and colour of the affected areas vary depending on the specific disorder and plant species and such symptoms aid in accurate diagnosis.

10.1.2.5 Stunting

- Reduced height: The plant appears shorter than its typical growth.
- Overall stunting: Stunted growth across the entire plant can indicate various nutrient disorders. This can be challenging to identify without comparison to normal plants of the same type and age.
- Specific growth patterns: Certain disorders may lead to acute stunting of growing points or young leaves, resulting in recognizable growth patterns.
- Zinc deficiency symptoms:
  - 'Little leaf' phenomenon: Some species particularly woody plants may exhibit reduced leaf size which is known as 'little leaf' syndrome.
  - Smaller young leaves: Younger leaves are notably smaller than usual.
  - Rosetting: Zinc deficiency may cause the stem to cease elongation behind the growing tip thus resulting in tightly bunched terminal leaves.
- Potential fatal deficiencies: B, Ca, K and occasionally Zn deficiencies can lead to death of growing points. This in woody plants can escalate to shoot dieback.

10.1.2.6 Root Stunting

- Root stunting is commonly associated with calcium deficiency, acidity, aluminum toxicity and copper toxicity.
- Certain plant species may exhibit similar symptoms when lacking boron.
- Affected roots appear shortened and thickened with stubby and peg-like laterals. Often, the entire root system may discolour ~ turning brown or gray.

10.1.2.7 Dieback

Dieback occurs when leaves or growing points of plants die rapidly and dry out. This condition may be attributed to deficiencies in boron or calcium levels within the plant.

10.1.2.8 Mottling

Mottling refers to the presence of irregular spotted patterns on the surface of leaves or other plant parts. It manifests as a blotchy pattern characterized by areas of both light and dark discolouration. Mottling is commonly associated with viral diseases in plants

10.1.2.9 Checkered Appearance (Reticulation)

The symptoms include a checkered appearance (reticulation) where small veins of the leaves retain their green colouration while the tissue between them turns yellow. This condition may indicate a manganese deficiency.

10.1.2.10 Spindle Formation

It is characterized by thin and succulent growth of stem and leaf petioles. This phenomenon is often associated with nitrogen deficiency.

10.1.3 Identifying and Addressing Nutrient Disorders

Identifying and addressing nutrient deficiencies and toxicities are crucial for optimizing plant health and maximizing yields in hydroponic systems. Here is how to recognize symptoms of nutrient imbalances and address them effectively:

10.1.3.1 Nitrogen Deficiency Symptoms

- Nitrogen deficiency manifests primarily in older leaves which may exhibit symptoms such as pale green or yellowing (chlorosis) due to a lack of chlorophyll, extending to newer growth over time. Eventually, affected leaves may wither and die.
- Stunted leaf growth and browning of leaf tips in older leaves are common signs of nitrogen deficiency.
- New growth may appear weak, spindly and less branched than usual.
- Overall growth including that of tops, roots and lateral shoots may be noticeably restricted.
- Some plants may develop purple discolouration in stems, petioles and lower leaf surfaces as a secondary symptom.
- Nitrogen deficiency usually results in uniform yellowing of entire leaves including the veins, without distinct interveinal striping or patches.
- Severe nitrogen deficiency may cause leaves to take on a yellowish-white hue.
- In certain cases, interveinal necrosis may replace chlorosis as the deficiency progresses.

Plants/Leaves Showing Nitrogen Deficiency.

Treatment

- **Adjustment of nutrient solution**: Increase of nitrogen concentration in the nutrient solution should be done by adding nitrogen-rich fertilizers or adjusting the ratio of nutrient solution to provide more nitrogen. Common nitrogen-rich fertilizers used in hydroponics include potassium nitrate, ammonium nitrate or urea.
- **Application of foliar sprays**: In addition to adjusting the nutrient solution, one should also apply foliar sprays containing nitrogen to provide immediate relief to the plants. Foliar sprays deliver nutrients directly to the leaves so bypassing any root uptake issues.

10.1.3.1.2 Phosphorus Deficiency Symptoms

- Initially manifests in older leaves.
- Symptoms are often subtle and challenging to discern.
- Plants exhibit stunted growth, lagging behind others in similar conditions.
- Leaves may darken to a dull green or gray with blotches.
- Overall plant growth is reduced with leaves appearing smaller and darker, sometimes with brown or purple spots.

Leaves Showing Phosphorus Deficiency.

- Bud and root development is hindered resulting in smaller *cum* underdeveloped structures.
- Petioles, leaf veins as well as stems may display purpling.
- Prolonged deficiency leads to poor root formation and delayed plant maturity.
- Certain plants such as tomatoes, lettuce, corn and brassicas may show distinct purpling of stems, petioles and leaf-undersides.
- Severe deficiency may cause leaves to develop a blue-gray lustre.
- Older leaves may develop a brown and netted veining pattern in advanced stages.

Treatment

- **Adjusting nutrient solution**: Phosphorus concentration in nutrient solution should be increased by using the fertilizer with higher phosphorus content or by adding a phosphorus-specific nutrient supplement to the reservoir.
- **Foliar sprays**: In severe cases of deficiency foliar application of phosphorus-containing fertilizers may be beneficial. This method allows direct absorption of phosphorus through leaves bypassing potential root uptake issues.

10.1.3.1.3 Potassium Deficiency Symptoms

- ✰ Results in taller plants having outwardly healthy appearance where initially older leaves turn yellow between veins, progressing dark yellow, brown and eventual curling or death.
- ✰ Small dead spots may emerge at leaf tips and between veins.
- ✰ Leaf margins cup downward with brown spots.
- ✰ Chlorotic older leaves develop dark necrotic spots (dead tissue) which may spread inward and upward to younger leaves as deficiency worsens.
- ✰ Weak and brittle stem and branches may occur.
- ✰ Leads to flower and fruit drop and poorly developed roots.

Leaves Showing Potassium Deficiency.

Treatment

- ✰ **Adjusting nutrient solution**: Increase of potassium concentration through addition of potassium sulphate or potassium nitrate to the reservoir as per manufacturer recommendations.
- ✰ **Supplement with foliar sprays**: In severe cases of potassium deficiency foliar application of potassium-containing fertilizers may be necessary to provide a quick boost to plant health. Diluted solution spray of potassium fertilizer directly on the leaves will be quite effective.

10.1.3.1.4 Magnesium Deficiency Symptoms

- ✰ Expression of symptoms is influenced by light intensity with higher exposure exacerbating symptoms.
- ✰ Initial effects are seen in older leaves.

Symptoms of Magnesium Deficiency Showing Yellowing in Older Leaves.

- Advanced interveinal chlorosis occurs with necrosis in severely affected tissue.
- Advanced deficiency may resemble potassium deficiency.
- Symptoms begin as mottled chlorotic areas in interveinal tissue of older leaves, progressing to younger leaves with severe deficiency.
- Interveinal tissue expands disproportionately, resulting in raised and puckered surfaces transitioning from chlorotic to necrotic tissue.
- Some plants such as *Brassica* species may exhibit orange, yellow or purple tints.

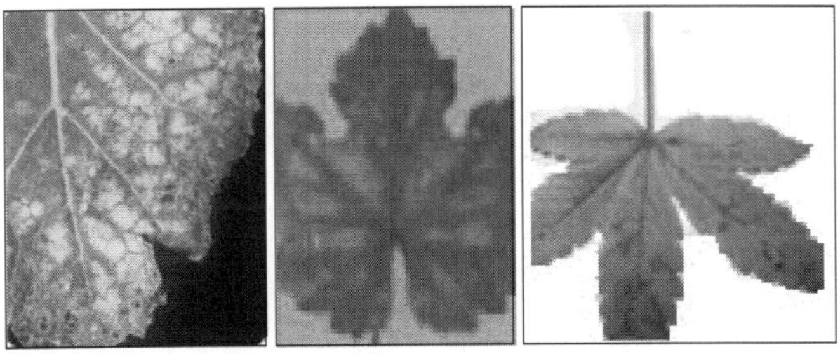

Symptoms of Magnesium Deficiency Showing Brown Spots on Leaves.

- Older leaves yellow from the centre outward while veins remain green.
- Brown spots develop on leaves.
- Leaf tips and edges may discolour or curl upward.
- Bud development is inhibited.
- In certain crops, interveinal yellow patches may progress to necrotic spots or patches and marginal leaf scorching.

Treatment

- ☆ **Adjusting nutrient solution**: Increase of Mg concentration in nutrient solution through Mg-containing fertilizers or supplements. Common sources of magnesium for hydroponics include magnesium sulphate (epsom salt) and magnesium nitrate.

10.1.3.1.5 Calcium Deficiency Symptoms

- ☆ Initial signs manifest in young growths characterized by tip and edge deficiencies in leaves, *vis-a-vis* lacking brittle tissue.
- ☆ Common symptoms include blossom-end rot in tomatoes, tip burn in lettuce, blackheart in celery and necrotic tissue in rapidly growing areas of various plants. This necrotic tissue is often attributed to poor calcium translocation rather than insufficient external calcium supply.

Calcium Deficiency Showing Blossom-End Rot of Tomato.

- ☆ New leaves exhibit stunted growth and dryness.
- ☆ Young leaves become distorted and chlorotic or develop irregular margins, spots or necrotic areas.

Leaves Showing Calcium Deficiency.

- ☆ Browning and blackening of new growth with potential death.
- ☆ Necrotic tissue remains slow to dry and does not crumble easily.
- ☆ Inhibited bud development.

- Prolonged deficiency leads to stunted growth and underdeveloped root formation.
- Fruits may suffer from stunting or deformities.
- Typically young leaves are affected first.
- Unlike B-deficiency, Ca-deficiency causes short internodes and does not promote lateral shoot growth.

Treatment

- **Use of Ca-rich nutrient formulations**: Choosing hydroponic nutrient formulations containing sufficient levels of calcium will be effective. Some formulations are specifically designed to address calcium deficiency and may include higher concentrations of calcium nitrate, calcium chloride or calcium carbonate.
- **Supplement with calcium additives**: If calcium levels in nutrient solution are inadequate this should be supplemented with calcium additives such as calcium chelates, calcium sulphate (gypsum) or calcium-based pH adjusters. Follow manufacturer recommendations for dosage and application.
- **Foliar spray application**: In addition to root uptake, calcium can be absorbed through the leaves *via* foliar sprays. Calcium-rich foliar sprays directly supply calcium to plant tissues particularly during periods of rapid growth or when root uptake is limited.

10.1.3.1.6 Sulphur Deficiency Symptoms

- Initial signs manifest in younger leaves.
- Symptoms may vary across plant species as in corn, S-deficiency results in interveinal chlorosis while wheat exhibits overall paleness particularly in younger leaves. Potatoes may develop leaf spotting.

Leaves Showing Sulphur Deficiency.

Sulphur Deficiency in Corn and Soybean.

- Young leaves become pale yellow with slowed growth.
- Veins turn yellow (chlorotic).
- Yellowing typically starts at the top of the plant and progresses downward affecting veins as well. Treatment mirrors that of magnesium deficiency.
- Leaf tips may yellow and curl downwards.
- Leaves may become brittle and narrower than usual.
- Similar to nitrogen deficiency, older leaves turn yellowish-green while stems become thin, hard and woody.
- Some plants exhibit orange and red hues rather than yellowing, particularly on the undersides of leaves. Petioles may also have a pinkish tone albeit less vivid than in nitrogen deficiency.
- Despite becoming hard and woody, stems elongate without an increase in diameter.

Treatment

- **Adjusting nutrient solution**: S-concentration in nutrient solution should be increased through addition of S-containing fertilizers or amendments. Common sulphur sources include magnesium sulphate (epsom salt) and potassium sulphate. Manufacturer's recommendations for application rates and frequency should be followed.
- **Considering foliar sprays**: In severe cases of sulphur deficiency, foliar application of S-containing fertilizers may provide a quick solution. Fertilizers should be diluted according to the manufacturer's instructions. Direct application should be made on the leaves ensuring thorough coverage.

10.1.3.1.7 Iron Deficiency Symptoms

- ☆ Its deficiency signs first appear in younger leaves.
- ☆ Its deficiency will mask Mg-deficiency.
- ☆ The newer growth will exhibit interveinal chlorosis typically differentiating it from Mg-deficiency.
- ☆ Interveinal chlorosis evolves into an overall chlorosis and ends as a totally bleached leaf that often develops necrotic spots.
- ☆ Its deficiency turns leaves pale yellow to white but veins remain green.
- ☆ Leaves dry out and turn crinkly.
- ☆ In severe cases, new leaves become completely lacking in chlorophyll but with little or no necrotic spots.
- ☆ The chlorotic mottling on immature leaves may start first near the bases of the leaflets so that in effect the middle of the leaf appears to have a yellow streak.

Leaves Showing Iron Deficiency.

Treatment

- ☆ **Selecting iron supplements**: A suitable iron supplement suitable for hydroponic system should be chosen. Common options include iron chelates such as iron EDTA, iron DTPA or iron EDDHA. Each type of chelate has different pH ranges in which they remain stable and available to plants. Iron EDDHA is most effective in high pH conditions while iron EDTA is suitable for low pH ranges.
- ☆ **Applying iron supplements**: Selected iron supplement should be added to the nutrient solution according to the manufacturer's recommendations. It should start with a low concentration with gradual increase as needed while monitoring plant response. Avoid overdosing, as excessive iron levels can lead to toxicity and may harm plants.

- **Considering foliar sprays**: In severe cases of iron deficiency, foliar application of iron supplements may provide a quick solution. Dilution of iron supplement as per manufacturer's instructions should be made and sprayed directly onto the leaves ensuring thorough coverage. This allows plants to absorb iron directly through their foliage.

10.1.3.1.8 Chlorine Deficiency Symptoms

- Chlorosis and wilting of young leaves. Chlorosis occurs on smooth flat depressions in interveinal areas of leaf blade.
- In more advanced cases, often there appears a characteristic bronzing on upper side of mature leaves.
- Wilted chlorotic leaves become bronze in colour.
- Roots become stunted and thickened near tips.

Leaves Showing Chlorine Deficiency.

Treatment

- **Confirming the deficiency**: Chlorine deficiency symptoms may include leaf yellowing, stunted growth and reduced vigour. However, these symptoms can also indicate other nutrient deficiencies or environmental stressors. Thorough assessment is made to confirm chlorine deficiency.
- **Checking water source**: Water source should be tested to determine its chlorine content. Tap water often contains chlorine added during municipal water treatment process. If water source lacks chlorine, use an alternative water source or water should be treated to ensure it contains adequate levels of chlorine.
- **Using chlorinated water**: In case water source lacks chlorine consider using chlorinated water or adding a chlorine-based disinfectant to the nutrient solution. Chlorine bleach (sodium hypochlorite) can be used to raise chlorine levels in the nutrient solution. However, use only proper dilution to avoid harming plants.
- **Considering foliar sprays**: In severe cases of chlorine deficiency, foliar application of chlorine-containing solutions may provide a quick remedy. Dilution of solution should be done according to manufacturer instructions and then this should be applied directly on leaves for ensuring thorough coverage.

10.1.3.1.9 Manganese Deficiency Symptoms

- ☆ Its deficiency signs first appear on younger leaves.
- ☆ Its deficiency exhibits a general chlorosis followed by yellowing or necrotic patches between veins of larger leaves.
- ☆ Symptoms appear as netted veins of mature leaves especially when these are viewed through transmitted light.
- ☆ Turns young leaves mottled yellow or brown and growth virtually stops.
- ☆ Interveinal chlorosis occurs on younger or older leaves followed by leaf shedding.
- ☆ In many plants it is indistinguishable from that of iron.
- ☆ On fruiting plants, the blossom buds often do not fully develop and turn yellow or abort.
- ☆ As the deficiency becomes more severe, the new growth becomes completely yellow.
- ☆ As the stress increases, the leaves take on a gray metallic sheen and develop dark spots and necrotic areas along the veins.
- ☆ A purplish lustre may also develop on the upper surface of the leaves.
- ☆ Grains such as wheat and barley are extremely susceptible to manganese deficiency. These develop a light chlorosis along with gray specks which elongate and coalesce, and eventually the entire leaf withers and dies.

Leaves Showing Manganese Deficiency.

Treatment

- **Adjusting nutrient solution**: Manganese concentration should be increased in nutrient solution through addition of Mn-containing fertilizers or chelated Mn-supplements. Manganese sulphate or chelated manganese are commonly used as potent sources. Manufacturer's instructions should be followed for application rates.
- **Monitoring other nutrients**: Mn-deficiency may occur alongside deficiencies in other micronutrients or imbalances in macronutrients. Regular monitoring of nutrient levels is important and in case there is deficiency, it should be adjusted to maintain a balanced ratio in the solution.
- **Considering foliar sprays**: In severe cases of Mn-deficiency, foliar application of Mn-containing fertilizers may provide a quick solution. Dilution of the fertilizer should be made as per manufacturer's instructions and then this should be sprayed directly onto the leaves ensuring thorough coverage.

10.1.3.1.10 Zinc Deficiency Symptoms

- In some plants, its deficiency first affects older leaves and in others it affects immature leaves.
- Its deficiency results in stunting, yellowing and downward curling of small leaves.
- Small young leaves with twisted leaflets, especially affecting growing points of all the plants.
- Plant's growth rate slows down or gets completely stopped.
- Chlorosis may accompany reduction of leaf size and a shortening between internodes.

Leaves Showing Zinc Deficiency.

- ☆ Leaf margins are often distorted or wrinkled.
- ☆ In some plants, the interveinal chlorotic mottling first appears on the older leaves though on others on immature leaves.
- ☆ The interveinal chlorotic mottling may be the same as that for iron and manganese except for the development of exceptionally small leaves.
- ☆ If onset of zinc deficiency is sudden such as left out of the nutrient solution the chlorosis can appear identical to that of iron and manganese without little leaf.
- ☆ As the deficiency progresses, these symptoms develop into an intense interveinal necrosis but the main veins remain green, as in the symptoms of recovering iron deficiency.
- ☆ In many plants, especially trees, the leaves become very small and the internodes shorten thus producing a rosette like appearance.

Treatment

- ☆ **Adjusting nutrient solution**: Judicious increase of zinc concentration in the nutrient solution should be made through addition of Zn-containing fertilizers or amendments. Common zinc sources include zinc sulphate, zinc chelates and zinc oxide. Manufacturer's recommendations should be followed for application frequency and rates.
- ☆ **Considering foliar sprays**: In severe cases of zinc deficiency, foliar application of Zn- containing fertilizers may provide a quick solution. Fertilizer should be diluted as per manufacturer's instructions. These should be applied directly on the leaves ensuring thorough coverage.

10.1.3.1.11 Copper Deficiency Symptoms

- ☆ Its deficiency signs first appear in younger leaves.
- ☆ Copper deficiency may be expressed as overall light chlorosis along with the permanent loss of turgor in the young leaves.
- ☆ Its deficiency causes wilting of new growth, *vis-a-vis* irregular growth.
- ☆ Growing tips die off first along with the crispy leaves.
- ☆ The new shoots turning brown or even white and start dieing first from the tips and margins.
- ☆ Young leaves often become dark green and twisted. They may die off or just exhibit necrotic spots.
- ☆ Recently matured leaves show netted *cum* green veining with areas bleaching to a whitish gray.

Leaves Showing Copper Deficiency.

- ✰ Leaves on top half of plant may show unusual puckering with veinal chlorosis.
- ✰ Absence of a knot on leaf where petiole joins the main stem of plant beginning about 10 or more leaves below growing point.
- ✰ Trees under chronic copper deficiency develop a rosette form of growth. Leaves are small and chlorotic with spotty necrosis.

Treatment

- ✰ **Adjusting nutrient solution**: Copper concentration in the nutrient solution should be increased through addition of Cu-containing fertilizers or amendments. Common sources of copper include copper sulphate and copper chelates. Manufacturer's recommendations for application rates and frequency should be followed.
- ✰ **Avoiding excess copper**: While addressing copper deficiency, it is important not to over-apply Cu-containing fertilizers as excessive copper levels can be toxic to plants so for this one should follow recommended dose guidelines and monitor copper levels in the solution to prevent toxicity.
- ✰ **Considering foliar sprays**: In severe cases of copper deficiency, foliar application of copper-containing fertilizers may provide a quick solution. Dilution of fertilizer according to manufacturer's instructions should be followed and applied directly to the leaves ensuring thorough coverage.

10.1.3.1.12 Boron Deficiency Symptoms

- ✰ Its deficiency sign first appears in young growth.
- ✰ Its deficiency may result in limited budding, bud break, distorted shoot growth, short internodes, increased branching, falling of flower buds and inhibition of fruit and seed development.

Boron Deficiency Symptoms in Pepper Leaves.

- Growing tips (leaves and petioles) may become light green to yellow.
- Rosetting of terminal growth occurs due to shortening of internodes.
- Terminal buds may die and new growth may form at lower leaf axils.
- Its deficiency may show slight chlorosis, brown to black scorching of new leaf tips and die-back of the growing points similar to calcium deficiency.
- Leaves show various symptoms which include drying, thickening, distorting, wilting and chlorotic or necrotic spotting.
- Also the occurrence of brown and black tissue die-back which dries up quite slowly, so that it can be crumbled easily.
- Both the pith and epidermis of stems may be affected as exhibited by hollow to roughened and cracked stems.
- Root tips often become swollen and discoloured.
- Internal tissues may rot and become host to fungal diseases.

Treatment

- **Adjusting nutrient solution**: Boron concentration in the nutrient solution should be increased through addition of boron-containing fertilizers or amendments by following manufacturer's instructions. Common boron sources include boric acid and borax.
- **Applying foliar sprays**: In severe cases of boron deficiency, foliar application of boron-containing fertilizers may provide a quick solution. Fertilizer should be diluted according to the manufacturer's instructions and its application should be directly on the leaves ensuring thorough coverage.

10.1.3.1.13 Molybdenum Deficiency Symptoms

- ☆ Its deficiency signs first appear on older leaves.
- ☆ Mottling of older leaves with veins remaining light green.
- ☆ Leaves showing molybdenum deficiency

Leaf Margins become Necrotic and may Curl Upward.

- ☆ Exhibits yellowing (necrotic) lesions on leaves from the tips inward.
- ☆ A very distinctive feature is that it occurs in the middle of the plant. It also may spread to the rest of the plant if not stopped, eventually killing the plant.
- ☆ Its deficiency causes leaves to turn pale and fringes to appear scorched.
- ☆ There may be irregular growth of leaves.
- ☆ Often interveinal chlorosis which occurs first on older leaves, then progressing to the entire plant.
- ☆ Younger leaves twist and may die.
- ☆ Older leaves show interveinal chlorotic blotches, become cupped and thickened.
- ☆ Chlorosis continues upward to younger leaves as deficiency progresses.
- ☆ In case of cauliflower, the lamina of the new leaves fail to develop resulting in a characteristic whiptail appearance.
- ☆ In many plants there is an upward cupping of the leaves and mottled spots developing into large interveinal chlorotic areas under severe deficiency.

Treatment

- ☆ **Adjusting nutrient solution**: Molybdenum concentration in the nutrient solution should be increased by adding molybdenum-containing fertilizers or amendments but as per manufacturer's recommendations. Sodium molybdate is a common source of hydroponic systems.

☆ **Considering foliar sprays**: In severe cases of molybdenum deficiency, foliar application of molybdenum-containing fertilizers may provide a quick solution. Dilute the fertilizer according to the manufacturer's instructions and apply it directly on the leaves ensuring thorough coverage.

10.1.3.1.14 Nickel (Ni) Deficiency Symptoms

Nickel deficiency in hydroponically grown plants is relatively rare but can occur particularly in crops that heavily rely on nickel-dependent enzymes for nitrogen metabolism such as legumes. Since some plant species are more sensitive to nickel deficiency than others, growers must be aware of the specific requirements of the plants one is growing.

☆ Nickel deficiency symptoms may manifest similarly to nitrogen deficiency including chlorosis (yellowing) of leaves, stunted growth and poor nitrogen uptake.

Treatment

☆ **Testing nutrient solution**: Analysis of the nutrient solution should be done to assess nickel levels. Low concentrations of nickel may indicate deficiency.

☆ **Supplementing with nickel:** If nickel deficiency is confirmed, it should be supplemented through nutrient solution with a source of soluble nickel, *e.g.* nickel sulphate ($NiSO_4$) or nickel chloride ($NiCl_2$). One of these can be dissolved in the trace amounts in water thoroughly.

10.1.3.1.15 General Treatment Strategies for Nutrient Deficiencies

☆ **Confirming deficiency**: Using of diagnostic tools such as nutrient solution testing kits or leaf tissue analysis to confirm the specific nutrient deficiency and rule out other potential issues. Measuring the nutrient levels in nutrient solution using a reliable testing kit or device. This helps determine the severity of the deficiency and guides the appropriate corrective action.

☆ **Electrical conductivity (EC) monitoring:** Regularly EC of the nutrient solution should be monitored to ensure adequate nutrient levels without causing nutrient imbalances.

☆ **Application rate**: Manufacturer's instructions should be followed or expert consultation should be made to determine the appropriate dosage of fertilizer based on the severity of the deficiency and the specific requirements of the crop.

☆ **Maintaining balanced nutrition**: While addressing specific nutrient deficiency, it is essential to maintain a balanced nutrient solution to prevent imbalances or deficiencies in other essential nutrients. Monitoring the nutrient levels regularly and adjust the solution accordingly to meet the plant's requirements.

- **Monitoring and adjusting pH levels**: Checking and adjusting pH of nutrient solution should be done regularly to ensure that it remains within the optimal range for nutrient uptake by the plants. Nutrient availability can be affected by pH levels so maintaining appropriate pH range (typically between 5.5 and 6.5 for most hydroponic crops) is crucial.
- **Optimizing watering and feeding**: It should be ensured that plants are receiving adequate water and nutrients. Maintaining a consistent watering schedule and avoiding overwatering should be standard practice as excessive moisture can inhibit nutrient uptake. Nutrient feeding schedules are strictly followed to provide a steady supply of nutrients throughout the growth cycle.
- **Environmental factors:** Factors such as temperature, humidity and light intensity can influence nutrient uptake and plant response to nutrient deficiency. Optimize environmental conditions to support plant health.
- **Optimizing temperature**: Maintain appropriate temperature range for nutrient uptake typically between 18 to 24°C. Extreme temperatures can affect nutrient absorption.
- **Optimizing root health**: Ensure that the root system is healthy and functioning properly to maximize nutrient uptake. Maintaining proper aeration, avoiding overwatering and regularly checking for any signs of root rot or other issues that may hinder nutrient absorption.
- **Considering microbial inoculants**: Introducing beneficial microbes such as nitrogen-fixing bacteria to the root zone can enhance nutrient availability and uptake as these can convert atmospheric nitrogen into a form that plants can utilize.
- **Monitoring progress**: Continuously monitor the plants for improvements in growth and leaf colouration after implementing nutrient supplementation. Adjust nutrient levels and other factors as needed to maintain optimal plant health.
- **Prevention**: To prevent future deficiencies, ensure proper nutrient management practices including balanced fertilization, regular pH monitoring and maintaining optimal growing conditions for plants.
- **Regular maintenance:** Keep the hydroponic system clean and well-maintained to prevent nutrient deficiencies and other problems.
- **Maintaining proper environmental conditions**: Ensure that environmental conditions such as temperature, humidity and airflow

are optimal for calcium uptake and transpiration in plants. Stressful conditions such as high temperatures or low humidity can hinder calcium absorption and exacerbate deficiency symptoms.

- ☆ **Seeking expert advice**: If you encounter persistent or severe magnesium deficiency despite corrective measures, seek advice from hydroponic specialists or agronomists to assess and address any underlying issues.

10.1.3.1.16 Foliar Feeding

Foliar feeding is a method of nourishing plants by applying liquid fertilizer directly onto their foliage. This involves spraying water-dissolved fertilizers onto the leaves either as a blend of elements or as a single nutrient to address specific deficiencies such as calcium. While hydroponics typically relies on a balanced nutrient solution delivered to the roots, foliar feeding can complement this regimen under certain circumstances:

- ☆ **Addressing deficiencies:** Foliar feeding can swiftly alleviate signs of nutrient deficiencies by targeting specific elements lacking in the plant. This provides a quick but temporary solution while adjusting the nutrient solution formula.
- ☆ **Combatting root diseases:** In instances where root diseases hinder nutrient absorption foliar feeding can deliver essential elements directly to the leaves thus reversing deficiencies temporarily as efforts are made to control the root disease.
- ☆ **Adapting to environmental factors:** Foliar feeding proves beneficial in environments with extreme humidity levels or conditions that impede nutrient uptake through the roots such as high or low pH levels in the nutrient solution, temperature stress, root diseases or the presence of pests affecting nutrient absorption.
- ☆ **Enhancing yield and quality:** During critical growth stages, foliar feeding can enhance both the yield and quality of crops.

10.1.3.1.16.1 Limitations of Foliar Feeding

However, foliar feeding has its limitations, and understanding these factors is crucial for utilizing foliar feeding effectively and avoiding potential drawbacks:

- ☆ **Incomplete nutrient supply:** While effective for targeted supplementation, foliar application alone cannot fulfill the entire nutrient requirements of crops.
- ☆ **Risk of leaf burn:** Applying high concentrations of nutrients through foliar feeding may lead to leaf burn as water evaporates and salts accumulate on the leaf surface.

10.1.3.1.16.2 Enhancing Foliar Feeding Efficiency

Several factors influence the effectiveness of foliar feeding:

- **pH of spray solution:** Typically an acidic pH enhances nutrient penetration through leaf surfaces. Moreover, pH impacts foliar nutrient absorption in several ways:
  - It influences the cuticle's charge thus affecting ion selectivity.
  - The ionic form of nutrients is pH-dependent so influencing penetration rates.
  - pH may influence the phytotoxicity of sprayed compounds.
- **Use of surfactants (adjuvants):** Surfactants promote uniform foliage coverage enhancing spray solution retention by reducing droplet surface tension.

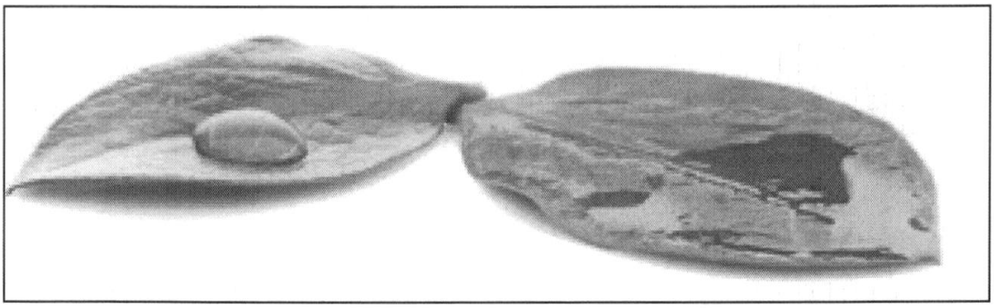

Without Surfactant **With Surfactant**

- **Timing:** Optimal foliar feeding occurs during early morning or late evening when stomata are open. Avoiding foliar feeding during temperatures exceeding 27°C is advisable.
- **Droplet size:** Smaller droplets offer broader coverage and increase foliar application efficiency. However, excessively small droplets (under 100 microns) may cause drift.
- **Spray volume:** Adequate spray volume is essential to cover plant canopy fully without excessive runoff.

By considering and optimizing these factors, foliar feeding can be made more efficient and effective in delivering nutrients to plants.

10.1.3.1.17 Root Drenching

Root drenching in hydroponics involves applying a liquid solution directly to the root zone of plants grown in a soilless medium such as rockwool, perlite or coco coir. This technique allows for targeted delivery of nutrients, water or other substances to the root system so promoting healthy growth and maximizing nutrient uptake. Root drenches can be used for various purposes including supplemental nutrition, correcting nutrient deficiencies or treating root diseases. Here is a brief overview of root drenches in hydroponics:

10.1.3.1.17.1 Purpose of Root Drenches

i) **Nutrient delivery:** Root drenches can supply essential nutrients directly to the root zone thus ensuring optimal nutrient levels for plant growth and development.

ii) **Watering:** Root drenching is an effective method for watering plants especially during periods of drought or when root system requires additional moisture.

iii) **pH and EC adjustment:** Root drenches can help adjust the pH and EC of root zone by applying pH-adjusted water or nutrient solutions.

iv) **Disease management:** Root drenches can be used to administer fungicides or beneficial microorganisms to control root diseases and promote root health.

10.1.3.1.17.2 Application of Root Drenches

i) **Preparation:** Prepare desired solution for root drenching such as a nutrient solution, water with pH adjusters or a solution containing additives for disease management.

ii) **Application:** Carefully pour the solution directly onto the root zone of the plants thus ensuring thorough coverage of the root system.

iii) **Timing:** Root drenches can be applied as needed based on the specific requirements of plants such as during watering or as part of a regular maintenance schedule.

iv) **Monitoring:** Monitor the plants closely after root drenching to assess their response and adjust future applications as necessary.

10.1.3.1.17.3 Benefits of Root Drenches

i) **Efficient nutrient uptake:** Direct application to the root zone ensures efficient nutrient uptake by the plants so minimizing wastage.

ii) **Targeted delivery:** Root drenches deliver substances directly to the root system thus bypassing potential barriers in the growing medium.

iii) **Quick response:** Plants can respond rapidly to root drenches especially when addressing nutrient deficiencies or root diseases.

iv) **Minimized risk of foliar damage:** Unlike foliar sprays, root drenches minimize the risk of leaf burn or damage so making them suitable for sensitive plants.

In summary, root drenches are a valuable technique in hydroponics for providing targeted nutrition, managing root diseases and optimizing plant health *cum* performance. Proper application and monitoring are essential to ensure successful outcomes and to minimize risks.

10.1.3.2 Nutrient Toxicities

Nutrient toxicities in hydroponics can occur when concentration of certain nutrients in nutrient solution exceeds the levels that plants can tolerate. Here are some common symptoms and treatments for nutrient toxicities in hydroponics:

10.1.3.2.1 Nitrogen Toxicity

Symptoms

- **i)** Dark green leaves.
- **ii)** Leaf curling or wilting.
- **iii)** Reduced fruit/flower production.
- **iv)** Stunted growth.
- **v)** Higher susceptibility to diseases and pests.

Treatment

- **i)** Flush the system with fresh pH-balanced water to remove excess nitrogen.
- **ii)** Adjust the nutrient solution to a lower nitrogen concentration for subsequent feedings.
- **iii)** Monitor and adjust feeding schedules to prevent over-fertilization with nitrogen-rich nutrients.

10.1.3.2.2 Phosphorus toxicity

Its excess is rare and usually buffered by pH limitations. Excess phosphorus can interfere with the availability of other elements such as iron, manganese, copper and zinc:

Symptoms

- **i)** Brown or black leaf tip necrosis.
- **ii)** Dark green leaves.
- **iii)** Reduced growth.
- **iv)** Delayed flowering.

Treatment

- **i)** Adjust pH of nutrient solution to optimize phosphorus uptake.
- **ii)** Flush the system with fresh water to remove excess phosphorus.

iii) Dilute the nutrient solution with water to lower phosphorus concentration.

iv) Choose plant varieties less sensitive to phosphorus toxicity.

10.1.3.2.3 Potassium toxicity

Excess potassium can aggravate the uptake of magnesium, manganese, zinc and iron though its excess is rare:

Symptoms

i) Leaf burn or necrosis (brown or yellow spots on leaves).

ii) Curling or wilting of leaves.

iii) Reduced growth.

iv) Reduced fruit/flower quality.

Treatment

i) Flush the system with fresh water to remove excess potassium.

ii) Adjust the nutrient solution to a lower potassium concentration.

iii) Monitor and adjust feeding schedules to prevent over-fertilization with potassium-rich nutrients.

10.1.3.2.4 Calcium Toxicity

Difficult to distinguish visually. May precipitate with sulphur in solution and cause clouding or residue in tank:

Symptoms

i) Leaf burn or necrosis, especially at leaf margins.

ii) Brittle or distorted leaves.

iii) Blossom end rot in fruiting plants.

iv) Its abundance in early life of plants causes stunted growth.

Treatment

i) Adjust pH of nutrient solution to optimize calcium uptake.

ii) Flush the system with fresh water to remove excess calcium.

iii) Dilute the nutrient solution with water to lower calcium concentration.

iv) Ensure proper balance with magnesium and potassium levels.

10.1.3.2.5 Magnesium Toxicity

Its toxicity is rare and not generally exhibited visibly.

Symptoms

i) Leaf discolouration or chlorosis (yellowing).

ii) Leaf curling or twisting.

iii) Reduced growth.

Treatment

i) Flush the system with fresh water to remove excess magnesium.

ii) Adjust the nutrient solution to a lower magnesium concentration.

iii) Monitor and adjust feeding schedules to prevent over-fertilization with magnesium-rich nutrients.

10.1.3.2.6 Sulphur Toxicity

Sulphur toxicity in hydroponics is less commonly encountered compared to other nutrient imbalances. However, sulphur toxicity can indeed occur in hydroponic systems under certain conditions such as excessive sulphur concentration in the nutrient solution, poor drainage leading to sulphur accumulation in the root zone and high levels of sulphur in the water source.

Leaf Showing Sulphur Toxicity

Symptoms

i) Leaf yellowing (chlorosis) starting from the tips or margins and progressing inward.

ii) Leaf burn or necrosis particularly at the leaf tips.

iii) Stunted growth.

iv) Reduced root development.

v) Reduced yield.

Treatment

i) Flush the hydroponic system with fresh pH-balanced water to remove excess sulphur.

ii) Adjust the nutrient solution to a lower sulphur concentration for subsequent feedings.

iii) Monitor and adjust feeding schedules to prevent over-fertilization with sulphur-rich nutrients.

iv) Check the sulphur content of the water source and consider using alternative water sources if necessary.

v) Ensure proper drainage to prevent sulphur buildup in the root zone.

vi) Consider using sulphur-free or low-sulphur fertilizers in hydroponic nutrient mix.

10.1.3.2.7 Iron Toxicity

Its excess accumulation is rare but could cause bronzing or tiny brown spots on leaf surface.

Symptoms

i) Leaf discolouration or chlorosis (yellowing).

ii) Leaf burn or necrosis.

iii) Reduced growth.

Treatment

i) Adjust pH of nutrient solution to optimize iron availability.

ii) Flush the system with fresh water to remove excess iron.

iii) Dilute the nutrient solution with water to lower iron concentration.

iv) Ensure proper balance with other micronutrients.

10.1.3.2.8 Chlorine Toxicity

Chlorine toxicity is indeed a potential issue in hydroponics though it is quite less common because most growers use water sources that don't contain high levels of chlorine. However, if tap water or another water source treated with chlorine is being used, it is essential to be aware of its potential effects on plants. Plants are generally tolerant of chloride but some species such as avocadoes, stone fruits and grapevines are sensitive to chlorine and can show toxicity even at low chloride concentration in the soil.

Symptoms

i) Leaf burn or necrosis, especially at leaf tip or margins.

ii) Bronzing, yellowing and leaf splitting.

iii) Reduced growth or stunted plants.

iv) Brown roots or root damage.

Treatment

i) Before using water in hydroponic systems, it should be dechlorinated using methods such as aging (letting it sit for 24-48 hours), using activated carbon filters or through dechlorinating agents available at gardening stores.

ii) Consider using RO water which removes chlorine and other impurities for hydroponic systems.

iii) There are commercial products available that neutralize chlorine in water, making it safe for plants.

iv) Chemicals like sodium thiosulphate can be used to neutralize chlorine or chloramines present in tap water.

v) Chlorine can affect pH of nutrient solution. Ensure that pH is within the optimal range for plant growth after dechlorinating the water.

vi) Regularly test water quality for chlorine levels especially when using tap water to ensure it meets the requirements for hydroponic cultivation.

10.1.3.2.9 Copper, Zinc, Manganese or Boron Toxicity

Symptoms

i) Leaf discolouration or chlorosis (yellowing).

ii) Leaf burn or necrosis.

iii) Reduced growth.

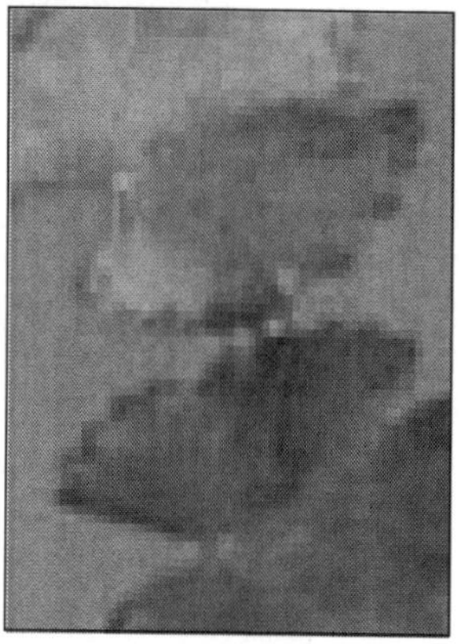

Leaves Showing Boron Toxicity

Treatment

i) Adjust pH of nutrient solution to optimize micronutrient availability.

ii) Flush the system with fresh water to remove excess micronutrients.

iii) Dilute the nutrient solution with water to lower micronutrient concentration.

iv) Monitor and adjust micronutrient supplementation carefully.

10.1.3.2.10 Nickel Toxicity

Nickel toxicity in hydroponically grown plants is relatively uncommon but can occur if nickel concentrations in the nutrient solution exceed tolerable levels.

Symptoms

- ☆ Nickel toxicity can lead to a range of adverse effects on plant growth and development including leaf necrosis, wilting, reduced growth and impaired nutrient uptake.

10.1.3.2.11 General Treatment Strategies for Nutrient Toxicities

- ☆ Some plant species may have higher tolerance to particular nutrient toxicity than others. Growers must be aware of the specific requirements and sensitivities of plants one is growing.
- ☆ Regularly monitor nutrient solution composition and plant health.
- ☆ Use diagnostic tools such as nutrient solution testing kits or leaf tissue analysis to confirm the specific nutrient toxicity and rule out other

potential issues. Measure the nutrient levels in the nutrient solution using a reliable testing kit or device. This helps determine the severity of the toxicities and guides the appropriate corrective action.

- Ensure that the nutrient solution's EC/CF and pH levels are within the appropriate range for the specific plant variety and growth stage.
- Implement a balanced nutrient feeding schedule.
- Monitor and maintain water temperature to support healthy nutrient uptake.
- Adjust greenhouse or room temperature as needed, considering the requirements of the particular crop being grown.
- Check the watering system to ensure that plants receive the correct amount of water. Overly wet conditions can inhibit root development and contribute to root rot. Maintain a moist but not saturated growing medium, allowing sufficient time between watering cycles for the medium to partially dry out.
- Flush the system periodically with fresh pH-balanced water.
- Adjust nutrient solution concentration based on plant growth stage and nutrient uptake.
- Choose plant varieties suited to hydroponic cultivation and less prone to nutrient imbalances.

By promptly identifying and addressing nutrient toxicities, one can help maintain optimal plant health and maximize yields in the hydroponic system. If symptoms persist or worsen despite treatment efforts, consult with experienced hydroponic growers or agricultural experts for further assistance.

10.1.4 Effect of Plant Nutrition on Plant Diseases

The resistance of plants to diseases is primarily governed by genetical factors yet the expression of genetic resistance can be influenced by plant nutrition. Certain nutrients play critical roles in enhancing plant health and resilience to diseases as calcium aids in formation of robust cell walls thus serving as a physical barrier against fungal infections. Adequate nutrition supports the synthesis of natural defense compounds such as antioxidants and phytoalexins.

However, imbalances in nutrient levels can have adverse effects so potentially exacerbating disease susceptibility. Insufficient levels of certain nutrients may lead to release of compounds that attract pathogens so facilitating their invasion into plant tissues. Calcium is known for its ability to inhibit enzymes released by bacteria thereby impeding their spread within the plant.

Moreover, some pathogens exploit specific vulnerabilities in plant anatomy or physiology to proliferate. Bacteria, for instance, can spread within plants *via* xylem vessels causing blockages that result in wilting and death. Certain nutrients can mitigate this risk by suppressing formation of bacterial slime within the vessels.

Furthermore, viruses are often transmitted to plants by sucking insects and fungi. While silicon is not considered a plant nutrient, it can deter the feeding activities of aphids and other pests, thereby reducing the likelihood of virus transmission.

Understanding the intricate relationship between plant nutrition and disease dynamics is crucial for implementing effective disease management strategies in agricultural practices. By optimizing nutrient levels and promoting plant health growers can bolster their crops' defenses and minimize the risk of disease outbreaks.

10.2 Algae

Algae, characterized by their greenish vegetative growth, are plant-like organisms devoid of stems or leaves. Their proliferation in hydroponic systems presents a formidable challenge impacting water quality, nutrient accessibility and overall well-being of plants. Usually these cover surface of plant media and infiltrate nutrient reservoirs. Their rampant growth is spurred by stagnant water, excessive moisture and light exposure. While algae themselves may not directly harm plants but their presence can lead to nutrient depletion and fluctuations in EC, thereby undermining optimal growing conditions.

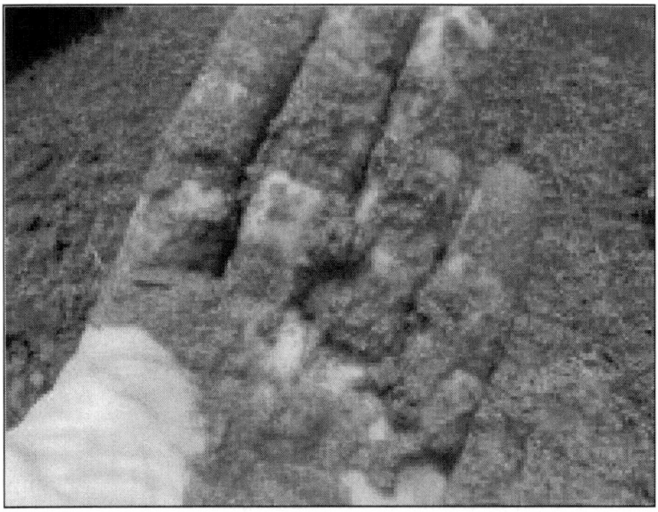

Algae

Preventing and treating algae growth in nutrient solutions is essential for maintaining plant health and optimizing yields. Following are some strategies to prevent and address this issue:

10.2.1 Prevention of Algal Growth

10.2.1.1 Light Management

Algae are photosynthetic organisms that utilize light energy to produce organic matter so thriving in nutrient-rich water environments. In hydroponic systems, algae can colonize nutrient solutions, growing media and surfaces exposed to light thus competing with plants for resources and potentially compromising crop health. Excessive algae growth can lead to clogged irrigation systems, reduced oxygen levels and nutrient imbalances thus impacting plant growth and productivity negatively.

Different species of algae exhibiting varying light preferences and tolerances. Most of the algae thrive in moderate to high light conditions with wavelengths in blue and red spectrum being particularly conducive to growth. Algae can adapt to fluctuating light intensities and may proliferate rapidly in environments with inconsistent lighting.

To control algal growth and maintain optimal growing conditions in hydroponic systems, growers can employ the following light management strategies:

i) **Light exclusion:** Minimize light exposure to nutrient solutions and growing media by using opaque materials for reservoirs, tubing and containers. Light-proofing hydroponic systems prevent algae from accessing the energy source needed for growth.

ii) **Light blockers:** Install light-blocking materials such as reflective films, covers or coatings to shield hydroponic components from external light sources. Light blockers prevent ambient light from penetrating the system thus reducing the likelihood of algae colonization.

iii) **Lighting schedule:** Implement a consistent lighting schedule for indoor hydroponic setups ensuring that grow lights are only operational during designated periods. Timed lighting regimes mimic natural daylight cycles while minimizing excess light exposure that can promote algal growth.

iv) **Light spectrum:** Select grow lights with spectra optimized for plant growth such as full-spectrum LEDs or fluorescent tubes. These lights emit wavelengths tailored to plant photosynthesis while minimizing spectral ranges favourable to algae.

v) **Light intensity:** Adjust light intensity levels to meet the specific requirements of cultivated crops while limiting excess light that can stimulate algal growth. Optimal light levels promote plant growth without providing excess energy for algal proliferation.

10.2.1.2 Aeration and Agitation

To mitigate algae growth and to maintain optimal growing conditions hydroponic growers can leverage aeration and agitation techniques. Aeration and agitation play crucial roles in disrupting algal growth and maintaining water quality in hydroponic systems:

Aeration introduces oxygen into the nutrient solution thus promoting aerobic conditions beneficial for plant roots while inhibiting anaerobic environments conducive to algal growth. Adequate oxygen levels enhance nutrient uptake by roots and support beneficial microbial populations contributing to overall health of the system.

Agitation techniques such as physical stirring, mixing or circulating the nutrient solution disrupt algal colonies and prevent their establishment on surfaces. Mechanical agitation helps distribute nutrients evenly thus preventing stagnation and reducing the accumulation of organic matter favourable to algal growth.

Strategies for aeration and agitation: Hydroponic growers can employ various strategies to incorporate aeration and agitation into their systems effectively:

i) **Air stones and diffusers:** Install air stones or diffusers in nutrient reservoirs to create fine bubbles that oxygenate the solution and promote nutrient circulation. Air stones enhance dissolved oxygen levels throughout root zone so inhibiting algal growth and supporting healthy root development.

ii) **Water pumps and circulation:** Use water pumps or circulation systems to agitate nutrient solution therefore maintaining continuous movement and preventing stagnation. Recirculating solution helps distribute oxygen and nutrients evenly while disrupting algal growth on surfaces.

iii) **Mixing and stirring devices:** Utilize mixing or stirring devices to agitate the nutrient solution periodically thus promoting oxygen exchange and preventing sedimentation. Mechanical agitation prevents stratification and ensures uniform nutrient distribution thus reducing the risk of algal colonization.

iv) **Timed agitation cycles:** Implement timed agitation cycles to disrupt the nutrient solution intermittently, simulating natural water movement and preventing algal attachment. Controlled agitation intervals maintain water quality without excessive energy consumption.

10.2.1.3 Regular Cleaning

Hydroponic growers must clean and disinfect nutrient solution reservoirs, growing containers, and irrigation systems regularly to remove algal spores and to prevent their buildup. Regular cleaning plays a vital role in preventing and controlling algal growth by:

i) **Removing organic debris:** Organic matters such as dead plant material and accumulated debris provide nutrients for algal growth. Regular cleaning removes organic debris thus depriving algae of essential nutrients and reducing their proliferation.

ii) **Disrupting algae colonies:** Mechanical removal of algal colonies through physical cleaning disrupts their growth and prevents their establishment on surfaces. Regular cleaning inhibits algae colonization and propagation which in turn also breaks down its biofilms thus removing even their spores.

iii) **Maintaining water quality:** Cleaning maintains water quality by removing contaminants, sediment and excess nutrients that contribute to algal growth. Clean water promotes optimal nutrient uptake by plants and prevents nutrient imbalances associated with algae-infested solutions.

iv) **Preventing algae build-up:** Regular cleaning prevents algal build-up on hydroponic equipment such as reservoirs, tubing and growing containers. By keeping surfaces clean and algae-free, growers can minimize the risk of clogged irrigation systems and nutrient delivery disruptions.

Hydroponic growers can implement following cleaning strategies to prevent and control algal growth effectively:

i) **Reservoir cleaning:** Regularly empty and clean nutrient reservoirs to remove accumulated debris, sediment and algal biofilms. Use mild cleaning agents or hydrogen peroxide solutions to disinfect reservoirs and to prevent algal regrowth.

ii) **Equipment maintenance:** Inspect equipments regularly for algal build-up and address any issues promptly. Clean and disinfect hydroponic equipment including pumps, tubing and irrigation components to prevent algal colonization and ensure optimal system performance.

iii) **Growing medium maintenance:** Replace or refresh growing medium periodically to remove accumulated algae and organic matter. Rinse inert growing media such as rockwool or perlite to remove algal spores and to prevent contamination.

10.2.1.4 Using Dry Barriers

For fine and porous media such as coco coir or perlite mixes addition of 2.5–5cm layer of gravel atop each pot serves as a dry barrier. This not only secures the lighter media but also deters algae and fungus gnats. Here is how it works and why it is effective:

i) **Physical barrier:** Placing a layer of dry gravel or similar material on top of the porous medium acts as a physical barrier that blocks light from reaching the underlying substrate. Algae require light to photosynthesize and grow so depriving them of light effectively inhibits their growth.

ii) **Moisture regulation:** The dry barrier helps regulate moisture levels within the growing medium by reducing evaporation and maintaining optimal moisture content. This prevents excessive moisture accumulation which can create favourable conditions for algal growth.

iii) **Aeration and drainage:** The presence of dry barrier promotes better aeration and drainage within the hydroponic system. It allows air to circulate around the root zone so preventing waterlogging and providing oxygen to plant roots while discouraging anaerobic conditions that may favour algal growth.

iv) **Substrate stability:** The layer of gravel or similar material helps stabilize the porous medium thus preventing it from shifting or becoming displaced during irrigation or maintenance activities. This ensures consistent coverage and maximizes the effectiveness of the barrier in blocking light and preventing algal growth.

v) **Aesthetic appeal:** In addition to its functional benefits a dry barrier can enhance the visual appeal of the hydroponic setup. The layer of gravel or decorative stones adds texture and visual interest to the growing environment thus creating a more appealing and aesthetically pleasing display.

When implementing a dry barrier to prevent algal growth it is important to ensure proper installation and maintenance:

- **Uniform coverage:** Ensure that dry barrier is evenly distributed across the surface of porous medium thus providing complete coverage and blocking all potential areas where light could penetrate.
- **Regular inspection:** Periodically inspect the dry barrier for signs of algal growth or debris accumulation. Remove any algae or debris promptly to maintain the effectiveness of the barrier in preventing algal proliferation.
- **Replacement:** Over time, the dry barrier may become compacted or contaminated with organic matter thus reducing its effectiveness. Replace the gravel or similar material as needed to maintain optimal coverage and prevent algae growth.

By incorporating a dry barrier into the hydroponic system growers can effectively prevent algae growth in porous media like coco coir or perlite mixes thus creating a cleaner and healthier growing environment for their plants.

10.2.1.5 Hygiene Practices

To prevent algal growth and to maintain a clean hydroponic environment, growers should adhere to the following hygiene practices:

i) Practice good personal hygiene including washing hands thoroughly before handling hydroponic equipment or interacting with plants to prevent introducing contaminants.

ii) Avoid introducing organic matter or debris into hydroponic systems as these can serve as nutrients for algal growth and compromise water quality.

10.2.2 Treatment of Algae Growth

Treatment strategies for algae growth: Hydroponic growers can employ several strategies to effectively treat algal growth and restore water quality in their systems:

10.2.2.1 Chemical Treatments

10.2.2.1.1 Algaecides

Algaecides are chemical agents specifically designed to combat algae infestations by inhibiting algal growth or killing existing algae populations. These are available in various formulations including liquid concentrates, powders and granules ~ each containing active ingredients that target algae cells through different mechanisms. Here are some commonly used algaecides:

i) **Copper-based algaecides:** Copper-based algaecides such as copper sulphate or copper chelates are effective at controlling algae in hydroponic systems. Copper ions disrupt algae cell membranes and inhibit photosynthesis which lead to algae cell death. However, copper-based algaecides should be used with caution as excessive copper levels can be toxic to plants and aquatic organisms.

ii) **Peroxide-based algaecides:** Peroxide-based algaecides such as hydrogen peroxide are oxidizing agents that effectively kill algae upon contact. Hydrogen peroxide breaks down into water and oxygen leaving no harmful residues behind. Peroxide-based algaecides are safe to use in hydroponic systems but should be applied carefully to avoid damaging plant tissues.

iii) **Quaternary ammonium compounds (QACs):** QACs such as benzalkonium chloride or alkyldimethylbenzylammonium chloride are surfactants that disrupt algae cell membranes and inhibit growth. QACs are effective against a wide range of algae species and are commonly used in commercial algaecide formulations. However, QACs should be used with caution as these can be toxic to aquatic organisms and may accumulate in the environment.

iv) **Bleach** or **sodium hypochlorite**: A 10 per cent bleach solution can be used as a chemical control for algae in hydroponic systems. Bleach is a potent disinfectant that can effectively kill algae upon contact. When properly diluted to a 10 per cent concentration (1 part bleach to 9 parts water), bleach can be applied to surfaces, equipment or nutrient solutions to sanitize and control algae growth. After applying bleach, surfaces or equipment must be thoroughly rinsed with clean water to remove any residual bleach. Residual bleach can be harmful to plants and may disrupt the balance of the hydroponic system.

v) **Herbicides:** Some herbicides such as diquat or glyphosate can be effective at controlling algae in hydroponic systems. Herbicides work by disrupting algae metabolism or photosynthesis leading to cell death. However, herbicides should be used sparingly and with caution as these can be toxic to plants and aquatic organisms.

vi) **Biological algaecides:** Biological algaecides contain naturally occurring microorganisms or enzymes that target and control algal growth. These products work by outcompeting algae for nutrients or by producing compounds that inhibit algae growth. Biological algaecides are generally safe to use in hydroponic systems and do not harm plants or aquatic organisms.

When selecting an algaecide for hydroponic use, it is important to choose a product that is safe, effective and compatible with specific needs of the system. Always follow manufacturer instructions carefully and use algaecides in accordance with proper safety protocols to minimize risks to plants, aquatic organisms and the environment. Consider integrated pest management strategies and non-chemical control methods as part of a comprehensive approach to algae management in hydroponics.

Benefits of Algaecides

- **Effective algae control:** Algaecides provide rapid and targeted control of algae populations, restoring water clarity and preventing further algae proliferation.
- **Versatility:** Algaecides can be used in various hydroponic setups including nutrient reservoirs, irrigation systems and growing containers so making these suitable for diverse applications.
- **Ease of application:** Most algaecides are simple to apply, requiring dilution in water according to manufacturer instructions before application to affected areas.
- **Compatibility:** Many algaecide formulations are compatible with hydroponic systems and plant growth thus posing minimal risk to crops when used as directed.

Considerations for Use

- **Selection of algaecide:** Choose an algaecide formulation suitable for specific algae species and hydroponic setup. Consider factors such as algae type, concentration and sensitivity to active ingredients when selecting an appropriate product.
- **Dosage and application:** Follow manufacturer instructions carefully regarding algaecide dosage, application frequency and safety precautions. Overdosing can harm plants and beneficial microorganisms while underdosing may be ineffective against stubborn algae infestations.

- ☆ **Avoid plant contact:** Algaecide can be damaging to plant tissues if applied directly. Take care to avoid contact with plant roots, leaves or stems when using bleach solutions.
- ☆ **Safety precautions:** Wear appropriate personal protective equipment (PPE) such as gloves and eye protection when handling algaecides to minimize exposure to skin and eyes. Store algaecides securely in a well-ventilated area away from direct sunlight and heat sources.

Best Practices for Application

i) **Prepare diluted solution:** Dilute the algaecide as per recommended instructions ensuring thorough mixing to achieve uniform distribution of the active ingredient.

ii) **Apply directly to affected areas:** Apply diluted algaecide directly to areas with visible algae growth such as nutrient reservoirs, tubing or growing containers. Use a spray bottle, watering can or pump sprayer for precise application.

iii) **Monitor effectiveness:** Monitor algae populations regularly following algaecide application to assess effectiveness. Repeat treatments as necessary until algae growth is adequately controlled.

10.2.2.1.2 pH Adjusters

Algae thrive in environments with favourable pH conditions typically preferring neutral to alkaline conditions for optimal growth. Adjusting pH of nutrient solution to slightly acidic levels (around 5.5-6.0) can create an inhospitable environment for algae so inhibiting their proliferation and promoting healthier plant growth. By manipulating pH levels growers can control algae infestations effectively and maintain water quality in hydroponic setups.

By utilizing acidic *e.g.* phosphoric acid and citric acid or basic *e.g.* potassium hydroxide and sodium hydroxide pH adjusters appropriately and maintaining pH within the desired range, growers can effectively treat algae infestations and ensure clean and thriving hydroponic environments. Thoughtful application of pH adjusters coupled with regular monitoring and maintenance is essential for successful algae control and optimal crop production in hydroponic systems.

Application of pH Adjusters

To treat algal growth effectively using pH adjusters in hydroponic systems, growers should follow these guidelines:

i) **Monitoring pH levels:** Regularly monitor the pH of the nutrient solution using a pH meter or test kit to determine if adjustments are necessary. Aim for slightly acidic pH levels (around 5.5-6.0) to inhibit algal growth while maintaining optimal nutrient availability for plants.

ii) **Adjusting pH levels:** Based on pH monitoring results, use the appropriate pH adjuster to modify pH levels as needed. Add acidic pH adjusters to lower pH and create an acidic environment inhospitable to algae or add basic pH adjusters to raise pH and discourage algal growth.

iii) **Dosage and application:** Follow manufacturer instructions and recommended dosage rates when applying pH adjusters to the nutrient solution. Gradually introduce pH adjusters to avoid drastic pH fluctuations that can stress plants or disrupt microbial balance in the system.

iv) **Regular maintenance:** Incorporate pH adjustment into regular maintenance routines to prevent algal regrowth and to maintain water quality in hydroponic systems. Monitor pH levels consistently and adjust as necessary to sustain optimal growing conditions.

10.2.2.2 Mechanical Removal

Mechanical removal offers a practical and immediate solution for controlling algal growth in hydroponic systems thus restoring system cleanliness and promoting optimal growing conditions. By incorporating targeted removal techniques into routine maintenance practices, growers can effectively manage algae infestations and maintain a healthy hydroponic environment for successful crop production. Thoughtful implementation of mechanical removal techniques enhances system hygiene and supports sustainable hydroponic cultivation practices.

10.2.2.2.1 Physical Scrubbing

Physical scrubbing involves manual removing of algal colonies and biofilms from hydroponic equipment and surfaces using gentle abrasive action. This targeted approach effectively disrupts algal growth and prevents further proliferation.

Benefits: Physical scrubbing offers several advantages for algae treatment in hydroponic systems:

i) **Localized treatment:** Physical scrubbing allows growers to target specific areas affected by algal growth such as reservoirs, tubing and growing containers. By focusing on problem areas growers can effectively remove algae colonies and prevent their spread to other parts of the system.

ii) **Immediate results:** Unlike chemical treatments or biological controls, physical scrubbing provides immediate results visibly removing algae and restoring system cleanliness. This instant gratification can boost growers' confidence in their ability to manage algae infestations effectively.

iii) **Minimal environmental impact:** Physical scrubbing is a chemical-free and environmental-friendly method for algae treatment thus minimizing the risk of introducing harmful substances into the hydroponic system. This eco-friendly approach aligns with sustainable growing practices and promotes health and longevity of the system.

Practical Considerations for Physical Scrubbing

To implement physical scrubbing as an algal treatment method in hydroponic systems effectively growers should consider the following practical considerations:

i) **Selection of tools:** Choose appropriate tools for mechanical removal such as soft-bristled brushes and sponges or scrapers to avoid damaging hydroponic equipment or surfaces. Abrasive materials can scratch or degrade plastic components thus compromising their integrity over time.

ii) **Gentle abrasive action:** Apply gentle pressure when scrubbing algae-infested surfaces to avoid causing damage or dislodging sensitive components. Gradually increase pressure as needed to remove stubborn algae biofilms without causing any harm.

iii) **Thorough rinsing:** After scrubbing, thoroughly rinse the equipment and surfaces with clean water to remove loosened algae cells and debris. Rinse repeatedly to ensure complete removal and prevent algal regrowth.

10.2.2.2.2 Filtration

Filtration plays a crucial role in algae control by removing algae cells and other particles from the nutrient solution so restoring water clarity and preventing further proliferation.

Benefits

Effective filtration techniques offer several benefits for algae control in hydroponic systems:

i) **Physical removal:** Filtration physically removes algae cells, debris and other particles from the nutrient solution so reducing algae population and restoring system cleanliness. This targeted approach helps maintaining water quality and promoting healthy plant growth.

ii) **Prevention of clogging:** Filtration systems prevent algal cells and debris from clogging irrigation components, such as pumps, tubing and emitters so ensuring uninterrupted nutrient delivery to plants. By preventing clogs, filtration systems help optimize system performance and minimize maintenance requirements.

iii) **Improved oxygenation:** Filtration systems enhance oxygenation of the nutrient solution by promoting water movement and aeration. Increased oxygen levels inhibit algal growth and support beneficial microbial populations thus contributing to overall system health.

iv) **Enhanced nutrient uptake:** Clean nutrient solutions facilitate optimal nutrient uptake by plant roots thus promoting vigorous growth and development. By removing algae-derived competition for nutrients, filtration systems improve nutrient availability and maximize plant productivity.

Effective Filtration Strategies

Hydroponic growers can implement various filtration strategies to control algae growth effectively and maintain water quality:

i) **Mechanical filters:** Install mechanical filters such as sediment filters, mesh screens or filter socks to trap algae cells and debris physically from the nutrient solution. Mechanical filters remove particles of varying sizes from large algae colonies to fine sediment so ensuring thorough filtration.

ii) **UV sterilization:** Incorporate UV sterilizers into hydroponic system to kill algae cells and to prevent their reproduction. UV radiation disrupts DNA of algae cells so rendering these unable to proliferate and causing them to die off. UV sterilizers offer continuous chemical-free algae control and promote water clarity.

10.2.2.3 Biological Controls

Biological controls for algae removal involve the use of natural organisms or compounds to inhibit algae growth and promote a healthy balance within aquatic environment. It is important to note that biological controls may require careful monitoring and management to prevent overpopulation or imbalance within the hydroponic system. Biological controls are often most effective when combined with other algae control methods such as proper nutrient management, light exclusion and regular cleaning and maintenance practices. By incorporating biological controls into a comprehensive algae management strategy, hydroponic growers can effectively control algae growth and maintain a healthy growing environment. Here are some common biological control methods used in hydroponic systems:

10.2.2.3.1 Beneficial Microorganisms and Biological Additives

Beneficial microorganisms and biological additives can be valuable tools for controlling algae growth in hydroponic systems. These organisms can help establish a balanced microbial community, outcompete algae for nutrients and promote overall health of the system. However, these differ in their nature, composition and mode of action. Here is a breakdown of differences between beneficial microorganisms and biological additives in hydroponics:

i) Beneficial microorganisms

- ☆ **Nature:** Beneficial microorganisms are naturally occurring bacteria, fungi or other microbial species that contribute to health of overall ecosystem in hydroponic systems.
- ☆ **Composition:** Beneficial microorganisms include a diverse range of species each with unique metabolic functions and interactions within the hydroponic environment.
- ☆ **Mode of action:** Beneficial microorganisms work by outcompeting algae for nutrients, metabolizing organic matter and maintaining microbial

balance in the system. These can help inhibit algae growth indirectly by consuming nutrients that algae rely on for growth.

Beneficial microorganisms commonly used in hydroponics include certain strains of bacteria (*e.g.*, *Bacillus* spp., *Pseudomonas* spp.) and fungi (*e.g.*, *Trichoderma* spp., *vis-a-vis* mycorrhizal fungi).

ii) Biological Additives

- **Nature:** Biological additives are commercially produced products containing a concentrated blend of beneficial microorganisms, enzymes and other natural compounds designed to enhance microbial activity and promote system health.
- **Composition:** Biological additives typically contain specific strains of beneficial bacteria, fungi, enzymes and other ingredients selected for their ability to support plant growth and control algae.
- **Mode of action:** Biological additives function similarly to naturally occurring beneficial microorganisms but are formulated to provide a standardized and concentrated dose of microbial activity. These can help improve nutrient cycling, breaking down of organic matter and suppressing algae growth.

Biological additives may include products such as microbial inoculants, bio-stimulants or enzyme formulations specifically designed for hydroponic applications.

Here are some beneficial microorganisms and biological additives commonly used for algae control in hydroponics:

i) **Beneficial bacteria:** Certain strains of bacteria such as *Bacillus* spp. and *Pseudomonas* spp. can help inhibit algal growth by consuming organic matter and nutrients that algae rely on for growth. These bacteria compete with algae for resources thus reducing nutrient availability and creating an unfavourable environment for algal proliferation.

ii) **Nitrogen-fixing bacteria:** Nitrogen-fixing bacteria such as *Rhizobium* spp. and *Azospirillum* spp. convert atmospheric nitrogen into forms that plants can readily use. By enhancing nitrogen availability for plants, these bacteria promote healthy plant growth and reduce nutrient availability for algae thereby controlling algae proliferation indirectly.

iii) **Beneficial fungi:** Mycorrhizal and other beneficial fungi can form symbiotic relationships with plant roots so enhancing nutrient uptake and promoting plant health. By improving nutrient availability to plants, beneficial fungi can suppress algal growth indirectly thus contribute to overall system balance.

iv) **Mycorrhizal inoculants:** Mycorrhizal inoculants contain beneficial mycorrhizal fungi that form symbiotic relationships with plant roots thus improving nutrient uptake and enhancing plant growth. By promoting plant health and vigour, mycorrhizal fungi can indirectly inhibit algae growth and maintain system balance.

v) **Beneficial enzymes:** Enzyme-based products containing beneficial enzymes such as cellulases and proteases can help break down organic matter and reduce nutrient availability for algae. These enzymes accelerate the decomposition of organic debris and detritus, promoting water clarity and inhibiting algae growth.

vi) **Competitive exclusion organisms:** Some beneficial microorganisms such as certain strains of yeasts and fungi can actively compete with algae for space and resources thus excluding algae from the environment effectively. By establishing themselves within the hydroponic system, these organisms create a competitive advantage that limits algae growth.

vii) **Probiotic solutions:** Probiotic solutions containing a diverse mix of beneficial microorganisms including bacteria, fungi and yeasts can help establish a balanced microbial community in hydroponic systems. These solutions promote beneficial microbial activity, suppress harmful pathogens and contribute to overall system resilience.

viii) **Humic substances:** Humic substances derived from organic sources such as leonardite or lignite can improve water quality and promote microbial balance in hydroponic systems. Humic substances contain natural compounds that can stimulate beneficial microbial activity and enhance nutrient availability to plants thus suppressing algal growth indirectly.

ix) **Biostimulants:** Biostimulant products containing beneficial microorganisms along with other growth-promoting compounds can enhance plant health and resilience while suppressing algal growth indirectly. These products stimulate plant growth and vigour thus improving nutrient uptake efficiency and reducing nutrient availability for algae.

When incorporating beneficial microorganisms for algae control in hydroponic systems, it is essential to choose products specifically formulated for hydroponic applications and follow manufacturer recommendations for application rates and timing. Regular monitoring of water quality and plant health is crucial to assess the effectiveness of microbial treatments and for ensuring a balanced ecosystem within the hydroponic system.

10.2.2.3.2 Algae-eating Organisms

Algae-eating organisms can be beneficial additions to hydroponic systems for controlling algae growth and maintaining water quality. Here are some common algae-eating organisms used in hydroponics:

i) ***Daphnia*:** Also known as water fleas, *Daphnia* is a small freshwater crustacean that feed on algae, bacteria and other organic matter suspended in the water. These are efficient grazers and can consume significant quantities of algae thus helping to control algae population in hydroponic systems.

ii) ***Ameca splendens***: These small freshwater fish are voracious algae eaters so can help control algal growth in hydroponic systems. These are particularly effective at grazing on algae biofilms and can help maintain water clarity and quality.

iii) **Siamese algae eater (*Crossocheilus siamensis*):** This fish species is known for its appetite for algae therefore it can be effective at controlling algal growth in hydroponic systems. Siamese algae eaters feed on various types of algae including green algae and filamentous algae thus helping to keep surfaces clean and free from algae buildup.

iv) **Amano shrimp (*Caridina multidentata*):** Amano shrimps are popular freshwater aquarium inhabitants known for their algae-eating capabilities. These feed on a variety of algae species including green algae, diatoms and hair algae thus making them valuable additions to hydroponic systems for algae control.

v) **Nerite snails (*Neritina* spp.):** Nerite snails are small algae-eating snails commonly used in aquaria and hydroponic systems. These have a voracious appetite for algae and can help keep surfaces clean by grazing on algae biofilms and algal growth on rocks, substrate and equipment.

When introducing algae-eating organisms into hydroponic systems, it is essential to ensure compatibility with existing plant and fish populations and to provide appropriate environmental conditions for their well-being. Monitoring and maintaining proper water parameters including temperature, pH and oxygen levels are crucial for health and productivity of algae-eating organisms in hydroponic environments.

10.2.2.3.2.1 Waste or Excretion of Algae-Eating Organisms

When using algae-eating organisms such as fish, snails or shrimp in hydroponic systems for algae control, their waste or excretion can impact overall nutrient dynamics and ecosystem balance within the system. Here are some considerations regarding waste or excretion of these organisms:

i) **Nutrient recycling:** The waste produced by algae-eating organisms contains organic matter and nutrients including nitrogen and phosphorus which can serve as a source of nutrients for plants in the hydroponic system. This process effectively recycles nutrients within the system thus contributing to overall nutrient availability for plant growth.

ii) **Nutrient management:** However, excessive accumulation of waste can lead to nutrient imbalances or fluctuations in the hydroponic solution affecting plant health potentially. It is important for growers to monitor nutrient levels and adjust accordingly to maintain optimal nutrient concentrations for plant growth.

iii) **Biological filtration:** Algae-eating organisms can also contribute to biological filtration within the hydroponic system by consuming organic matter and detritus so helping to maintain water clarity and quality. Their activity can help prevent the buildup of organic debris and algae biomass thus reducing the likelihood of clogs or system issues.

iv) **Oxygen demand:** The metabolic processes of algae-eating organisms including respiration and waste production can contribute to oxygen demand within the hydroponic system. Adequate aeration and oxygenation are essential to support health and activity of these organisms and to prevent oxygen depletion in the water.

v) **Waste management:** To manage waste accumulation, growers can implement strategies such as regular water changes, filtration systems or use of additional biological controls to help break down organic matter and to maintain water quality. Removing excess waste and detritus can help prevent nutrient buildup and ensure a healthy growing environment.

Overall, while the waste or excretion of algae-eating organisms can influence nutrient dynamics and ecosystem balance in hydroponic systems, proper management and monitoring can help optimize their beneficial contributions while minimizing potential drawbacks. By integrating algae-eating organisms into a well-balanced and carefully managed hydroponic setup, growers can effectively control algae growth and maintain a healthy growing environment for their plants.

10.2.2.3.3 Barley Straw

Barley straw is a natural and chemical-free method for controlling algal growth in hydroponic systems. When barley straw decomposes in water it releases compounds that inhibit algal growth. Barley straw contains allelopathic substances that disrupt algal reproduction and growth thus providing a natural and chemical-free method of algae control. Here is how barley straw can be used for algae control in hydroponics:

i) **Preparation:** Start by obtaining barley straw which is available at garden centres or aquatic supply stores. Cut or break the barley straw into smaller pieces to increase its surface area and to facilitate the release of algae-inhibiting compounds.

ii) **Placement:** Place the barley straw pieces directly into the hydroponic reservoir or nutrient solution. Ensure that the straw is submerged and distributed evenly throughout the system to maximize its effectiveness.

iii) **Monitoring:** Monitor the water clarity and algal growth in the hydroponic system regularly. Over time, the barley straw will begin to decompose thus releasing compounds that inhibit algal growth. Observe any changes in algae populations and adjust the quantity of barley straw as needed.

iv) **Replacement:** As the barley straw decomposes it will gradually lose its effectiveness. Replace the barley straw periodically to maintain continuous algae control. Remove any decomposed straw and replenish with fresh pieces to ensure consistent algae inhibition.

It is important to note that barley straw may take some time to exert its effects on algae and the results may vary depending on factors such as water temperature, circulation and severity of algae problem. While barley straw offers a natural and environment friendly approach to algae control, it may not provide immediate or complete eradication of algae populations. However, when used as part of a comprehensive algae management strategy, barley straw can be an effective tool for maintaining water quality and promoting healthy plant growth in hydroponic systems.

10.3 Insect-Pests and Diseases

While hydroponically grown plants are generally shielded from soil-borne pests and diseases, certain vulnerabilities persist even within the meticulously controlled environment of hydroponic systems. Beyond the usual suspects of insect pests, diseases such as mildew, moulds and viruses can still emerge. These maladies often stem from a range of factors:

- ☆ Imbalanced pH levels
- ☆ Stagnant air or insufficient aeration around the roots and plant environment
- ☆ High humidity
- ☆ Over-watering or poor drainage leading to stagnant water
- ☆ Inadequate lighting conducive to algal growth
- ☆ Nutrient imbalances causing toxicity or deficiencies
- ☆ Introduction of diseases *via* contaminated tools, soils or water sources

In recirculating hydroponic setups, the reuse of nutrient solutions can exacerbate root diseases and nutrient deficiencies over time. Consequently, regular reconstitution, filtration, sterilization and even replacement of the nutrient solution may be necessary throughout the growing period.

In fully controlled aeroponic environments, disease transmission is rare due to minimal plant-to-plant contact. Each sterile spray pulse mitigates the risk and any plant exhibiting disease symptoms is promptly removed to prevent further spread.

By diligently monitoring and addressing these factors, hydroponic growers can effectively manage plant pests and diseases, ensuring the health and productivity of their crops.

10.3.1 Mites and other Pests

While hydroponic gardens generally offer protection against soil-borne pests, harmful insects can still infiltrate the growing environment. These pests pose a threat by feeding on plant foliage leading to discolouration and premature leaf drop. Certain insects such as fungus gnats target and damage roots while others may attack fruits.

Insects can also serve as vectors for disease transmission, transferring pathogens from one plant to another through sap exchange. To mitigate these risks it is crucial for growers to conduct regular inspections for signs of pests and diseases.

One effective monitoring method involves using brightly coloured sticky strips strategically placed throughout the garden. These strips attract and trap insects, allowing growers to assess their presence and monitor population growth with the aid of a magnifying glass.

Common indicators of pest infestation include the presence of white spots on leaves, webbing and overall yellowing of foliage. By staying vigilant and proactive in pest detection and management, hydroponic growers can safeguard their crops and ensure healthy thriving plants.

Some harmful insect-pests that are usually seen in the garden are:

10.3.1.1 Spider Mites

Minute spider mites nearly invisible to the naked eye are arachnids known for their ability to spin fine webs resembling cobwebs. These come in various hues of red, yellow or green and sustain themselves by extracting sap from plant tissues. These thrive in low humidity levels making humidity adjustment a key control measure. These mites exhibit dormancy during winter and re-emerge with warmer temperatures.

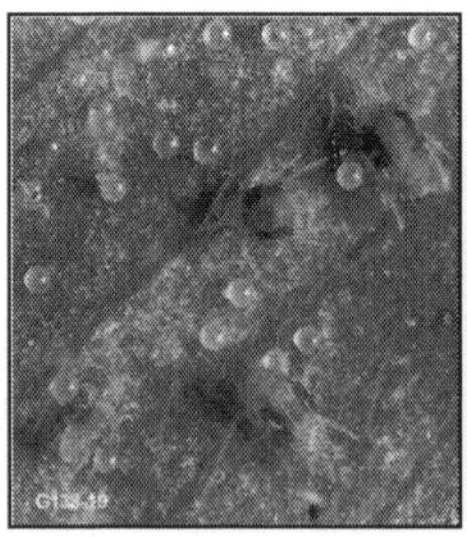
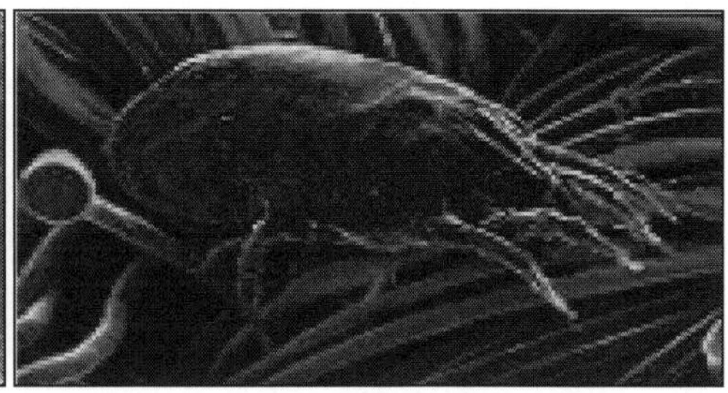

Feeding on leaves involves puncturing the leaf surface and drawing out cell contents thus resulting in the formation of tiny holes. This feeding pattern leads to a characteristic mottling of leaves which gradually turn brown and eventually drop from the plant. Plants infested with spider mites often exhibit a sickly appearance accompanied by telltale signs such as white spots, webbing and overall yellowing of foliage.

The two-spotted spider mite is one of the most widespread which is distinguished by its pair of dark spots. Despite their diminutive size, these pests can inflict significant damage to crops so making early detection *cum* proactive management is crucial for maintaining plant health in hydroponic systems.

Treatment for affected plants requires a multi-faceted approach beginning with isolation to prevent spread of the pests. These are controlled by washing the leaves with warm soapy water to physically remove mites and their eggs. Following this, botanical sprays can be applied regularly for ongoing control, however, reserving the use of insecticides for severe infestations only. Incorporating predatory mites into the environment is another effective strategy. These beneficial mites prey on spider mites thus reducing their populations over time effectively. Introducing ladybugs can prove highly beneficial as these are natural predators known for their voracious appetite for spider mites.

10.3.1.2 Aphids

Aphids are small slow-moving insects characterized by their tiny pear-shaped bodies which can vary in colour from green to yellow or some other hues. These pests feed by piercing plant tissues and sucking sap from leaves leading to symptoms such as curling, puckering and discolouration. Infested leaves may also exhibit a sticky residue known as honeydew left behind by the aphids.

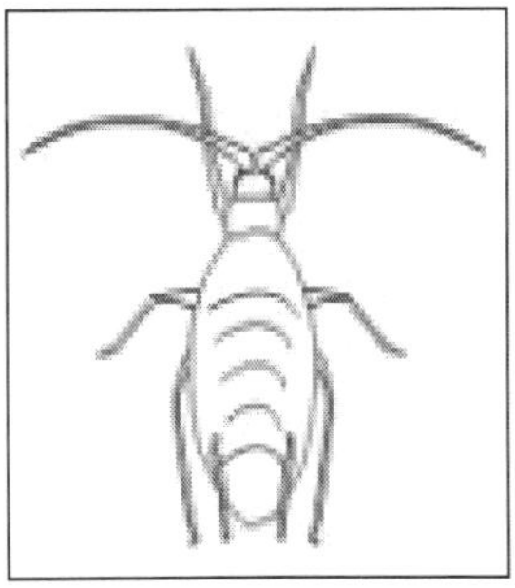
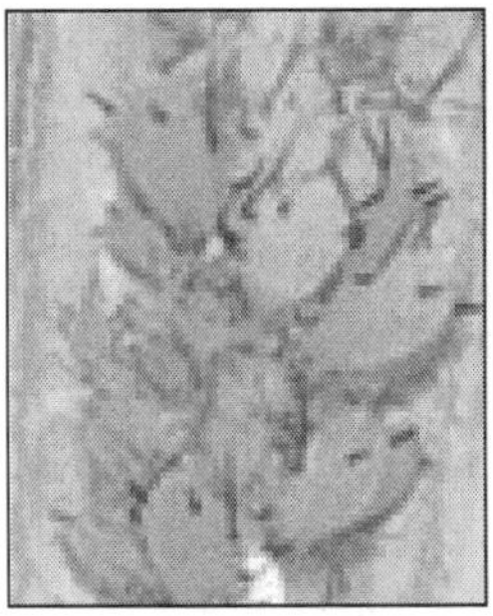
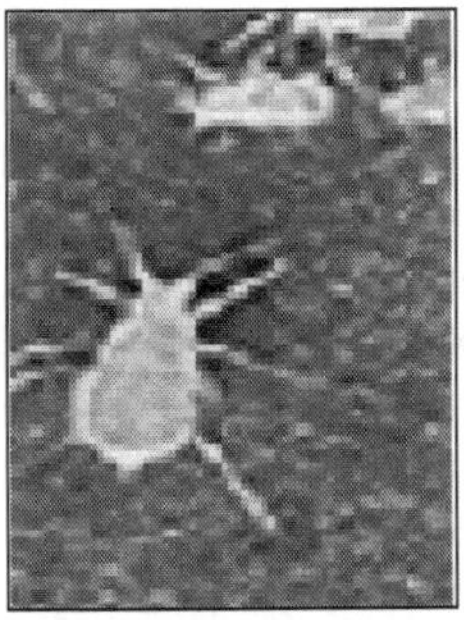

Aphids thrive in environments with lower humidity levels so making humidity control a key strategy in managing their populations. Remarkably adaptable aphids can enter a dormant state during winter and re-emerge once temperatures rise again. Adjusting humidity levels and understanding their seasonal behaviour are essential in managing aphid infestations effectively.

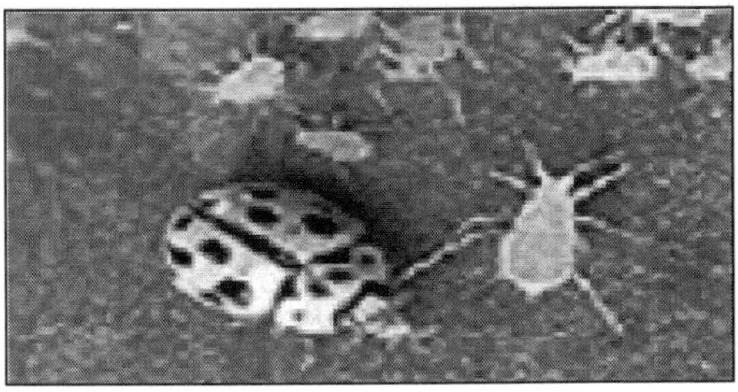

Ladybug Eating Aphids

Treatment for aphid infestations involves a multi-step approach. It should start by carefully removing damaged leaves, then proceeding to cleanse the remaining foliage with warm soapy water over a span of several days and then application of a targeted spray of insecticidal soap to control aphid populations effectively. Encouraging the presence of natural predators is also key to managing aphid infestations in an eco-friendly manner. Beneficial insects such as ladybugs, lacewings and specific parasitic wasps like *Aphidius colemani* (in certain species) and *Aphidoletes aphidimyza* are effective predators of aphids. By fostering these beneficial insects, growers can naturally regulate aphid populations thus minimizing the reliance on chemical interventions.

10.3.1.3 Fungus Gnat

While fungus gnats are not as common in hydroponic gardens as in soil-based systems but it is essential for growers to remain vigilant and address any signs of infestation promptly to safeguard the health and vitality of their plants.

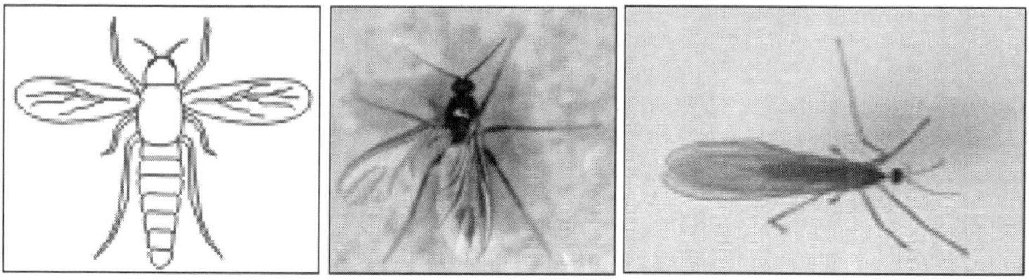

Fungus gnat

Fungus gnats are small pests resembling tiny fruit flies, measuring approximately 2mm in length. Typically these sport a black colouration with two wings and prominent antennae. The larvae of these pests resembling small worms reside in the top inch or two of the growing medium where these continue feeding on organic matter such as debris, fungi and algae. However, these can also pose a threat to young seedlings by nibbling on their roots.

Indications of a fungus gnat infestation include the presence of small black flies hovering around the surface of the growing medium. Affected plants may exhibit symptoms such as yellowing of leaves and development of brown spots so signalling potential nutrient deficiencies and root rot.

The growers for its treatment can employ bright yellow sticky strips to trap these effectively and to monitor pest activity. It is essential to remove any damaged plant and apply suitable insecticidal spray to control pest populations promptly. Furthermore, the grower may consider incorporating natural enemies such as *Steinernema feltiae* and *Steinernema carpocapsae* nematodes into the hydroponic system. These nematodes prey on harmful insects thus providing an eco-friendly and sustainable approach to pest management.

10.3.1.4 Whitefly

These small whiteflies are white flying insects measuring approximately

2mm in length resembling as miniature moths with four wings. These pests feed on plant sap causing foliage to yellow and wilt. Extracting significant amounts of sap, whiteflies excrete honeydew so leaving the leaves sticky and prone to fungal infections and decay. In severe cases, the fungal growth and decay associated with honeydew can decimate entire crop. Whiteflies typically rest on plant leaves and swiftly take flight when disturbed. Shiny sticky leaves serve as telltale signs of whitefly infestation. Moreover, whiteflies pose a considerable threat to plant health as these can transmit numerous plant viruses.

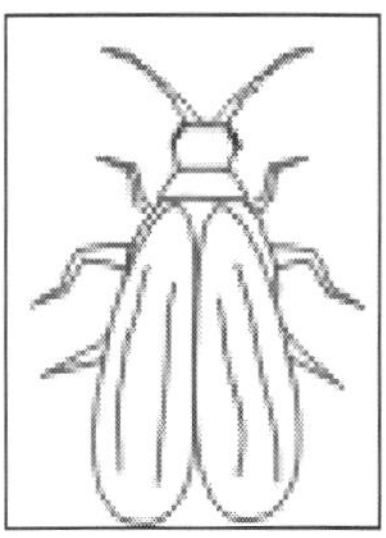

Treatment options include spraying affected plants with insecticide or stronger alternatives ensuring that surrounding plants are also treated. Yellow sticky traps are effective in capturing whiteflies. Several natural enemies can aid in controlling whitefly populations. Predatory insects such as green lacewings and ladybugs prey on whiteflies. *Encarsia formosa* ~ a parasitic wasp lays its eggs inside whitefly larvae and their feeding turn these black and easily identifiable. Adult *Encarsia formosa* feed on honeydew and bodily fluids of whitefly larvae so further suppressing their population.

10.3.1.5 Thrips

Thrips are characterized by their diminutive size (1-3mm) and slender bodies with hairy wings. These pose a challenge to hydroponic growers. The greenhouse thrips sustain themselves by feeding on plant sap often leaving behind telltale signs of their presence.

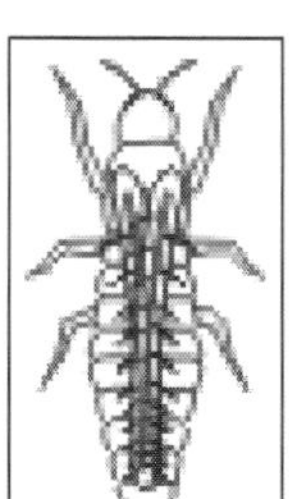

Symptoms of thrips infestation manifest as fine yellow spots speckling the leaves particularly on inner foliage and fruit. As the infestation progresses, affected leaves may develop a silvery-metallic sheen but only in some cases as not all the thrips species exhibit this trait. Scattered black specks as thrips faecal material may be observed on leaf surfaces.

Thrips infestations can easily be overlooked due to their small size and inconspicuous behaviour. Therefore, diligent observation and close inspection are essential for early detection and effective pest management in hydroponic systems.

Flower thrips are known to feed on buds, flowers and leaves inflicting shallow yet numerous punctures that result in a characteristic shrunken appearance of the affected tissue. These pests target the thick and fleshy petals, pistils and stamens of flowers. Consequently, the areas they feed upon may exhibit symptoms such as browning, yellowing, blackening, shrivelling and premature dropping.

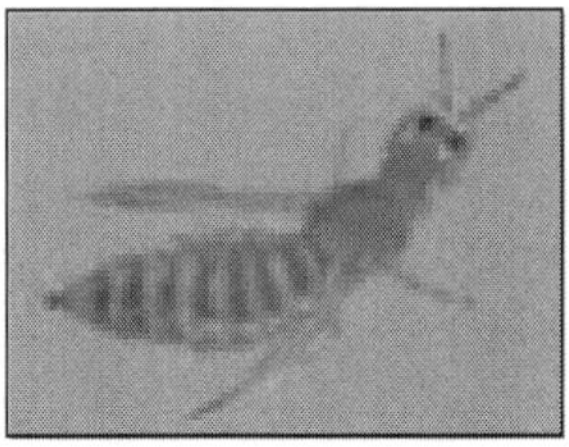

Flower Thrips

Flower thrips thrive under specific environmental conditions. These favour lower humidity levels and thrive in high-temperature environments. These conditions provide an ideal habitat for their activity and reproduction so exacerbating the damage they cause to plants.

Treatment options for insect infestations include manual removal of thrips followed by thorough leaf washing with water and application of insecticide. Preventive measures such as dusting bulbs and corms can help deter flower thrips infestations in subsequent growing seasons.

In terms of biological control, deploying natural enemies of pests can be highly effective. Predator nematodes target immature thrips while predatory mites particularly suited to greenhouse environments with elevated humidity levels prey on adult thrips. Other beneficial insects such as green lacewings, pirate bugs and ladybugs also serve as effective predators against thrips infestations. Integrating these natural enemies into pest management strategies can help maintain a balanced ecosystem and minimize reliance on chemical interventions.

10.3.1.6 Scale

Scale insects are characterized by their small, brown, circular to slightly oval and hard-bodied appearance resembling oyster shells when attached to leaves and stems. Affected plants typically exhibit symptoms such as stunted growth, yellowing and distortion. Initially, shiny honeydew may appear covering the leaves which may indicatie scale infestation. These pests feed on plant sap from both leaves and stems.

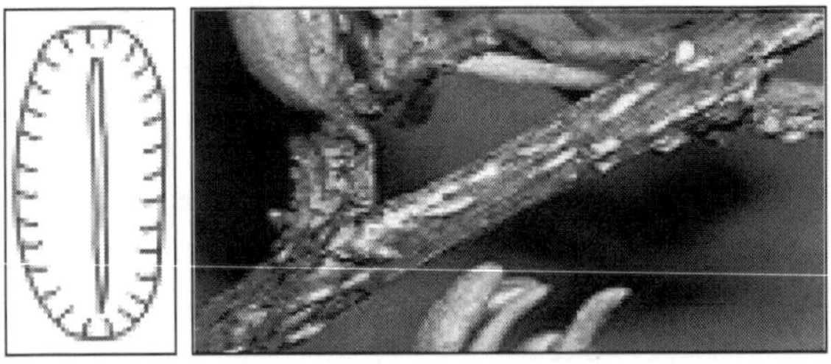

Treatment involves manual scraping off the scales and if possible isolating the affected plant. Washing the leaves with warm soapy water can help removing the scales. Insecticide application may be necessary for severe infestations. Natural predators such as green lacewings target the crawler stage of scale insects while mealybug destroyers can also feed on scales when mealybug populations decrease. However, effectiveness of scale insect control using natural predators can vary.

10.3.1.7 Mealybugs

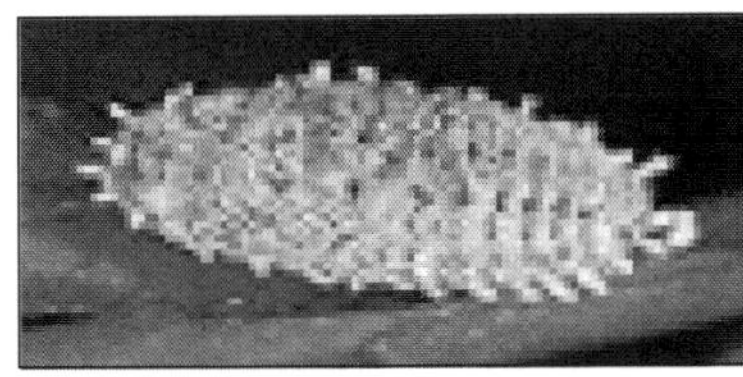

Mealybugs are characterized by their small size and soft bodies coated with a fluffy *cum* waxy substance. These are sluggish insects often found congregating at plant joints. Their presence manifests in powdery masses and these feed on plant juice so leading to distorted leaves and weakened plants. Moreover, affected plants may become covered in shiny honeydew which can attract sooty mould which threatens survival of the plants ultimately.

To combat mealybug infestations, effective treatment methods include scraping off the insects and applying insecticide. Alternatively, washing leaves with water or soapy water can help remove the pests with additional insecticide application if necessary. Natural predators such as green lacewings can serve as beneficial allies in controlling mealybug populations which can offer a sustainable and environment friendly approach to pest management.

10.3.1.8 Some other Harmful Pests

Insect	*Description*	*Best controlled with*
Earwigs 	Small dark brown centipede-like insects with a pincer tail. These usually come out at night.	Pick off by hand.
Cabbage worm 	Small velvety green caterpillars that eat plant leaves.	Spray with insecticidal soap.

Insect	*Description*	*Best controlled with*
Mexican bean beetle 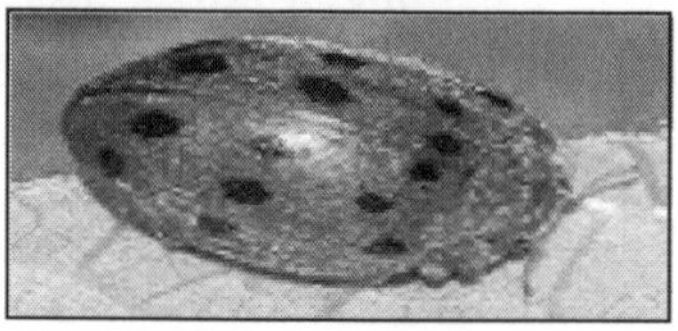	Small 1.25cm beetles eat on roots and leaves leaving only the veins.	Wash with soap.
Caterpillars 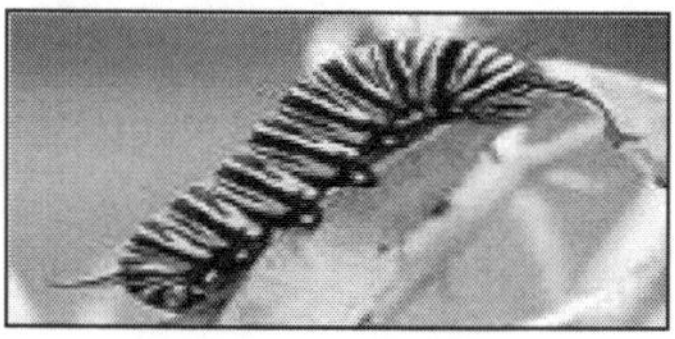	Most kinds of caterpillars are considered plant pests and will eat the leaves	Pick off by hand.
European corn borers 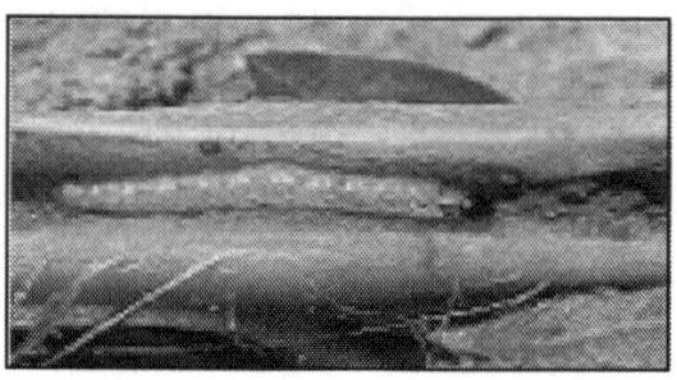	Small 2.5cm long caterpillars that bore into plant stems and eat them from inside out.	Spray with insecticide.
Cutworms 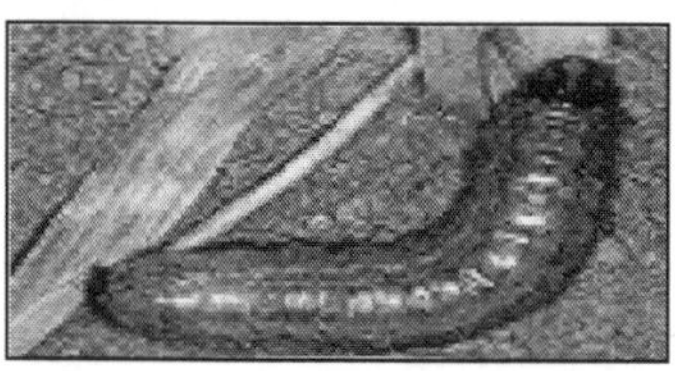	Small worm-type insects which when touched curl up into a ring shape. These eat through plant roots and topple the plant.	Not too common in hydroponics but when found spray with insecticide.
Colorado potato beetle 	2.5cm long striped beetles and their larvae will eat everything on a plant.	Pick off by hand and kill or spray with insecticide.
Tarnished plant bugs 	Small beetles about 6mm long with tarnish-like markings on its back. These insects feed on plants with a substance that deforms leaf tips and stem joints.	Clean off all nearby organic debris and spray with insecticide.

Insect	*Description*	*Best controlled with*
Striped cucumber beetle 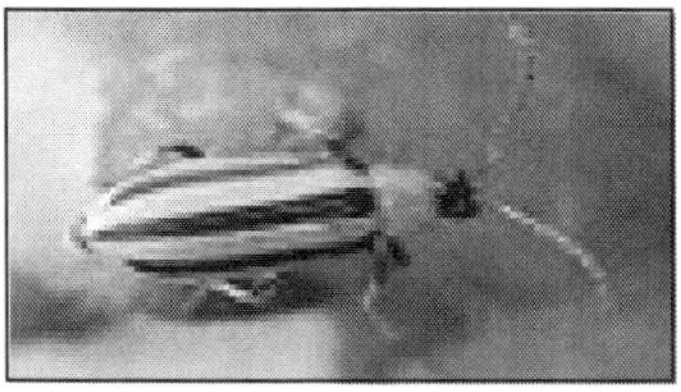	Small, very destructive, striped back beetles 6mm long. As adults, these plant pests eat leaves and its larvae eat on roots.	Spray with insecticide.
Root maggots 	Fly larvae that hatch at the base of the plants and eat the roots.	Not too common in hydroponics gardens. If found spray with insecticide.
Leaf hoppers 	Small, 3mm long, wedge shaped insects that suck the sap through the plant leaves.	Pick off by hand and apply insecticide if needed.

10.3.2 Fungal Diseases

The main causes of fungal diseases are high humidity and poor air circulation. Fungal diseases may be controlled through use of fungicides and other agricultural practices, however, new races of fungi often evolve that are resistant to various fungicides. Fungal problems are best dealt with prevention approaches such as:

☆ By maintaining good sanitation in and around the greenhouse hydroponics.

- ❖ The greenhouse floor should be kept clean by removing any debris.
- ❖ If possible, greenhouse floor can be covered using a white plastic groundcover. This will accomplish several things:
    - Reduce pest and pathogens which reside in soil.
    - Reflect additional light to the plants.
    - Decrease humidity by preventing soil moisture from evaporating up from the warm greenhouse floor.

- The plant debris should be composted far from the greenhouse air intakes.
- Weeds should not be allowed to grow in or near the greenhouse.
- Any affected plants should be burnt.

- All growing media should be sterilized before use.

Some common fungal diseases are explained below:

10.3.2.1 Anthracnose

It is also known as 'black dot root rot' and it appears as small sunken spots on ripe fruits. Leaves become dark and dry. In some plants such as cucumbers, pale green ulcers appear near the top of the fruit but afterwards become covered with mould and turn pink, then black and are finally covered with white mould. As the disease progresses the fruits become yellow and die. Anthracnose usually develops under high humid *cum* temperature conditions and may develop due to over-watering. To prevent it, the substrate for seedlings must be sterilized and room must be regularly ventilated. It can be controlled by:

- Removing or cutting off damaged foliage and fruits.
- Spraying with a mild fungicide.

10.3.2.2 Gray Mould (*Botrytis*)

It is a sooty grayish-white and furry growth on the green fruits (*e.g.* tomato) and stems. It is the most common fungal disease which is usually aggravated in cool and humid conditions with stale air. The infection usually enters the plant through damaged or dead parts. If not detected at early stage it may prove to be fatal. If it gets detected at an initial stage, it can be controlled to a great extent without using pesticides but by following the common precautions such as:

- By cleaning it using a soft-dry cloth (at initial stage) and then washing the plant gently.
- By decreasing the humidity and increasing air circulation and temperature.
- Removing and destroying the affected plants.
- The dead matter and debris should be removed from the system completely.
- If required, first mild and natural pesticides should be used and harsh pesticides should be used only if mild pesticides are not effective.

10.3.2.3 Powdery Mildew (*Erysiphe*)

It first appears as small white or grayish spots or patches on the tops of the leaves, and then progressing to a fine pale gray-white powder all over. Growth of infected plants become slow; leaves turn yellow and the plant dies. It thrives in cool temperature, dim light and high humidity. Its treatment includes:

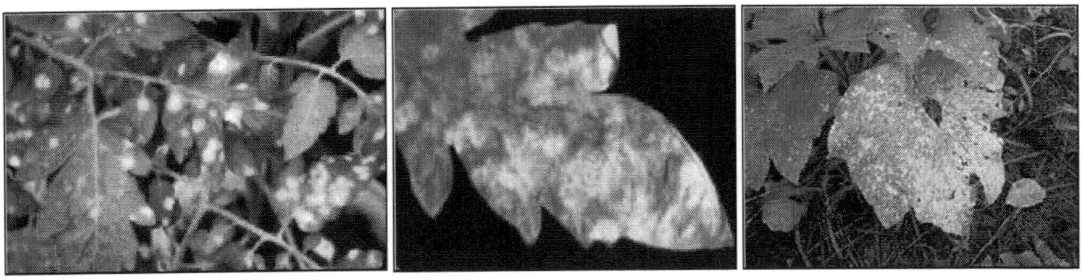

- ☆ Decrease of humidity
- ☆ Increase in air circulation and ventilation
- ☆ Increase in light intensity
- ☆ Use of pesticides finally

10.3.2.4 Buckeye Rot

Appears as large dark brown soft spots on fruits. Also can cause damping-off of seedlings, crown and root rot. Moderate moisture and temperatures favour this disease. Fungicides can help to control buckeye rot.

10.3.2.5 Damping-off (*Pythium, Rhizoctonia, Phytophthora, etc.*)

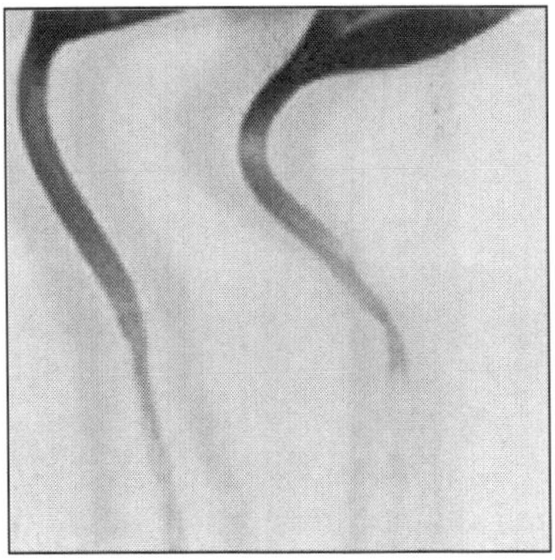

It is caused by soil-borne fungi under too-moist planting media and usually affects the newly planted seedlings. It affects the base of the stem and causes toppling over and death of the plants. Since there is no cure therefore situations causing such conditions should be prevented:

- ☆ Only sterile and fast draining media should be used for seedlings.
- ☆ The damaged sections must be removed.

10.3.2.6 Root Rot

The root rot is common in both indoor and outdoor plants ~ more serious being in indoor plants with poor drainage. A common belief that hydroponics growing method would exclude soil borne pathogens is not correct. Microorganisms similar to those found in soils rapidly establish themselves in hydroponic systems. Certain species of water mould genera of *Phytophthora* and *Pythium* from parasitic Oomycetes and *Fusarium* cause many of the root rot problems destroying the plant from the bottom. Among the species of *Phytophthora*, the most aggressive one is *P. cinnamomi*. *Phytophthora* species are generally host-specific whereas *Pythium* species can infect a large range of hosts. Most species of *Pythium* are plant parasites but *Pythium insidiosum* is an important pathogen of animals. The three most commonly encountered species of *Pythium* are *P. aphanidermatum*, *P. irregulare* and *P. ultimum*. *P. aphanidermatum* and *P. irregulare* may occur in 'ebb and flow' systems because they form a swimming spore stage that can move in water. This is likely to occur only if irrigation times are long (30 minutes or longer) or if pots sit in puddles of water because the bench or floor does not drain completely. If *Pythium* infects a cutting bed or if contaminated water is used in propagation large losses occur. *Pythium ultimum* is primarily associated with soil and sand. Almost all plants are susceptible to *Pythium* root rot. Root tips are very important in taking up nutrients hence these are attacked and killed first. *Pythium* can also rot the base of cuttings. Usually a plant with root rot does not survive but through vegetative propagation it can be regenerated. The airborne spores from root rot causing agents and which are also carried by insects and other organisms in the system may contaminate other plants therefore it is suggested to remove and destroy such plants immediately. Root rot that occurs in hydroponics is usually due to poor aeration in nutrient solution resulting in root zone oxygen starvation. Oxygen level available to the roots greatly affects the nutrient uptake of the plant. Root rot is a condition when the roots of the plant rot or decay mainly due to soggy root environment, overwatering and poor drainage. Excess water makes it very difficult for roots to get the air that they need resulting in decay. Aeration in nutrient solutions can be accomplished using air pumps, air stones, air diffusers and by adjusting the frequency and length of watering cycles. Overly warm nutrient or waterlogged media greatly affect the oxygen content in nutrient solution. As the water warms up it loses its capacity to hold oxygen. The nutrient temperature should be below 25°C. If a plant is affected with root rot these become stunted, pale and die and by mid-day the infected plants may show wilting but recovering in the night. On uprooting root tips exhibit browning and death, outer root tissues also show browning which can easily be pulled off leaving a strand of vascular tissues exposed, and under microscope the cells of such roots show round and thick-walled spores. To prevent this disease, all the bench surfaces, reservoirs of nutrient solution, all the equipments, *vis-a-vis* drain floors should regularly be cleaned and disinfected, proper temperature and aeration in nutrient solution should be maintained, damaged roots should be sheared off and dipped in some effective fungicide, only clean water free from pathogens especially RO or UV-treated water, chlorinated, ozonised or heated *cum* cooled ones should be used in nutrient solution to avoid contamination, no contaminated debris should be allowed to fall in the nutrient solution, nutrient solution should be reconstituted and/or replaced within the growing period and after every crop and propagation *cum* growing media should also be properly sterilized. Addition of hydrogen peroxide to nutrient solution will destroy single-celled organisms by liberating oxygen. Some biological control agents (micro-organisms) that may prove effective are **yeast** ~ *Candida oleophila*, **bacteria** ~ *Bacillus subtilis* and *Streptomyces* *griseoviridis*, **fungi** such as *Gliocladium catenulatum*, *Trichoderma harziamum* and *T. virens*. However, *Trichoderma* may enhance biomass production as is helpful in promoting root development.

10.3.2.7 Pythium Rot

Though it is same damping off but its attack is at a later stage of the plant. Its infection appears as a sunken brown spots on fruit (as in tomatoes) outlined with concentric rings. It can be controlled through:

- ☆ Using low concentrations of surfactants that prove to be effective in killing the mobile zoospores in re-circulating (closed) hydroponic systems thus preventing spread of the disease.
- ☆ However, surfactants are not effective in curing diseased plants.

10.3.2.8 Wilts (*Verticullum* and *Fusarium*)

Wilts initially appear as small spots on the leaves of infected plants such as tomato, peppers or eggplants. Lower leaves start to curl up, dry out and wilt. Sometimes, portions of the plant may wilt suddenly. *Fusarium* wilt causes an overall yellowing of the plant. Sometimes, it starts with older leaves. Browning of vascular tissue in stem indicates the fungus is clogging those tissues thus causing wilting. Roots turn brown and taproot may rot away. These wilts can be prevented:

- ☆ By choosing wilt resistant varieties.
- ☆ By destroying all infected vines/plants.
- ☆ By using only fresh and clean medium for each planting.
- ☆ By removing the infected plant from the system.

10.3.2.9 Early Blight

Irregular dark brownish spots and patches can be seen on lower leaves or stems of infected plants. These spots may enlarge and create a pattern of bull's-eye. Its symptoms on green fruits may not appear at initial stages but as the infection increases symptoms become visible. Fungicides may not be very effective. It is best treated with harsh chemicals. Its preventive measures are:

- ☆ Use of resistant varieties.
- ☆ Decreasing humidity and increasing air circulation.
- ☆ Removal of older leaves.

10.3.2.10 Leaf Mould

It spreads rapidly especially in greenhouses. Its symptoms include gray or brown spots on lower side of the leaf and pale areas on upper surface. It can be controlled by:

- ☆ Decreasing humidity and increasing ventilation.
- ☆ Fungicides are also effective.

10.3.2.11 Late Blight

Its effect can be seen as irregular greenish-black water-soaked spots on older leaves with downward bending of petioles which lead to leaf drop. Under humid conditions, a white fuzzy growth may give plants a frosted look. Fruit lesions are brown, firm and irregular. It is most common when day temperatures are high and night temperatures are low. It can be controlled with fungicides.

10.3.2.12 Rust

This powdery fungus is often found on the lower surface of leaf with uneven red patch that may be slightly raised. It thrives at high levels of humidity. It is usually cured by using harsh chemicals in small amounts.

- ☆ It will be better if plants with root rot are removed and destroyed.

10.3.2.13 Club Root

This disease transforms previously healthy roots into balls or clumps of club-shaped tubers which results into stunted growth. Fungicides can be used for its treatment.

10.3.2.14 Crown and Stem Rot

Crown and stems of the plants become pulpy followed by subsequent rot.

- ☆ Remove rotted sections of the plant and spray the fungicide

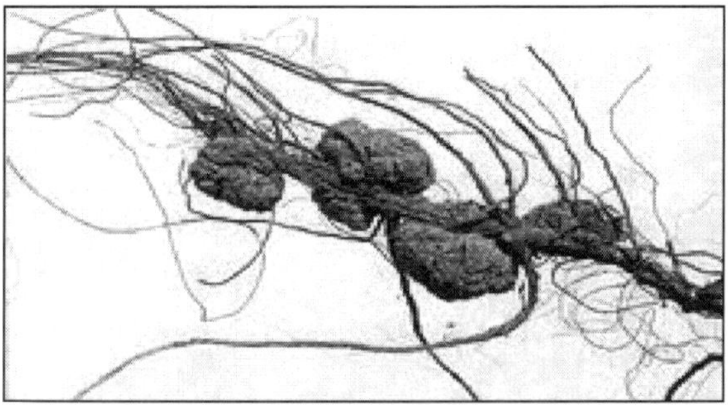

Crown Gall Disease is Caused by *Agrobacterium*.

10.3.3 Bacteria

Bacterial diseases are much common in sub-tropical and tropical regions of the world. Most of the bacteria that are associated with plants do not actually harm the plant itself. However, a small number of known species are able to cause diseases. Bacterial and viral diseases spread at a very high speed especially in a closed hydroponics system where nutrient solution is re-circulated. Some bacterial diseases that may infect hydroponics plants are:

10.3.3.1 Bacterial Canker

It is mainly caused by infected seeds and thrives well in moderate temperatures and high humidity. This disease first appears on older leaves causing unilateral wilting on one side of the leaf which later on dies. Often, the petioles remain on the plant which help distinguishing bacterial canker from other diseases. Small raised white spots with centre-browning may appear on fruits. Once the plants are infected by bacterial canker it becomes difficult to cure. The spread of the disease can be controlled to a great extent by removing infected plant and sterilizing all equipment and media.

10.3.3.2 Bacterial Spot

Initially, small and dark water-soaked spots appear on leaves which progress to dried and cracked lesions surrounded by yellow colouration. This spreads rapidly *via* wounds such as pruning or sucking insects. This disease flourishes in warm *cum* moist temperatures. The spread of this disease can be controlled at initial stage through copper sprays. Also, all the dead matters and debris should be destroyed.

10.3.3.3 Bacterial Wilt

This disease is highly contagious and can cause severe damage to the garden as there is no known effective control for it. Symptoms first appear as wilting of older leaves, followed by whole plant. It can be quickly diagnosed by placing a freshly cut infected stem in a glass of water. A white to milky stream of bacterial ooze flowing from the cut confirms the presence of this disease.

10.3.4 Viruses

There are many types of plant viruses causing great loss to crop yield. It is not economically viable to control it effectively with exception of perennial species such as fruit trees. Plant viruses are mostly transmitted from plant to plant often through a vector, especially aphids or through tools though certain other insects, fungi, nematodes or protozoa also spread the viruses. In many cases, the insects and viruses are specific for virus transmission such as the beet leafhopper that transmits the curly top virus causing disease in several crop plants.

Some precautionary measures that should be taken to prevent bacterial and viral infections in a hydroponics system are:

- Water being used for nutrient solution should be clean and free from pathogens.
- Though UV and ozone water purifiers can be used in hydroponic systems but these can be very expensive and have adverse effects on the minerals in the solution.
- Removal of any weed that may be susceptible to diseases.
- The clothes and shoes of persons entering the greenhouse grow-room should be free from soil.
- If possible, shallow trays of bleach solution should be kept at the entrance of the greenhouse so that the workers can clean off their shoes before entering the growing areas.
- Hands and tools must be cleaned regularly to prevent spread of disease within a greenhouse.
- Tobacco or tobacco products should be strictly kept away from greenhouse because these products may contain various tobacco viruses *i.e.* tobacco mosaic virus.
- Smoking or chewing tobacco in the greenhouse must be strictly prohibited.
- After handling tobacco or its products, workers must wash their hands properly.
- Aphids can transmit viruses, so control of insect-pests will reduce the risk of viral problems.

Some viral diseases that may infect hydroponics plants are:

10.3.4.1 Tobacco Mosaic Virus (ToMV)

ToMV which is quite difficult to eradicate remains alive for years on tobacco leaves and can badly affect many hydroponics crops such as cucumbers, peppers and tomatoes. Its symptoms appear as disfigurement of the leaves and stunted growth. This virus is usually transmitted through hands, the working tools and through sucking insects in the greenhouse. The best practices to prevent this virus are:

- Using resistant cultivars.
- Keeping harmful insects away from the greenhouse.
- Tobacco or tobacco products should be strictly kept away from greenhouse because these products may contain tobacco mosaic virus.
- Smoking or chewing tobacco in the greenhouse must be strictly prohibited.

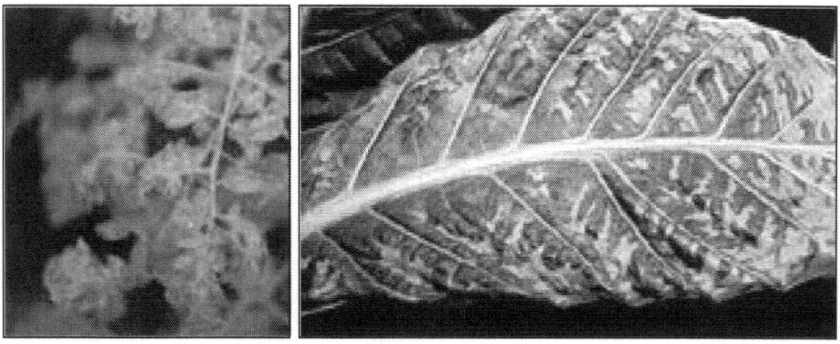

10.3.4.2 Tomato Yellow Leaf Curl Virus (TYLCV)

It is transmitted by tobacco whitefly, *Bemisia tabaci* which is a serious problem in the Middle East. It causes interveinal chlorosis and upward curling of leaf edges. Fruit set is greatly reduced. Particularly destructive if young plants are attacked as production of fruit is drastically affected. Its preventive method is a complete control of whitefly.

10.3.4.3 Cucumber Mosaic Virus

Dark green blotches appear on short deformed fruits. The green skin becomes covered with white, yellow and dark-green spots. The symptoms become more evident as the temperature rises inside the greenhouse. To prevent it, the soil or compost must be sterilized.

10.3.4.4 Other Viruses

Other mosaic viruses, such as **common mosaic**, **aucuba mosaic**, and **cucumber mosaic** can all infect tomatoes and various other crops. Symptoms include mottled areas of the leaves, yellowing and stunted growth.

10.3.5 Protozoa

Protozoans are very small organisms which often live inside larger animals or plants. There are a few examples of plant diseases caused by protozoa. These are transmitted as zoospores that are very durable and may be able to survive in a resting state in the soil for many years. These have also been shown to transmit plant viruses. When zoospores come into contact with a root hair these produce a plasmodium and invade the roots. It can be prevented by maintaining good sanitation in and around the hydroponics system.

10.3.6 Nematodes

Nematodes are microscopic, multi-cellular and worm-like organisms most of them living freely in the soil. Some nematodes are even beneficial parasites that make no harm to plants or earthworms, instead these can be introduced into the garden to control most of the soil-dwelling insects such as moths, fleas, loopers, weevils and borers. These enter the body of its prey and destroy the bad bug. Some of the species of nematodes can parasitize plant roots. These are a problem in tropical and subtropical regions of the world where these may infect crops. Nematodes have also been shown to transmit plant viruses. It can be prevented by maintaining good sanitation in and around the hydroponics system, especially when a soil plant is being transplanted in hydroponics system.

Root-Knot Nematode Galls

10.4 General preventive measures of pathogen contamination

i) **Water quality:** Start with high quality clean water sources to minimize the risk of pathogen contamination. Use filtered or sterilized water whenever possible.

ii) **Sanitation:** Practice strict sanitation measures to prevent introduction and spread of pathogens. Clean and disinfect all equipment, tools and surfaces regularly.

iii) **Hygiene:** Minimize contact with plants and nutrient solutions to reduce the risk of pathogen transmission. Avoid introducing pathogens through contaminated hands, tools or equipment.

iv) **Quarantine:** Quarantine new plants or growing media before introducing them into the hydroponic system to prevent the spread of pathogens.

10.5 Insect Control

The best way to get rid of unwanted insects is their detection and identification. Once the insects in the garden are identified correctly, these can be controlled easily using one or more of the following options:

i) **Bright sticky trap strips:** Sticky strips provide a safe method of trapping insects. The insects are attracted to the bright colour of the sticky strip and when once these land are immediately stuck. When the strips are full, simply discard and replace with new ones. These sticky traps can be made at home as:

   a) Spray paint a cardboard with bright yellow colour (for insects such as aphids, fungus gnats and whiteflies) or with bright blue colour (for insects such as thrips).

   b) Let it dry.

   c) Coat with vaseline (petroleum jelly).

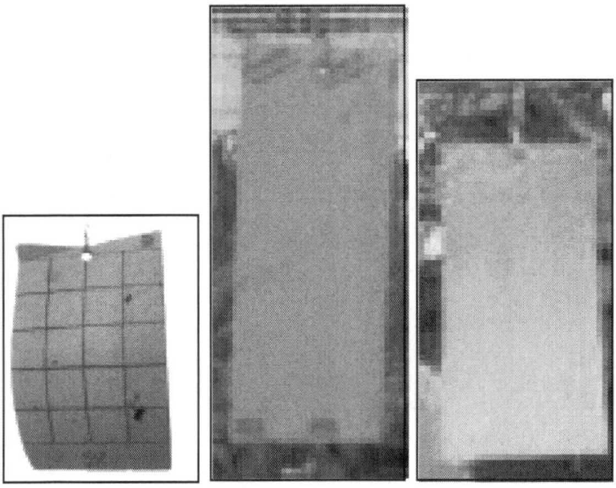

Bright Sticky Trap Strips.

ii) **Physical control methods**:

a) Sometimes a strong blast of water from a hose may kill and wash away the pests.

b) Pests like caterpillars can be hand-picked and dropped into soapy water.

c) Potato beetles can be removed by shaking off the plants onto a sheet.

d) Aphids, cabbage worms and white flies are attracted to yellow colour so these can be trapped by covering a piece of yellow poster board with sticky glue and hanging it among the infested plants.

iii) **Predators and parasites**: A predator is an animal that kills and eats other animals. A parasite is a small animal or plant that lives on or inside a larger animal or plant and gets its food from it. Almost all the insects have predators and/or parasites. These enemies of harmful pests can be purchased from the predators and parasites garden supply centres or insectaries (facilities that raise insects) that breed and sell beneficial insects. These beneficial insects can be introduced in the gardens to eradicate harmful pests. These beneficial insects are shipped as eggs, larvae or adults. Some such beneficial insects are:

a) **Ladybugs beetles**: These are small spotted insects that love to eat aphids and are effective in controlling many other insect-pests (*e.g.* thrips, mites, scale insects and small caterpillars) in the larval stage as well as the adult.

 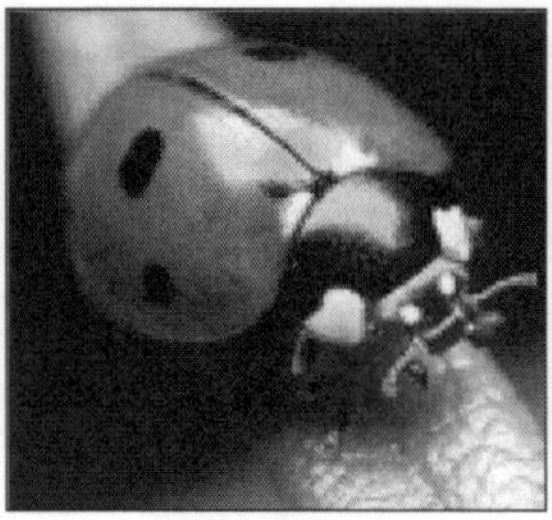 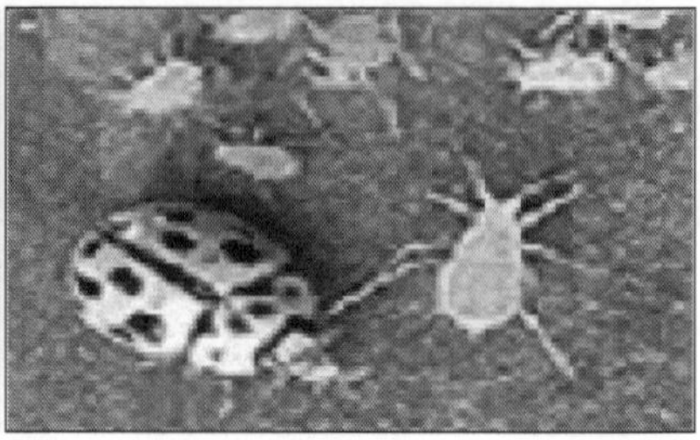

b) **Lacewings**: These are known for their voracious appetite for various insects [*e.g.* aphids, thrips, whiteflies, mealybugs and scales (only if mealybugs run low)] in the larval stage as well as the adult ones.

c) **Praying mantis:** Effective in controlling various insects in the larval stage as well as at the adult stage. It produces egg cases which contain approximately 200 babies. The babies are also deadly predators.

d) ***Trichogramma* wasp**: This wasp feeds on more than 200 types of insects (such as tomato hornworms, cabbageworms, fruitworms, corn earworms, codling moths, cutworms, armyworms, cabbage loopers, corn borers and cane borers) but does no harm to other beneficial insects or people.

e) **Predatory polistes wasp**: Controls bollworms or other caterpillars on a cotton plant.

f) **Larvae of many hoverfly species**: Principally feed upon greenfly and also eat fruit tree spider mites and small caterpillars. Adults feed on nectar and pollen which they require for egg production.

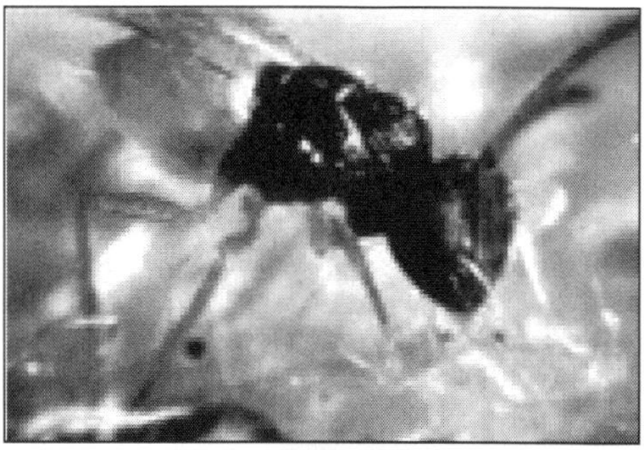

g) ***Encarsia formosa***: This tiny parasitic wasp is effective in some species of the whitefly (*e.g.* tobacco whitefly) and lays its eggs in their larvae. Parasitized larvae turn black and are easily recognized. Adult *Encarcia formosa* also feed on honeydew and the body fluids of whitefly larvae.

h) ***Eretomocerus californicus***: The tobacco whitefly can be parasitized by *Encarsia* but these are controlled better by *Eretomocerus* species.

i) ***Ichneumonid* wasps**: These are 5–10mm long parasitic wasps which prey mainly on caterpillars of butterflies and moths.

j) **Braconid wasps**: These are some 5mm long tiny parasitic wasps attacking caterpillars and a wide range of other insects including greenfly.

k) **Chalcid wasps**: These are among the smallest of insects (<3mm). These parasitize the eggs/larvae of greenfly, whitefly, cabbage caterpillars, scales and strawberry tortrix moth (*Acleris comariana*).

l) **Tachinid flies**: This parasitizes a wide range of insects including caterpillars, adult and larval beetles, true bugs and others.

m) ***Aphidius colemani***: This parasitic wasp is effective against some species of aphids. This parasitizes aphids form characteristic white mummies.

n) ***Aphidoletes aphidimyza***: Effective on a wide range of aphid species and lays its eggs in aphid colonies. The orange larvae hatch from these eggs and start feeding voraciously on aphids.

o) **Predatory mite** (*Phytoseiulus persimilis*): Used for control of the two-spotted spider mite.

Phytoseiulus persimilis

p) ***Verticillium lecanii***: A common soil borne fungus which affects various kinds of insects. It is widespread in temperate and tropical areas but does not harm birds, fish, mammals or plants. The fungal spore germinates and begins to grow on the honeydew secretion on the whitefly body and the whitefly dies from infection before the fungus even becomes visible. Since the fungus is not mobile and cannot seek its host, it is only effective in very high densities of whitefly so repeated applications are necessary.

q) **Black and yellow garden spider**: These feed on juice of various harmful as well as beneficial insects.

iv) Entomopathogenic organisms: Entomopathogenic organisms can play a valuable role in managing insect pests in hydroponic systems. While hydroponic environments may not naturally support as diverse a population of these organisms as soil-based systems, these can still be introduced and utilized effectively. Here are some examples:

a) **Entomopathogenic Fungi**: Fungi that cause disease in insects are known as entomopathogenic fungi. A remarkable additional feature of some fungi is their effect on plant fitness.

i) ***Beauveria bassiana***: This fungus infects a wide range of insect-pests including aphids, thrips, whiteflies, weevils and spider mites.

ii) ***Metarhizium anisopliae***: Another versatile entomopathogenic fungus that can target various pests like aphids, whiteflies and thrips.

iii) ***Lecanicillium lecanii***: Effective against whiteflies, aphids and other soft-bodied insects commonly found in hydroponic systems.

b) **Entomopathogenic or parasitic nematodes** (microscopic worms): These destroy harmful insects such as grubs, fleas, gnats, flies, cutworms, billbugs, ants and Japanese beetles. These control harmful soil-dwelling insects by feeding on their larvae though not harming to other beneficial insects.

i) ***Steinernema feltiae*** **and** ***S. carpocapsae***: These nematodes are effective against soil-dwelling pests such as fungus gnats, thrips pupae and some species of root aphids.

ii) ***Heterorhabditis bacteriophora***: Another species of entomopathogenic nematode that can target soil-dwelling pests, including various larvae and pupal stages of insects.

iii) Phasmarhabditis hermaphrodita**:** It is a microscopic nematode that kills slugs, thereafter feeding and reproducing inside.

c) **Entomopathogenic bacteria:** While fungi and nematodes are more commonly associated with biological control in hydroponics, certain bacteria can also play a role. Bacteria used for biological control infest insects *via* their digestive tracts so insects with sucking mouth parts such as aphids and scale insects are difficult to control. There is only one example such as:

i) ***Bacillus thuringiensis*** **(Bt)**: Bt is the most widely applied species of bacteria used for biological control. This bacterium produces toxins that are lethal to certain insect-pests particularly caterpillars, beetles and mosquitoes. The bacteria are available in sachets of dried spores which are mixed with water and sprayed onto vulnerable plants such as brassicas and fruit trees or incorporated into the growing medium in hydroponic systems to target specific pests. Bt has also been incorporated into crops making them resistant to these pests and thus reducing the use of pesticides.

v) **Homemade organic pesticides**: Homemade organic pesticides explained earlier can be used.

vi) **Mild organic pesticides**: Horticultural oils, insecticidal soaps and botanical insecticides all are organic pesticides. Botanicals are non-toxic soaps which include plant compounds combined into efficient formulae for easy use. Most are safe to use on food crops because these break down quickly and have no long-lasting effect on the environment. However, these are not only toxic to the pests and diseases but can also be toxic to birds and soil microorganisms that are part of a healthy garden.

vii) **Harsh pesticides** should only be used if all the above methods are ineffective.

10.6 Disease Control

See 'Nutrient disinfectants (sterilizing agents)' and 'Microbial inoculants containing beneficial microorganisms' in 'Additives' section of 'Factors affecting nutrient solution'.

10.7 Monitoring and Management

i) Regular monitoring of nutrient solution parameters, plant health and system cleanliness to detect signs of algal growth or pathogen contamination early.

ii) Implement IPM strategies to prevent and manage algal growth and pathogen outbreaks effectively.

iii) Maintain detailed records of sanitation practices, treatment protocols and any other observed issues to guide future management decisions.

By implementing these strategies for preventing and treating algae growth and pathogen contamination, growers can maintain a healthy and productive hydroponic system thus maximizing crop yields. Regular monitoring, proper sanitation and proactive management practices are key to preventing and mitigating issues related to algae and pathogens in hydroponic cultivation.

Chapter 11

Advanced Techniques and Innovations

11.1 Automated Nutrient Delivery Systems

Hydroponic agriculture has gained prominence due to its ability to efficiently produce high-quality crops in controlled environments. Automated nutrient delivery systems represent a cornerstone of modern hydroponic agriculture offering precision, efficiency and scalability for growers. With ongoing advancements in technology and increased accessibility, these systems are poised to play an even more significant role in shaping the future of hydroponic farming, enabling sustainable food production and agricultural innovation.

11.1.1 Components of Automated Nutrient Delivery Systems

Apart from hydroponic components mentioned in "Chapter 2", automated nutrient delivery system comprises 'Automation and Control Interface' with sensors, controllers and software that automate nutrient delivery processes and monitor environmental conditions. It allows growers to set and adjust nutrient schedules, monitor system performance and receive alerts for any deviations from set parameters. The key components in this system are designed to efficiently manage the distribution of nutrients to plants in hydroponic setups. These components work together seamlessly to ensure precise nutrient delivery and optimal plant growth. The main components of such systems typically include:

11.1.1.1 Sensors

Automated nutrient delivery systems in hydroponics often rely on sensors and monitoring devices to ensure precise control over nutrient levels, pH, temperature and other environmental parameters. Sensors play a crucial role in automated nutrient delivery systems by providing real-time data on environmental parameters such as temperature, humidity and nutrient levels. This information allows growers to make informed decisions and adjustments to optimize plant growth and performance. Here are some common sensors and monitoring devices used in hydroponic setups:

11.1.1.1.1 pH Meter

The pH meter serves as an indispensable tool within hydroponic systems, offering continuous monitoring of the delicate balance between acidity and alkalinity within the nutrient solution. Its primary function revolves around ensuring that the pH levels remain within the optimal range conducive to robust plant growth and nutrient absorption. By gauging and regulating pH levels meticulously, growers can fine-tune nutrient solution to provide an environment where plants thrive.

With precision and accuracy, the pH meter acts as a sentinel, diligently overseeing the chemical composition of the nutrient solution. Its role extends beyond mere measurement ~ serving as a guardian against fluctuations that may impede plant health and productivity. Through real-time monitoring, growers gain invaluable insights into the pH dynamics within the hydroponic system thus empowering them to make informed adjustments as needed.

Moreover, pH meter embodies efficiency and reliability ~ offering a steadfast companion in the quest for optimal plant nutrition. Its continuous monitoring capabilities provide growers with a comprehensive understanding of the intricate interplay between pH levels and nutrient uptake. By maintaining pH within ideal range growers unlock the full potential of hydroponic cultivation ensuring that plants receive the essential nutrients they need for vigorous growth and development.

11.1.1.1.2 EC/TDS Meter

Electrical conductivity and total dissolved solids meter is an essential tool in hydroponic agriculture, offering growers invaluable insights into the composition and strength of their nutrient solutions. By measuring EC of solution these meters provide a quantitative assessment of the concentrations of dissolved salts, minerals and nutrients in hydroponic water. These data serve as a critical indicator of the overall nutrient strength, allowing growers to fine-tune and optimize their nutrient formulations to meet the precise requirements of their crops.

Utilizing advanced sensing technology, EC/TDS meters accurately assess the conductivity of the solution which is directly correlated with concentration of ions dissolved within it. This includes essential macronutrients such as nitrogen, phosphorus and potassium as well as micronutrients and trace elements vital for plant growth and development. By analyzing the EC/TDS readings, growers can ensure that their plants receive the appropriate balance of nutrients, promoting healthy root development, vigorous growth and optimal yield.

Furthermore, EC/TDS meters enable growers to monitor changes in nutrient levels over time, facilitating proactive adjustments to nutrient dosing regimes based on crop growth stages, environmental conditions and other factors. This dynamic approach to nutrient management allows for precise control and optimization of nutrient delivery thus minimizing the risk of nutrient deficiencies or imbalances that could compromise plant health and productivity.

11.1.1.1.3 Nutrient Sensors

Nutrient sensors serve as invaluable tools in hydroponic agriculture, playing a pivotal role in the meticulous monitoring of essential nutrients within the nutrient solution. These sensors are calibrated to detect concentration levels of vital nutrients crucial for plant health and growth. By providing real-time data on nutrient levels growers can precisely adjust and fine-tune their nutrient formulations ensuring that plants receive an optimal balance of essential elements for robust and healthy growth.

These sensors act as guardians of plant nutrition, offering growers critical insights into the nutrient status of their hydroponic systems. Armed with this data, growers can make informed decisions regarding nutrient dosing, supplementation and management strategies. Whether adjusting nutrient levels to meet specific crop requirements or addressing deficiencies to prevent stunted growth, nutrient sensors empower growers to tailor nutrient solutions to the precise needs of their plants.

Furthermore, nutrient sensors play a fundamental role in optimizing resource efficiency and minimizing waste in hydroponic systems. By monitoring nutrient levels accurately, growers can avoid over- or under-fertilization, thereby conserving valuable resources and reducing environmental impact. Nutrient sensors contribute to overall crop health and productivity by ensuring that plants receive a steady supply of essential nutrients, promoting vigorous growth, and maximizing yield potential.

11.1.1.1.4 Dissolved Oxygen (DO) Meter

DO meter serves as a pivotal tool in hydroponic cultivation, tasked with measuring oxygen content dissolved in nutrient solution. This measurement is of paramount importance as directly it influences the maintenance of aerobic conditions vital for thriving ecosystem within hydroponic systems. Adequate levels of dissolved oxygen are indispensable in preventing suffocation of roots and fostering robust root development which is crucial for overall health and productivity. By providing growers with precise insights into oxygen levels, DO meters play a fundamental role in ensuring optimal growing conditions, thereby contributing to the success and sustainability of hydroponic agriculture.

11.1.1.1.5 Temperature Sensor

Temperature sensors play a crucial role in hydroponic systems by monitoring both ~ the temperature of nutrient solution and ambient air. This dual monitoring ensures that environmental conditions remain within the optimal range conducive to plant growth and metabolic activity. In the nutrient solution maintaining an ideal temperature is essential as it directly impacts nutrient uptake, root health and overall plant physiology. Fluctuations outside the optimal temperature range can hinder nutrient absorption and lead to stress or even damage the root system. Similarly, monitoring ambient air temperature is vital as it affects transpiration rates, photosynthesis and overall plant metabolism. Extreme temperatures can impact plant health negatively so causing wilting, leaf burn or slowed growth.

By utilizing temperature sensors growers can assess environmental conditions accurately and make informed adjustments to maintain a stable and favourable growing environment for their hydroponic crops. This proactive approach to temperature management contributes to ensuring success of hydroponic cultivation endeavours, promoting plant health and maximizing crop yields.

11.1.1.1.1 Water Level Sensor

Maintaining precise levels of nutrient solution in the system is critical for optimal plant growth and health. Water level sensors serve as invaluable tools as these are designed to detect and monitor nutrient solution level within reservoirs or individual growing containers. By measuring the height (depth) of the nutrient solution continuously these sensors provide growers with real-time feedback thus allowing them to prevent both overfilling and depletion scenarios. Overfilling can lead to waterlogged roots and nutrient imbalances while depletion can result in inadequate hydration and nutrient uptake.

11.1.1.2 Controllers

Control systems receive data from sensors and use algorithms to analyze information and make real-time adjustments to nutrient delivery. These systems can regulate nutrient dosing pumps, valves and other components to maintain optimal nutrient levels and solution parameters. Advanced control systems may include programmable logic controllers (PLCs), microcontrollers or computer-based control systems for precise regulation and automation. Here are some common controlling devices used in hydroponic setups:

11.1.1.2.1 Dosing Pumps and Valves

Dosing pumps and valves serve as critical components in the orchestration of nutrient delivery within hydroponic systems so ensuring that plants receive the precise amounts of essential nutrients, pH adjusters and supplements required for optimal growth and development. These systems operate in tandem with sensor readings and programmed parameters to achieve precision and consistency in nutrient dosing.

Whether peristaltic or diaphragm dosing pumps are integral to the accurate dispensation of liquid nutrients. By precisely metering out specified volumes of nutrient solutions, pH adjusters and supplements dosing pumps enable growers to maintain desired composition of nutrient solution. This meticulous dosing is essential for preventing nutrient imbalances and ensuring that plants receive appropriate nourishment at all stages of growth.

Furthermore, automated valves play a pivotal role in regulating the flow of nutrient solutions and water throughout the hydroponic system. These valves provide capability for precise adjustments and mixing thus allowing growers to tailor nutrient solution to specific requirements of their crops. By controlling the flow rates and proportions of different components, automated valves facilitate optimal nutrient uptake by plants while minimizing waste and maintaining system efficiency.

11.1.1.2.2 Irrigation System

It is a fundamental component within hydroponic setups serving as the conduit through which vital nutrient solutions are transported to plant roots. It employs various delivery mechanisms such as drip lines, misters or other specialized systems tailored to specific needs of the crops being cultivated. These delivery mechanisms ensure precise distribution of nutrient solution ~ allowing each plant to receive necessary hydration and essential nutrients directly at root zone. By optimizing delivery of nutrients *cum* moisture, the irrigation system plays a critical role in promoting healthy plant growth and maximizing crop yield. The efficiency and reliability of irrigation system are paramount in maintaining consistent growing conditions and mitigating the risk of nutrient imbalances or water-related issues that could adversely affect plant health and productivity. Thus, meticulous attention to the design, operation and maintenance of the irrigation system is essential for success of hydroponic cultivation endeavours.

11.1.1.2.3 pH and EC Controllers

pH and EC controllers are fundamental components of hydroponic systems as these ensure the precise management of nutrient solutions to support optimal plant growth. pH controllers are responsible for monitoring and modulating the acidity or alkalinity of the nutrient solution while EC controllers regulate strength of nutrient solution by measuring its electrical conductivity.

In automated hydroponic setups, these controllers work in tandem to maintain ideal conditions for plant development. pH controllers continuously monitor pH levels of nutrient solution and make adjustments as necessary to keep these within the optimal range. This is achieved through utilization of pH up or pH down solutions along with dosing pumps that precisely administer these solutions to achieve desired pH levels.

Similarly, EC controllers play a crucial role in ensuring appropriate nutrient concentrations in the solution. By measuring electrical conductivity of the solution,

EC controllers provide valuable insights into nutrient strength. Automated systems then adjust EC levels by regulating the dosing of concentrated nutrient solutions or water so ensuring that plants receive correct balance of essential nutrients for healthy growth.

Together, pH and EC controllers form the backbone of nutrient management in hydroponic systems, facilitating the maintenance of stable and optimized growing conditions. Their precise monitoring and adjustment capabilities contribute to the overall success and efficiency of hydroponic cultivation allowing growers to achieve superior crop yields and quality.

11.1.1.2.4 Automation Software

Automation software is a critical component within hydroponic systems ~ serving as the central hub for integrating data from sensors and orchestrating the operation of various components such as dosing pumps, controllers and irrigation systems. By leveraging predefined settings, this software ensures effecient functioning of the hydroponic system in accordance with desired parameters. In addition to its control functions, automation software also facilitates comprehensive data logging capabilities. This feature enables the recording of sensor readings, nutrient dosing rates and system parameters over time thus providing growers with a wealth of historical data for analysis and optimization purposes.

Furthermore, automation software often includes sophisticated data analysis tools. These tools process the collected data to identify trends, patterns and correlations empowering growers to optimize nutrient delivery strategies, adjust system settings and diagnose any issues or irregularities promptly.

In essence, automation software serves as the backbone of modern hydroponic operations ~ enabling growers to manage their systems with precision, efficiency and informed decision-making. Through seamless integration of data collection, control and analysis functions automation software maximizes the potential for successful crop cultivation and yield optimization in hydroponic environments.

11.1.1.3 Data Logging and Control System

Data logging and control systems in hydroponics serve as vital infrastructure for collecting real-time data from sensors and managing various aspects of growing environment. These systems are primarily responsible for capturing and storing data related to key parameters such as pH levels, EC, temperature, humidity and nutrient concentrations. Data logging systems ensure that growers have access to comprehensive information about their hydroponic setups so allowing them to monitor system performance over time and identify trends or anomalies. Control systems integrated within these setups enable growers to make adjustments to nutrient dosing, pH levels and environmental conditions based on collected data. By automating routine tasks and providing growers with greater control over their systems, data logging and control systems facilitate efficient management of hydroponic operations, optimizing plant growth and maximizing crop yield.

11.1.1.4 Remote Monitoring and Control Systems

On the other hand, remote monitoring and control systems represent a more advanced layer of technology that allows growers to access and manage their hydroponic systems remotely. These systems leverage connectivity and internet-enabled devices to provide growers with real-time access to system data and controls from anywhere with an internet connection. Remote monitoring functionality enables growers to track various aspects of system performance in real-time including environmental conditions, nutrient levels and plant health. This capability allows growers to identify potential issues early on and take proactive measures to address them even when they are not physically present in the growing facility. Moreover, remote control functionality empowers growers to make real-time adjustments to system settings and parameters as needed, ensuring optimal growing conditions and maximizing crop productivity. Overall, remote monitoring and control systems offer growers unparalleled flexibility and convenience in managing their hydroponic operations, contributing to increased efficiency and improved crop outcomes.

11.1.1.5 Integrated Data Logging and Remote Monitoring Systems

Data logging and control systems along with remote monitoring and control capabilities are integral components of modern hydroponic agriculture thus offering growers unparalleled insight, control and flexibility in managing their systems. These systems work synergistically to collect real-time data from sensors, provide comprehensive insights into system performance and enable growers to make informed decisions and take proactive measures to support plant growth and development.

By seamlessly integrating data logging and control functionalities growers can remotely monitor and adjust various aspects of their hydroponic systems from anywhere with an internet connection. This integration empowers growers to track critical parameters such as pH, EC, temperature, humidity and nutrient levels in real-time so gaining valuable insights into the status of their hydroponic operations. Armed with this information, growers can identify potential issues early on and take prompt action to optimize growing conditions and maximize crop yields.

Moreover, data logging and control systems are equipped with customizable alerts and automation capabilities ~ further enhancing operational efficiency and resource utilization. These systems streamline operations by automating routine tasks such as nutrient dosing and environmental control while also providing growers with timely alerts for any deviation from pre-defined parameters. Whether it is a sudden drop in pH, a spike in temperature or a nutrient imbalance growers receive instant notifications enabling them to respond promptly and prevent potential crop damage.

In addition to monitoring, remote control functionality empowers growers to make real-time adjustments to system settings and parameters as needed. Whether it is fine-tuning nutrient dosing, adjusting pH levels or modifying environmental conditions growers can remotely intervene to maintain optimal growing conditions and ensure the health and vitality of their crops. This level of control enables growers to adapt to changing conditions and to optimize crop production efficiently, ultimately contributing to the success and sustainability of hydroponic farming operations.

11.1.2 Advancements in Automated Nutrient Delivery Systems

11.1.2.1 Integration with 'Internet of Things' (IoT) Technology and Cloud-Based Platforms

Integration with Internet of Things (IoT) technology and cloud-based platforms represents a groundbreaking advancement in the realm of hydroponic agriculture. By merging IoT sensors, devices and data seamlessly with cloud-based infrastructure, modern hydroponic systems transcend geographical boundaries offering growers unparalleled access and control over their operations.

At the heart of this integration lies IoT technology comprising a network of interconnected sensors strategically deployed throughout hydroponic environment. These sensors serve as the eyes and ears of the system capturing real-time data on crucial parameters such as temperature, humidity, pH levels, nutrient concentrations and more. Through their continuous monitoring efforts, IoT sensors generate a wealth of valuable information so providing growers with insights into the dynamic conditions within their hydroponic setups.

Complementing this sensor network is the utilization of cloud-based platforms which serve as centralized hub for data storage, analysis and management. Leveraging the power of cloud computing, these platforms offer growers the ability to access, analyze and act upon their hydroponic data from virtually anywhere with an internet connection. Whether from the comfort of their homes, on the go *via* mobile devices or across different geographical locations seamlessly the growers can tap into their hydroponic systems and make informed decisions in real-time.

Integration of IoT technology and cloud-based platforms enables a wide array of functionalities including remote monitoring, data analysis and control of hydroponic systems. Growers can remotely monitor environmental conditions, track nutrient levels, detect anomalies and receive alerts or notifications when predefined thresholds are exceeded. Moreover, through sophisticated data analysis techniques such as machine learning and predictive analytics growers can gain deeper insights into crop performance, growth trends and potential issues thus empowering them to optimize growing conditions and maximize yields.

Furthermore, cloud-based platforms facilitate seamless control and management of hydroponic systems allowing growers to remotely adjust parameters, schedule tasks and automate processes as needed. Whether fine-tuning nutrient dosing, regulating pH levels or adjusting environmental conditions growers have the flexibility and agility to respond promptly to changing conditions and optimize crop growth with precision.

In essence, the integration of IoT technology and cloud-based platforms revolutionizes hydroponic agriculture so enabling growers to transcend physical limitations and harness the power of data-driven insights and remote control. Through this transformative synergy, hydroponic farming enters a new era of efficiency, sustainability and innovation thus paving the way for a more connected and resilient agricultural future.

11.1.2.2 Machine Learning (ML) and Artificial Intelligence (AI)

ML and AI represent cutting-edge technologies that are being integrated increasingly into hydroponic systems to revolutionize crop management and maximize yields. These advanced systems leverage sophisticated algorithms and computational techniques to process vast amounts of sensor data so enabling them to perform complex analyses and make data-driven real-time decisions.

In the context of hydroponics, ML and AI algorithms are employed to analyze sensor data collected from various environmental and nutrient monitoring devices deployed throughout the growing environment. These algorithms have capability to identify patterns, correlations and trends within the data thus allowing growers to gain deeper insights into the factors influencing plant growth and health.

One of the key applications of ML and AI in hydroponics is the prediction of plant growth patterns. By analyzing historical data on environmental conditions, nutrient levels and crop performance these algorithms can forecast future growth trajectories with remarkable accuracy. This predictive capability enables growers to anticipate changes in crop development, optimize resource allocation and proactively address potential issues before these arise.

Furthermore, ML and AI algorithms play a crucial role in optimizing nutrient dosing strategies. Continuously by analyzing sensor data and monitoring plant responses these algorithms can adjust nutrient concentrations dynamically in the hydroponic solution to meet the evolving needs of the plants. This adaptive approach ensures that plants receive optimal balance of nutrients throughout their growth cycle thus leading to healthier plants, higher yields and improved resource efficiency.

Overall, the integration of ML and AI technologies into hydroponic systems represents a significant advancement in precision agriculture. By harnessing the power of data analytics and computational intelligence growers can unlock new insights, optimize growing conditions and achieve unprecedented levels of productivity *cum* sustainability in hydroponic crop production.

11.1.3 Customization and Flexibility

In contemporary hydroponic setups, growers benefit from remarkable customization and flexibility afforded by modern systems. These systems are designed with a diverse array of customizable settings and flexible programming options thus enabling growers to fine-tune nutrient delivery and environmental control according to the precise requirements of their crops and the unique conditions within their growing environments.

The customization capabilities of modern hydroponic systems extend across various aspects of cultivation. Growers can adjust nutrient formulations, pH levels and EC concentrations to align with the specific needs of different plant varieties and growth stages. This tailored approach ensures that plants receive the optimal balance of essential nutrients so promoting robust growth and maximizing yield potential.

Moreover, the flexibility inherent in these systems empowers growers to adapt to changing conditions and optimize performance over time. By fine-tuning parameters such as irrigation schedules, light cycles and CO_2 supplementation growers can respond dynamically to environmental fluctuations and seasonal variations thus optimizing plant health and productivity throughout the cultivation cycle.

Furthermore, modern hydroponic systems offer versatile programming options that enable growers to automate tasks and create customized cultivation protocols. Whether it is scheduling nutrient dosing cycles, adjusting environmental parameters based on sensor data or implementing precise irrigation regimes, growers have the freedom to tailor system operations to suit their specific preferences and objectives.

In essence, the customization and flexibility inherent in modern hydroponic systems empower growers to optimize crop production with precision and efficiency. By tailoring nutrient delivery and environmental control to the unique requirements of their crops and growing conditions, growers can achieve superior results, maximize resource utilization and unlock the full potential of hydroponic agriculture.

11.1.4 Improved User Interfaces

Incorporating user-friendly interfaces and intuitive mobile applications has significantly elevated the accessibility and usability of hydroponic system management. These sophisticated interfaces are designed to streamline the monitoring and control processes offering growers an enhanced level of control and visualization over critical system parameters. Here is an elaboration on how improved user interfaces are revolutionizing hydroponic agriculture:

i) **Intuitive control features**: User-friendly interfaces present growers with clear and intuitive control features, allowing them to adjust settings and parameters effortlessly to meet specific needs of their crops. Intuitive controls streamline the process of managing nutrient dosing, pH adjustments and environmental conditions so ensuring precise control with minimal effort.

ii) **Enhanced visualization**: Advanced user interfaces provide growers with comprehensive visualization tools offering real-time data displays and graphical representations of system parameters. These visualizations enable growers to quickly assess the health and status of their crops, identify trends and make informed decisions about system adjustments or interventions.

iii) **Mobile accessibility**: Mobile applications extend the functionality of hydroponic system management by enabling growers to monitor and control their systems remotely from any location with internet connectivity. With mobile accessibility, growers can stay connected to their hydroponic setups at all times receiving alerts, notifications and updates on system performance directly to their smartphones or tablets.

iv) **Customization options**: Improved user interfaces often include customization options that allow growers to tailor the interface to their preferences and workflows. Customizable dashboards, layouts and alerts empower growers to create personalized monitoring and management setups that align with their specific needs and priorities.

v) **User support and training**: In addition to intuitive design user-friendly interfaces may include built-in support resources and training materials to assist growers in mastering the functionality of the system. User guides, tutorials and troubleshooting resources provide growers with the knowledge and confidence they need to effectively utilize the features of the interface.

In essence, improved user interfaces in hydroponic agriculture are designed to simplify system management, enhance visualization capabilities and empower growers with greater control and accessibility. By leveraging these advanced interfaces growers can optimize their hydroponic setups with confidence, efficiency and precision so ultimately leading to improved crop health, productivity and profitability.

11.1.5 Benefits of Automated Nutrient Delivery Systems

i) **Precision and consistency**: Automated systems ensure precise nutrient delivery, pH regulation and environmental control thus optimizing growing conditions for maximum plant health and yield.

ii) **Resource efficiency**: By dosing nutrients and water accurately, automated systems minimize waste and optimize resource use leading to reduced costs and environmental impact.

iii) **Labour savings**: Automated systems automate routine tasks such as nutrient dosing and pH adjustment so reducing the need for manual labour and freeing up time for other activities.

iv) **Real-time monitoring and control**: Sensors provide real-time data on nutrient levels, pH and environmental conditions thus allowing growers to monitor and adjust system parameters remotely.

v) **Scalability**: Automated systems can be scaled to accommodate small-scale hobbyists or large commercial operations thus providing flexibility for growers to expand their hydroponic operations.

11.2 Environmental Monitoring and Control Systems

Hydroponic agriculture relies on precise control of environmental conditions to optimize plant growth and yield. Environmental monitoring systems play a crucial role in providing real-time data on key parameters thus allowing growers to make informed decisions and maintain optimal growing conditions. Here we will explore the importance, components and benefits of environmental monitoring systems in hydroponics.

11.2.1 Sensors

11.2.1.1 Temperature Sensors

Temperature sensors play a crucial role in hydroponic systems by monitoring the ambient temperature continuously within the growing environment. These sensors are strategically placed to provide real-time data on temperature fluctuations thus ensuring that plants are not subjected to extremes of heat or cold that could compromise their growth and development.

By measuring the surrounding temperature accurately these sensors enable growers to maintain optimal growing conditions for their plants. Temperature extremes can have detrimental effects on plant metabolism, nutrient uptake and overall health. High temperatures can accelerate water loss through transpiration leading to dehydration and wilting while low temperatures can impede metabolic processes and stunt growth.

The data collected by temperature sensors allow growers to detect and respond to temperature variations promptly. If temperatures rise above or fall below the desired range automated systems can be activated to adjust environmental controls such as ventilation, heating or cooling to bring temperatures back within the optimal range.

In addition to ensuring plant health and vitality, temperature sensors also contribute to the overall efficiency and sustainability of hydroponic operations. By preventing temperature extremes growers can minimize stress on plants, reduce the risk of crop loss and optimize resource use. Ultimately, temperature sensors serve as indispensable tools for maintaining a stable and conducive growing environment, supporting healthy plant growth and maximizing yields in hydroponic systems.

11.2.1.2 Humidity Sensors

Humidity sensors serve as indispensable tools in hydroponic systems diligently monitoring the relative humidity levels within the growing environment. By assessing the moisture content in the air continuously these sensors play a crucial role in averting potential issues that could compromise plant health and vitality.

Excessive moisture, *vis-a-vis* dryness in the air can have detrimental effects on plants so disrupting their physiological processes and impeding growth. Humidity sensors act as vigilant guardians thus alerting growers to deviations from optimal humidity ranges. In conditions of excessive moisture these provide early warnings so helping growers mitigate the risk of fungal diseases such as powdery mildew and prevent issues such as root rot. Conversely, in environments characterized by dryness, humidity sensors prompt growers to take corrective action to avoid desiccation stress and ensure adequate water uptake by plants.

By providing real-time data on relative humidity levels these sensors empower growers to maintain a balanced and conducive growing environment for their crops. This proactive approach to humidity management contributes to the overall health, vigour and productivity of plants in hydroponic systems thus fostering optimal conditions for robust growth and bountiful yields.

11.2.1.3 Light Sensors

Light sensors play a pivotal role in hydroponic cultivation by measuring both the intensity and duration of light exposure precisely within the growing environment. These sensors serve as indispensable tools for growers so enabling them to fine-tune and optimize artificial lighting systems to meet specific needs of their plants.

By gauging light intensity accurately growers can ensure that plants receive an optimal amount of illumination for photosynthesis ~ the fundamental process by which plants convert light energy into chemical energy. This ensures that plants have access to the necessary energy to fuel growth, development and ultimate yield.

Moreover, light sensors provide valuable insights into duration of light exposure thus allowing growers to establish appropriate photoperiods tailored to the requirements of different plant species and growth stages. By controlling the duration of light exposure, growers can mimic natural daylight cycles, promoting healthy growth patterns and physiological responses in their crops.

Overall, light sensors empower growers to create and maintain ideal lighting conditions within their hydroponic systems thus ensuring that plants receive precise amount and duration of light necessary for robust growth and development. By leveraging this technology growers can optimize photosynthetic efficiency, maximize crop yields and ultimately achieve success in hydroponic cultivation endeavours.

11.2.1.4 CO_2 Sensors

CO_2 sensors play a pivotal role in hydroponic agriculture by monitoring CO_2 levels continuously in the surrounding air. This real-time data is invaluable for growers as it provides critical insights into the atmospheric CO_2 concentrations within the growing environment. By measuring CO_2 levels accurately these sensors enable growers to make informed decisions regarding CO_2 supplementation ~ a key factor in optimizing plant growth and development.

The data obtained from CO_2 sensors allow growers to precisely regulate CO_2 supplementation thus ensuring that plants receive the optimal amount of this essential gas for photosynthesis. By fine-tuning CO_2 levels based on sensor readings growers can create an environment conducive to maximum photosynthetic efficiency and plant productivity.

Furthermore, CO_2 sensors contribute to overall optimization of plant growth by facilitating the adjustment of environmental conditions to meet specific requirements of different crops. By monitoring and controlling CO_2 levels the growers can tailor the growing environment to promote vigorous growth, healthy foliage development and high crop yields.

In summary, CO_2 sensors are indispensable tools in hydroponic agriculture, providing growers with the data needed to optimize CO_2 supplementation and create an ideal growing environment for plants. By leveraging the insights provided by these sensors, growers can enhance photosynthetic rates, improve plant health and ultimately maximize the productivity and quality of their hydroponic crops.

11.2.2 Environmental Controllers

Environmental controllers represent a sophisticated and integral component of modern hydroponic systems, orchestrating a harmonious interplay between various environmental parameters to create and maintain optimal growing conditions for plants. These controllers serve as the nerve centre of the hydroponic setup, seamlessly integrating data from a network of sensors strategically placed throughout the growing environment.

Once armed with real-time data on critical variables such as temperature, humidity, CO_2 levels and light intensity environmental controllers spring into action orchestrating a symphony of automated responses to fine-tune the growing environment. These wield control over a suite of environmental control systems including heaters, humidifiers, fans and CO_2 injectors among others.

Through precise modulation of these environmental factors, environmental controllers orchestrate an environment tailored to the specific needs of the plants being cultivated. For example, in cooler climates or during periods of low light intensity heaters may be activated to maintain optimal temperature levels conducive to plant growth. Conversely, in humid environments, humidifiers may be engaged to prevent excessive moisture buildup which may reduce the risk of fungal diseases.

Moreover, environmental controllers play a pivotal role in optimizing CO_2 levels within the growing space ~ a key determinant of photosynthetic efficiency and overall plant productivity. By synchronizing CO_2 injectors with sensor data, controllers ensure that plants receive an adequate supply of carbon dioxide essential for robust growth and development.

In essence, environmental controllers act as the guardians of hydroponic ecosystem, vigilantly monitoring environmental conditions and orchestrating precise adjustments to create an environment conducive to plant flourishing. By automating the management of environmental parameters, these controllers empower growers to maintain optimal growing conditions, maximize plant health and ultimately achieve bountiful yields in their hydroponic endeavours.

11.2.3 Data Logging and Analysis Software

Data logging and analysis software play a pivotal role in modern hydroponic agriculture by facilitating the collection, organization and interpretation of data generated by sensors deployed within the system. This software serves as a comprehensive platform for capturing real-time measurements of environmental parameters such as temperature, humidity, pH levels, nutrient concentrations and light intensity.

Once the data are collected the software employs sophisticated algorithms and analytical tools to analyze trends, identify patterns and extract actionable insights. By processing vast quantities of sensor data, growers gain valuable intelligence into the dynamic behaviour of their hydroponic systems over time. This insight allows for informed decision-making and proactive management strategies to optimize system performance and maximize crop yields.

Furthermore, data logging and analysis software empower growers to monitor key metrics and track changes in environmental conditions with precision. By visualizing trends and anomalies through graphical interfaces and customizable dashboards, growers can quickly identify areas of concern and take timely corrective action as needed. Whether adjusting nutrient dosing schedules, fine-tuning environmental controls or implementing targeted interventions the software provides growers with the tools they need to maintain optimal growing conditions and support healthy plant growth.

In essence, data logging and analysis software serve as invaluable assets for hydroponic growers, enabling them to harness the power of data-driven insights to optimize crop production, enhance resource efficiency and ensure the long-term success of their operations. By embracing these advanced technologies growers can stay ahead of challenges, maximize productivity and unlock the full potential of hydroponic agriculture.

11.2.4 Benefits of Environmental Monitoring Systems

i) **Enhanced plant health**: Environmental factors such as temperature, humidity, light intensity and CO_2 levels directly influence plant growth, development and productivity in hydroponic systems. By maintaining optimal environmental conditions, monitoring systems promote healthy plant growth, vigour and resistance to pests and diseases.

ii) **Disease prevention**: Monitoring environmental conditions helps detect early signs of disease or pest infestation thus allowing the growers to take timely action and prevent the spread of pathogens.

iii) **Resource efficiency**: By optimizing environmental conditions growers can minimize resource use, reduce energy consumption and maximize crop yield in hydroponic setups.

iv) **Increased yield and quality**: Precise control of environmental parameters leads to improved crop yield, quality and consistency thus maximizing profitability for growers.

v) **Reduced risk of crop loss**: Early detection of environmental issues allows growers to take corrective actions promptly thus minimizing the risk of crop loss due to adverse conditions.

vi) **Data-driven decision making**: Real-time data provided by environmental monitoring systems enables growers to make data-driven decisions thus leading to improved crop management and higher profitability.

vii) **Remote monitoring and control**: Advanced monitoring systems offer remote access capabilities thus allowing growers to monitor and control environmental conditions from anywhere with an internet connection, enhancing flexibility and convenience.

11.3 Disease and Pest Management

In the realm of disease and pest management in hydroponics, the integration of sensors stands as a pivotal advancement. These sophisticated devices are instrumental in detecting the earliest indications of disease outbreaks or pest infestations within the hydroponic environment. By monitoring key parameters continuously such as temperature, humidity and nutrient levels the sensors have the capability to identify subtle deviations that may signal the presence of pathogens or pests.

The ability to detect these early warning signs empowers growers to take swift and targeted action to mitigate the risk of disease spread and minimize potential crop damage. Armed with real-time data and insights provided by sensors, growers can implement preventive measures such as adjusting environmental conditions, implementing biological controls or applying targeted treatments to contain the spread of pathogens or pests effectively.

Furthermore, sensors enable growers to adopt a proactive approach to disease and pest management rather than reacting to outbreaks after these have already occurred. This proactive stance allows for more efficient use of resources and reduces the likelihood of extensive crop losses. By leveraging sensor technology in disease and pest management strategies, hydroponic growers can cultivate healthier and more resilient crops while optimizing productivity and sustainability in their operations.

11.4 Synergistic Integration of Automated Systems

Seamless integration of automated nutrient delivery systems with other critical components including environmental controls *cum* disease and pest management systems marks a significant advancement in modern hydroponic agriculture. This integration fosters a holistic approach to cultivation wherein various elements work in harmony to create an optimal growing environment and maximize crop productivity.

One key aspect of this integration is the coordination between automated nutrient delivery systems and environmental control systems. By integrating these components growers can create a unified growing environment where adjustments to nutrient delivery are synchronized with changes in temperature, humidity and lighting conditions. This coordinated approach ensures that plants receive tailored conditions precisely to support their growth and development at every stage of the growing cycle. For instance, during periods of high light intensity, nutrient delivery systems may adjust nutrient concentrations to meet increased metabolic demands of plants while simultaneously optimizing temperature and humidity levels to prevent stress and promote optimal physiological function. Conversely, during low light periods or periods of high humidity nutrient delivery may be adjusted to prevent nutrient imbalances or excesses that could lead to plant health issues.

Furthermore, integration allows for real-time monitoring and data exchange between different systems thus enabling automated responses to changing environmental conditions or pest and disease threats. For example, if sensors detect signs of pest infestation automated pest management systems can deploy targeted treatments while adjusting nutrient delivery simultaneously to support plant recovery and resilience.

Ultimately, the integration of automated nutrient delivery, environmental controls and disease *cum* pest management systems creates a synergistic growing environment that optimizes plant growth and resource efficiency. By harnessing the power of automation and coordination growers can achieve higher yields, superior crop quality and greater sustainability in their hydroponic operations.

11.5 Future Directions: Nanotechnology and Biofortification

Emerging trends in hydroponic nutrient solutions are focussed on leveraging advanced technologies such as nanotechnology and biofortification to enhance nutrient delivery, plant health and crop productivity. Here is how these trends are shaping future of hydroponic agriculture:

11.5.1 Nanotechnology in Nutrient Delivery

Nanotechnology heralds a paradigm shift in nutrient delivery within hydroponic systems ~ harnessing the manipulation of materials at the nanoscale level spanning from 1 to 100 nanometers to engineer sophisticated structures and functional systems tailored explicitly for plant nutrition.

At its core, nanotechnology offers a transformative approach to enhancing nutrient uptake efficiency, bolstering nutrient solubility and refining nutrient release dynamics to align with the intricate demands of plant roots precisely. Through utilization of nano-scale delivery systems nutrients can be encapsulated within nanoparticles thus providing a protective shield against degradation while facilitating direct and targeted delivery to plant cells thereby optimizing nutrient absorption and utilization.

A diverse array of nanostructured materials including nanocarriers, nanogels and nanofertilizers are currently undergoing rigorous development for specialized nutrient delivery and controlled release within hydroponic environments. These cutting-edge materials epitomize precision in nutrient delivery thus empowering growers to tailor nutrient formulations to the exacting needs of specific plant species and growth stages. By leveraging nanotechnology, hydroponic agriculture stands poised to achieve unprecedented levels of crop productivity, resource efficiency and sustainability thus ushering in a new era of innovation and advancement in plant nutrition management.

11.5.2 Biofortification for Enhanced Nutrient Content

Biofortification stands as a pioneering strategy aimed at bolstering the nutrient content of crops through a myriad of techniques including traditional breeding methods, genetic engineering and targeted nutrient supplementation. This concerted effort is geared towards addressing malnutrition and improving human health outcomes by fortifying staple crops with essential nutrients crucial for overall well-being.

Within the realm of hydroponics, biofortification finds an ideal avenue for implementation, leveraging the precision and flexibility offered by nutrient solutions to fine-tune the nutrient profile of crops. In hydroponic systems, nutrients can be meticulously calibrated and tailored to provide plants with optimal levels of essential micronutrients such as iron, zinc, selenium and vitamins. This targeted approach ensures that plants receive the necessary nutrients in readily absorbable forms thus facilitating enhanced nutrient uptake and accumulation.

Biofortified crops cultivated in hydroponic environments hold immense potential in combatting malnutrition and addressing specific nutrient deficiencies prevalent in certain populations. By imbuing crops with elevated levels of key micronutrients, biofortification endeavours aim to bolster dietary intake and promote better health outcomes particularly in vulnerable communities.

In pursuit of biofortification goals, cutting-edge genetic engineering techniques including gene editing and transgenic approaches are being harnessed to enhance nutrient uptake and accumulation in crops cultivated in hydroponic systems. These innovative methodologies enable researchers to manipulate the genetic makeup of plants precisely thus facilitating the development of varieties with enhanced nutrient-absorption capacities and heightened nutrient content.

In essence, biofortification represents a pivotal strategy in the quest to alleviate malnutrition and improve public health with hydroponic systems serving as a promising platform for advancing these efforts through targeted nutrient enrichment and genetic enhancement of crops.

11.5.3 Smart Nutrient Formulations

In the realm of hydroponics, the evolution of nutrient formulations towards smarter and more responsive solutions heralds a new era of precision agriculture. This paradigm shift is fuelled by integration of cutting-edge technologies such as smart materials, nanosensors and responsive polymers which collectively empower growers with unprecedented control over nutrient management in hydroponic systems.

- **Advanced formulations**: Smart nutrient formulations represent a fusion of traditional nutrient solutions with innovative materials and technologies. By incorporating smart materials and responsive polymers these formulations possess the capability to dynamically adapt and respond to the ever-changing needs of plants and their surrounding environment.
- **Real-time monitoring and adjustment**: One of the defining features of smart nutrient solutions is their ability to provide real-time monitoring and adjustment of nutrient levels. Nanosensors embedded within the nutrient solution continuously scan for changes in nutrient concentrations, pH levels and other critical parameters. These real-time data are then utilized to orchestrate automated adjustments ensuring that nutrient levels remain finely tuned to meet the precise requirements of plants at every growth stage.
- **Dynamic response to plant needs**: Smart nutrient solutions go beyond static formulations offering a dynamic response to the nuanced needs of plants. By harnessing the intelligence embedded within these formulations growers can optimize nutrient delivery in real-time by tailoring it to match the specific demands of individual plants, environmental conditions and growth stages. This adaptive approach not only maximizes nutrient uptake efficiency but also minimizes waste so enhancing overall resource utilization and sustainability.

In essence, the emergence of smart nutrient formulations represents a transformative leap forward in hydroponic agriculture. By harnessing the power of advanced materials and nanotechnology, these formulations pave the way for a more responsive and efficient approach to nutrient management ultimately driving higher yields, superior crop quality and greater environmental stewardship in hydroponic systems.

11.5.4 Microbial Biostimulants and Biocontrol Agents

In the dynamic world of hydroponic agriculture harnessing the power of beneficial microorganisms has emerged as a promising strategy to optimize plant growth, bolster resilience and fortify crop productivity. At the forefront of this approach lie plant growth-promoting rhizobacteria (PGPR) and mycorrhizal fungi revered for their multifaceted roles as microbial biostimulants and biocontrol agents within hydroponic systems.

Incorporating these beneficial microorganisms into hydroponic nutrient solutions represents a paradigm shift in agricultural practices as these function as potent allies in promoting plant vigour and combating deleterious pathogens. PGPR and mycorrhizal fungi exert their beneficial influence through a myriad of mechanisms including enhanced nutrient uptake, stimulation of root growth and suppression of pathogenic microbes. By fostering symbiotic relationships with plant roots these microorganisms not only augment nutrient acquisition but also fortify the natural defenses against disease and stress of the plants.

As hydroponic systems continue to evolve so too is the formulations of microbial inoculants tailored for these unique growing environments. Advanced formulations are meticulously crafted to ensure compatibility with hydroponic conditions, optimizing factors such as pH, nutrient availability and oxygen levels to maximize microbial activity and efficacy. Moreover, these formulations are engineered to provide sustained benefits so delivering long-lasting improvements in plant health and productivity throughout the growth cycle.

In essence, the integration of microbial biostimulants and biocontrol agents into hydroponic agriculture represents a synergistic convergence of nature and technology unlocking new avenues for sustainable crop production and resilience in the face of environmental challenges. By harnessing the transformative potential of beneficial microorganisms, hydroponic growers can cultivate more robust and healthier plants while minimizing reliance on conventional agrochemicals thus fostering a greener and more resilient agricultural future.

11.5.5 Sustainable Nutrient Sources

Innovations in sustainable nutrient sources are heralding a transformative shift in hydroponic agriculture as growers increasingly turn to organic-certified fertilizers, plant-based extracts and waste-derived nutrients to meet the nutritional needs of their crops. These alternative sources of nutrients are rapidly gaining traction for their capacity to provide essential elements while aligning with principles of environmental stewardship and resource conservation.

By embracing renewable and environment-friendly nutrient sources, growers can reduce their reliance on synthetic fertilizers significantly which often carry environmental drawbacks such as soil degradation, water pollution and greenhouse gas emissions. Instead, organic-certified fertilizers harness the power of natural ingredients such as compost, seaweed extracts and animal manures to deliver a rich blend of nutrients that nourish plants while enhancing soil health and fertility.

Similarly, plant-based extracts derived from sources like alfalfa, kelp and legumes offer a sustainable alternative for enriching nutrient solutions in hydroponic systems. These extracts are prized for their bioactive compounds, vitamins and minerals which promote robust plant growth and vitality without introducing harmful chemicals or synthetic additives into growing environment.

Furthermore, the utilization of waste-derived nutrients presents an innovative approach to sustainability in hydroponics. By repurposing organic waste streams such as composted food scraps, agricultural residues and byproducts from biofuel production growers can harness valuable nutrients that would otherwise be destined for disposal. Through processes like composting, anaerobic digestion and bioconversion these waste materials are transformed into nutrient-rich amendments that replenish soil fertility and support healthy plant growth in hydroponic systems.

In essence, the adoption of sustainable nutrient sources represents a fundamental shift towards environmentally responsible farming practices in hydroponics. By prioritizing the use of organic-certified fertilizers, plant-based extracts and waste-derived nutrients growers not only safeguard the health of their crops and ecosystems but also contribute to the broader goal of building resilient and sustainable food systems for future generations.

These emerging trends in hydroponic nutrient solutions hold promise for revolutionizing modern agriculture by providing innovative solutions to enhance nutrient delivery, improving crop quality and addressing global food security challenges. As research and development in these areas continue to advance, hydroponic farming is poised to play a significant role in sustainable food production for the future.

Chapter 12

Hydroponics Step by Step

12.1 Plant Selection

A diverse array of vegetables and ornamental plants can thrive in hydroponic systems, offering efficiency and control in cultivation. Selecting plants suitable for hydroponics is critical, focussing on varieties that excel in this environment. Common choices include lettuce, spinach, basil, parsley, tomatoes, peppers, cucumbers and strawberries. However, factors like available space, lighting conditions and nutrient needs must be carefully considered. This ensures optimal growth and success in the hydroponic setup.

The initial phase of establishing a hydroponic system involves thoughtful selection of plants suited for this method. Space requirements are crucial as larger plants need adequate room to avoid overcrowding and promote proper development. Profitability, market potential and indoor or outdoor implementation influence plant selection.

While a wide range of plants can thrive indoors, outdoor hydroponic cultivation faces challenges due to fluctuating weather conditions. Furthermore, different plants have varying light level preferences with some requiring high-intensity illumination and others thriving in lower light conditions.

After selecting plants, meticulous attention is given to documenting their requirements. Key considerations include nutrient strengths, pH levels, nutrient solution temperature, light intensity preferences, identification of deficiency symptoms and anticipation of potential diseases and pathogens that may affect plant health.

12.2 Direct Seeding

Direct seeding or direct planting entails placing propagules such as seeds or cuttings directly into the hydroponic growing medium such as perlite, rockwool or coco coir at the suitable depth and spacing. This method eliminates the need for starting plants in a separate nursery environment, allowing them to germinate and mature directly within the hydroponic system. By bypassing the initial step of raising seedlings in separate containers, growers can save time and resources while promoting efficient resource utilization. Advantages of direct sowing include:

i) **Simplified process**: Direct seeding eliminates the need for starting seeds indoors or in seedling trays, streamlining the planting process.

ii) **Cost-effectiveness**: It reduces the costs associated with seedling trays, growing medium and labour required for transplanting.

iii) **Reduced transplant shock**: By avoiding the transplanting process, direct seeding minimizes the risk of transplant shock which occurs when seedlings are moved from one environment to another.

iv) **Precise control over environmental conditions**: Direct establishment in the hydroponic system allows for precise control over environmental conditions such as moisture levels and nutrient availability right from start of life cycle of the plant.

v) **Suitable for specific crops**: The crops such as lettuce, spinach and herbs as well as basil and cilantro adapt well to direct seeding in hydroponic systems.

vi) **Uniform growth**: When properly spaced and managed, direct seeding can result in uniform plant growth thus facilitating easier management and harvesting.

However, there are some considerations and challenges associated with direct seeding:

i) **Germination rate**: Direct seeding may result in lower germination rates compared to starting seeds indoors under controlled conditions. Factors such as temperature, humidity and seed quality can affect germination success.

ii) **Variability**: Achieving uniform plant spacing and growth can be challenging with direct seeding leading to variability in plant size and maturity.

iii) **Environmental conditions**: Direct-seeded crops are more susceptible to fluctuations in environmental conditions such as temperature and moisture levels which can affect germination and early growth.

In conclusion, direct seeding can be a viable option for certain crops in hydroponic systems offering simplicity, cost-effectiveness and reduced transplant shock. However, careful attention to seed quality, environmental conditions and weed management is essential to ensure successful germination and uniform plant growth.

In hydroponics, some plants may not transplant well or may prefer direct seeding due to their specific growth characteristics and sensitivity to transplant shock and these include:

i) **Leafy greens**: Plants like lettuce, spinach, kale and Swiss chard often prefer direct seeding in hydroponic systems. These have delicate root systems and can be sensitive to disturbance during transplanting.

ii) **Herbs**: Culinary herbs such as basil, cilantro, parsley and dill are typically well-suited to direct seeding in hydroponic setups. Transplanting can disturb their root systems and affect their growth.

iii) **Radishes**: Radishes are fast-growing root vegetables that prefer direct seeding. Transplanting can cause root damage and may result in stunted growth or misshapen roots.

iv) **Microgreens**: Microgreens are harvested at a very young stage often within a few weeks of germination. Direct seeding ensures uniform growth and minimizes the risk of damage during transplanting.

v) **Certain flowers**: Some flowering plants, such as marigolds and nasturtiums are best suited to direct seeding in hydroponic systems. Transplanting can disrupt their growth and flowering patterns.

vi) **Root vegetables**: Plants with taproots or deep root systems such as carrots and beets may not transplant well in hydroponics. Direct seeding allows their roots to develop without disturbance so leading to better overall growth.

vii) **Fast-growing crops**: Plants with rapid growth rates such as radishes, arugula and mustard greens often perform better when directly seeded. Transplanting can delay growth and reduce yields.

Overall, while many plants can be successfully transplanted in hydroponic systems with proper care, direct seeding is often preferred for crops that are sensitive to transplant shock or have specific requirements for uninterrupted root development. Growers should consider the unique needs of each plant species when deciding whether to transplant or directly seed in hydroponics.

12.3 Raising or Obtaining Seedlings

The process of raising and transplanting seedlings in a hydroponic system is pivotal for successful soil-less cultivation, fostering the growth of healthy and robust plants. Seedlings can be initiated from seeds, cuttings or through other vegetative materials using seed trays or peat pellets or obtained from reputed nurseries or garden centres.

Should a grower choose to transfer plants from soil to a hydroponic environment? Prior to transplantation it is crucial to cleanse the roots of the plant gently and thoroughly to remove soil particles. Subsequently, the root zone undergoes a washing procedure achieved by dipping and shaking it in a bucket of purified water. Whether sourced from seeds, cuttings or nurseries careful consideration must be given to selecting seeds or seedlings well-suited for hydroponic cultivation, ensuring they are free from pests or diseases. This attention to detail sets the stage for healthy and successful growth within the hydroponic system.

12.3.1 Growing Medium for Raising Seedlings

To ensure optimal conditions for seed germination and production of healthy, pest- and disease-free seedlings it is crucial to select the growing medium carefully. The chosen medium should strike a balance between fertility, drainage, water retention, aeration, pH stability and freedom from pests and diseases. Various options are available for raising seedlings or rooting planting materials including 'old coir-dust,' 'carbonized rice husk,' 'fine sand,' rockwool, peat, coco-peat, perlite and vermiculture.

Before use, it is essential to sterilize the growing medium properly to prevent contamination. For instance, hydrated lime can be added to coir-dust to neutralize its pH (approximately 100-250g of hydrated lime is added per 5 kg coir-dust block).

While many growers traditionally opt for rockwool due to its excellent water and nutrient retention properties, it has drawbacks such as environmental concerns, non-biodegradability and potential health risks if inhaled or when is in skin contact.

The growing mix for starting the seeds can be easily prepared at home by mixing equal amounts of fine peat moss, perlite and vermiculture in a bucket. Modern growers are increasingly turning to a mixture of 3 parts coco-peat, 1 part perlite and 1 part vermiculite. To neutralize the acidity of the peat moss a $1/4^{th}$ teaspoon of lime can be added. The ratio of peat moss can be increased to increase the moisture holding capability of the mixture. Note that the ratio of lime is directly proportional to that of fine peat moss. The sand is also a good medium for seedlings. In some cases the seeds can be directly placed in the grow media.

Coco-peat undergoes a weathering process to degrade it and remove excess salts. It boasts excellent porosity, facilitating air movement and drainage, and is typically pest- and pathogen-free. Perlite aids in aeration and drainage due to its heat-expanded aluminum silicate composition. However, its low weight can cause it to float when watered. Vermiculite ~ a heat-expanded mica offers good water retention capacity and is available in various size grades. Both perlite and vermiculite are neutral in reaction and contribute to the overall balance of the growing medium.

Root cubes, also known as root plugs or propagation cubes are individual blocks or plugs designed to support seed germination or rooting cuttings. These serve as a medium for seeds or cuttings to develop roots. Root cubes are typically made from materials like rockwool, peat moss or coco coir. Seeds are planted directly into pre-made holes or depressions in the cubes, providing optimal seed-to-medium contact for germination. Root cubes are often used in hydroponic systems or as a starting medium for transplants.

Root Cubes.

12.3.2 Sponge Nursery Technique

'Sponge nursery technique' is an alternative method for seed germination and early-stage plant development. Here is how it works:

Instead of traditional nursery media, sponge pieces serve as an effective medium. Utilizing 2.5cm cube sponge blocks seeds are placed at the centre, nestled within a cut made on the topside of the sponge. This setup ensures direct contact between the seeds and the nurturing sponge material. Maintenance of the sponge nursery aligns with standard nursery practices. However, nutrient supply should commence once the first true leaf begins to unfold ~ kickstarting the seedlings' growth phase.

One of the technique's advantages is its versatility. Depending on the chosen cultivation method, seedlings can seamlessly transition into hydroponic systems with the sponge block intact. As the plants mature, the sponge block can be delicately removed ensuring minimal damage to the developing roots.

12.3.3 Seedling Containers/Trays

Growers can select a hydroponic system suitable for starting seedlings. Options include 'Deep water culture' (DWC), NFT or a simple tray system with a nutrient solution. Seedling containers or trays play a crucial role in the early stages of plant growth providing a stable environment for germination and initial root development. Here are some common types of seedling containers/trays:

i) **Plastic seedling trays**: Plastic seedling trays are typically made from lightweight but durable materials such as polypropylene or polystyrene. These materials are resistant to moisture and provide excellent structural integrity thus ensuring long-lasting performance in various growing conditions. Seedling trays come in a variety of designs and configurations to suit different growing needs. These feature a grid-like structure with multiple compartments or cells each intended to hold a single seed or seedling. Common configurations include trays with 72, 128 or more cells per tray with cells arranged in rows and columns. Size and shape of the cells in seedling trays can vary depending on the type of plants being grown and the specific requirements of the propagation process. Cells may be square, round or hexagonal in shape with dimensions ranging from small compartments for delicate seedlings to larger cells for more robust plants.

ii) **Peat pots or pellets**: Peat pots are biodegradable containers made from compressed peat moss. These provide a natural option for starting seeds and can be planted directly into the soil once the seedlings have matured thus reducing transplanting shock.

iii) **Styrofoam trays**: Styrofoam trays are lightweight and inexpensive options for starting the seed. These are commonly used in commercial nurseries and can be reused multiple times with proper cleaning and sterilization.

iv) **Plug trays**: Plug trays are typically made of durable plastic and come in various sizes and configurations. These feature a grid-like structure with individual cells arranged in rows and columns. The size of the cells can vary depending on the type of plants being grown and the specific requirements of the propagation process. Plug trays offer a convenient and efficient way to organize and transport seedlings or root cubes/plugs. These help maximize space utilization so allowing growers to propagate a large number of plants in a relatively small area. Plug trays also facilitate easy handling and transplanting of seedlings thus streamlining the propagation process.

Plug Tray

v) **Recycled containers**: Many gardeners repurpose household items such as egg cartons, yogurt cups or plastic food containers as seedling containers. It should be ensured that these containers are clean and have adequate drainage holes before use.

vi) **Hydroponic net pots**: For hydroponic seed starting net pots are commonly used. These pots have a mesh bottom that allows roots to grow through and come into direct contact with the nutrient solution.

vii) **Biodegradable seedling trays**: It is made from materials like moulded pulp or bamboo. Biodegradable seedling trays offer an eco-friendly alternative to traditional plastic trays. These can be planted directly into the soil along with the seedlings thus reducing waste and environmental impact.

When selecting seedling containers or trays, growers must consider factors such as the type of seeds to start, the space available for seedling growth and the preferred transplanting method. They should ensure that containers have adequate drainage to prevent waterlogging and promote healthy root development. Each container should be labelled with the type of seed and date of planting to track growth progress accurately.

12.4 Propagation

12.4.1 Germinating Seeds

One common method of propagation for hydroponic systems is seed germination. Though seeds can be directly sown in the main hydroponic system but these are typically started in seed trays or peat pellets where these are provided with optimal conditions such as moisture, warmth and light to encourage sprouting. Once germinated, seedlings can be transplanted into the hydroponic system for further growth.

Growers should diligently adhere to the guidelines provided with their selected growing medium to ensure successful seed germination. This process typically entails following specific instructions such as pre-soaking the medium to optimize moisture levels, planting the seeds at the recommended depth and meticulously maintaining ideal conditions for germination including precise moisture levels and temperature control. By adhering to these instructions meticulously, growers can establish an optimal environment for seed germination so setting the stage for healthy and vigorous plant growth.

When starting seedlings in a plug tray, it is crucial to use a sterile fine-growing medium specifically tailored for seed starting. This medium should strike a balance ~ it must be light enough to allow seedlings to easily push their roots and tops through together with maintaining a consistent moisture level ~ just moist enough but not overly wet. It should possess excellent drainage properties to prevent the occurrence of damping off ~ a fungal disease that can affect seeds in overly moist conditions. Furthermore, the medium should be easily removable from the roots during transplanting into the hydroponic system, facilitating a smooth transition for the seedlings.

To ensure proper hydration it is essential that the bottom of the root cubes or plug tray comes into contact with water. This ensures consistent moisture uptake by the seedlings. As the roots begin to grow and extend beyond the planting cones, maintaining water levels at a height where these just touch the bottom of the plastic cones is sufficient.

For seeding, it is advisable to place one seed per block filled with the growing medium at the correct depth within the pots or trays. Provide optimal conditions for seed germination including maintaining appropriate moisture, temperature and humidity levels. Covering germination trays with wet papers or cloth can help maintain an adequate temperature until seeds sprout; however, these coverings should be removed promptly upon seedling emergence. Throughout the germination process, meticulous attention in maintaining the moisture level of the medium is crucial to ensure uniform germination and robust seedling growth.

12.4.2 Propagation using Vegetative Parts

Another method of propagation is through cuttings or clones. This involves taking a portion of a healthy plant such as stem or leaf and encouraging it to develop roots. It is imperative to select only those vegetative parts that are devoid of pests, disease-causing organisms and nematodes to ensure successful propagation. To initiate this process, individual containers or trays filled with a suitable growing medium such as rockwool or perlite are utilized for rooting these vegetative parts. Once rooted, the cuttings can be transferred to the hydroponic system for continued growth.

For instance, specific materials are employed for the propagation of different crops:

- **Strawberry**: Plantlets or runners that have been separated from the mother plants are commonly utilized for propagation.
- **Gerbera**: Plantlets that have been detached from the mother plants are ideal candidates for propagation purposes.
- **Mint**: Semi-hardwood stem cuttings approximately 10-12cm in length serve as effective planting materials for mint propagation.

12.4.3 Nutrient Supply

To ensure optimal growth and development, a strategic approach to nutrient application is crucial during early stages of plant growth. Here is a comprehensive guide:

12.4.3.1 Nutrient Application during Early Growth Stages

i) **Initial phase**: Until the emergence of the first two true leaves it is recommended to irrigate with clean water devoid of any nutrients. This ensures that seedlings receive the necessary hydration without overwhelming them with nutrients.

ii) **Transition phase**: As the seedlings develop and the first true leaves unfold nutrient supply should commence gradually. Since growing medium initially contains minimal plant nutrients, it is essential to introduce nutrients gradually to support healthy growth. It is advised to use a balanced nutrient formula designed for seed starting or follow recommendations for young plants. Ensure that pH of the nutrient solution is within the optimal range for seedling growth (typically between 5.5 and 6.5).

iii) **Fertilizer selection**: Fertilizer mixtures designed for hydroponic plants can also be used for nursery plants during this stage. Dilute the nutrient solution as per recommended ratios and apply it daily to ensure consistent and balanced nutrient uptake by the seedlings.

iv) **Hydration technique**: During early stages, trays or pots containing seedlings should be placed in shallow containers filled with nutrient solution. This setup allows the lower portion of the trays or pots to be submerged in the solution thus facilitating nutrient uptake through capillary action. Similarly, vegetative parts for propagation when planted in individual containers or trays should also be positioned in shallow nutrient solution containers mimicking the setup of seedling trays. This ensures that propagation materials receive adequate hydration and nutrient uptake during rooting process.

12.4.4 Growing Conditions for Seedlings

Once the seedlings are established these require optimal conditions for growth. This typically involves providing adequate light, moisture and temperature levels to support healthy development.

12.4.4.1 Optimal Light Conditions for Indoor Cloning and Seed Starting

In hydroponic systems, the initial stages of seedling growth are crucial for establishing healthy plants. Typically, hydroponic seedlings are first placed in a dark environment at around 21°C to encourage germination. This darkness provides optimal conditions for seeds to sprout and initiate growth.

Proper lighting is essential throughout the seedling stage to prevent issues such as stretching or leggy growth which can occur if seedlings do not receive enough light. By providing consistent and adequate light growers can foster healthy development and strong growth in their hydroponic seedlings setting the stage for success in their hydroponic gardens.

Achieving the right light conditions is crucial for healthy growth of seedlings and clones during their early stages. Here is what growers need to know:

i) **Light spectrum**: Seedlings and clones thrive under bright light preferably from special spectrum fluorescent lights tailored to their growth needs. These lights provide a balanced spectrum that promotes vigorous growth and sturdy development ensuring a healthier start for the plants.

ii) **Fluorescent lights**: High-power fluorescent bulbs are often utilized to provide necessary light for seedling growth in hydroponic systems. These fluorescent bulbs emit a spectrum of light that is suitable for promoting plant growth including wavelengths necessary for photosynthesis.

iii) **Placement of fluorescent lights**: By positioning the bulbs at the appropriate distance above the seedlings growers can ensure that plants receive sufficient light intensity for optimal growth. The lights should be placed no more than 16cm above the containers to ensure that seedlings and clones receive adequate light intensity for optimal growth without causing stretching or weak stems.

iv) **Alternative lighting options**: While fluorescent lights are popular for seed starting and cloning, some growers opt for HID (high-intensity discharge) lights. If using HID lights these should be hung higher ~ approximately 0.9-1.2m away from the plants to prevent burning. Proper distance ensures that the plants receive sufficient light without being subjected to excessive heat or light intensity.

v) **Light duration**: Most clones and seedlings benefit from 16 to 18 hours of light per day during their early growth stages. Providing extended periods of light helps promote photosynthesis and healthy plant development. However, it is crucial to allow for a period of darkness to avoid stressing the plants.

vi) **Light intensity and uniformity**: In addition to duration of light exposure growers should also consider the intensity and uniformity of light distribution. Ensuring consistent light levels across all plants helps prevent uneven growth and ensures that every seedling or clone receives adequate illumination for optimal development.

12.4.4.2 Temperature Requirement for Clones/Seedlings

Maintaining the appropriate temperature is crucial for healthy development of clones and seedlings. Here is what growers need to know:

i) **Optimal temperature range**: Clones and seedlings thrive in a temperature range typically between 21 to 27°C. This range provides ideal conditions for root development, nutrient uptake and overall plant growth.

ii) **Consistency**: It is essential to maintain consistent temperatures to avoid stressing the plants. Fluctuations in temperature can slow down growth and lead to health issues such as wilting or stunted growth.

iii) **Day/night temperature differential**: Providing a slight temperature drop during night (around 5-6°C) can mimic natural environmental conditions and promote robust growth. However, avoid drastic temperature fluctuations that may shock the plants.

iv) **Monitoring**: Regularly monitor temperatures using a reliable thermometer placed at plant level to ensure these remaining within optimal range. Adjust heating or cooling systems as needed to maintain consistent conditions.

v) **Seasonal considerations**: In colder climates, supplemental heating may be necessary to maintain temperatures within the optimal range especially during winter months. Conversely, in warmer climates adequate ventilation and shading may be required to prevent overheating during summer.

12.4.4.3 Humidity Level for Clones/Seedlings

Maintaining appropriate humidity levels is essential for successful growth of clones and seedlings. Here is what growers need to know:

i) **Optimal humidity range**: Clones and seedlings typically thrive in a humidity range of 50 to 70 per cent. This range provides necessary moisture for root development and foliar growth without promoting fungal diseases or excessive moisture-related issues.

ii) **Early growth stage**: During early stages of growth when roots are developing and foliage is delicate, higher humidity levels within the range of 65 to 70 per cent are beneficial. This helps prevent desiccation and promotes healthy growth.

iii) **Later growth stage**: As seedlings mature and develop stronger root systems, humidity levels can gradually be reduced to around 50 to 60 per cent. This encourages stronger stem development and prepares the plants for transplantation or further growth.

iv) **Monitoring and adjustments**: Regularly monitor humidity levels using a reliable hygrometer placed near the plants. Adjust humidity levels as needed by employing techniques such as misting, ventilation or using a humidifier or dehumidifier.

v) **Preventing moisture-related issues**: Be cautious not to allow humidity levels to remain too high for prolonged periods as this can increase the risk of fungal diseases such as damping off. Adequate air circulation and ventilation are essential for maintaining optimal humidity levels while preventing moisture-related issues.

vi) **Seasonal considerations**: Humidity levels may vary depending on environmental factors such as seasonality and geographic location. During colder months or in drier climates supplemental humidity may be necessary to maintain optimal levels.

12.4.5 Monitoring Seedling Growth

Monitoring seedling growth is a critical aspect of hydroponic cultivation ensuring the optimal development of plants. It is essential for growers to observe the progress of seedlings closely taking note of factors such as light exposure, water levels and nutrient availability. Adequate light is necessary for photosynthesis and healthy growth, ensuring that seedlings receive sufficient light intensity and duration. Moreover, maintaining optimal water levels in the hydroponic system is crucial to prevent dehydration or waterlogging which can hinder seedling development. Regular checking the nutrient levels in the solution is also important to ensure that the seedlings have access to essential minerals for growth. Any signs of nutrient deficiencies such as yellowing leaves or stunted growth should be addressed promptly by adjusting the nutrient solution accordingly. Vigilant monitoring for pests or diseases is necessary to prevent any potential damage to the seedlings. By diligently monitoring seedling growth and addressing any challenges that arise growers can foster the growth of strong and healthy plants in their hydroponic systems.

12.5 Preparing Hydroponic System for Transplanting

In selecting a hydroponic system one should consider factors such as available space, budget constraints and the specific requirements of the plants being grown. There are various types of hydroponic systems available either for retail purchase or for self-construction as most hydroponic systems are relatively simple to build and can be cost-effective. The choice of a hydroponic system depends on what is being grown, where it is being grown and the scale of production. Any water culture or medium culture hydroponic system discussed previously can be selected according to the specific requirements. Common options include 'deep water culture' (DWC), NFT, ebb and flow and drip systems. Following either the manufacturer's instructions or DIY plans is crucial when setting up the hydroponic system ensuring that reservoirs, growing containers, pumps and lighting are placed correctly for effective operation and plant growth.

Selecting a suitable location is crucial for setting up a hydroponic system. Factors to consider include whether it will be indoors or outdoors, available space and environmental conditions such as light and temperature. Once the location is determined appropriate hydroponic system should be chosen based on available resources and preferences. Gather all necessary equipment and supplies including reservoirs, growing trays, pumps and lighting. Follow step-by-step assembly instructions to install the system correctly ensuring proper placement and alignment of components.

12.5.1 Building a Simple Ebb and Flow System

i) Container Selection

- ☆ Choose a suitable growing container either rectangular or square with a flat bottom and sufficient depth to accommodate the desired plants. Consider the size of the container in relation to the nutrient tank to ensure adequate water volume.
- ☆ Select a nutrient reservoir which can be any appropriately sized bucket or container. For instance a 3-5 gallon bucket is suitable for a growing container measuring 24” × 16” × 6” deep.
- ☆ Consider the drain setup which can be fitted at the bottom or side of the container. Side drains may alleviate issues when placing the container on shelves and prevent underwater fittings from leaking.

ii) Planning Placement of Nutrient Reservoir

Determine where to position the nutrient reservoir to guide drilling holes in the container.

iii) Establishing Maximum Water Level

- ☆ Determine the maximum water level typically about three-quarters up the pot where plants will be grown. Mark this level on the side of the container.
- ☆ Place a top fill hole 2.5 to 5.0cm above this marked water level and drill two holes (approximately 5/8”) at these points for installing fill and drain flow tubes.
- ☆ Attach hoses to these holes, necessitating some plumbing work. The same tubes can be utilized for flooding the grow bed and draining the solution back to the reservoir.

iv) Building a Nutrient Solution Reservoir

- ☆ Select a sturdy and waterproof container for nutrient reservoir, ensuring it has sufficient capacity to supply nutrient solution to the plants.
- ☆ Size the reservoir larger than the growing container to account for water evaporation and maintain pH stability. Cover the reservoir with a lid to minimize evaporation and prevent debris contamination.
- ☆ If the nutrient tank is transparent, cover or paint it to prevent algal growth.

v) Installing the Water Pump

- ☆ Choose a submersible pump suitable for your hydroponic setup. It can be placed directly in the nutrient solution tank or be an external pump.
- ☆ Connect the pump supply hosepipe to pump and ensure there are no kinks in the hoses.

- ☆ If your system has separate paths for filling and draining, bring one end of the drain hose into the nutrient tank.
- ☆ Control the water pump with a timer to regulate flood times. Adjust the timer settings based on factors such as growing medium, plant types and growth stages.

vi) Installing the Air Stone and Air Pump

- ☆ Purchase an air stone and air pump from an aquarium store or hydroponic supplier.
- ☆ Place the air stone in the nutrient tank to oxygenate the nutrient solution to promote healthy plant growth.
- ☆ Connect air pump to air stone and ensure it is securely positioned in the nutrient tank.
- ☆ Turn on the air pump to create bubbles that will oxygenate nutrient solution thus enhancing plant growth and overall system health.

12.6 Selecting the Growing Media

Selecting the right hydroponic growing media is crucial for providing support, oxygen and carbon dioxide to plants. It should ensure maximum water availability, good nutrient retention, moisture holding, eco-friendliness and pH stability. Popular options include hydroton, cocofibre, rockwool, lava rock, perlite and vermiculite. Avoid sharp-edged media to prevent root damage and prioritize quality to maintain pH stability. However, solution culture hydroponic systems do not require growing media.

12.7 Preparation and Addition of Nutrient Solution in Reservoir

Nutrient solution preparation is a critical step in ensuring the success of hydroponic system. Growers can either prepare the nutrients using fertilizer salts or opt for pre-prepared hydroponic nutrients available in stores. Regardless of the method chosen, it is essential to provide plants with necessary nutrients at each stage of their growth including vegetative, flowering and fruiting stages.

When preparing nutrient solution it is important to refer to the product label of the chosen fertilizer or hydroponic nutrient solution. This label typically contains a list of ingredients and their percentages including macro elements such as N, P and K as well as minor elements. Ensure that the selected nutrient solution contains all the essential nutrients required for the specific plants one is growing in hydroponic system.

Following points must be considered while preparing nutrient solution:

i) **Nutrient requirements**: Understanding the specific nutrient requirements of plants at different growth stages and adjusting nutrient levels accordingly to optimize growth, flowering and fruiting.

ii) **Water quality**: Selection of fertilizers and their concentration in hydroponic nutrient solution depends greatly on the quality of the raw water as the raw water for hydroponic nutrient solution may contain minerals. Therefore, it is necessary to test the raw water prior to deciding on a fertilizer formula. Also, water for nutrient solution should be free from pathogens.

iii) **Balanced formulation**: Growers should aim for a balanced nutrient formulation that meets the needs of their plants; ensuring that the nutrient solution contains the appropriate ratios of macro and micronutrients for optimal plant health and growth.

iv) **Mixing instructions**: The mixing instructions provided on the fertilizer or nutrient solution packaging should be followed carefully. The appropriate amount of fertilizer salts or liquid concentrate should be measured and mixed thoroughly with water to achieve desired concentration.

v) **EC/PPM monitoring**: The EC or ppm of the nutrient solution must also be measured using a conductivity meter and nutrient concentration should be adjusted as needed to maintain the desired EC/PPM levels for the plants being grown.

vi) **pH adjustment**: A major factor in determining ability of the plants to uptake hydroponic nutrients is the relative acidity or pH of the hydroponic growing medium or hydroponic solution from which they feed. The pH level of the nutrient solution should be monitored and adjusted to ensure it falls within the optimal range for hydroponic cultivation. The ideal pH range typically falls between 5.5 and 6.5 for most hydroponic systems.

vii) **Nutrient solution temperature**: For healthy growth of the plants, a proper temperature of nutrient solution is necessary. An aquarium heater can be used to warm nutrient solution in the winter and chillers can be used to cool nutrient solution in the summer (if high nutrient temperature becomes a problem).

Plants tolerate a range of nutrients. It is the grower's responsibility to read the plant and if necessary modify the nutrient to suit the environment. The grower must regularly check the plants for nutrient deficiencies and toxicity symptoms. A close observation should be made regularly for any symptom of pests and diseases. Water quality, temperature, humidity, light intensity, EC and pH are some other factors that should be regularly monitored and managed. Hence, any nutrient solution formula used for any crop in one system may not necessarily be suitable for the same crop in another system and may require some modifications for that system.

12.8 Transplanting Seedlings

Once the hydroponic system becomes ready, the seedlings have developed strong roots and several sets of leaves, they are ready for transplanting into the main hydroponic system or into individual net pots filled with a growing medium. The seedlings should be handled carefully to avoid damaging the delicate roots. If transplanting seedlings into a different hydroponic system, growers should gradually acclimatize them to the new environment by increasing the nutrient concentration slowly and adjusting light levels over a few days.

12.9 Providing Lighting and Environmental Conditions

Grow lights are essential for plant development, providing necessary light for photosynthesis alongside carbon dioxide and water. While outdoor hydroponic systems may rely on natural sunlight indoor systems may require artificial grow lights, chosen based on factors like plant type, size, growth stage and system type. Options include fluorescent, metal halide, high-pressure sodium and LED grow lights. Timers automate nutrient pump and grow light schedules streamlining hydroponic gardening and allowing for precise control over feeding and lighting cycles. Alarms can also be installed to monitor nutrient reservoir pH and ppm levels.

Maintaining proper temperature and humidity levels within the growing area is also necessary to promote plant health and productivity.

12.10 Ongoing Care and Maintenance

After transplantation, growers must maintain careful oversight of seedling growth to ensure optimal conditions for plant development. This involves regular monitoring and adjustment of nutrient levels, pH and water temperature to promote healthy growth. Growers should engage in pruning activities to remove dead or diseased foliage and to encourage bushier growth if necessary. Regular inspection for signs of pests, diseases or nutrient deficiencies is essential with prompt action taken to address any issues that arise. For vining plants like tomatoes or cucumbers training techniques such as trellising or staking should be employed to support their growth and optimize space utilization.

12.11 Harvesting

After meticulous tending to the hydroponic garden and ensuring optimal growing conditions growers should be ready to harvest mature plants according to their specific growth patterns and harvesting guidelines. Employing clean scissors or shears is essential to prevent any harm to the plants during harvesting process. Once the harvest is complete, growers can relish in the satisfaction of their efforts by enjoying the fresh and nutritious produce cultivated in their hydroponic garden. Growing lettuce in a non-circulating (static) hydroponic system follows several steps for successful cultivation as given under:

i) **Preparing the seedlings (starter plants):** Begin by preparing the starter plants 12-16 days before planting. Place the lettuce seeds in root cubes ensuring only one seed per cube. Submerge cubes slightly in water in a dish to facilitate sprouting ensuring these never dry out.

Root Cubes in a Dish

ii) **Selecting and preparing planting container:** Utilize a dark five-gallon plastic bucket or similar container. If the container is transparent, it is advisable to paint it with a dark colour or cover it using dark plastic or tin foil.

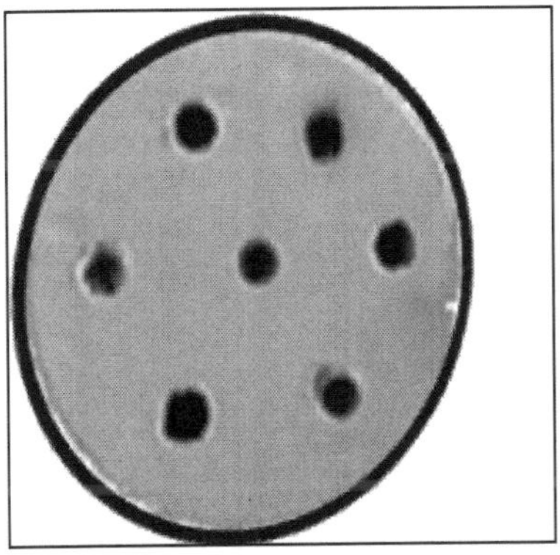

Container Cover with Evenly Spaced Holes.

Five Gallon Bucket with Cover.

iii) **Preparing a cover for container:** Create a cover for the container from plastic, plywood board, or foam board ensuring that it covers the surface area completely. Make evenly spaced holes around the cover slightly away from the rim to suspend the plastic planting cones (tubes) for mature plants.

iv) **Plastic tubes (preferably cone-shaped):** Suspend a plastic tube in each hole on cover of the container.

v) **Preparing nutrient solution:** Dissolve and add necessary nutrients to the water in container. Dissolve each chemical in water separately including 2 teaspoons of water-soluble hydroponic fertilizer (8-15-36 N-P-K percentage), 1 teaspoon of epsom salt (magnesium sulphate) and 2 teaspoons of calcium nitrate. Add the dissolved chemicals to a 5-gallon water in the container.

vi) **Placing cover with planting tubes:** Position the cover containing the planting tubes on the container and secure it with duct tape.

vii) **Transplanting the seedlings:** Insert a root cube with the seedling into each planting cone.

viii) **Monitoring for insects:** Keep a close watch on the plants for insects such as grasshoppers, caterpillars and slugs.

ix) **Sunlight requirements:** Lettuce prefers about equal amounts of direct and indirect sunlight. Ensure the hydroponic system is placed in an area with ample sunlight while being protected from rain.

Following these steps meticulously will facilitate the successful growth of lettuce in a non-circulating hydroponic system.

Chapter 13

Growing Tomatoes in Hydroponics System

pH			CF	EC	ppm
Min	Max	Optimum			
>=5.0	<=6.5	5.5-6.0	20-50	2.0-5.0	1400-3500

13.1 Introduction

Tomatoes, scientifically classified as fruits, are commonly prepared and served as vegetables. Tomato is scientifically known as *Lycopersicon esculentum* which grows to heights of 1–3 meters with a weak stem that often sprawls over the ground and vines over other plants. While perennial in its native habitat, tomatoes are often grown outdoors in temperate climates as annuals. There exist thousands of tomato varieties varying in shape, size, and colour with important varieties including 'Arka Vardan', 'Arka Vishal', 'GS-600', 'Naveen', 'N.S. 646', 'Nun 7711', 'Pusa Ruby' and 'Pusa Sheetal' among others. Cherry tomato varieties like 'BR-126', 'HA-818' and 'T-56' are also notable.

Tomatoes can be classified as determinate, vigorous determinate, semi-determinate or indeterminate. Determinate tomatoes grow to a compact height and stop growing once fruit sets on the top bud with all tomatoes ripening around the same time. These require minimal staking and are favoured by commercial growers. Indeterminate varieties develop into vines and continue producing fruits until frost, requiring support and management throughout the season. Vigorous determinate or semi-determinate tomatoes resemble determinate varieties but produce a second crop after the initial harvest.

Hydroponic systems accommodate a vast array of tomato varieties with many tailored specifically for controlled environments. Given their size, tomato plants necessitate substantial support such as trellises or vertically-tied strings to ensure proper growth. Alternatively, ample space should be provided to accommodate their expansive growth. Crops with extended growing seasons such as tomatoes and cucumbers require ample root space to thrive.

13.2 Cultivation

13.2.1 Propagation and Transplanting

Hydroponic tomatoes flourish in controlled environments, requiring support like trellises or strings for robust growth. Ample root space is essential particularly when grown alongside cucumbers. Propagation begins with selecting the tomato variety based on factors such as space, market potential and system location. Tomato seeds are sown 0.6 to 1cm deep with uniform watering through overhead, flooding and drain systems. Growers can sprinkle a thin layer of vermiculite over the seeds or cover the germination cubes or pots with a large piece of clear plastic to conserve moisture at the surface. It is advised to avoid use of plastic if cubes receive direct sunlight as the temperature may get too hot for good germination. The plastic must be removed as soon as emergence begins.

After emergence, seedlings are transferred to larger blocks and spaced evenly for optimal light exposure. Transplantation into final media occurs before flowering, ensuring proper leaching and moistening of the growing medium followed by immediate irrigation with nutrient solution. Denser spacing is maintained in hydroponic systems, *i.e.* some 60cm^2 per plant with adjustments for lower light conditions as needed.

13.2.2 Environmental Condition

13.2.2.1 Light

Adequate light is essential for plant growth and photosynthesis. Insufficient light, high humidity, or water stress can decrease photosynthesis resulting in reduced sugar production, *vis-a-vis* affecting fruit quality and size. Supplementary artificial lighting such as cool white fluorescent or high intensity discharge lamps can be beneficial when sunlight is lacking but is not as absolute substitute. While artificial lighting is generally not cost-effective for vegetable crops, it may be necessary for seed production and during winter months in northern latitudes. Providing shade to tomato plants in high-temperature and humidity areas can prevent issues like yellow or green shoulders and cracking in fruits, *vis-a-vis* improving overall fruit quality.

13.2.2.2 Temperature

Tomato plants thrive in day temperatures of 21-26°C and night temperatures of 16-18.5°C with some varieties tolerating minimal day-night temperature differences. Ideal growing temperatures vary by variety and can be obtained from the seed supplier. For seedlings, maintain a constant temperature of 20-22°C, gradually acclimating them to diurnal fluctuations before transplanting. Temperature control in greenhouses can be achieved using electrical heaters or solar energy for warming, and fans, evaporative pads or fogging for cooling.

13.2.2.3 Humidity

Ideal humidity levels for tomatoes range from 80-90 per cent during the day and 65-75 per cent at night. High humidity impedes transpiration leading to water and nutrient uptake issues thus resulting in nutrient deficiencies such as blossom end rot. It can also reduce leaf size, quality and yields, *vis-a-vis* increasing flower and fruit abortion and may cause glassiness *cum* gold fleck in fruit. Conversely, low humidity accelerates water and nutrient flow through plants leading to nutrient deposition in leaves rather than fruit, *vis-a-vis* also causing deficiencies. Misting and fogging systems increase humidity while exhaust fans and ventilation reduce it.

13.2.2.4 Carbon Dioxide

Carbon dioxide is essential for plant growth and in closed environments like greenhouses or grow rooms plants can quickly deplete the available CO_2. Optimal CO_2 levels for tomatoes range from 1000 to 1500ppm during daylight hours which may lead to substantial yield increases. To achieve this bottled CO_2 with a regulator and solenoid valve can be utilized to inject CO_2 into the greenhouse. Specialized CO_2 controllers are available to monitor the levels ensuring that this remains within optimal range. Greenhouses should be sealed during injection to prevent CO_2 loss with fan operation resuming after absorption. After a designated period to allow for CO_2 absorption, greenhouse fans can be activated once again.

13.2.3 Preparing Nutrient Solution

Selection and concentration of fertilizers in hydroponic nutrient solutions are contingent upon the quality of the raw water which may contain pathogens, pesticides and minerals. Hence, it is imperative to conduct thorough testing and treatment of the raw water with preference given to RO water if feasible. Raw water should be analyzed before determining the fertilizer formula. Macronutrient concentrations in tomato nutrient solutions vary throughout maturity of the crop while micronutrient levels remain consistent. Adapted from Jensen and Malter (1995), the following tables illustrate macronutrient concentrations during initial growth stage (from seedlings to 60cm tall plants with initial fruit) and the subsequent growth stage (from fruit set to harvest).

Nutrient Requirements for Tomatoes
(Adapted from Jensen and Malter, 1995)

Nutrient	*Seedlings to First Fruit Set (ppm or mg/l)*	*Fruit Set to Harvest (ppm or mg/l)*
	Macronutrients	
Magnesium (Mg)	50	50
Potassium (K)	199	199
Phosphorus (P)	62	62
Nitrogen (N)	113	144
Calcium (Ca)	122	165
	Micronutrients	
Iron (Fe)	2.5	2.5
Boron (B)	0.44	0.44
Copper (Cu)	0.05	0.05
Chlorine (Cl)	0.85	0.85
Manganese (Ma)	0.62	0.62
Molybdenum (Mo)	0.06	0.06
Zinc (Zn)	0.09	0.09

Preparation of Macronutrient and Iron Solutions for Tomato
(Adapted from Jensen and Malter, 1995)

Chemical Compound (Fertilizer Grade)	*Seedlings to First Fruit Set (g/1000 litres)*	*Fruit Set to Harvest (g/1000 litres)*
Magnesium sulphate (epsom salts)	500	500
Monopotassium phosphate (0-22.5-28)	270	270
Potassium nitrate (13.75-0-36.9)	200	200
Potassium sulphate (0-0-43.3)	100	100
Calcium nitrate (15.5-0-0)	500	680
Chelated iron	25	25

The above table is based as per standard practice but certain newer tomato varieties may have different nitrogen and potassium requirements. It is advisable to consult the seed company for recommended nutrient formulae tailored to specific tomato variety being grown. As the crop matures and lighting conditions change, optimizing the nitrogen-to-potassium (N:K) ratio becomes crucial. Higher light levels typically increase nitrogen usage while lower light levels necessitate more potassium.

Preparation of Micronutrient Stock Solution for Tomatoes

Fertilizer Salt	*Grammes of Chemical in 450 ml Stock Solution*
Boric acid	7.50
Manganous chloride	6.75
Cupric chloride	0.37
Molybdenum trioxide	0.15
Zinc sulphate	1.18

Add 250ml of the micronutrient stock solution to every 1000 litres of the previously prepared nutrient solution (Jensen and Malter, 1995). If using a concentrated stock solution for the macronutrients ensure that calcium salts are kept separate in their own solution.

13.2.4 The pH

Recommended pH range for hydroponic tomato nutrient solution is 5.5 to 6.5. Below 5.0 there is a risk of nutrient toxicity while above 7.0 nutrient deficiency may arise. pH adjustment can be done using nitric or phosphoric acid as needed.

13.2.5 Symptoms of Nutrient Deficiencies and Toxicities

- **Nitrogen deficiency:** Older leaves turn yellow, new growth is restricted and fruits are small. Treat with a foliar spray of 0.25-0.5 per cent urea solution. Toxicity symptoms include dark green plants with reduced root growth.
- **Phosphorus deficiency:** Plants are stunted with delayed maturity and fruit-set. Red/purple colours may appear on undersides of leaves but no toxicity symptoms.
- **Potassium deficiency:** Older leaves turn chlorotic with dead spots and the fruit ripening becomes uneven. Treat with a foliar spray of 2 per cent potassium sulphate. No toxicity symptoms.
- **Sulphur deficiency:** Rare but symptoms include yellowing in young leaves and purple stems coupled with stunted growth.
- **Magnesium deficiency:** Interveinal chlorosis on older leaves. Treat with a foliar spray of 10 per cent magnesium sulphate. No toxicity symptoms.

- ☆ **Calcium deficiency:** Causes blossom end rot, leaf yellowing and curling. Treat with a foliar spray of 0.75-1 per cent calcium nitrate solution. No toxicity symptoms.
- ☆ **Iron deficiency:** Stunted growth, interveinal chlorosis on young leaves and aborted flowers. Treat with a foliar spray of 0.2-0.5 per cent iron chelate. Not usually toxic.
- ☆ **Chlorine deficiency:** Though rare but can cause leaf wilting and chlorotic bronzing.
- ☆ **Manganese deficiency:** Leaves light green with dead patches *cum* interveinal chlorosis on older leaves. Treat with a foliar spray of 1 per cent manganese sulphate. Not usually toxic.
- ☆ **Boron deficiency:** Causes withering of growing points and brittle leaves. Treat with a foliar spray of 0.1-0.25 per cent borax. Not usually toxic.
- ☆ **Zinc deficiency:** Reduces internodal length and causes puckered leaf margins. Treat with a foliar spray of 0.1-0.5 per cent zinc sulphate. Commonly accompanies iron chlorosis.
- ☆ **Copper deficiency:** Young leaves are dark green and misshapen. Treat with a foliar spray of 0.1-0.2 per cent copper sulphate with 0.5 per cent hydrated lime. Not usually toxic.
- ☆ **Molybdenum deficiency:** Leaves turn golden yellow. Causes interveinal chlorosis on older leaves. Treat with a foliar spray of 0.07-0.1 per cent molybdenum solution.

When addressing deficiencies adjust the nutrient solution concentration accordingly and consider foliar sprays for faster response but be cautious of potential plant burning.

13.2.6 Pollination

Pollination is crucial for tomato plants in hydroponic systems so relying solely on natural air movements may not suffice. Growers have several options to ensure effective pollination such as artificial wind, gentle shaking/vibrating of flower clusters or introducing bumblebee hives into the greenhouse. However, bumblebee pollination is more suitable for large greenhouse setups rather than smaller home systems or outdoor locations. Mechanical methods such as shaking should be performed every two days usually under optimal humidity and temperature conditions or even under less favourable conditions. For larger operations, utilizing commercially available pollinators or bumblebee hives is recommended with one hive typically sufficient for around 0.2 hectares of tomato cultivation.

13.2.7 Diseases and Pests

In hydroponic tomato cultivation diseases and pests remain a concern despite efforts to maintain sanitation and vigilance. Common diseases include various fungal infections like 'Gray mould' (*Botrytis*), mildews and blights as well as *Verticillium* and *Fusarium* wilt. Nematodes, bacteria such as *Clavibacter michiganensis* and *Xanthomonas vesicatoria*, and viruses like 'tobacco mosaic virus' and 'tomato yellow leaf curl virus' (TYLCV) threaten the crop. Insect-pests such as whiteflies (*Trialeurodes vaporariorum*, *Bemisia tabaci*), tomato fruit worm (*Helicoverpa armigera*), leaf miner, tomato pinworm and aphids among others can inflict damage.

Here is an overview of common pests and diseases encountered in hydroponic tomato systems:

13.2.7.1 Pests

i) **Whiteflies**: These tiny sap-sucking insects can quickly infest tomato plants causing leaf yellowing, stunted growth and reduced fruit yield. Whiteflies also secrete honeydew thus promoting growth of sooty mould.

ii) **Aphids**: Aphids feed on plant sap so causing leaf curling, yellowing and distortion. These reproduce rapidly forming large colonies and spreading viruses to tomato plants.

iii) **Thrips**: Thrips feed on plant tissues causing silvery stippling on leaves and distorted growth. These can also transmit viruses which affect tomato plant health and productivity.

iv) **Spider mites**: These minuscule pests feed on plant cells causing stippling, webbing and leaf discolouration. Severe infestations can weaken plants and reduce fruit quality.

v) **Caterpillars (tomato hornworm)**: Caterpillars can devour tomato leaves and fruits so causing significant damage if not controlled timely. Handpicking and biological control methods are often used to manage caterpillar infestations.

13.2.7.2 Diseases

i) **Powdery mildew**: Powdery mildew is a fungal disease that appears as white powdery patches on tomato leaves, stems and fruit. It thrives in warm and humid conditions which reduces plant vigour and yield drastically.

ii) **Early blight**: Early blight is a fungal disease characterized by dark brown lesions on lower tomato leaves and further spreading to stems and fruits. It thrives in warm *cum* moist environments and can lead to defoliation and reduced fruit quality.

iii) **Late blight**: Late blight is also a fungal disease that causes dark water-soaked lesions on tomato leaves and fruits. It spreads rapidly in cool and wet conditions and can decimate entire tomato crops if not managed effectively.

iv) **Bacterial wilt (*Ralstonia solanacearum*)**: It is a bacterial wilt causing plant yellowing, wilting and eventual death of tomato plants. It is a soil-borne disease which can spread rapidly through hydroponic systems *via* contaminated water or equipment.

v) **Fusarium and verticillium wilts**: These soil-borne fungal diseases infect tomato plants through roots and cause yellowing, vascular discolouration and wilting. These can persist in hydroponic systems leading to recurring infections in successive crops.

Controlling diseases and pests in hydroponic tomato cultivation requires a multifaceted approach. Sanitation measures such as maintaining a clean growing environment and sterilizing equipment can help prevent disease spread. Regular monitoring of plants for signs of infection or infestation is crucial for early detection. IPM strategies including biological controls like beneficial insects or nematodes can help manage pest populations. Cultural practices such as proper spacing and airflow can reduce disease risk. In case it is very necessary then chemical control methods like fungicides or insecticides may be employed following label instructions carefully to ensure safety and efficacy.

Chapter 14

Growing Cucumber in Hydroponics System

pH			*CF*	*EC*	*ppm*	*Temperature*	*Humidity*
Min	*Max*	*Optimum*					
>=5	<=6	5.5-6.0	17-25	1.7-2.5	1190-1750	26.7-29.5°C	75 – 85 per cent

EC and light intensity

- ☆ Preferable EC during low light intensity: 2.2
- ☆ Preferable EC as light intensity increases: 2.5

14.1 Introduction

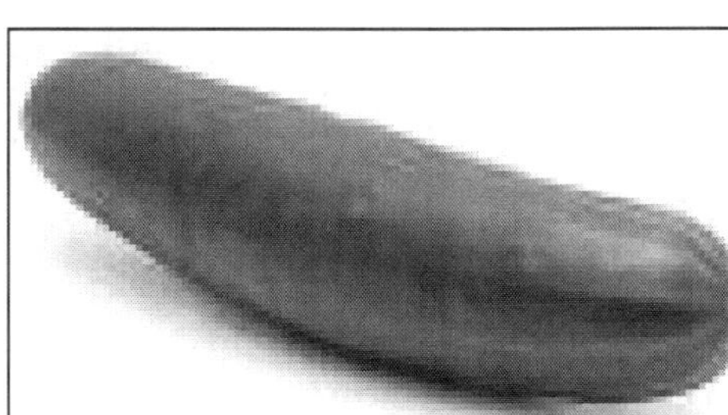

Cucumber (*Cucumis sativus* L.) is an annual herbaceous vining plant belonging to the family of Cucurbitaceae. Despite being botanically classified as fruits, cucumbers are typically consumed raw or as salad vegetables. These warm-season crops feature creeping vines that can reach up to 2 metres in length, often climbing trellises or other supporting structures using spiralling tendrils. The stems are characterized by a covering of white hairs. Certain varieties produce hermaphrodite flowers that develop into long round fruits.

While conventional cucumbers can thrive in hydroponic systems, parthenocarpic cucumbers which produce fruits without pollination are favoured among growers worldwide especially in controlled environments. Notable parthenocarpic (gynoecious) varieties include 'Alamir', 'Bronco', 'Dinar', 'Hasan', 'Kian', 'Muhasan', 'Mustang', 'Nun-9729', 'Sarig' and 'Satis'. Monoecious varieties like 'Japanese Long Green', 'Priya', 'Poinsett' and 'Pusa Sanyog' require pollination for seed development. If pollination occurs in parthenocarpic cucumbers it may lead to distorted fruit shapes and a bitter taste.

Cucumber varieties can vary in several aspects including being profiled or hybrid, bee-pollinated, partly self-pollinated or self-pollinated parthenocarpic fruits as well as being categorized as bunch-forming (short-vined) or long-vined and having normal or clustered ovaries (bucketed).

Optimal for controlled environments, self-pollinated (parthenocarpic) varieties are particularly well-suited for hydroponic production. Given their substantial size, cucumber plants necessitate sturdy support structures such as trellises or vertically-tied strings. Alternatively, ample space should be provided for their unobstructed growth. It is essential to accommodate the expansive root systems of large crops like tomatoes and cucumbers.

14.2 Cultivation

14.2.1 Propagation and Transplanting

Cucumber propagation begins with selecting the appropriate variety based on space, profitability, and indoor or outdoor cultivation. Parthenocarpic cucumbers not requiring pollination are ideal for hydroponic systems to avoid fruit distortion and bitterness. Seeds germinate rapidly at temperatures around 29°C taking 2-3 days for emergence. At 21°C it takes 5-6 days for germination while at 16°C 9-10 days. Once the seed has germinated, the temperature should be raised to 25°C. Optimal watering and nutrient levels are crucial during seedling and transplant production stages to prevent stress. During seeding and transplant production stages the plants should never be stressed for water or nutrients.

If rockwool is chosen as the growing medium, it should be soaked before transplanting cucumber seedlings. In case any other media such as perlite is used this should be properly sterilized, water-soaked and drained out before transplanting. Few growers prefer direct seeding into soilless media such as perlite, coconut fibre or composted pine bark and at the same time seeded a few transplants to use as replacement afterwards in case of casualty.

Cucumber seedlings with 3-4 true leaves are ready for transplanting with an emphasis on maintaining an upright growth habit. Adequate spacing between cucumber plants is essential following instructions on seed packets. Hydroponics/greenhouse cucumber plants require ample sunlight and space, ideally about 2m per plant under good sunlight conditions with adjustments based on lighting conditions. The exact spacing between rows and plants within a row depends on the chosen training and production system. Utilizing trellises to train cucumber vines not only conserves space but also enhances fruit quality and yield.

14.2.2 Environmental Condition

14.2.2.1 Light

A mature cucumber crop thrives with higher natural light levels, given adequate water, nutrients, carbon dioxide and optimal air temperature. Inadequate lighting during early growth stages can result in elongated and weakened plants. Indoor cultivation of cucumbers necessitates a minimum of 14-16 hours of daily light exposure thus requiring artificial grow lights.

14.2.2.2 Temperature

Greenhouse cucumbers especially seedless varieties are more sensitive to temperature fluctuations than tomatoes. Sustained maximum production requires minimum temperatures above 18°C while temperatures exceeding 35°C can hinder fruit production and quality. Optimal conditions for cucumbers include warm temperatures ranging from 26-30°C during the day and at least 20°C at night, preferably between 22°C and 23°C. Some varieties may perform well with minimal temperature fluctuations. Germination typically occurs at 15-30°C with an optimum temperature of around 28°C. Stem extension and leaf development are best at 18-27°C with reduced growth rates observed above 28°C. Flowering and pollination occur optimally at 18-21°C. Fruit production peaks at daytime temperatures of 20-22°C and nighttime temperatures of 19-20°C. Root survival requires nutrient solution temperatures of 25-30°C. Optimum growth of cucumber is obtained when greenhouse temperatures are 20-26°C though nutrient solution temperatures of 22-25°C. Temperature regulation can be achieved using electrical furnaces or solar energy for heating, and fans, evaporative pads or fogging for cooling. It is important to note that cool and cloudy growing seasons may result in bitter fruits.

14.2.2.3 Humidity

Cucumbers thrive in a humid environment ~ optimal humidity levels varying throughout their growth stages. During daytime humidity should range between 75 and 85 per cent while at night 60 and 75 per cent. Effective management of substrate humidity is crucial with different stages requiring specific levels:

- ☆ **Seedling stage**: 50-70 per cent
- ☆ **From seedling/planting to fruit formation**: 70-80 per cent
- ☆ **Mass-flowering stage**: Substrate humidity can be reduced to 55-60 per cent
- ☆ **From fruit formation to first harvest**: 75-85 per cent
- ☆ **From first harvest to the end of the growth cycle**: 85-95 per cent. Persistent humidity exceeding 95 per cent for 7-10 days may lead to onset of *Ascochyta* leaf spot. Misting and fogging systems are useful for

increasing humidity and lowering temperatures while exhaust fans and proper ventilation help in humidity reduction.

14.2.2.4 Carbon Dioxide

Cucumbers exhibit optimal growth when exposed to elevated CO_2 concentrations, typically ranging between 1200 to 1800ppm during daylight hours. Increased CO_2 levels reaching 2-5 times the ambient levels have been observed enhancing cucumber production significantly.

14.2.3 Preparing Nutrient Solution

Preparing the nutrient solution for hydroponic cultivation is a critical step that demands careful consideration. The choice of fertilizers and their concentrations hinges on the quality of the raw water which may harbour pathogens, pesticides and minerals. Therefore, it is imperative to test and treat the raw water thoroughly though RO water is most preferred one. Prior to formulating fertilizer mix, conducting comprehensive tests on the raw water is essential.

Nutrient concentration should be adjusted based on plant variety and growth stage. At the initial stages when root system is still developing it is advisable to maintain a lower nutrient concentration. As the plants progress through different growth and fruit-bearing stages the nutrient concentration should be gradually increased. Consulting the seed company for recommended nutrient formulae tailored to the specific cucumber variety being grown hydroponically is recommended for optimal results. Various experiments have demonstrated promising outcomes with following nutrient solution composition.

Component	*Minimum*	*Optimum*	*Maximum*
$N-NO_3$	120	160	190
$N-NH_4$	-	7	20
P	30	38	50
K	200	240	270
Ca	120	140	190
Mg	20	25	60
S	25	35	60
Fe	0.4	0.6	2.0
Mn	0.3	0.6	1.0
Zn	0.1	0.3	1.0
B	0.1	0.2	0.5
Cu	0.02	0.03	0.04
Mo	0.03	0.05	0.08
EC	1.5	2.0	2.5

Concentration of nutrient solution in hydroponic cucumber cultivation is influenced not only by the cucumber variety and growth stages but also by various environmental factors such as light intensity, pH, temperature and humidity. Growers must closely monitor plant health for signs of deficiency or toxicity and adjust nutrient solution concentration accordingly. In case of deficiency symptoms, the nutrient solution should be replenished with an increased concentration of the deficient element (25 to 30 per cent), gradually restoring it to slightly elevated levels once the deficiency is corrected. While foliar sprays offer a quicker response, caution must be exercised to avoid burning of plants.

14.2.4 pH

Optimal pH range for hydroponic cucumber nutrient solutions is typically between 5.5 and 6.0. When pH falls below 5.0, there is a risk of nutrient absorption becoming excessive so leading to toxicity issues. Conversely, if pH exceeds 6.5 certain nutrients may become inaccessible to the plants resulting in deficiencies. To adjust pH levels downward, nitric or phosphoric acid can be utilized.

14.2.5 Pests and Diseases

In hydroponic cucumber cultivation, like any other crop growers face challenges from pests and diseases. Warm temperatures and high humidity found in greenhouse environments create favourable conditions for development of diseases in cucumber plants. Various diseases, insect-pests and nematodes can pose threats to greenhouse cucumbers.

Here are some common pests and diseases encountered in hydroponic cucumber systems:

14.2.5.1 Pests

i) **Aphids**: These small sap-sucking insects can quickly colonize cucumber plants causing damage to leaves and transmitting viral diseases.

ii) **Whiteflies**: Whiteflies are tiny moth-like insects that feed on cucumber plants, particularly on the underside of leaves. These can cause leaf yellowing, stunting and spread of viral diseases.

iii) **Spider mites**: These arachnids feed on plant sap and causing yellowing, stippling and webbing on leaves. Spider mite infestations can weaken cucumber plants and reduce yield.

iv) **Thrips**: Thrips are slender winged insects that feed on cucumber leaves causing silvering, stippling and distortion of foliage. These can also transmit viral diseases.

v) **Fungus gnats**: These small flying insects lay their eggs in the growing medium and their larvae feed on plant roots leading to stunted growth and root damage.

14.2.5.2 Diseases

i) **Powdery mildew**: Powdery mildew is a fungal disease that appears as a white powdery growth on the leaves, stems and fruits of cucumber plants. It can reduce photosynthesis and weaken plants leading to decreased yield.

ii) **Downy mildew**: Downy mildew is another fungal disease that causes yellowing and browning of cucumber leaves often accompanied by a fuzzy growth on underside of leaves. It can spread rapidly under humid conditions.

iii) ***Pythium* and *Phytophthora* rots**: These soil-borne pathogens can cause root rot in cucumber plants leading to wilting, yellowing and eventual plant death.

iv) **Bacterial wilt**: This bacterial disease causes yellowing and wilting of cucumber leaves. It spreads through cucumber beetles and certain other insects.

v) **Cucumber mosaic virus (CMV)**: CMV is a viral disease transmitted by aphids and certain other insects. Infected cucumber plants show mosaic patterns, yellowing, and distortion of leaves leading to reduced yield and quality.

Effective control measures include selecting resistant varieties, maintaining greenhouse sanitation and applying pesticides at appropriate time. Proper sanitation and sterilization of hydroponic media, containers and equipment help prevent most fungus and virus diseases which can also be managed with fungicides.

Chapter 15

Growing Lettuce in Hydroponics System

15.1 Introduction

Lettuce stands out as one of the most extensively cultivated vegetable crops worldwide, particularly in hydroponic systems under greenhouses. Varieties of lettuce include 'Romaine' such as 'Parris Island' (green), 'Freckles' (red'), 'Oakleaf' (including 'Cocarde', 'Berenice' (green), 'Oscarde', and 'Dano' (Red') as well as red curly types such as 'Lolla Rosa', 'Ruby', 'Red Sails', 'New Red Fire' and 'Brunia'.

Majority of hydroponically grown lettuce belongs to the bibb types. Under the longer days of summer these reach maturity usually in 40-48 days. Seedlings are nurtured for 12-18 days before transplantation, followed by an additional 28 to 32 days to reach full maturity.

Green vegetables such as lettuce, spinach, watercress and certain herbs thrive in water-based hydroponic systems like NFT and 'raft culture'.

15.2 Cultivation

15.2.1 Propagation and Transplanting

Lettuce seed viability diminishes rapidly with age hence growers must ensure proper date-labelling for all the seeds. Storage in a cold and dry atmosphere such as refrigerator preserves seed quality for up to or more than 6 months.

Propagation and transplanting of lettuce involve several critical steps for successful establishment and healthy growth. It begins with meticulous seed selection and sowing. High-quality lettuce seeds are chosen and sown in a sterile well-draining growing medium including coco coir, peat moss, rockwool cubes, oasis cubes and multipacks with vermiculite medium.

Sowing lettuce seeds by hand can be cumbersome due to their small size. To simplify this, growers can opt for 'pelletized' lettuce seeds ~ encapsulated in an inert clay-like material so that these become easy to handle and sow manually. For large-scale operations investing in an automatic seeding system becomes necessary.

The growing medium must remain consistently moist to facilitate germination. Thinning becomes necessary as seedlings emerge and develop true leaves to ensure proper spacing. Only properly growing strongest growing seedlings are retained while others are thinned out. Lettuce seeds germinate optimally within a temperature range of 21 to 24°C. Placing seeded trays at 4.5°C for 1 to 2 days (stratification) aids uniform germination. Subsequently, seedlings should be transferred to warmer greenhouse environments at 15-18°C during daytime and 13-16°C at night.

To ensure optimal growth conditions, it is recommended to maintain a relative humidity of 60-80 per cent, carbon dioxide enrichment at 1000 ppm during daytime, solution pH 5.5-6.0 and EC from 1.0 to 2.3mS depending on light conditions. Supplementary lighting may be necessary in northerly latitudes during winter while shade curtains can regulate sunlight in high solar light regions to prevent leaf burn.

Seedlings are typically grown for 14 to 21 days before transplanting to the hydroponic production area. Seedlings are transplanted when these reach 2-3 leaf stage and are sturdy enough to handle and have developed a robust root system. The transplanting site or containers are prepared with a nutrient-rich growing medium. Carefully, the seedlings are removed from their original containers taking care not to damage the delicate roots, and transplanted into the prepared site or containers at the same depth as they were in the nursery. Immediate watering helps the seedlings settle into their new environment.

Transplanting lettuce in late afternoon is preferable to minimize stress from high daytime temperatures. Transplanting of seedlings during cooler evening hours help these getting rid of the uprooting shock and settling in their new surroundings through adjustment of the roots into the nutrient solution. When transplanting ensure that base of plants makes contact with the nutrient solution flow below. This facilitates moisture absorption, keeping the base of the cube moist and encouraging root extension into the solution. Within a few days, roots will extend from the cubes into the solution so promoting vigorous plant growth.

Finally, post-transplant care is crucial for well-being of the seedlings. Consistent proper watering is provided to keep growing medium evenly moist, however, overwatering may lead to root rot. Gradual acclimation to the final growing conditions is essential particularly if transitioning from indoor to outdoor environments or from controlled environments to greenhouses. Protection from extreme temperatures, pests and diseases ensures the continued health and vigour of the transplanted lettuce seedlings.

Regardless of operation size continuous lettuce production is essential requiring daily sowing, transplanting, harvesting and system cleanup.

15.2.2 Environmental Conditions

15.2.2.1 Light

Lettuce thrives with 12-16 hours of daylight, necessitating supplementary artificial lighting during short winter days to expedite the cropping period. Due to its leafy nature, metal halide (MH) lighting is optimal for lettuce cultivation.

15.2.2.2 Temperature

Maintaining stable air and water temperatures preferably below 26.5°C day temperature as well as pH levels is crucial to prevent stress.

15.2.2.3 Humidity

It is 60-70 per cent RH within the lettuce head quite satisfactory. Calcium deficiency can contribute to tip burn.

15.2.2.4 Carbon Dioxide

Lettuce requires 800-1000ppm CO_2 levels during daytime for proper growth to support optimal growth. Consistency in environmental conditions ensures healthy lettuce development which even minimizes the risk of stress-related issues like tip burn.

15.2.3 Nutrient Solution

A comprehensive nutrient formulation supplies all necessary elements essential for plant growth. Initially, lettuce seedlings are nourished with a half-strength solution until transplanting. This solution comprises approximately half the concentration of macro-elements while maintaining full concentration of micro-elements. The choice of specific formulation depends on factors such as temperature, day length and sunlight intensity. During summer when sunlight is abundant and days are long, higher nitrogen levels can accelerate plant growth. Conversely, under low light conditions, potassium and nitrogen levels should be reduced to optimize growth.

A typical nutrient formulation might include:

Nutrient	*Seedlings to First Fruit Set (ppm or mg/l)*
Macronutrients	
Magnesium (Mg)	40-50
Potassium (K)	200-210
Phosphorus (P)	50-60
Nitrogen (N)	140-160
Calcium (Ca)	175-195
Micronutrients	
Iron (Fe)	2.5-3
Boron (B)	0.44 – 0.5
Copper (Cu)	0.1
Manganese (Mn)	0.5 - 0.62
Molybdenum (Mo)	0.05 - 0.06
Zinc (Zn)	0.09 – 0.1

Growers should opt for top-grade highly soluble fertilizers known for their exceptional purity. It is important to maintain optimal pH for lettuce cultivation which falls within the range of 5.5 to 5.8. Most lettuce formulations typically have an electrical conductivity ranging from 1.5 to 2.0.

15.2.4 Nutritional Disorders

The prevalent nutritional disorder in lettuce cultivation is tip burn resulting from excessive water loss from the leaves coupled with insufficient water uptake by the roots. Effective control measures involve ensuring proper nutrition, providing an adequate water supply and fostering healthy root development. Maintaining sufficient oxygen levels in the nutrient solution is crucial for sustaining robust root systems. It is imperative to note that excessive EC levels should be avoided to mitigate the risk of tip burn.

15.2.5 Insect-Pests and diseases

Lettuce is susceptible to various pests commonly found in other crops including thrips, whiteflies and larvae from moths and butterflies. Effective management strategies for these pests often involve utilizing biological agents or homemade organic sprays. Among diseases, *Pythium* poses one of the most significant threats to lettuce cultivation.

Here are some of the most common insect-pests and diseases:

15.2.5.1 Insect-Pests

i) **Aphids**: These small sap-sucking insects can colonize lettuce plants causing stunted growth, distorted leaves and transmission of viral diseases.

ii) **Thrips**: Thrips are tiny slender insects that feed on lettuce leaves causing silvering or bronzing of foliage and deformed growth.

iii) **Whiteflies**: Whiteflies are small white insects that feed on plant sap and can transmit viruses to lettuce plants leading to yellowing, wilting and reduced vigour.

iv) **Spider mites**: These microscopic pests feed on the undersides of lettuce leaves causing stippling, webbing and leaf damage that reduces plant growth and vigour.

15.2.5.2 Diseases

i) **Downy mildew**: Downy mildew is a fungal disease that thrives in cool humid conditions. It appears as yellow or brown lesions on lettuce leaves, eventually leading to wilting and plant death if left untreated.

ii) **Powdery mildew**: Powdery mildew is another fungal disease characterized by white powdery patches on lettuce leaves. It can reduce photosynthesis and overall plant health.

iii) **Botrytis**: *Botrytis* (gray mould) is a fungal disease causing brown lesions and fuzzy gray growth on lettuce leaves especially under humid conditions. It can lead to rotting and decay of plant tissue.

iv) ***Pythium* and *Phytophthora***: These are water moulds that cause root rot in hydroponic lettuce systems leading to stunting, yellowing and wilting.

One can get rid of diseases through:

- Implementing strict sanitation measures to prevent the introduction and spread of pests and diseases.
- Regular monitoring for signs of pests and diseases and taking prompt action if detected.
- Using biological controls such as predatory insects or mites to manage pest populations.
- Providing adequate air circulation and ventilation to reduce humidity levels and minimize the risk of fungal diseases.
- Maintaining proper nutrient and water management to promote healthy plant growth and vigour.
- Utilizing disease-resistant lettuce varieties when possible to reduce susceptibility to common pathogens.

By being proactive in insect-pest and disease management and employing integrated pest management strategies, growers can minimize the impact of pests and diseases on hydroponic lettuce crops, thus ensuring a healthy and productive growing environment.

Chapter 16

Growing Capsicum (Bell Pepper) in Hydroponics System

Temperature Requirements (until the end of crop)

Nutrient Solution			*Air*		
Min	*Max*	*Ideal*	*Min. night*	*Max. night*	*Min. day*
18°C	28°C	22-25°C	15°C	28°C	18°C

Nutrient Solution's pH and CF Requirements

pH			*CF*		
Min	*Max*	*Optimum*	*CF at Planting*	*CF at Cropping*	*Ppm*
>=6.0	<=6.5	6.2-6.3	15-18	18-22	1260-1540

16.1 Introduction

Capsicum (family: Solanaceae) varieties come in diverse shapes, sizes and colorus with their plants typically growing bushy reaching the heights of about 60–80cm. While capsicums are semi-perennials, these are commonly cultivated as annuals. Their fruit range from small tapered or ball-shaped varieties to long cones and large round bells. Capsicums offer versatility in culinary applications ~ being enjoyed fresh or dried, whole or ground and used alone or in combination with various flavouring agents. These are commonly incorporated into salads, baked dishes, stuffed dishes, salsa, pizzas and more. Capsicums with thick flesh are well-suited for roasting and cooking while thinner-skinned varieties are ideal for raw consumption. Varieties mostly with 4-lobed or square shapes fetch higher prices in the market.

In India, important capsicum varieties include 'Mazurka', 'Parker', 'Torkel' (red), 'Fiesta' (yellow); 'HA-988', 'HA-1038', 'HA-1195', 'HA-1589', 'HA-1931' (Israeli); 'Bharat', 'Bombay', 'Heera', 'Indira', 'Mahabharat', 'Natasha' (red), 'Golden Summer', 'Orobellee', 'Swarna', and 'Tanvi'.

Capsicums thrive in various hydroponic systems like NFT or media-based systems. It is crucial to maintain the nutrient solution's oxygenation, temperature, conductivity (CF) and pH levels at desirable ranges.

- ☆ In NFT systems, gullies should not exceed 15 meters in length and should have a minimum fall of 1 in 40 down the 15 meters. The nutrient flow rate down the gully should be around 1 litre of nutrient per minute.
- ☆ In media-based systems, capsicums can be cultivated in various media such as rockwool, perlite or pumice. Any inert and sterile material with a grain size allowing air permeation to the roots is suitable. Composted pine bark has been successfully used in hydroponic systems under open shade structures in places like Allahabad.

16.2 Open Shade Structure

Capsicum production in open shade structures offers several advantages over traditional open field production systems. It can result in significantly higher yields with the potential for a harvest season extended by 3-4 months. This extended season is particularly beneficial in temperate regions where harvest can be prolonged from June to November up to the first frost. The use of open shade structures promotes production of large and high-quality fruits throughout entire growing season thus further enhancing the overall productivity and profitability of the operation.

Open Shade Structure

16.3 Cultivation

16.3.1 Propagation and Transplanting

Propagation Environment					
Water Quality			***Media Type***	***Media Temp.***	***Plants per Square metre***
Before Germination	***At the Two Leaf Stage (until 20 cm tall)***	***Until 30 to 35 cm Tall***			
Clean water to maintain a damp media should be used until the seed has germinated.	Should be fed with a balanced nutrient at 5-10 CF	When plants can be planted onto the system, the CF should be increased to 12-15 and then to 15-18.	Any suitable medium such as rockwool, perlite or vermiculite	21-25°C	2.5- 3 plants per square metre

High-quality capsicum seeds suitable for hydroponic cultivation can be raised in plugs in any suitable, sterile and well-draining growing media such as perlite, vermiculite, rockwool cubes, oasis cubes or coco coir. Care is taken to maintain proper spacing between seeds to prevent overcrowding to avoid hinderance in growth and development. Capsicum seeds can also be sown directly into the growbeds or into the netpots in growbeds.

Optimal propagation media temperatures ranging between 21 to 25°C are crucial for successful seed germination, typically occurring within 2 to 3 weeks under these conditions. During this period it is essential to maintain a consistently damp medium using clean water until germination occurs. Once the seeds germinate and develop true leaves, attention turns to seedling care. Optimal temperature levels around 24-27°C and RH around 70-80 per cent are maintained to promote healthy seedling growth. Adequate lighting either natural sunlight or artificial grow lights is provided to ensure robust photosynthesis and sturdy seedling development. The growing medium is kept consistently moist but never waterlogged to facilitate continued growth and root development. Once seedlings reach two-leaf stage these should be introduced to a balanced nutrient solution with an EC of 5 to 10 CF (conductivity factor). As the seedlings grow to a height of around 20cm gradually the CF is increased to 12-15 to support their nutrient needs. When seedlings reach a height of 30-35cm these are ready to be transplanted onto the hydroponic system. Planting density should be approximately 2.5 to 3 plants per square metre to ensure adequate spacing and optimal growth conditions.

Transplanting is the next crucial step in the process. Once the seedlings are strong enough to handle these are carefully transplanted into the hydroponic system. The system is prepared with a nutrient solution tailored to specific requirements of capsicum plants. Seedlings are gently removed from their original containers ensuring that delicate roots are not damaged in the process. These are then transplanted into the hydroponic system ensuring that roots are fully submerged in nutrient solution. Proper spacing between seedlings is maintained to avoid competition for resources and for promoting healthy growth.

After transplanting, diligent post-transplant care is essential. Environmental conditions such as temperature, humidity and light levels are monitored closely to ensure optimal growing conditions for transplanted seedlings. The nutrient solution is regularly adjusted to provide necessary nutrition for capsicum plants with pH and EC levels monitored and maintained within appropriate ranges. Regular inspection for insect-pests, diseases or nutrient deficiencies is conducted and appropriate action is promptly taken to address any issues that arise. Structural support such as trellising may be provided for proper growth support and to prevent any breakage.

16.3.2 Environmental Condition

Creating optimal environmental conditions is crucial for successful capsicum cultivation in hydroponics.

16.3.2.1 Light

Lighting is another critical factor to consider. Capsicum plants require abundant light for optimal growth and fruit production. Providing at least 8-12 hours of bright, direct sunlight or supplemental artificial lighting with high-quality grow lights ensures sufficient light intensity. Metal halide or high-pressure sodium lights are commonly used for indoor hydroponic setups.

16.3.2.2 Temperature

Temperature plays a vital role with capsicum plants thriving in warm conditions. The ideal range for daytime temperatures should be 21-29°C while during nighttime it should be 15-21°C. Consistent temperatures within this range promote healthy growth and fruit development.

16.3.2.3 Humidity

Maintaining moderate humidity levels is essential for capsicum plants typically ranging from 60-70 per cent. High humidity can increase the risk of fungal diseases while low humidity can lead to flower drop and fruit abortion. Proper ventilation and airflow within the growing area help regulate humidity levels and prevent disease development.

Good air circulation is vital to prevent the buildup of heat and humidity around the plants. Installing fans or ventilation systems helps maintaining proper airflow within the growing environment thus promoting healthy plant growth and reducing the risk of diseases.

16.3.2.4 Carbon Dioxide

Increasing CO_2 levels in the growing environment can enhance capsicum growth and productivity. CO_2 enrichment systems can be used to maintain CO_2 levels between 800-1200ppm during daytime thus promoting vigorous plant growth and increased yields.

16.3.3 Plant Training Systems

Many growers employ a single string to an overhead wire for young plants ~ a common practice for initial support. Once the plant establishes itself it is typically pruned at the head allowing for growth of four lateral branches. These branches are then attached to strings leading to the top wire for additional support.

In our method we utilize 1.52-1.55m long bamboo poles placed in each pot to secure the plants to the poles for stability. Metal fence posts are strategically positioned down the middle of the double row every 3.0 meters. These bamboo poles are then fastened to the metal posts using horizontal polypropylene strings. As the plants continue to grow, additional horizontal strings are added to accommodate the increasing weight of the branches so that optimal support throughout the growth cycle is ensured.

16.3.4 Nutritional Requirements

Capsicum plants require a well-balanced nutrient solution to support their growth and fruit development. These have diverse mineral nutrient needs encompassing nitrogen, potassium, phosphorous, calcium, magnesium, sulphur, iron, manganese, boron, copper, zinc and molybdenum. The nutrient solution should contain essential macro- and micro-nutrients in appropriate concentrations. These are particularly susceptible to over-feeding, though these actually do not die but just look ridiculous.

To promote quality growth, especially during adverse conditions like low light in winter, growth promoters can be incorporated into the nutrient solution. It is imperative to tailor nutrient formulations to the water supply necessitating a water mineral analysis for optimal formulation.

For hydroponic cultivation of shadehouse capsicums in a composted pine bark media, specific nutrient concentrations in ppm (mg/l) are recommended at different growth stages:

Nutrient Concentration (ppm or mg/l)

Nutrient	*Transplant to First Flower*	*After First Flower*
N	75	125
P	45	45
K	110	180
Ca	140	140

Nutrient	Transplant to First Flower	After First Flower
Mg	35	35
S	45	50
Fe	2.6	2.6
Cu	0.2	0.2
Mn	0.7	0.7
Zn	0.3	0.3
B	0.6	0.6
Mo	0.05	0.05

Adjusting pH and EC levels based on the growth stage of the plants is crucial. The pH of nutrient solution should be maintained between 5.5-6.5 while EC adjusted based on the growth stage though typically ranging from 1.5-2.5mS/cm. Monitoring and adjusting these levels regularly help prevent nutrient deficiencies and toxicities.

16.3.5 Pollination

In hydroponic capsicum cultivation, pollination plays a crucial role in setting of fruits and subsequent yield. Unlike some other crops that rely on wind or insects for pollination, capsicum plants are typically self-pollinating. However, in a controlled environment such as a hydroponic greenhouse where natural pollinators may be limited, growers often need to facilitate pollination manually to ensure optimal fruit production.

There are several methods for pollinating hydroponic capsicum plants:

i) **Hand pollination**: This method involves manual transfer of pollen from the male flower to the female using a small brush or cotton swab. Growers can identify male and female flowers by their physical characteristics; male flowers typically have a long slender stem while females a small bulbous structure at the base which will develop into the fruit if pollination is successful. By brushing gently the inside of the male flower to collect pollen and then transferring it to the stigma of the female flower can easily be performed by the growers.

ii) **Vibration**: Another method for promoting pollination in hydroponic capsicum plants is to use vibration. This can be achieved by gently tapping or shaking the plant which helps dislodging the pollen from the male flowers to fall onto the female flowers.

iii) **Electric toothbrush**: Some growers use an electric toothbrush to mimic the vibration caused by wind or insects to facilitating pollination. By holding the toothbrush carefully near the flowers growers can transfer pollen effectively from male to female flowers.

iv) **Commercial pollination products**: Also there are commercial products available specifically designed to aid in pollination in hydroponic systems. These products may include vibrating wands or devices designed to disperse pollen more effectively.

Regardless of the method used, consistent and timely pollination is essential for maximizing fruit set and yield in hydroponic capsicum cultivation. Growers should monitor flowering patterns closely and intervene as needed to ensure occurrence of adequate pollination for optimal crop production.

16.3.6 Insect-Pests and Diseases

Hydroponic capsicum cultivation offers numerous advantages but growers must remain vigilant against pests and diseases that can impact plant health and reduce yields. Here are some common pests and diseases encountered in hydroponic capsicum systems:

16.3.6.1 Insect-Pests

i) **Whiteflies**: These small flying insects feed on plant sap causing yellowing, reduced vigour and wilting in capsicum plants. These also excrete honeydew leading to growth of sooty mould.

ii) **Aphids**: Aphids are another sap-sucking insects that can infest capsicum plants. These reproduce rapidly and can cause distortion of leaves, stunted growth and transmission of viral diseases.

iii) **Thrips**: Thrips feed on leaves and flowers causing stippling, silvering and distortion of plant tissues. These can also transmit viral diseases to capsicum plants.

iv) **Spider mites**: These tiny arachnids feed on plant sap causing yellowing, stippling and webbing on leaves. Heavy infestations can lead to defoliation and reduced yields.

v) **Leaf miners**: The larvae of leaf miner flies tunnel through the leaves leaving distinctive trails or mines. Severe infestations can weaken plants and reduce photosynthetic capacity.

16.3.6.2 Diseases

i) **Powdery mildew**: Powdery mildew is a fungal disease characterized by white powdery growth on leaves, stems and flowers. It thrives in warm and humid conditions and can lead to reduced photosynthesis and premature defoliation.

ii) ***Botrytis* (Gray mould)**: *Botrytis* is a fungal disease that causes grayish-brown lesions on leaves, stems and fruits. It thrives in cool and moist environments and can lead to fruit rot.

iii) **Bacterial leaf spot**: Bacterial leaf spot causes water-soaked lesions on leaves which later on turn brown and necrotic. It can spread rapidly in humid conditions and through contaminated water or tools.

iv) ***Phytophthora* and *Pythium***: These soil-borne pathogens can cause root rot in capsicum plants leading to yellowing, wilting and eventual death. In hydroponic systems, these can be introduced through contaminated water or growing media.

To manage insect-pests and diseases in hydroponic capsicum cultivation, growers should implement IPM strategies including cultural practices, biological controls and judicious use of pesticides. Sanitation, regular monitoring and proper crop rotation can also help prevent outbreaks and to maintain plant health in hydroponic systems.

Chapter 17

Growing Chrysanthemum in Hydroponics System

17.1 Introduction

Chrysanthemum (family Asteraceae) boasts a diverse array of flower colours, types and plant sizes making them versatile for various applications. Some are prized for their suitability in cut flower production while others thrive in outdoor settings often referred to as hardy or fall mums. Flowers themselves exhibit a wide range of forms from daisy-like blooms to decorative pompons or buttons offering an abundance of choices for floral arrangement and landscaping.

Within this genus numerous hybrids and cultivars have meticulously been developed to meet horticultural demands. Notable among these are several important cut flower varieties each prized for its unique characteristics. These include 'Barliz' (spider form), 'Dark Splendour' (red), 'Greeno' (green), 'Paragon' (white), 'Reagan Cherry' (red), 'Reagan Lemon' (yellow), 'Reballet' (pink), 'Sunny' (yellow) and 'Tigro' (light orange) among others. Export varieties are 'Ajay', 'Anne', 'Birbal Sahni', 'Chandrama', 'Nanako' and 'Sonar Bangla' which are highly sought after for their quality and appeal. In all there are many thousand cultivars.

National Botanical Research Institute (NBRI), Lucknow has developed chrysanthemum cultivars engineered for year-round sequential flowering ~ a departure from the typical blooming season of November to December. These specially bred varieties such as 'Himanshu', 'Jwala', and 'Jyoti' for April to June flowering; 'Phuhar' for July to August; 'Ajay' and 'Sharda' for September to October; 'Makhmal', 'Megami', 'Mohini' and 'Sharad Bahar' for October to November; and 'Maghi' for February to March flowering ensure a continuous and vibrant display of blooms throughout the year.

17.2 Cultivation

17.2.1 Propagation

Dendranthema grandiflora has again been renamed as *Chrysanthemum morifolium*. It is revered as the national flower of Japan. Its flowers have versatile applications in horticulture ~ serving as cut flowers, loose flowers, potmums, bedding plants, border accents and hanging basket adornments. As a quintessential short-day plant, it thrives under critical day lengths of less than 9½ hours ~ a characteristic essential for its optimal growth and flowering. Renowned for its extensive spectrum of captivating flower colours and sizes as well as its remarkable vase life, chrysanthemum boasts diverse classes including Single, Anemone, Korean, Double, Decorative, Pompon, Incurved, Incurving, Reflexed, Reflexing, Quilled, Spoon and Spider varieties among others.

While chrysanthemums can be propagated *via* seeds (primarily for bedding types and select pompons), plant divisions, cuttings or tissue culture, however, commercial cultivation predominantly relies on cuttings. Growers typically obtain cuttings from healthy plants in their gardens or procure rooted or unrooted cuttings from specialized propagators and breeders. These propagators meticulously maintain true-to-type mother stock, employ culture-virus indexing to eradicate diseases, and cultivate extensive stocks to meet market demands.

For commercial chrysanthemum cultivation, propagation options include:

i) **Rooted cuttings**: This method is favoured by small-scale growers with limited propagation space. While it offers convenience but the cost per cutting may be higher where growers have limited control over cutting quality. Rooting facilities necessitate reduced light conditions with elevated humidity typically achieved through periodic misting for 2-3 days until root initiation. Upon arrival, rooted cuttings from propagators require prompt inspection and transplantation into final containers. If immediate planting is not feasible, cuttings can be refrigerated at 0.6 – 4.5°C for several days.

ii) **Unrooted cuttings**: Unrooted cuttings are directly placed into final containers or propagated in specialized cells, cubes or trays so offering a flexible option for growers.

Selecting the appropriate propagation method is crucial for ensuring successful establishment and robust growth of chrysanthemum crops in commercial settings, ultimately contributing to the quality and marketability of the final product.

17.2.1.1 Propagation Environment

Creating the optimal environment for chrysanthemum, propagation is essential for successful rooting and healthy plant development. Here are key factors to consider:

i) **Relative humidity**: Maintain relative humidity close to 100 per cent in the propagation area to promote moisture retention and prevent desiccation of cuttings. Ensure the propagation area is adequately isolated to control humidity levels effectively. Larger unused areas may complicate humidity control.

ii) **Sanitation**: Maintain strict sanitation protocols to prevent spread of pathogens. Disinfect all surfaces and equipment that may come in contact with cuttings to minimize the risk of contamination. Keep bench surfaces and floors clean and free of weeds to ensure a hygienic environment.

iii) **Temperature**: Maintain an optimal air temperature ranging from 21 to 29.5°C to promote rooting. During winter, it should be ensured that temperatures do not drop below 20°C. Providing bottom heat to maintain temperature of propagation medium between 22 to 24°C can significantly accelerate rooting. During flowering the night temperatures should be 10 to 16°C and daytime temperatures 18 to 21°C.

iv) **Light**: Provide adequate light intensity ranging from 3200 to 3800 foot-candles (fc) for optimal rooting conditions. During summer, shading can be employed to reduce light intensity and to control heat. Night-break incandescent lighting is essential to simulate long days. Supplemental HID lighting may be necessary during periods of low natural light. For flowering, provide lighting with wavelengths between 600 and 800 nm and light intensity of 1.23 to 1.6 MJ/m2/day. Flowering is induced when critical day length is less than 9½ hours per day.

v) **Mist**: Ensure uniform misting to keep cuttings turgid during rooting. Excessive misting can leach nutrients, oversaturate the medium and may cause stretching of cuttings. Conversely, inadequate misting can lead to wilting and necrosis. Use programmable misting systems to regulate frequency and duration based on environmental conditions.

vi) **Growing medium**: Choose a well-drained coarse medium such as rockwool or foam strips for rooting chrysanthemum cuttings. Maintain proper spacing, with cuttings typically placed 2.5cm apart in rows and 2.5 to 5.0cm between rows.

vii) **Rooting hormone**: Apply rooting hormones containing IBA or NAA to enhance rooting uniformity. Use liquids or talcs containing 1500ppm IBA applied to the lower end of the cutting.

viii) **Growth retardants**: Utilize growth retardants on vigorous cultivars during warm periods to prevent excessive stretching and promote compact growth. Treat cuttings with B-Nine at 1000ppm and place them in plastic bags in a cooler overnight before sticking.

ix) **Nutrition**: Initiate fertilization once callus forms (4-7 days) using a balanced fertilizer at 200-250ppm N. Some growers opt for a dilute solution of potassium nitrate through the mist system during initial days of propagation.

By managing these environmental factors carefully, growers can optimize chrysanthemum propagation and ensure healthy and robust plant growth.

17.2.1.2 Direct stick

Many growers opt to insert unrooted cuttings directly into the final container ~ a method that offers several advantages. By bypassing the transplanting step, this approach reduces labour costs and often yields a more uniform product. Production time is typically shortened. However, implementing this method necessitates a larger and more controlled propagation area. Moreover, the failure of even a single cutting in a pot can compromise the marketability of entire 15cm pot ~ underscoring the importance of careful monitoring and management. Furthermore, ensuring uniformity in the size of cuttings placed in each pot is essential for achieving a consistent final product.

Containers designated for direct sticking should be filled nearly to the brim with growing medium and watered prior to insertion. To facilitate rooting, one or two lower leaves should be removed from the base of the cuttings, allowing them to be inserted approximately 1.25-2.5cm deep into the medium. When executed meticulously thios method can streamline the propagation process and enhance overall efficiency in plant production.

17.2.2 Environmental Condition after Transplanting

After transplanting chrysanthemums, maintaining optimal environmental conditions is crucial to ensure their successful establishment and continued growth. Here are the key environmental factors to consider:

17.2.2.1 Light

Chrysanthemums require ample sunlight to thrive. After transplanting, ensure that the plants receive at least 6 to 8 hours of direct sunlight per day. If growing indoors or in a greenhouse there should be supplemental natural light with artificial grow lights to maintain adequate light levels.

17.2.2.2 Temperature

Maintaining specific temperature ranges is crucial for optimal growth and flowering of chrysanthemum plants. It is established that once the root system becomes established, maintaining daytime temperatures between 22-26°C and nighttime temperatures between 18-22°C is optimal for promoting healthy growth and flowering in chrysanthemums.

Lowering temperatures below these ranges such as maintaining daytime temperatures around 18°C and nighttime temperatures around 14°C can indeed prolong the time to flowering. This is because cooler temperatures slow down the metabolic processes within the plant including flowering initiation. Conversely, exposing chrysanthemum plants to high temperatures such as daytime temperatures of 30°C and nighttime temperatures of 24°C can negatively impact flower quality and yields. High temperatures can lead to stress in the plants so causing issues such as reduced flower size, discolouration and decreased flower longevity.

Therefore, maintaining optimal temperature levels within the recommended ranges is essential for maximizing flower quality, yield and overall plant health in chrysanthemum cultivation.

17.2.2.3 Humidity

Maintain moderate humidity levels around 50 to 60 per cent to prevent moisture-related issues such as fungal diseases and dehydration. Adequate air circulation can help regulate humidity levels and prevent stagnant air around the plants.

17.2.2.4 Air Circulation

Good air circulation is essential to prevent buildup of humidity and to reduce the risk of fungal diseases. Ensure proper spacing between plants to allow for adequate airflow especially in greenhouse or indoor growing environments.

17.2.2.5 CO_2 Enrichment

CO_2 enrichment to a level of 500-1,000 ppm will accelerate growth rates.

17.2.4 Nutrients Requirements

Chrysanthemums exhibit a significant demand for nutrients with a particular emphasis on nitrogen and potassium. During initial stages of growth a high nitrogen supply is essential to support robust vegetative development. However, as the plants progress and bud formation commences it is advisable to reduce nitrogen levels gradually to encourage optimal flowering. Concurrently, potassium supplementation should be heightened to support bud development and enhance flower quality.

In media-based systems, maintaining EC within the range of 16-22 is recommended to ensure adequate nutrient availability for chrysanthemum plants throughout their growth cycle. By managing nutrient levels carefully, growers can promote vigorous growth, abundant flowering and overall plant vitality in chrysanthemum cultivation.

Nutrients	*ppm*	*Nutrients*	*ppm*
Nitrogen	150-160	Potassium	120-140
Phosphorus	60-72	Magnesium	50-60
Calcium	170-180	Sulphur	70-80

Trace elements are required in standard ranges.

17.2.5 Insect-Pests and Diseases

Hydroponically grown chrysanthemums though are generally less susceptible to soil-borne insect-pests and diseases but still can be affected by a range of issues. Here are some common insect-pests and diseases encountered in hydroponic chrysanthemum cultivation:

17.2.5.1 Insect-pests

i) **Aphids**: These small sap-sucking insects can infest chrysanthemum plants causing leaf curling, yellowing and distortion. Aphid infestations can lead to reduced plant vigour and flower quality.

ii) **Thrips**: Thrips are tiny and slender insects that feed on plant tissues causing stippling, discolouration and distortion of leaves and flowers. Severe thrips infestations can result in reduced plant growth and flower production.

iii) **Whiteflies**: Whiteflies are small winged insects that feed on plant sap causing stunting, yellowing and plant wilting. Heavy infestations can lead to spread of viral diseases and significant damage to chrysanthemum crops.

iv) **Spider mites**: Spider mites are microscopic arachnid pests that feed on plant cells causing stippling, webbing and leaf discolouration. Severe spider mite infestations can weaken plants and reduce flower quality.

v) **Fungus gnats**: These small flying insects lay their eggs in moist growing media where larvae feed on plant roots. Fungus gnat larvae can cause root damage and increase the risk of root diseases.

17.2.5.2 Diseases

i) **Powdery mildew**: Powdery mildew is a fungal disease characterized by white powdery patches on leaves, stems and flowers. It thrives in warm and humid conditions and can reduce plant vigour and flower quality.

ii) ***Botrytis* blight (Gray mould)**: *Botrytis* blight is a fungal disease that causes brownish-gray lesions on leaves, stems and flowers. It thrives in cool and humid environments and can lead to floral rot and plant collapse.

iii) ***Pythium* root rot**: *Pythium* and other root rot pathogens can infect chrysanthemum roots in hydroponic systems causing root decay and stunted growth. Excessive moisture and poor drainage can exacerbate the problem.

iv) ***Fusarium* wilt**: *Fusarium* wilt is a soil-borne fungal disease that affects the vascular system of chrysanthemum plants causing yellowing, wilting and eventual plant death. It can spread rapidly in hydroponic systems if contaminated water or equipment is used.

To manage insect-pests and diseases in hydroponic chrysanthemum cultivation, growers should implement IPM strategies including regular monitoring, sanitation, use of biological control agents and cultural practices. Maintaining proper growing conditions such as adequate airflow, temperature and humidity control can help prevent insect-pest and disease outbreaks and promote healthy plant growth.

Chapter 18

Homegrown Hydroponics: Cultivating a Bounty with Advanced Methods

We have successfully cultivated a variety of crops including tomatoes, eggplants, cucumbers, lettuce, bell peppers, strawberries, broccoli (as well as a few other cole crops), roses and chrysanthemums in our small hydroponic garden. Details regarding the hydroponic techniques, nutrient composition, pH and EC of nutrient solution that were utilized are provided below.

18.1 Hydroponic Techniques and Substrates Used

The hydroponic techniques we have used for these crops are as following:

Crops	*Hydroponic Methods Used*
Tomatoes	Dutch buckets
Cucumbers	Dutch buckets and Passive drip system
Bell peppers	Kratky method
Bok choy	Kratky method
Lettuce	DWC, Kratky, NFT, static solution culture
Strawberries	Kratky method
Eggplants	Ebb and flow system

Crops	Hydroponic Methods Used
Broccoli and some other cole crops	Ebb and flow system
Roses	Drip irrigation system
Chrysanthemums	Drip irrigation system
Substrates: Coco coir has been employed in the majority of these methods. However, for bell peppers in the Kratky system, hydroton has been utilized instead.	

While we have implemented these methods it is important to recognize that the suitability of each hydroponic approach can vary depending on factors such as space availability, budget and individual grower preferences. Considering the specific requirements of each plant species is essential when selecting the most suitable hydroponic system. Moreover, diligent monitoring and adjustment of environmental conditions are paramount for achieving success in hydroponic cultivation.

It is important to note that regardless of the hydroponic method chosen, proper monitoring of pH, EC, temperature and humidity is essential for optimal growth of all the crops. Selecting appropriate nutrient solution tailored to specific needs of each crop is crucial for success in hydroponic gardening.

18.2 Nutrient Composition, pH and EC

The composition of macro and micronutrients in hydroponically grown plants can vary depending on various factors such as growth stage of the plant, specific nutrient solution formulation and environmental conditions. Details regarding nutrient composition, pH and EC utilized and those giving best results are provided below.

18.2.1 Tomatoes

Nutrient		Seedling Stage	Vegetative Stage	Flowering and Fruit Set	Fruit Development
Macronutrients	Nitrogen (N)	75-100 ppm	100-200 ppm	150-250 ppm	200-300 ppm
	Phosphorus (P)	50-75 ppm	50-100 ppm	100-150 ppm	150-200 ppm
	Potassium (K)	75-100 ppm	150-250 ppm	200-300 ppm	250-400 ppm
	Calcium (Ca)	Throughout growth	100-200 ppm		
	Magnesium (Mg)		50-100 ppm		
	Sulphur (S)		50-100 ppm		

	Nutrient	*Seedling Stage*	*Vegetative Stage*	*Flowering and Fruit Set*	*Fruit Development*
Micronutrients	Iron (Fe)	Throughout growth	2-5 ppm		
	Manganese (Mn)		0.5-2 ppm		
	Zinc (Zn)		0.1-0.5 ppm		
	Copper (Cu)		0.05-0.1 ppm		
	Molybdenum (Mo)		0.01-0.05 ppm		
	Boron (B)		0.5-1 ppm		
	Chlorine (Cl)		20-100 ppm		

pH levels:

- ☆ **Seedling stage:** pH 5.5-6.0
- ☆ **Vegetative stage:** pH 5.8-6.3
- ☆ **Flowering and fruit setting stage:** pH 6.0-6.5

Electrical conductivity:

- ☆ **Seedling stage:** EC 1.2-1.8 mS/cm (milliSiemens per centimeter)
- ☆ **Vegetative stage:** EC 1.5-2.0 mS/cm
- ☆ **Flowering and fruit setting stage:** EC 2.0-2.5 mS/cm

18.2.2 Eggplant (*Solanum melongena*)

	Nutrient	*Seedling Stage*	*Vegetative Stage*	*Flowering and Fruit Setting Stage*
Macronutrients	Nitrogen (N)	100-150 ppm	200-250 ppm	250-300 ppm
	Phosphorus (P)	75-100 ppm	150-200 ppm	200-250 ppm
	Potassium (K)	100-150 ppm	250-300 ppm	300-350 ppm
	Calcium (Ca)	Throughout growth	100-200 ppm	
	Magnesium (Mg)		50-100 ppm	
	Sulphur (S)		20-50 ppm	
Micronutrients	Iron (Fe)		2-5 ppm	
	Manganese (Mn)		1-2 ppm	
	Zinc (Zn)		0.5-1 ppm	
	Copper (Cu)		0.1-0.5 ppm	
	Molybdenum (Mo)		0.01-0.05 ppm	
	Boron (B)		0.5-1 ppm	
	Chlorine (Cl)		20-100 ppm	

<table>
<tr><th>Nutrient</th><th>Seedling Stage</th><th>Vegetative Stage</th><th>Flowering and Fruit Setting Stage</th></tr>
<tr><td colspan="4">pH level:
☆ Seedling stage: Maintain pH between 5.5 and 6.0.
☆ Vegetative stage: Keep pH between 5.8 and 6.2.
☆ Flowering and fruit setting stage: Maintain pH between 6.0 and 6.5.</td></tr>
<tr><td colspan="4">Electrical conductivity:
☆ Seedling stage: Aim for an EC between 0.8 and 1.2 mS/cm (milliSiemens per centimeter).
☆ Vegetative stage: Maintain EC between 1.2 and 1.8 mS/cm.
☆ Flowering and fruit setting stage: Keep EC between 1.8 and 2.5 mS/cm.</td></tr>
</table>

18.2.3 Cucumbers

<table>
<tr><th colspan="2">Nutrient</th><th>Seedling Stage</th><th>Vegetative Stage</th><th>Flowering and Fruit Setting Stage</th></tr>
<tr><td rowspan="6">Macronutrients</td><td>Nitrogen (N)</td><td>100-150 ppm</td><td>200-250 ppm</td><td>250-300 ppm</td></tr>
<tr><td>Phosphorus (P)</td><td>50-75 ppm</td><td>100-150 ppm</td><td>150-200 ppm</td></tr>
<tr><td>Potassium (K)</td><td>100-150 ppm</td><td>200-300 ppm</td><td>300-400 ppm</td></tr>
<tr><td>Calcium (Ca)</td><td rowspan="10">Throughout growth</td><td colspan="2">100-200 ppm</td></tr>
<tr><td>Magnesium (Mg)</td><td colspan="2">50-100 ppm</td></tr>
<tr><td>Sulphur (S)</td><td colspan="2">20-50 ppm</td></tr>
<tr><td rowspan="7">Micronutrients</td><td>Iron (Fe)</td><td colspan="2">2-5 ppm</td></tr>
<tr><td>Manganese (Mn)</td><td colspan="2">1-2 ppm</td></tr>
<tr><td>Zinc (Zn)</td><td colspan="2">0.5-1 ppm</td></tr>
<tr><td>Copper (Cu)</td><td colspan="2">0.1-0.5 ppm</td></tr>
<tr><td>Molybdenum (Mo)</td><td colspan="2">0.01-0.05 ppm</td></tr>
<tr><td>Boron (B)</td><td colspan="2">0.5-1 ppm</td></tr>
<tr><td>Chlorine (Cl)</td><td colspan="2">20-100 ppm</td></tr>
<tr><td colspan="5">pH level:
☆ Seedling stage: 5.8 - 6.2
☆ Vegetative stage: 5.8 - 6.2
☆ Flowering and fruit setting stage: 5.8 - 6.2</td></tr>
<tr><td colspan="5">Electrical conductivity:
☆ Seedling stage: 1.2 - 1.6 mS/cm (milliSiemens per centimeter)
☆ Vegetative stage: 1.6 - 2.0 mS/cm
☆ Flowering and fruit setting stage: 2.0 - 2.5 mS/cm</td></tr>
</table>

18.2.4 Lettuce

<table>
<tr><th colspan="2">Nutrient</th><th>Seedling Stage</th><th>Vegetative Stage</th><th>Heading Stage</th></tr>
<tr><td rowspan="6">Macronutrients</td><td>Nitrogen (N)</td><td>75-100 ppm</td><td>150-200 ppm</td><td>100-150 ppm</td></tr>
<tr><td>Phosphorus (P)</td><td>50-75 ppm</td><td>100-150 ppm</td><td>75-100 ppm</td></tr>
<tr><td>Potassium (K)</td><td>75-100 ppm</td><td>150-200 ppm</td><td>100-150 ppm</td></tr>
<tr><td>Calcium (Ca)</td><td rowspan="10">Throughout growth</td><td colspan="2">100-200 ppm</td></tr>
<tr><td>Magnesium (Mg)</td><td colspan="2">50-100 ppm</td></tr>
<tr><td>Sulphur (S)</td><td colspan="2">20-50 ppm</td></tr>
<tr><td rowspan="7">Micronutrients</td><td>Iron (Fe)</td><td colspan="2">2-5 ppm</td></tr>
<tr><td>Manganese (Mn)</td><td colspan="2">1-2 ppm</td></tr>
<tr><td>Zinc (Zn)</td><td colspan="2">0.5-1 ppm</td></tr>
<tr><td>Copper (Cu)</td><td colspan="2">0.1-0.5 ppm</td></tr>
<tr><td>Molybdenum (Mo)</td><td colspan="2">0.01-0.05 ppm</td></tr>
<tr><td>Boron (B)</td><td colspan="2">0.5-1 ppm</td></tr>
<tr><td>Chlorine (Cl)</td><td colspan="2">20-100 ppm</td></tr>
</table>

pH levels:

- ☆ **Seedling stage:** 5.5 - 6.0
- ☆ **Vegetative stage:** 5.5 - 6.5
- ☆ **Flowering stage (if applicable):** 5.5 - 6.5

Electrical conductivity:

- ☆ **Seedling stage:** 0.8 - 1.2 mS/cm (milliSiemens per centimeter)
- ☆ **Vegetative stage:** 1.2 - 1.8 mS/cm
- ☆ **Flowering stage (if applicable):** 1.2 - 1.8 mS/cm

18.2.5 Bell Pepper

<table>
<tr><th colspan="2">Nutrient</th><th>Seedling Stage</th><th>Vegetative Stage</th><th>Flowering and Fruit Setting Stage</th></tr>
<tr><td rowspan="6">Macronutrients</td><td>Nitrogen (N)</td><td>75-100 ppm</td><td>150-250 ppm</td><td>200-300 ppm</td></tr>
<tr><td>Phosphorus (P)</td><td>50-75 ppm</td><td>100-150 ppm</td><td>150-200 ppm</td></tr>
<tr><td>Potassium (K)</td><td>100-125 ppm</td><td>200-250 ppm</td><td>250-350 ppm</td></tr>
<tr><td>Calcium (Ca)</td><td rowspan="3">Throughout growth</td><td colspan="2">100-200 ppm</td></tr>
<tr><td>Magnesium (Mg)</td><td colspan="2">50-100 ppm</td></tr>
<tr><td>Sulphur (S)</td><td colspan="2">20-50 ppm</td></tr>
</table>

Nutrient		Seedling Stage	Vegetative Stage	Flowering and Fruit Setting Stage
Micronutrients	Iron (Fe)	Throughout growth	2-5 ppm	
	Manganese (Mn)		1-2 ppm	
	Zinc (Zn)		0.5-1 ppm	
	Copper (Cu)		0.1-0.5 ppm	
	Molybdenum (Mo)		0.01-0.05 ppm	
	Boron (B)		0.5-1 ppm	
	Chlorine (Cl)		20-100 ppm	

pH levels:

- ☆ **Seedling stage:** pH 5.5 - 6.0
- ☆ **Vegetative stage:** pH 5.5 - 6.5
- ☆ **Flowering and fruit setting stage:** pH 6.0 - 6.5

Electrical conductivity levels:

- ☆ **Seedling stage:** 1.2 - 1.6 mS/cm (milliSiemens per centimeter)
- ☆ **Vegetative stage:** 1.6 - 2.0 mS/cm
- ☆ **Flowering and fruit setting stage:** 2.0 - 2.4 mS/cm

18.2.6 Strawberries

Nutrient		Seedling Stage	Vegetative Stage	Flowering and Fruit Setting Stage
Macronutrients	Nitrogen (N)	100-150 ppm	150-200 ppm	200-250 ppm
	Phosphorus (P)	50-75 ppm	75-100 ppm	100-150 ppm
	Potassium (K)	100-125 ppm	150-200 ppm	200-250 ppm
	Calcium (Ca)	Throughout growth	100-200 ppm	
	Magnesium (Mg)		50-100 ppm	
	Sulphur (S)		20-50 ppm	
Micronutrients	Iron (Fe)		2-5 ppm	
	Manganese (Mn)		1-2 ppm	
	Zinc (Zn)		0.5-1 ppm	
	Copper (Cu)		0.1-0.5 ppm	
	Molybdenum (Mo)		0.01-0.05 ppm	
	Boron (B)		0.5-1 ppm	
	Chlorine (Cl)		20-100 ppm	

<table>
<tr><th>Nutrient</th><th>Seedling Stage</th><th>Vegetative Stage</th><th>Flowering and Fruit Setting Stage</th></tr>
<tr><td colspan="4">pH:
☆ Seedling stage: 5.5 - 6.0
☆ Vegetative and fruit development stages: 5.8 - 6.5</td></tr>
<tr><td colspan="4">Electrical conductivity:
☆ Seedling stage: 0.8 - 1.2 mS/cm (milliSiemens per centimeter)
☆ Vegetative and fruit development stages: 1.2 - 2.0 mS/cm</td></tr>
</table>

18.2.7 Broccoli, Brussels Sprout, Savoy Cabbage, Kale and Bok Choy

<table>
<tr><th colspan="2">Nutrient</th><th>Seedling Stage</th><th>Vegetative Stage</th><th>Flowering and Fruit Setting Stage</th></tr>
<tr><td rowspan="6">Macronutrients</td><td>Nitrogen (N)</td><td>100-150 ppm</td><td>200-250 ppm</td><td>250-300 ppm</td></tr>
<tr><td>Phosphorus (P)</td><td>50-75 ppm</td><td>100-150 ppm</td><td>150-200 ppm</td></tr>
<tr><td>Potassium (K)</td><td>100-150 ppm</td><td>200-300 ppm</td><td>300-400 ppm</td></tr>
<tr><td>Calcium (Ca)</td><td rowspan="10">Throughout growth</td><td colspan="2">100-200 ppm</td></tr>
<tr><td>Magnesium (Mg)</td><td colspan="2">50-100 ppm</td></tr>
<tr><td>Sulphur (S)</td><td colspan="2">20-50 ppm</td></tr>
<tr><td rowspan="7">Micronutrients</td><td>Iron (Fe)</td><td colspan="2">2-5 ppm</td></tr>
<tr><td>Manganese (Mn)</td><td colspan="2">1-2 ppm</td></tr>
<tr><td>Zinc (Zn)</td><td colspan="2">0.5-1 ppm</td></tr>
<tr><td>Copper (Cu)</td><td colspan="2">0.1-0.5 ppm</td></tr>
<tr><td>Molybdenum (Mo)</td><td colspan="2">0.01-0.05 ppm</td></tr>
<tr><td>Boron (B)</td><td colspan="2">0.5-1 ppm</td></tr>
<tr><td>Chlorine (Cl)</td><td colspan="2">20-100 ppm</td></tr>
<tr><td colspan="5">pH:
☆ Seedling stage: Maintain pH levels between 5.5 and 6.0.
☆ Vegetative stage: Maintain pH levels between 5.8 and 6.2.
☆ Flowering and fruit setting stage: Maintain pH levels between 6.0 and 6.5.</td></tr>
<tr><td colspan="5">Electrical conductivity:
☆ Seedling stage: Maintain EC levels between 1.0 and 1.5 mS/cm (milliSiemens per centimeter).
☆ Vegetative stage: Maintain EC levels between 1.5 and 2.0 mS/cm.
☆ Flowering and fruit setting stage: Maintain EC levels between 2.0 and 2.5 mS/cm.</td></tr>
</table>

Broccoli is a unique vegetable which does not produce traditional flowers and fruits as many other plants do. Instead, broccoli plants form edible flowerheads which are harvested before these bloom fully. However, for purposes of hydroponic nutrient management, we can consider the stage when broccoli heads begin to form as an equivalent to the 'Flowering and Fruit Setting Stage' found in other plants.

During this stage, broccoli plants allocate significant resources towards development of central edible flowerhead. Therefore, it is crucial to provide adequate nutrients to support this growth phase. Adjusting nutrient levels, pH and EC appropriately during this stage can help optimize the size, quality and yield of broccoli heads. So, while broccoli doesn't produce traditional flowers and fruits, the stage when the central edible head begins to form can be considered analogous to the flowering and fruit setting stage in other plants.

18.2.8 Rose

<table>
<tr><th colspan="2">Nutrient</th><th>Seedling Stage</th><th>Vegetative Stage</th><th>Flowering Stage</th></tr>
<tr><td rowspan="6">Macronutrients</td><td>Nitrogen (N)</td><td>150-200 ppm</td><td>200-250 ppm</td><td>150-200 ppm</td></tr>
<tr><td>Phosphorus (P)</td><td>50-75 ppm</td><td>100-150 ppm</td><td>100-150 ppm</td></tr>
<tr><td>Potassium (K)</td><td>150-200 ppm</td><td>200-300 ppm</td><td>300-400 ppm</td></tr>
<tr><td>Calcium (Ca)</td><td rowspan="10">Throughout growth</td><td colspan="2">100-200 ppm</td></tr>
<tr><td>Magnesium (Mg)</td><td colspan="2">50-100 ppm</td></tr>
<tr><td>Sulphur (S)</td><td colspan="2">20-50 ppm</td></tr>
<tr><td rowspan="7">Micronutrients</td><td>Iron (Fe)</td><td colspan="2">2-5 ppm</td></tr>
<tr><td>Manganese (Mn)</td><td colspan="2">1-2 ppm</td></tr>
<tr><td>Zinc (Zn)</td><td colspan="2">0.5-1 ppm</td></tr>
<tr><td>Copper (Cu)</td><td colspan="2">0.1-0.5 ppm</td></tr>
<tr><td>Molybdenum (Mo)</td><td colspan="2">0.01-0.05 ppm</td></tr>
<tr><td>Boron (B)</td><td colspan="2">0.5-1 ppm</td></tr>
<tr><td>Chlorine (Cl)</td><td colspan="2">20-100 ppm</td></tr>
</table>

pH levels:

- ☆ **Propagation stage (Rooting):** pH should be maintained between 5.5 and 6.0.
- ☆ **Vegetative growth stage:** pH should be maintained between 5.5 and 6.5.
- ☆ **Flowering stage:** pH should be maintained between 5.5 and 6.5.

Electrical conductivity levels:

- ☆ **Propagation stage (Rooting):** EC should be maintained between 0.5 and 1.0 mS/cm (milliSiemens per centimeter).
- ☆ **Vegetative growth stage:** EC should be maintained between 1.5 and 2.5 mS/cm.
- ☆ **Flowering stage:** EC should be maintained between 1.5 and 2.5 mS/cm.

18.2.9 Chrysanthemums

<table>
<tr><th colspan="2">Nutrient</th><th>Seedling Stage</th><th>Vegetative Stage</th><th>Flowering Stage</th></tr>
<tr><td rowspan="6">Macronutrients</td><td>Nitrogen (N)</td><td>100-150 ppm</td><td>200-250 ppm</td><td>150-200 ppm</td></tr>
<tr><td>Phosphorus (P)</td><td>50-75 ppm</td><td>100-150 ppm</td><td>75-100 ppm</td></tr>
<tr><td>Potassium (K)</td><td>100-150 ppm</td><td>200-250 ppm</td><td>250-300 ppm</td></tr>
<tr><td>Calcium (Ca)</td><td rowspan="10">Throughout growth</td><td colspan="2">100-200 ppm</td></tr>
<tr><td>Magnesium (Mg)</td><td colspan="2">50-100 ppm</td></tr>
<tr><td>Sulphur (S)</td><td colspan="2">20-50 ppm</td></tr>
<tr><td rowspan="7">Micronutrients</td><td>Iron (Fe)</td><td colspan="2">2-5 ppm</td></tr>
<tr><td>Manganese (Mn)</td><td colspan="2">1-2 ppm</td></tr>
<tr><td>Zinc (Zn)</td><td colspan="2">0.5-1 ppm</td></tr>
<tr><td>Copper (Cu)</td><td colspan="2">0.1-0.5 ppm</td></tr>
<tr><td>Molybdenum (Mo)</td><td colspan="2">0.01-0.05 ppm</td></tr>
<tr><td>Boron (B)</td><td colspan="2">0.5-1 ppm</td></tr>
<tr><td>Chlorine (Cl)</td><td colspan="2">20-100 ppm</td></tr>
</table>

pH levels:1.

- ☆ **Seedling stage:** pH should be maintained between 5.5 and 6.0.**2.**
- ☆ **Vegetative growth stage:** pH should be maintained between 6.0 and 6.5.**3.**
- ☆ **Flowering and late flowering stages:** pH should be maintained between 6.0 and 6.5.

Electrical Conductivity:

- ☆ **Seedling stage:** Keep EC relatively low, typically between 0.6 and 0.8 mS/cm
- ☆ **Vegetative stage:** Increase gradually, typically ranging from 1.2 to 1.8 mS/cm
- ☆ **Flowering stage:** Maintain similar to the vegetative stage, between 1.2 and 1.8 mS/cm
- ☆ **Late flowering and pre-harvest stage:** Reduce EC slightly to around 1.0 to 1.5 mS/cm to promote maturation and flavour development

18.2.10 General guidelines

- ☆ These nutrient concentrations are general guidelines and may vary based on factors such as tomato variety, environmental conditions and specific nutrient requirements.
- ☆ It is crucial to monitor nutrient levels regularly in the hydroponic solution and adjust as needed based on plant growth and development.
- ☆ Consider using a balanced hydroponic nutrient solution formulated specifically for tomatoes or adjust nutrient formulations based on laboratory analysis or recommendations from agricultural experts.

- Always follow manufacturer instructions and guidelines when preparing and applying nutrient solutions to avoid over or under-fertilization.

18.2.11 pH and EC

- pH and EC requirements may vary slightly depending on factors such as varieties, growing media and environmental conditions.
- It's essential to monitor pH and EC levels regularly using appropriate meters or test kits and adjust them as needed to maintain optimal nutrient uptake and prevent nutrient imbalances.
- Most plants are sensitive to fluctuations in pH and EC so strive for stability in these parameters to promote healthy growth and flowering.
- Use pH-adjusting solutions (pH up or down) as necessary to maintain desired pH range.
- EC levels can be adjusted by adding or diluting nutrient solutions based on growth stage of plants and nutrient requirements.

18.3 Optimal Growing conditions

Optimal environmental conditions including nutrient solution temperature, air temperature, light and humidity play a crucial role in growth and development of hydroponically grown crops. In our hydroponic garden we have followed specific growing conditions tailored to different growth stages of crops. Details regarding growing environment utilized by us are provided below.

18.3.1 Tomatoes

Growing conditions that we followed at different growth stages during cultivation of tomatoes:

Seedling Stage

- **Nutrient solution temperature:** Maintained between 18 to 22°C to foster healthy root development without causing stress to young seedlings.
- **Air temperature:** Kept between 20 to 25°C during daytime and slightly cooler at night.
- **Light:** Provided 16 to 18 hours of light per day using fluorescent or LED grow lights to promote strong and compact seedling growth.
- **Humidity:** Maintained high humidity levels around 70 to 80 per cent to prevent desiccation of seedlings, utilizing a humidity dome or misting system if necessary.

Vegetative Growth Stage

- **Nutrient solution temperature:** Maintained 18 to 25°C temperature for optimal nutrient uptake and plant metabolism.

- **Air temperature:** Temperature between 21 to 27°C during daytime and slightly cooler at night supports vigorous vegetative growth.
- **Light:** High intensity lighting from 14 to 16 hours per day using metal halide or high-pressure sodium (HPS) grow lights promote leaf and stem development.
- **Humidity:** Maintaining the moderate humidity levels around 60 to 70 per cent reduces risk of fungal diseases and promotes transpiration.

Flowering and Fruit Development Stage

- **Nutrient solution temperature:** Maintaining temperatures between 20 to 25°C supports nutrient uptake and flowering.
- **Air Temperature:** Air temperature should be maintained between 21 to 27°C during daytime and slightly cooler at night. Temperature fluctuations should be avoided to prevent blossom drop.
- **Light:** Per day high light intensity should be provided from 12 to 14 hours using HPS grow lights to promote flowering and fruit set. Providing supplementary lighting with red and blue LED lights enhances fruit development.
- **Humidity:** There should be moderate humidity levels around 50 to 60 per cent to reduce the risk of diseases and to ensure proper fruit development.

Ripening Stage

- **Nutrient solution temperature:** It should be between 20 to 25°C to support nutrient uptake and fruit ripening.
- **Air temperature:** Air temperature should be maintained between 20 to 25°C during daytime and slightly cooler at night to promote even ripening and to prevent fruit cracking.
- **Light:** It should be provided 10 to 12 hours of moderate-intensity lighting per day using HPS or LED grow lights to support ripening without causing stress to the plants.
- **Humidity:** It should maintain moderate humidity levels around 40 to 50 per cent to prevent mould growth and to ensure proper fruit ripening.

18.3.2 Eggplant

We have implemented tailored growing protocols for different growth stages during eggplant cultivation:

Seedling Stage

- **Nutrient solution temperature:** Temperature of nutrient solution should be kept between 20 to 25°C to foster root development and for nutrient uptake.

- **Environmental temperature:** Air temperature is kept between 22 to 26°C during daytime and slightly lower at night to support healthy seedling growth.
- **Light requirements:** It should be provided with 16 to 18 hours of light per day using fluorescent or LED grow lights suited for seedling growth.
- **Humidity:** High humidity levels around 70 to 80 per cent should be maintaied to prevent seedlings from desiccating. If needed, a humidity dome or misting system may be installed.

Vegetative Stage

- **Nutrient solution temperature:** To sustain temperature of nutrient solution between 20 to 25°C will facilitate robust vegetative growth.
- **Environmental temperature:** Air temperature between 24 to 28°C during daytime and slightly lower at night should be provisioned to promote healthy leaf and stem development.
- **Light requirements:** Using high-intensity discharge (HID) lamps or full-spectrum LED grow lights, it should be supplied with 14 to 16 hours of light daily to support vegetative growth.
- **Humidity:** A moderate humidity levels around 60 to 70 per cent should be maintained to mitigate fungal diseases and to ensure optimal transpiration.

Flowering and Fruit Development Stage

- **Nutrient solution temperature:** Nutrient solution temperature between 20 to 25°C should be maintained to encourage flowering and fruit set.
- **Environmental temperature:** Air temperature should be maintained between 22 to 28°C during daytime and slightly lower at night to facilitate flower development and for fruit formation.
- **Light requirements:** Through high intensity lamps or LED grow lights rich in red and blue lighting should be used to provide 12 to 14 hours of light per day to stimulate flowering and fruiting.
- **Humidity:** Humidity should slightly be reduced to around 50 to 60 per cent to deter excessive moisture around flowers and fruits to avoid fungal issues.

Harvesting Stage

- **Nutrient solution temperature:** Nutrient solution temperature should be maintained between 20 to 25°C to support nutrient uptake until the end of the growing cycle.
- **Environmental temperature:** Air temperature should be maintained between 22 to 28°C to ensure optimal fruit ripening.

- **Light requirements:** Consistent light levels should be maintained as provided during flowering and fruiting stages to support fruit development and ripening.
- **Humidity:** Moderate humidity levels should be maintained to prevent moisture-related issues during fruit ripening while avoiding excessive drying out.

18.3.3 Cucumber

Growing conditions that we followed at different growth stages during cucumber cultivation are as under:

Seedling Stage

- **Nutrient solution temperature:** Between 18 to 22°C.
- **Ambient temperature:** Air temperatures around 20 to 25°C.
- **Light requirements:** 16 to 18 hours of light per day using full spectrum grow lights or natural sunlight.
- **Humidity:** High humidity levels around 70 to 80 per cent to prevent seedlings from drying out.

Vegetative Growth Stage

- **Nutrient solution temperature:** Between 20 to 25°C.
- **Ambient temperature:** Air temperatures around 22 to 28°C during daytime and slightly cooler at night.
- **Light requirements:** 14 to 16 hours of light per day with a combination of blue and red spectrum grow lights.
- **Humidity:** Around 60 to 70 per cent.

Flowering and Fruit Development Stage

- **Nutrient solution temperature:** Between 22 to 26°C.
- **Ambient temperature:** Air temperatures around 24 to 28°C during daytime and slightly cooler at night.
- **Light requirements:** 12 to 14 hours of light per day with a combination of blue, red and far-red spectrum grow lights to promote flowering and fruit set.
- **Humidity:** Slightly reduced to around 50 to 60 per cent to prevent fungal diseases and promote fruit development.

Late Fruit Development and Ripening Stage

- **Nutrient solution temperature:** Between 22 to 26°C.

- **Ambient temperature:** Air temperatures around 22 to 26°C during daytime and slightly cooler at night.
- **Light requirements:** 10 to 12 hours of light per day with a balanced spectrum to support fruit ripening.
- **Humidity:** Moderate humidity levels around 50 to 60 per cent.

18.3.4 Lettuce

Growing conditions that we followed for different growth stages during the course of lettuce cultivation are as under:

Seed Germination and Seedling Stage

- **Nutrient solution temperature:** Between 21 to 24°C to promote rapid germination and seedling establishment.
- **Environmental temperature:** The ambient air temperature around 20 to 24°C to support healthy seedling growth.
- **Light:** 16 to 18 hours of continuous lighting using fluorescent or LED grow lights with a colour spectrum suitable for seedling growth.
- **Humidity:** High humidity levels around 70 to 80 per cent to prevent desiccation of seedlings. Using a humidity dome or misting system can help maintain humidity.

Vegetative Growth Stage

- **Nutrient solution temperature:** Between 18 to 22°C to support nutrient uptake and vegetative growth.
- **Environmental temperature:** The ambient air temperature around 18 to 22°C during vegetative stage.
- **Light:** 14 to 16 hours of light per day using full-spectrum LED grow lights or natural sunlight supplemented with artificial lighting, if necessary.
- **Humidity:** There should be moderate humidity levels (50 to 70 per cent) to prevent wilting and promote healthy leaf expansion.

Pre-harvest Stage (Head formation)

- **Nutrient solution temperature:** There should be 18 to 20°C temperature to support nutrient uptake and head formation.
- **Environmental temperature:** The ambient air temperature should be 18 to 20°C during pre-harvest stage.
- **Light:** There should be 12 to 14h of light per day using full-spectrum LED grow lights or natural sunlight supplemented with artificial lighting, if necessary.

- **Humidity:** Humidity levels should be noderate, *i.e.* 50 to 60 per cent to prevent disease development and to ensure optimal head formation.

Harvest Stage

- **Nutrient solution temperature:** Between 18 to 20°C to support nutrient uptake and post-harvest quality.
- **Environmental temperature:** Ambient air temperature around 18 to 20°C during harvesting and post-harvest handling.
- **Light:** 10 to 12 hours of light per day during harvest stage.
- **Humidity:** Moderate humidity levels around 50 per cent to 60 per cent to prevent wilting and maintain post-harvest quality.

18.3.5 Bell Pepper

Optimal conditions for hydroponically grown bell peppers vary throughout different growth stages. Here is a breakdown of recommended temperature, light and humidity requirements for each stage:

Seedling Stage

- **Nutrient solution temperature:** 18-22°C to promote healthy root development and nutrient uptake.
- **Environmental temperature:** 21-26°C to facilitate seed germination and early growth.
- **Light:** Seedlings require indirect 14-16h of bright,light; per day using fluorescent or LED grow lights to promote strong and healthy growth.
- **Humidity:** RH of 70-80 per cent is necessary to prevent seedlings from drying out. Use of a humidity dome or misting system may be necessary to increase humidity.

Vegetative Growth Stage

- **Nutrient solution temperature:** 18-22°C to support vigorous vegetative growth.
- **Environmental temperature:** 21-28°C to encourage healthy foliage development and stem elongation.
- **Light:** 14-16 hours of high-intensity light per day is beneficial. Growers can use full-spectrum grow lights or natural sunlight to ensure robust growth and leaf expansion.
- **Humidity:** 60-70 per cent RH may be necessary to prevent moisture-related diseases and promote optimal transpiration.

Flowering and Fruit Development Stage

- ☆ **Nutrient solution temperature:** 18-22°C to support flower formation and fruit set.
- ☆ **Environmental temperature:** 21-26°C to facilitate pollination and fruit development.
- ☆ **Light:** 12-14 hours of high-intensity light per day. Growers should ensure consistent light levels to support flower production and fruit maturation.
- ☆ **Humidity:** There should be 50-60 per cent RH. Maintaining slightly lower humidity levels during flowering and fruiting is desired to reduce the risk of fungal diseases and to promote fruit quality.

Late Growth Stage and Ripening

- ☆ **Nutrient solution temperature:** 18-22°C to support fruit ripening and nutrient uptake.
- ☆ **Environmental temperature:** A moderate temperature of 18-24°C will promote uniform ripening and fruit quality.
- ☆ **Light:** It requires 10-12 hours of light per day. Slightly reduced light exposure during late growth stage will promote fruit ripening and colour development.
- ☆ **Humidity:** An average RH level of 40-50 per cent is required during fruit growth and ripening stages. Lower humidity levels prevent fruit rot and maximize shelf life during storage.

18.3.6 Strawberries

Seedling Stage

- ☆ **Nutrient solution temperature:** A temperature range of 18 to 22°C is quite effective to promote root development and seedling establishment.
- ☆ **Environmental temperature:** An ambient air temperature between 18 to 25°C is essential to support healthy growth and to minimize stress on seedlings.
- ☆ **Light:** Seedlings require high light levels, preferably 12 to 16 hours of direct or indirect sunlight per day or supplemental artificial lighting equivalent to sunlight.
- ☆ **Humidity:** Around 70 per cent to 80 per cent RH is necessary to prevent desiccation of delicate seedlings.

Vegetative Growth Stage

- ☆ **Nutrient solution temperature:** A temperature range of 18 to 22°C is useful to support vigorous vegetative growth and nutrient uptake.

- **Environmental temperature:** An ambient air temperature between 18 to 25°C is essential during vegetative stage for optimal growth and development.
- **Light:** Per day 14 to 16 hours of light with a balance of red and blue wavelengths is beneficial to promote vegetative growth and leaf expansion.
- **Humidity:** RH of around 60 to 70 per cent is essential to prevent moisture-related diseases and to ensure adequate transpiration.

Flowering and Fruit Development Stage

- **Nutrient solution temperature:** The temperatures at 18 to 22°C supports healthy flowering and fruit set.
- **Environmental temperature:** An ambient air temperatures between 15 to 22°C during flowering and fruit development prevent heat stress and ensure proper pollination.
- **Light:** 12 to 14 hours of lighting per day with a balance of red and blue wavelengths along with supplementary far-red light during flowering stage promote flower induction and fruit development.
- **Humidity:** Around 60 to 70 per cent RH prevents moisture-related diseases and maintains optimal fruit quality.

Ripening Stage

- **Nutrient solution temperature**: A temperature range of 18 to 22°C supports continued nutrient uptake and ripening processes.
- **Environmental temperature:** An ambient air temperature between 15 to 22°C during ripening stage promotes gradual fruit ripening and optimal flavour development.
- **Light:** A light regime of 10 to 12 hours per day during ripening stage reduces light intensity to prevent stress on ripening fruit.
- **Humidity:** Relative humidity levels around 60 to 70 per cent prevent moisture-related diseases and maintain fruit quality during ripening.

18.3.7 Broccoli, Brussels Sprout, Savoy Cabbage, Kale and Bok Choy

Seedling Stage

- **Nutrient solution temperature:** A temperature regime of 18 to 22°C promotes healthy root development.
- **Environmental temperature:** Air temperature around 20 to 25°C during daytime and slightly lower at night is quite favourable for seedling growth.

- **Light requirements:** 14 to 16 hours of lighting per day through fluorescent or LED grow lights encourages vigorous growth.
- **Humidity:** High humidity levels of around 70 to 80 per cent prevent seedlings from drying out.

Vegetative Growth Stage

- **Nutrient solution temperature:** Temperature regime of 18 to 22°C favours optimal nutrient uptake.
- **Environmental temperature:** Air temperature regime of 18 to 22°C during daytime and slightly lower at night supports vegetative growth.
- **Light requirements:** Using high-intensity grow lights such as metal halide or LED for 14 to 16 hours per day promote leafy growth.
- **Humidity:** Moderate to high humidity levels between 60 to 70 per cent prevent wilting and promote healthy transpiration.

Flowering/Bolting Stage (for broccoli and Brussels sprouts)

- **Nutrient solution temperature:** During flowering, a temperature regime of 18 to 22°C supports nutrient uptake during flowering.
- **Environmental temperature:** Air temperature around 15 to 20°C during daytime and slightly lower at night prevents premature bolting.
- **Light requirements:** Cooler spectrum LED grow lights for 12 to 14 hours per day discourage bolting and promote head development.
- **Humidity:** Moderate humidity levels between 50 and 60 per cent prevent excessive moisture-related diseases.

Maturation/Harvest Stage

- **Nutrient solution temperature:** A temperature regime of 18 to 22°C ensures continued nutrient uptake during maturation.
- **Environmental temperature:** Air temperature regime of around 15 to 20°C during daytime and slightly lower at night favours proper maturation and flavour development.
- **Light requirements:** Cooler spectrum LED grow lighting for 10 to 12 hours per day favours maintaining growth and quality during maturation.
- **Humidity:** Moderate humidity levels between 50 and 60 per cent prevent moisture-related issues during harvest.

18.3.8 Rose

Seedling Stage

- **Nutrient solution temperature:** Nutrient solution temperature of 18-22°C promotes root development without causing any stress.

- **Environmental temperature:** Ambient air temperature around 20-24°C during daytime and slightly cooler at night is quite favourable.
- **Light:** Provision of 16-18 hours of indirect bright light per day using fluorescent or LED grow lights is most suitable at seedling stage.
- **Humidity:** Maintaining high humidity levels of 70-80 per cent prevent desiccation of delicate seedlings.

Vegetative Growth Stage

- **Nutrient solution temperature:** Maintaining nutrient solution temperature between 18 and 22°C ensures optimal nutrient uptake and root health.
- **Environmental temperature:** Ambient air temperature around 20-24°C during daytime and slightly cooler at night are quite favourable for plant growth at vegetative phase.
- **Light:** Provision of 12-14 hours per day of direct bright sunlight or artificial light ensures a light intensity of 10,000-20,000 lux.
- **Humidity:** Maintaining moderate humidity levels of 60-70 per cent prevents leaf wilting and dehydration.

Flowering Stage

- **Nutrient solution temperature:** Maintaining temperature of nutrient solution at 18-22°C supports nutrient uptake and flower development.
- **Environmental temperature:** An ambient air temperature around 18-22°C during daytime and slightly cooler at night supports flowering.
- **Light:** Provilion of 12 hours per day of direct bright sunlight or artificial lighting ensures a light intensity of 10,000-20,000 lux.
- **Humidity:** Maintaining moderate humidity levels around 50-60 per cent prevent fungal diseases and ensure flower quality.

Pre-harvest Stage

- **Nutrient solution temperature:** Nutrient solution temperature between 18 to 22°C supports nutrient uptake and maintains plant health.
- **Environmental temperature:** Ambient air temperatures around 18-22°C during daytime and slightly cooler at night support pre-harvest stage.
- **Light:** Provision of 10-12 hours per day of indirect bright sunlight or artificial light supports this stage.
- **Humidity:** Maintaining a moderate humidity levels around 50-60 per cent prevent mould and maintain flower quality.

18.3.9 Chrysanthemums

For hydroponically grown chrysanthemums, optimizing environmental conditions such as nutrient solution temperature, air temperature, light and humidity are crucial for achieving healthy growth and maximizing yield. Given below are recommended conditions for different growth stages:

Seedling Stage

- ✰ **Nutrient solution temperature:** Maintaining temperature of nutrient solution between 18 to 22°C promotes healthy root development.
- ✰ **Air temperature:** Air temperature between 20 to 25°C during daytime and slightly cooler at night supports seedling growth in chrysanthemum.
- ✰ **Light requirements:** Seedlings require 16 to 18 hours of lighting per day through fluorescent or LED grow lights with a spectrum suitable for vegetative growth.
- ✰ **Humidity:** High humidity levels of around 70 to 80 per cent facilitate preventing seedlings from drying out.

Vegetative Growth Stage

- ✰ **Nutrient solution temperature:** Nutrient solution temperature of 18 to 22°C supports nutrient uptake and plant growth.
- ✰ **Air temperature:** Air temperature regime of 20 to 25°C during daytime and slightly cooler at night encourages robust vegetative growth.
- ✰ **Light requirements:** It requires 14 to 16 hours of light per day using high-intensity discharge (HID) lights or LED grow lights with a spectrum optimized for vegetative growth.
- ✰ **Humidity:** Maintaining a humidity regime of 50 to 70 per cent reduces the risk of fungal diseases and ensures efficient transpiration.

Flowering Stage

- ✰ **Nutrient solution temperature:** It should be within the range of 18 to 22°C to support nutrient uptake and flower development.
- ✰ **Air temperature:** It should be a little cooler, *i.e.* 18-22°C during daytime and 15-18° at night to prevent heat stress and to promote flowering.
- ✰ **Light requirements:** Per day it should be 12 to 14 hours managed through HPS or LED grow lights with a spectrum optimized for flowering.
- ✰ **Humidity:** To prevent bud rot and maintain flower quality, it should be maintained between 50 to 60 per cent.

Late Flowering and Pre-harvest Stage

- **Nutrient solution temperature:** Temperature of nutrient solution should be maintained between 18 to 22°C to support nutrient uptake and proper development of flowers.
- **Air temperature:** Air temperature should be little cooler, *i.e.* 18 to 22°C during daytime and 15 to 18°C at night to promote proper floral development.
- **Light requirements:** Lighting through HPS or LED grow lights for 10 to 12 hours per day will maintain flower quality.
- **Humidity:** Bare minimum RH levels of 40 to 50 per cent will prevent fungal diseases, *vis-a-vis* flower quality.

18.4 Additional Considerations

- Sudden change in environmental conditions should be avoided and appropriate sensors should be installed for proper monitoring of temperature, light and humidity. To monitor humidity levels, hygrometer should be used.
- Under enclosed growing environments, there should be adequate ventilation *cum* airflow to prevent heat build up and for reduction of diseases, *vis-a-vis* proper CO_2 levels should be maintained for photosynthesis.
- There should be sufficient support to the plants to prevent lodging or breakage and to prevent them from becoming tangled or stressed.
- For optimal nutrient uptake, pH and EC of the solution should regularly be monitored.
- Regularly the plants should be inspected for signs of nutrient deficiencies, pests and diseases.
- For adequate light and airflow penetration in the planting proper spacing among the plants should be ensured.

Growers can use shade cloth and other shading methods to protect plants from excessive sunlight and heat during warmer months. It would be better to have a controlled environment with supplemental heating or cooling *cum* RH, *vis-à-vis* supplemental CO_2 levels.

Chapter 19

Organic Hydroponics

A traditional hydroponic system typically employs synthetic or inorganic nutrient solutions to provide essential elements to plants, sometimes incorporating organic compounds like iron chelates. These solutions are composed of a mixture of salts and minerals dissolved in water, ensuring a well-rounded diet for the plants. Inorganic hydroponic systems are commonly preferred for their precise formulation of nutrient solutions, which deliver consistent and readily available nutrients, thereby streamlining management.

However, organic hydroponics is gaining traction due to concerns regarding environmental impact, along with the growing demand for organic produce and sustainable agricultural practices. Organic hydroponics combines the principles of organic farming with hydroponic techniques, placing emphasis on the use of natural and sustainable inputs. In contrast to conventional hydroponics, which often relies on synthetic fertilizers and chemicals, organic hydroponic systems utilize organic nutrients, organic substrates, and biological controls to foster plant growth. Organic nutrient solutions are sourced from plant and animal materials or naturally mined substances (Atkin and Nichols, 2004).

Both organic and inorganic hydroponic systems possess distinct advantages and disadvantages. Organic systems are often lauded for their environmental friendliness and potential to yield crops with enhanced nutritional value. However, they can pose challenges in management and may lack precision in nutrient delivery. In contrast, inorganic systems provide precise control over nutrient levels, mitigating the risk of imbalances and potentially resulting in higher yields and faster growth rates. Yet, they may rely on synthetic chemicals and may not align with sustainability goals. Moreover, they necessitate vigilant monitoring and management to maintain optimal nutrient levels for plant health and productivity.

The selection between organic and inorganic hydroponic systems is frequently influenced by personal preferences, environmental considerations, and the specific needs of the cultivated plants.

Aquaponics exemplifies an organic system that integrates hydroponics with aquaculture, utilizing fish waste to nourish plants. Employing organic fish feed and natural filtration methods sustains water quality and promotes plant growth. Nevertheless, meticulous management is essential in organic hydroponic systems to regulate nutrient levels and mitigate organic matter buildup, preventing potential system clogs and other issues.

By integrating organic nutrient sources and sustainable cultivation techniques, growers can adopt organic hydroponics, producing high-quality, nutrient-rich crops while minimizing environmental impact and fostering ecological balance. This approach offers a viable and sustainable solution to meet the increasing demand for organic produce while supporting environment-friendly agriculture.

While organic fertilizers are commonly used in traditional soil-based agriculture, their effectiveness can be more challenging to leverage in hydroponic systems. Unlike soil, which harbors beneficial microbes aiding in the breakdown of organic fertilizers into accessible nutrients for plants, soilless hydroponic substrates lack such microbial support. Consequently, organic fertilizers applied in hydroponics may not undergo efficient breakdown, potentially hindering plant growth rather than enhancing it (Kawamura-Aoyama *et al.,* 2014). Moreover, organic hydroponics necessitates extensive knowledge, experience and investment compared to conventional agricultural practices (Phibunwatthanawong and Riddech, 2019).

19.1 Organic Certification

Organic certification for hydroponic producers entails strict adherence to organic standards, which can vary depending on the region and the certifying body involved. While traditional organic farming methods primarily focus on soil-based cultivation, hydroponic systems can also meet these rigorous criteria, albeit with some unique considerations. This process necessitates meticulous record-keeping, consistent adherence to organic farming practices, and regular inspections to ensure compliance.

To attain organic certification in hydroponics, growers must adhere to several key principles:

i) **Organic inputs**: Organic hydroponic operations must utilize organic-approved inputs, including nutrient solutions, growing media, seeds, and pest control products. These inputs must be derived from natural sources and free from synthetic chemicals, genetically modified organisms (GMOs), and irradiation.

ii) **Soilless growing media**: While soil is not used in hydroponic systems, organic-certified growing media must meet specific criteria to ensure they are free from prohibited substances and support plant health. Common organic-approved media include coconut coir, peat moss, perlite, and vermiculite.

iii) **Nutrient management**: Organic hydroponic nutrient solutions must be derived from organic sources such as compost teas, fish emulsions, seaweed extracts, and other natural amendments. These nutrient sources provide essential elements for plant growth without synthetic additives or chemical fertilizers.

iv) **Pest and disease management**: Organic hydroponic growers employ natural pest control methods such as beneficial insects, biological predators, botanical extracts, and cultural practices to manage pests and diseases. Synthetic pesticides, herbicides, and fungicides are strictly prohibited in organic production.

v) **Record-keeping and documentation**: Organic certification requires meticulous record-keeping to track all inputs, practices and activities related to hydroponic production. Detailed documentation ensures transparency, traceability, and compliance with organic standards during inspections and audits.

vi) **Certification process**: Growers seeking organic certification must undergo a thorough certification process conducted by accredited certifying agencies. This process involves completing an application, submitting documentation, undergoing on-site inspections and demonstrating compliance with organic standards.

vii) **Compliance with regulations**: Organic hydroponic operations must comply with national and international regulations governing organic agriculture such as the USDA National Organic Program (NOP) in the United States, the European Union Organic Regulation, and other regional standards.

viii) **Ongoing compliance and renewal**: Once certified, organic hydroponic growers must maintain ongoing compliance with organic standards and undergo regular inspections to retain certification status. Certification renewal typically occurs annually, requiring growers to demonstrate continued adherence to organic practices.

Overall, organic certification in hydroponics demonstrates a commitment to sustainable and environment-friendly agricultural practices, ensuring the production of organic crops that meet stringent quality and integrity standards. By following established organic protocols and maintaining transparency throughout the certification process, hydroponic growers can contribute to the growth and credibility of the organic farming movement. However, the certification of hydroponic systems as organic has sparked considerable debate within the agriculture industry.

In the United States, hydroponic production is the only exception where it can be certified as "organic." However, there are no explicit guidelines for organic hydroponics, leading to ambiguity and differing interpretations among stakeholders. This lack of clear regulations has raised concerns about the integrity of organic certification in hydroponics.

Conversely, in Canada, hydroponics and aeroponics are not eligible for organic certification due to their deviation from traditional soil-based farming methods. Canadian organic standards mandate that crops must be grown in soil to qualify for organic certification, excluding hydroponic systems from organic labeling and marketing.

Similarly, recent revisions to the European Union's organic regulations have stipulated that crops grown hydroponically cannot be certified as organic. This decision reflects the EU's commitment to soil-based agriculture and its skepticism towards soilless growing methods like hydroponics.

The exclusion of hydroponic systems from organic certification in various regions underscores the ongoing debate surrounding the compatibility of soilless growing methods with organic farming principles. While hydroponics offers advantages in resource efficiency and crop productivity, questions persist about its alignment with traditional notions of organic agriculture, particularly concerning soil health, biodiversity and sustainability. As the demand for organic produce continues to rise, reconciling these divergent perspectives on hydroponic certification remains a complex and contentious issue within the organic farming community.

19.2 Strategies for Organic Hydroponic Cultivation

Here are some strategies for organic hydroponic cultivation:

i) **Organic nutrient sources:** Instead of synthetic fertilizers, organic hydroponic systems use nutrient solutions derived from organic sources such as compost teas, seaweed extracts, fish emulsion and other natural inputs. Organic nutrient solutions can be prepared by composting, fermenting or steeping organic matter to extract nutrients. These organic nutrients provide essential elements and micronutrients to plants in a form that is readily available for uptake.

ii) **Solubility**: Organic nutrients may not always dissolve as readily as synthetic ones which could lead to clogging or buildup in the system. Ensuring that the organic nutrients are fully dissolved before adding them to the solution can help mitigate this issue.

iii) **Nutrient balance**: Organic nutrient solutions may not always provide exact balance of nutrients required by plants. It is essential to use a well-formulated organic nutrient solution or supplement it with additional nutrients as needed to ensure that plants receive all the essential elements for healthy growth.

iv) **Companion planting and crop rotation:** Companion planting and crop rotation can be implemented to enhance nutrient uptake, improve soil health (in soilless media) and prevent pest and disease issues. Certain plants have beneficial relationships when grown together such as nitrogen-fixing legumes.

v) **Beneficial microorganisms:** Mycorrhizal fungi and rhizobacteria have to be introduced into the hydroponic system to enhance nutrient availability, root health and overall plant vigour. Microbial inoculants can be applied to the root zone or incorporated into organic nutrient solutions to promote symbiotic relationships with plants.

vi) **Microbial activity, pH and EC control**: Maintaining optimal pH and EC levels is crucial for nutrient uptake in hydroponic plants. Organic nutrients can promote microbial activity in the nutrient solution which may lead to changes in pH or nutrient availability over time. Regular monitoring and adjustments may be necessary to maintain optimal conditions for plant growth. Organic acids and natural pH adjusters are used to regulate pH levels in organic hydroponic systems.

vii) **Water management**: Efficient water management techniques including recirculating systems and drip irrigation help conserve water and minimize wastage in organic hydroponic setups.

viii) **Organic pest and disease management:** IPM strategies must be employed using organic and natural methods to control pests and diseases. Physical barriers, beneficial insects, botanical insecticides, *e.g.*, neem oil and pyrethrin, and organic fungicides or herbicides are utilized to manage pest and disease outbreaks without synthetic chemicals.

ix) **Soilless media and substrates:** Growers must choose organic-certified or natural soilless media and substrates such as coconut coir, peat moss, perlite, vermiculite or organic compost blends for hydroponic cultivation. Synthetic materials, chemical additives and genetically modified organisms (GMOs) must be avoided in hydroponic substrates.

x) **Organic-certified seed and plant selection:** Organic-certified seeds, seedlings or starter plants should be obtained from reputable organic suppliers to ensure the integrity of organic production from the beginning. Plant varieties that are well-suited for hydroponic cultivation and are compatible with organic growing practices should be selected.

xi) **Sustainability:** Organic hydroponic systems often emphasize sustainability by incorporating practices such as water conservation, energy efficiency and minimizing waste generation.

Organic nutrients in hydroponics offer several benefits such as improved substrate fertility, enhanced nutrient uptake and reduced environmental impact compared to synthetic fertilizers. However, organic hydroponic systems may require more frequent monitoring and maintenance to prevent nutrient imbalances and ensure optimal plant growth. Organic nutrients may be more expensive and less readily available compared to synthetic fertilizers.

19.3 Advantages over Conventional Hydroponics

Organic hydroponics represents a sustainable and environment-friendly approach to food production offering numerous benefits for growers, consumers and the planet alike. By combining the best practices of organic farming with efficiency of hydroponic technology, organic hydroponic systems pave the way for a healthier and more sustainable food system.

Some key benefits include:

i) **Environmental sustainability**: Organic hydroponic systems prioritize sustainable practices such as water conservation, reduced chemical inputs and minimal environmental impact. By utilizing organic nutrients and natural pest control methods these systems minimize pollution and promote ecosystem health.

ii) **Healthier produce**: Organic hydroponic crops are cultivated without synthetic pesticides, herbicides or fertilizers so this results in produce that is free from chemical residues. This makes organic hydroponic fruits and vegetables healthier and safer for consumption.

iii) **Enhanced flavour and nutrition**: By prioritizing soil health and organic nutrient sources organic hydroponic systems can produce crops with superior flavour *cum* aroma and nutritional content compared to conventionally grown counterparts. The absence of synthetic chemicals allows plants to express their full genetic potential which may result in tastier and more nutrient-rich produce.

iv) **Reduced dependency on external inputs**: Organic hydroponic systems rely on natural and sustainable inputs such as compost teas, organic fertilizers and biological pest control agents. This reduces dependency on synthetic inputs and fossil fuels thus contributing to long-term agricultural resilience and self-sufficiency.

v) **Ecological benefits**: Methods of preparing organic nutrient solutions offer additional ecological benefits by requiring less energy compared to production of inorganic fertilizers. These also contribute to waste diversion from landfills where nutrients are not efficiently cycled so waste decomposition generates greenhouse gases (Kawamura-Aoyama *et al.*, 2014).

19.4 Challenges Compared to Conventional Hydroponics

Organic fertilizers can either supplement or fully substitute the inorganic compounds commonly found in traditional hydroponic solutions (Sholto Douglas,

1985; Benton, 2004). While organic hydroponics offers numerous benefits, it also presents several challenges compared to conventional hydroponic methods. Some of these challenges include:

i) **Variable nutritional composition:** Organic fertilizers exhibit high variability in nutritional compositions encompassing minerals, *vis-a-vis* various organic and inorganic species. Similar materials can vary significantly based on their sources as quality of manure fluctuates depending on diet of animals.

ii) **Lower crop yield:** Organic hydroponic systems often yield lower crop yields compared to conventional hydroponics as through a study when lettuce was grown in organic solutions exhibited slower growth rates and yielded less compared to that grown in conventional solutions (Atkin and Nichols, 2004).

iii) **Organic nutrient management:** Organic hydroponic systems rely on organic nutrient sources such as compost teas, fish emulsions and seaweed extracts. Managing nutrient levels and ensuring an adequate supply without synthetic additives can be complex so necessitating careful monitoring and adjustment. Biochemical degradation and conversion processes of organic materials make their mineral ingredients available to plants.

iv) **Particulate organic fertilizers:** Organic fertilizers may clog substrates or other growing equipment. Sieving or milling the organic materials to fine particles is therefore necessary.

v) **Disease transmission:** Organic fertilizers sourced from animal byproducts raise concerns about disease transmission, especially for plants grown for human consumption or animal forage.

vi) **Pest and disease management:** Organic hydroponic systems prioritize natural pest control methods such as biological predators, botanical extracts and cultural practices. However, organic pest management can be less effective than chemical pesticides thus necessitating more frequent monitoring and intervention.

vii) **Foul odours:** Some organic materials such as manures and offal can emit foul odours under anaerobic conditions.

viii) **Oxygen demand:** Many organic molecules such as sugars demand additional oxygen during aerobic degradation ~ essential for cellular respiration in plant roots.

ix) **Non-essential organic compounds:** Certain organic compounds like sugars and vitamins are not necessary for normal plant nutrition (Murashige and Skoog, 1962).

x) **Limited availability of organic inputs:** Organic hydroponic growers may face challenges sourcing certified organic inputs such as growing media, nutrients and pest control products so impacting feasibility and profitability.

xi) **Soil health and microbial diversity:** Organic hydroponic systems aim to mimic soil-based conditions so promoting soil health and microbial diversity in the root zone which requires specialized techniques and additives.

xii) **Risk of contamination:** Organic hydroponic systems are vulnerable to contamination from external sources so necessitating stringent hygiene protocols to safeguard plant health and food safety.

xiii) **Balancing nutrient levels:** Organic nutrient sources may vary in composition and availability thus making it challenging to achieve optimal nutrient balance and prevent deficiencies or excesses.

xiv) **Large pH fluctuations:** Organic hydroponic systems are prone to a large pH fluctuations with limited options to balance them (Williams and Nelson, 2016).

xv) **Growing medium selection:** Choosing an appropriate organic-approved growing medium that provides adequate support, aeration and water retention can be challenging.

Despite these challenges, organic fertilizers can successfully be used in hydroponics (Sholto Douglas, 1985; Benton, 2004). It is gaining popularity as growers seek sustainable alternatives to conventional agriculture. Overcoming these challenges requires innovation, research and collaboration within the organic farming community to develop effective solutions and best practices for organic hydroponic production.

19.5 Organic Nutrient Solution

Preparing an organic nutrient solution involves using natural and organic ingredients to provide essential nutrients to the plants. There are various liquid organic fertilizers available on the markets that are specifically formulated for hydroponic systems. These fertilizers typically contain a balanced blend of nutrients derived from natural sources such as plant extracts, animal manures and mineral deposits.

When assessing the efficacy of organic fertilizers, it is crucial to recognize significant variability in effectiveness among different types of fertilizers. Beyond variations inherent to fertilizer types, the mineral composition can fluctuate widely between batches due to the organic origins of the sources and the diverse nutritional requirements of different crops. Despite these challenges, there exists a range of promising organic fertilizer sources worth exploring. Here is a basic recipe to get started:

19.5.1 Sources of Organic Ingredients for Nutrient Solution

i) **Vermicompost leachate**: A potent fertilizer termed vermicompost leachate or worm compost tea can be crafted from high-grade food and animal waste. Its liquid form enables effortless integration into hydroponic setups (Kawamura-Aoyama *et al.*, 2014). Nutrient composition of vermicompost is subject to fluctuations influenced by variables such as feedstock variety, composting conditions and worm species. Thus the nutrient profile may differ based on production circumstances and feedstock utilized. For practical applications this requires laboratory assessments to ascertain precise nutrient levels.

ii) **Treated food waste**: A pioneering study carried out in Japan unveiled compelling findings by harnessing treated food waste as an innovative fertilizer solution. This groundbreaking method seamlessly integrated a fish-based soluble fertilizer and corn steep liquor derived from surplus resources in food production and this yielded remarkable outcomes (Kawamura-Aoyama *et al.*, 2014).

iii) **Organic molasses**: Molasses serve as a rich carbohydrate source ~ nurturing beneficial microorganisms within the soil or growing medium thereby enhancing the capacity of plants for nutrient absorption. In a pioneering study conducted in Thailand, researchers delved into a concoction comprising waste molasses, distillery slop and sugarcane leaves. Through meticulous experimentation involving diverse ratios of these constituents, a precise blend was pinpointed ~ showcasing growth rates akin to those facilitated by conventional chemical fertilizers (Phibunwatthanawong and Riddech, 2019).

iv) **Compost tea**: Organic compost tea represents a potent liquid fertilizer crafted from decomposed organic materials like composted manure, plant residue and assorted organic substances. This nutrient-rich elixir boasts a diverse array of essential nutrients and beneficial microorganisms crucial for fostering robust plant development. To concoct compost tea compost is immersed in water to initiate the extraction process for vital microorganisms and nutrients. Gardeners can produce their own compost tea by enclosing compost within a permeable pouch or container, submerging it in water for several days and subsequently filtering out the solids thus yielding a liquid supplement integral to nurturing plant vitality.

v) **Fish emulsion**: Derived from fish waste, fish emulsion stands as a natural fertilizer enriching plants with essential nitrogen, phosphorus, potassium and trace minerals. This organic nutrient source proves particularly beneficial for hydroponic cultivation.

vi) **Kelp meal ~ liquid seaweed extract**: An array of commercially available products derived from diverse species of seaweed offer a rich blend of nutrients such as nitrogen, potassium, and essential micronutrients. These seaweed extracts contain growth-promoting hormones and advantageous compounds that foster enhanced plant growth and development. Growers are encouraged to seek out products sourced from sustainably harvested seaweed for optimal environmental stewardship.

vii) **Bloodmeal**: It is a desiccated and neutral blood-derived powder which serves the dual roles ~ as an enriching high-nitrogen organic fertilizer and a protein-rich feed for animals. Renowned for its unparalleled non-synthetic nitrogen content this resource typically originates from slaughterhouse by-products of cattle or hogs.

viii) **Bone ashes**: It is a white substance derived from calcination of bones which exhibits a composition that fluctuates with the bone source. Typically represented by the formula $Ca_5(OH)(PO_4)3$, bone ash boasts a density of approximately 3.10g/ml and a melting point of 1670°C. Remarkably, even after calcination many bones retain their cellular structure.

ix) **Bonemeal**: It is derived from finely and coarsely ground animal bones and byproducts from slaughterhouses serving dual purposes in agriculture. Primarily recognized as a dietary supplement for monogastric livestock, it provides essential calcium and phosphorus in the form of hydroxyapatite. As a slow-release organic fertilizer, it enriches soil with phosphorus, calcium and a modest nitrogen content thus beneffiting plant growth and development.

x) **Hoof- or horn-meal**: The hoof ~ a distinctive feature found at the tip of an ungulate mammal's toe boasts a sturdy casing of thick and horny keratin. In contrast, a horn characterized by its permanent pointed projection on the head of diverse animals comprises of a keratinous sheath enveloping a core of living bone.

xi) **Fishmeal**: It is a commercially produced substance derived from a combination of whole wild-caught fish, bycatch and fish by-products.

xii) **Liquid compost of poultry and cattle manure**: Refined liquid compost derived from sheep, goat, horse or cow waste ~ meticulously sifted to eliminate solids and rigorously screened for pathogens.

xiii) **Guano**: It is derived from the Quechua term 'wanu', guano is a prized organic fertilizer harvested from the accumulated excrement of seabirds or bats. Renowned for its potency, guano boasts high levels of nitrogen, phosphate and potassium ~ essential nutrients crucial for optimal plant development. Its commercial availability and natural composition make it a highly effective option for enhancing soil fertility and promoting robust plant growth.

xiv) **Rock phosphate**: It is derived from natural deposits. Rock phosphate serves as a rich reservoir of phosphorus along with assorted trace minerals. When integrated into hydroponic setups it offers a gradual and sustained release of phosphorus thus contributing to sustained plant nourishment.

xv) **Other sources:** Wood ashes, cottonseed ashes, cottonseed meal, dried locust or grasshopper, leather waste milled to a fine dust, *etc.*

xvi) **Organic apple cider vinegar** (optional): Supplementing the nutrient solution with a small quantity of apple cider vinegar can effectively regulate its pH level when needed.

19.5.1.1 Nutritional Contents of above Organic Sources

The table below showcases a selection of organic materials along with their average nutritional contents, represented as percentages of dried mass (Sholto Douglas, 1985; Benton, 2004):

Organic Material	*Nitrogen (N)*	*Phosphorus pentoxide (P_2O_5)*	*Potassium Oxide (K_2O)*	*Calcium Oxide (CaO)*	*Magnesium Oxide (MgO)*	*Sulpur Dioxide (SO_2)*
Vermi-compost	1 per cent - 2.5 per cent	0.5 per cent - 1.5 per cent	0.5 per cent - 1.5 per cent	1.0 per cent - 2.5 per cent	0.3 per cent - 0.8 per cent	0.1 per cent - 0.3 per cent
Bloodmeal	13.0 per cent	2.0 per cent	1.0 per cent	0.5 per cent	–	–
Bone ashes	–	35.0 per cent	–	46.0 per cent	1.0 per cent	0.5 per cent
Bonemeal	4.0 per cent	22.5 per cent	–	33.0 per cent	0.5 per cent	0.5 per cent
Hoof/horn meal	14.0 per cent	1.0 per cent	–	2.5 per cent	–	2.0 per cent
Fishmeal	9.5 per cent	7.0 per cent	–	0.5 per cent	–	–
Wool waste	3.5 per cent	0.5 per cent	2.0 per cent	0.5 per cent	–	–
Wood ashes	–	2.0 per cent	5.0 per cent	33.0 per cent	3.5 per cent	1.0 per cent
Cottonseed ashes	–	5.5 per cent	27.0 per cent	9.5 per cent	5.0 per cent	2.5 per cent
Cottonseed meal	7.0 per cent	3.0 per cent	2.0 per cent	0.5 per cent	0.5 per cent	–

Organic Material	Nitrogen (N)	Phosphorus pentoxide (P_2O_5)	Potassium Oxide (K_2O)	Calcium Oxide (CaO)	Magnesium Oxide (MgO)	Sulpur Dioxide (SO_2)
Dried locust or grass-hopper	10.0 per cent	1.5 per cent	0.5 per cent	0.5 per cent	–	–
Leather waste milled to a fine dust (Benton, 2004)	5.5 per cent - 22 per cent	–	–	–	–	–
Kelp meal, liquid seaweed	1 per cent	–	12 per cent	–	–	–
Poultry manure (Sholto Douglas, 1985)	2 per cent - 5 per cent	2.5 per cent - 3 per cent	1.3 per cent - 3 per cent	4.0 per cent	1.0 per cent	2.0 per cent
Sheep manure (Sholto Douglas, 1985)	2.0 per cent	1.5 per cent	3.0 per cent	4.0 per cent	2.0 per cent	1.5 per cent
Goat manure (Sholto Douglas, 1985)	1.5 per cent	1.5 per cent	3.0 per cent	2.0 per cent	–	–
Horse manure (Sholto Douglas, 1985)	3 per cent - 6 per cent	1.5 per cent	2 per cent - 5 per cent	1.5 per cent	1.0 per cent	0.5 per cent
Cow manure (Sholto Douglas, 1985)	2.0 per cent	1.5 per cent	2.0 per cent	4.0 per cent	1.1 per cent	0.5 per cent

Organic Material	*Nitrogen (N)*	*Phosphorus pentoxide (P_2O_5)*	*Potassium Oxide (K_2O)*	*Calcium Oxide (CaO)*	*Magnesium Oxide (MgO)*	*Sulpur Dioxide (SO_2)*
Bat guano: high in micro-nutrients (Benton, 2004).	8.0 per cent	40 per cent	29 per cent	Trace	Trace	Trace
Bird guano: high in micro-nutrients (Benton, 2004)	13 per cent	8 per cent	20 per cent	Trace	Trace	Trace

19.5.1.2 Organically Sourced Micronutrients

Organic fertilizers can serve as sources of micronutrients too such as composted pine bark which is naturally rich in manganese and is occasionally incorporated into hydroponic solutions to meet this mineral requirement (Benton, 2004). Adhering to the standards set by National Organic Programs, growers may include pulverized but unrefined minerals such as gypsum, calcite and glauconite to fulfil the nutritional needs of plants.

19.5.2 Key Microorganisms in Organic Hydroponic Systems

In organic hydroponics, microorganisms play a crucial role in nutrient cycling and maintaining a healthy root environment for plant growth. Unlike conventional hydroponic systems that often rely on synthetic fertilizers, organic hydroponic systems utilize natural and organic inputs to support plant growth. Here are some key microorganisms commonly found in organic hydroponic systems:

i) **Nitrogen-fixing and other beneficial bacteria**: These microbes (*Bacillus, Pseudomonas* and *Rhizobium*) break down organic matter into forms that plants can readily absorb. These also help in nitrogen fixation thus making atmospheric nitrogen available to plants. Apart from these, *Azotobacter* and *Azospirillum* also convert atmospheric nitrogen into usable forms through nodulation.

ii) **Mycorrhizal fungi**: Mycorrhizae form symbiotic relationships with plant roots, extending the root system's reaching and aiding in nutrient uptake, especially phosphorus and micronutrients. These also contribute to soil structure and water retention. Species such as *Glomus* spp. and *Rhizophagus* spp. are commonly used in organic hydroponic systems.

iii) **Actinomycetes**: These are a type of beneficial bacteria that decompose complex organic compounds and in turn releasing nutrients for plant uptake. These also produce antibiotics that can suppress harmful pathogens. *Streptomyces* is a well-known genus of actinomycetes.

iv) **Protozoa and nematodes**: These microscopic organisms help in breaking down organic matter and releasing nutrients in a form accessible to plants. These also play a role in controlling populations of harmful pathogens. Predatory nematodes can feed on harmful nematodes that may attack plant roots.

v) **Composting microbes**: In systems where compost teas or organic compost extracts are used, a diverse range of microbial species including bacteria, fungi and protozoa contribute to nutrient cycling and disease suppression.

Maintaining a balanced and diverse microbial community is essential in organic hydroponic systems to support plant health and productivity. This can be achieved through use of organic amendments, compost teas and microbial inoculants as well as proper management practices that encourage the proliferation of beneficial microorganisms while minimizing conditions favouring harmful pathogens.

19.5.3 Introducing Microorganisms into Organic Hydroponic Systems

Introducing microorganisms into organic hydroponic systems can help promote plant growth, improve nutrient uptake and prevent disease. Here are some ways to introduce them:

i) **Composting microbes**: Compost tea is made by steeping compost in water and aerating it to encourage the growth of beneficial microorganisms. Once the tea is ready, it can be added to the hydroponic system as a nutrient-rich solution.

ii) **Microbial inoculants**: These are commercially available products containing beneficial microorganisms such as bacteria and fungi. These can be added directly to the hydroponic nutrient solution or applied to the growing medium.

iii) **Mycorrhizal fungi**: Mycorrhizae form symbiotic relationships with plant roots, extend the root system's reach and aid in nutrient uptake, especially phosphorus and micronutrients. These also contribute to soil structure and water retention. *Glomus* spp. and *Rhizophagus* spp. are commonly used in organic hydroponic systems.

iv) **Vermicompost**: Vermicompost or worm casting is rich in beneficial microorganisms. Addition of vermicompost to hydroponic system can help introduce these microorganisms and improve nutrient cycling.

v) **Biological additives**: Some hydroponic nutrient solutions contain biological additives that promote the growth of beneficial microorganisms. These additives can include beneficial bacteria, fungi and other microbes.

vi) **Rhizosphere colonization**: Introducing plants into the hydroponic system will naturally bring beneficial microorganisms into the root zone. These microorganisms through colonizing the rhizosphere (area around the plant roots) provide various benefits to plants.

When introducing microorganisms into a hydroponic system, it is essential to monitor the system's pH and nutrient levels to ensure optimal conditions for both ~ the plants as well as microorganisms. It is a good idea to start with a small amount of microorganisms and gradually increase the dosage to avoid any potential adverse effects.

19.5.4 Bio-enzymes

The need of bio-enzymes in addition to organic nutrients in organic hydroponic system depends on several factors including the specific setup, the types of plants being grown and grower's overall goals. Bio-enzymes help in breaking down organic matter into simpler forms that are more readily absorbed by plants. These enhance nutrient availability and microbial activity in the growing medium. If organic materials or compost is being used in the hydroponic system, bio-enzymes can aid in decomposing these materials and releasing nutrients.

While bio-enzymes assist in nutrient release, these may not provide all the essential nutrients which plants require. Depending solely on bio-enzymes may not supply sufficient levels of nitrogen, phosphorus, potassium and micronutrients required for optimal plant growth. Supplementing the hydroponic system with additional organic nutrients ensures that plants receive a balanced diet. Organic nutrients like compost tea, fish emulsion, seaweed extract and others provide a broader spectrum of essential nutrients that bio-enzymes alone may not supply in adequate quantities. It is advised to consider the specific nutrient needs of the plants being grown. Some plants may require higher levels of certain nutrients than others. Nutrient supplementation should be tailored accordingly to meet the requirements of the crops.

Maintaining a balance between bio-enzymes and other organic nutrients is essential. Too much reliance on one component over the other may lead to imbalances in the system. Regular monitoring of nutrient levels, pH and plant health can help adjusting nutrient regimen as needed.

In summary, while bio-enzymes can play a beneficial role in organic hydroponic systems by enhancing nutrient availability, these are typically used in conjunction with other organic nutrients to provide a comprehensive nutrient supply for optimal plant growth. Experimentation and observation of plant response can help determining most effective nutrient regimen for specific setup and crops.

19.5.4.1 Bio-enzyme Preparation

Bio enzymes can be beneficial for organic hydroponic systems as these help breaking down organic matter into simpler forms that are more readily available to plants. Here's a general guide on how to prepare bio-enzymes for use in organic hydroponic nutrient solutions:

i) **Ingredients:** Pesticide- or chemical-free fruit peels (lemons, oranges, *etc.*), brown sugar or jaggery and water.

ii) **Container selection:** A large airtight plastic container or glass jar with a tight-fitting lid is most suitable.

Fruit peels are chopped or blended into small pieces to increase surface area ~ allowing for better fermentation. These chopped fruit peels are placed into the container along with brown sugar or jaggery. Sugar acts as a food source for microorganisms that will ferment the mixture. Now water is poured into this container covering the fruit peels and sugar completely and leaving some space at the top as the mixture will expand during fermentation. The mixture is stirred well to ensure full dissolution of sugar. The container is sealed tightly with the lid and placed in a warm and dark place. Ideal temperature for fermentation is around 25-30°C. Mixture should be stirred daily to promote oxygenation and even distribution of microorganisms. Fermentation typically takes around 2-3 weeks but it can be allowed to ferment for longer for a stronger solution.

After completion of fermentation mixture is strained using a fine mesh strainer or cheesecloth to separate the liquid (bio-enzyme) from the solid residue. The strained bio-enzyme liquid should be stored in clean airtight containers away from direct sunlight. It can be stored for several months.

19.5.4.2 Bio-enzyme Application in Hydroponics

The bio-enzyme must be diluted with water before using it in hydroponic systems. The dilution ratio can vary though typically it is 1 part bio-enzyme to 10 parts water. Diluted bio-enzyme solution can be used as part of the nutrient solution regimen. It helps in breaking down organic matter and improves nutrient uptake by plants.

Regularly the pH and nutrient levels should be monitored in the hydroponic system so that this may maintain optimal conditions for plant growth.

19.6 Preparing Organic Nutrient Solution

A clean container should be filled with filtered or dechlorinated water to avoid harmful chemicals that can harm beneficial microorganisms in the nutrient solution. Growers can obtain organic compost tea, liquid seaweed extract, fish emulsion and molasses from hydroponic stores and add them to the water according to the manufacturer's instructions. These should be stirred or mixed well to ensure that all the ingredients are thoroughly combined. Growers can make the compost tea by steeping compost in water to extract beneficial microorganisms and nutrients;

aerating it encourages the growth of beneficial microorganisms. It would be better to place the compost in a breathable bag or container which should be soaked in water for several days and then after straining out the solids before using the liquid as part of the nutrient solution.

Amount of organic nutrient extracts to add to water in an organic hydroponic system can vary depending on several factors including the specific nutrient requirements of the plants, the growth stage, the concentration of nutrients in the extracts and overall nutrient balance in the system. As a general guideline, growers can start by adding around 5-10 per cent of compost tea to 10-litre water reservoir. Here are some general guidelines to help determining how much organic nutrient extract should be added:

i) **Follow manufacturer's recommendations**: If using commercially available organic nutrient extracts, the manufacturer will typically provide guidelines on how much to use for hydroponic applications. These recommendations should be followed as a starting point. Once the organic nutrient solution is prepared, it can be used to water the plants or to fill the reservoir of the hydroponic system according to regular watering schedule.

ii) **Start with dilute solutions**: If the grower is unsure about the strength of organic fertilizers, it is generally advisable to start with a more dilute solution (around 5 per cent initially) and gradually increasing the concentration based on plant response. This helps prevent over-fertilization which can damage plants.

iii) **Monitor plant response**: It is advised to observe the plants closely for signs of nutrient deficiencies or excesses such as leaf yellowing, leaf burn or stunted growth. The nutrient concentration must be adjusted accordingly based on plant response.

iv) **Consider growth stage**: Nutrient requirements of plants can vary depending on their growth stage as a plant may require higher levels of nitrogen during vegetative growth and higher levels of phosphorus and potassium during flowering or fruiting stages.

v) **Maintain balanced nutrition**: Growers must ensure that plants receive balanced level of nutrient (N, P, K, Ca, Mg, Fe, Mn, Zn, Cu, B and Mo) solution

vi) **pH and EC monitoring**: It is recommended to regularly monitor the pH and EC of nutrient solution using a pH testing kit and EC meter. The optimal pH for most plants is between 5.5 and 6.5. If necessary, the pH must be adjusted using organic apple cider vinegar, adding small amounts at a time until the desired pH level is reached. EC indicates the concentration of dissolved salts in the solution and can help assess nutrient levels.

vii) **Keeping records**: Keep detailed records of the nutrient concentrations which are being used and their effect on the health and growth of the plants. This will help fine-tuning the nutrient regimen over time for optimal results.

It is important to note that there is not a one-size-fits-all answer to how much organic nutrient extract to add to water in a hydroponic system. It may require some experimentation and adjustment based on the specific needs of your plants and the conditions of your growing environment.

19.6.1 General Guidelines for Dilution Ratio of Self Prepared Compost Tea

The amount of self-prepared compost tea to use in a 10 litre water solution for hydroponics can vary depending on factors such as strength of the compost tea and specific needs of plants. As a general guideline, one can start with a dilution ratio of around 1:10 to 1:20 (compost tea to water). Here's how it can be calculated:

i) **Determining dilution ratio**: While making 1:10 dilution ratio, this means that one part of compost tea should be added to every ten parts of water.

ii) **Calculating amount of compost tea needed**: Since there is a 10 liter (10,000ml) water solution, one needs to divide the total volume by the sum of the parts in the dilution ratio. Using 1:10 ratio as an example one would have to divide 10,000ml by 11 (1 part compost tea + 10 parts water) to get amount of compost tea needed.

So, for a 1:10 dilution ratio, one would use approximately 909 ml of self-prepared compost tea in 10 liters of water to prepare the organic nutrient solution. Adjust the amount based on the specific recommendations for compost tea and the needs of your plants.

19.7 pH Fluctuation and its Balancing

Maintaining proper pH balance is crucial in organic hydroponic systems just as it is in conventional hydroponics. However, achieving and managing pH levels in organic hydroponics can present some unique challenges due to the nature of organic inputs and the absence of synthetic pH stabilizers. Here are some key considerations for pH balance in organic hydroponic systems:

i) **Natural buffering capacity:** Organic nutrient solutions often have a natural buffering capacity due to presence of organic acids and other compounds. This can help stabilize pH levels to some extent but it may also make pH adjustments more gradual compared to conventional hydroponic solutions.

ii) **pH drift:** Organic hydroponic systems may experience pH drift over time as organic materials decompose and release acids or alkaline compounds into the nutrient solution. Regular monitoring and adjustment of pH levels are essential to prevent significant fluctuations that could affect plant health.

iii) **Organic pH adjusters:** While synthetic pH adjusters are not used in organic hydroponics, growers can employ organic alternatives such as citric acid, vinegar or certain types of organic acids to lower pH while potassium bicarbonate or calcium carbonate to raise the pH. These organic pH adjusters can help maintain pH within the optimal range for plant growth without compromising organic certification requirements.

iv) **Biological activity:** Microbial activity in organic hydroponic systems can influence pH levels as microorganisms metabolize organic matter and release acids or alkalis as by-products. Balancing microbial activity through proper nutrient management and aeration can help stabilize pH levels.

v) **Monitoring and adjustment:** Regular monitoring of pH levels using a reliable pH meter is essential in organic hydroponics. Growers should adjust pH as needed using organic pH adjusters and strive to keep pH within the optimal range for the specific crops being grown.

vi) **Nutrient interactions:** pH levels can affect nutrient availability and uptake by plants. Organic nutrients may exhibit different solubility and availability at various pH levels, so growers should consider how pH adjustments may impact nutrient uptake and plant health.

Overall, maintaining proper pH balance in organic hydroponic systems requires careful attention to the unique characteristics of organic inputs and the interplay between organic materials, microbial activity and nutrient dynamics. By monitoring pH levels regularly and making appropriate adjustments using organic pH adjusters, growers can create optimal growing conditions for healthy thriving plants in organic hydroponic setups.

Chapter 20

Aquaponics (Pisciponics)

Aquaponics revolutionizes food production by integrating aquaculture and hydroponics which results in a sustainable and environmental friendly system. In this symbiotic relationship, fish waste serves as a nutrient source for plants while plants act as natural filters ~ purifying the water for fish. Beneficial bacteria play a crucial role in converting fish waste toxins into nutrients usable by plants thus creating a self-sustaining cycle that requires minimal external inputs. As a result, aquaponic systems operate with little to no need for inorganic mineral nutrients or pesticides ~ making these almost entirely organic. This approach not only eliminates the need for synthetic fertilizers but also reduces the reliance on herbicides as plants are grown without soil. Whether for household use, community food production or commercial farming, aquaponics offers a scalable solution for cultivating fresh fish and vegetables sustainably ~ making it a versatile and efficient method of food production.

Aquaponics systems span a spectrum ranging from small-scale setups tailored for hobbyists or backyard food production to large-scale commercial operations focused on selling fish and plants. These systems operate as closed-loop facilities, meticulously treating and recycling water within the system. In aquaponics, a harmonious symbiosis unfolds among fish, microbes (beneficial bacteria) and plants each contributing to other's growth in a natural and organic manner. Water containing fish waste and leftover food undergoes circulation, propelled either continuously or on a predetermined schedule by a timer-controlled water pump ~ from the fish tank to the grow beds. These beds typically filled with expanded clay or similar growing media serve as the habitat for nitrifying bacteria which convert fish waste into essential nutrients for plant uptake. As the water flows through grow beds, plants filter out impurities aided by growing medium and possibly a mechanical filter if integrated into the system before cleansed water is returned *via* gravity or pump back to the fish tank. This cyclical process underscores the interdependence of plants and fish within the aquaponic ecosystem ~ the fish providing nutrients to the plants while, in turn, the plants ensure the clean water for inhabiting the fish. Aquaponics system relies solely on natural chemicals and fish food additives. Any additional substances introduced to enhance plant growth could jeopardize the delicate balance of the system, potentially harming the fish and disrupting the bacterial colony. Nonetheless, a few exceptions exist such as use of liquid seaweed, small doses of chelated iron and select minerals for pH adjustment which can be safely incorporated into the system without adverse effects on its inhabitants.

In aquaponic systems, microorganisms play a vital role in maintaining the ecological balance necessary for the system's success. Among these microorganisms are bacteria, fungi, phytoplankton, and zooplankton, each contributing to the system's functionality in unique ways. Bacteria, in particular are crucial as these break down fish waste and uneaten feed into essential nutrients for plant growth. These microscopic organisms drive essential processes that ensure the system's biological equilibrium, making the water habitable and conducive to plant and fish life. While invisible to the naked eye, bacteria constitute a significant portion of the system's biomass, utilizing nutrients like carbon, nitrogen, phosphorous and potassium to fuel their cellular processes. These derive energy by facilitating chemical reactions such as nitrification where ammonia is converted to nitrate. This process releases energy that sustains bacterial metabolism and indirectly benefits plant and fish health. However, like all aquatic organisms, bacteria rely on water's chemical composition for their well-being. Maintaining optimal water quality is thus essential to support a thriving bacterial community and by extension a healthy aquaponic system.

20.1 Aquaponics vs. Hydroponics

Comparison of Hydroponics and Aquaponics

Hydroponics	*Aquaponics*
Plants thrive in meticulously optimized conditions where factors such as light, temperature, humidity and nutrient levels are carefully regulated. To further enhance plant growth and resilience against pests and diseases, beneficial bacteria can be introduced into the growing environment. These beneficial microbes play a crucial role in supporting plant health by promoting nutrient uptake, suppressing harmful pathogens and enhancing overall plant vigour.	Aquaponics creates a harmonious ecosystem where plants, fish, bacteria and worms coexist in a finely tuned symbiotic relationship. This intricate system operates on interconnected components where each element influences and supports the others thus resulting in a balanced and thriving environment.

Hydroponics	*Aquaponics*
Food	
Hydroponic systems exclusively yield vegetarian produce; no fish are cultivated or harvested in these setups.	Aquaponics yields both plant-based produce and fish or other aquatic animals for consumption.
Nutrients/Fertilizers	
Managing nutrients in hydroponics poses a more intricate and demanding challenge compared to aquaponics.	In aquaponics, achieving equilibrium within the ecosystem is key to effective management.
Hydroponic nutrients are typically administered in the form of mineral salts.	Nourished by the organic waste generated by the fish which contains minimal salts.
In hydroponics, it's essential to maintain a balance in the nutrient solution which often requires regular replacement to ensure optimal plant health and growth.	Regular water maintenance in the fish tank primarily involves topping up the water level as needed to compensate for evaporation. Complete water changes are typically reserved for addressing severe or unexpected issues as the system is designed to maintain a balanced and stable aquatic environment without frequent water replacement.
Over time, the nutrient concentration in the solution may become imbalanced.	As time goes on, a well-maintained aquaponics system tends to thrive and improve continuously.
Each crop has unique nutritional requirements, necessitating specific compositions of nutrients for optimal growth and development.	Fish waste serves as a nutrient source, supplying nitrates for all crops.
Excessive fertilization can lead to rapid damage to the plants, resulting in burning.	In a fully cycled and well-established system, the risk of over-fertilization is minimized or eliminated altogether.
Optimal pH	
5.5 to 6.5	Maintaining optimal pH levels is crucial for the health and productivity of the aquaponics system, ensuring harmonious interaction between plants, fish and beneficial bacteria. Typically, the pH range is maintained between 6.4 and 7.0 to create an ideal environment for all components of the system to thrive.
Water temperature	
The temperature is carefully regulated to meet the specific needs of each plant variety and its growth stage.	Choose fish and plants that thrive in similar optimal temperature ranges.

Hydroponics	*Aquaponics*
EC	
Regular monitoring and adjustment of electrical conductivity levels is essential.	Only nitrates are to be measured. If nitrates are low (close to zero), more fish should be added to the system. If nitrates are high (above 50) more grow beds and/or plants should be added.
Use of pesticides	
Pesticides containing toxic chemicals can be employed for pest control.	The application of pesticides within the system poses a significant risk, potentially leading to the demise of both the fish population and the beneficial bacteria essential for the system's balance and functionality.
Diseases and pests	
Hydroponics operates within a controlled and sterile environment.	Aquaponics embraces all micro-organism as each of these play an important part in the growing process. Aquaponics tends to have less diseases and pest problem.
System start time	
The system can be initiated promptly upon completion of installation and the addition of plants and nutrients.	The setup process for an aquaponics system typically spans from a few days to about a month as it involves development of nitrifying bacteria through a crucial phase known as 'cycling'. During this period, beneficial nitrifying bacteria establish themselves within the system converting the ammonia present in fish waste into nitrates which serve as essential nutrients for the plants.
Flood and drain cycle	
The frequency of nutrient solution replenishment can vary, occurring either continuously or at intervals ranging from every 3 to 8 hours. This timing is contingent upon factors such as water retention capacity of growing media, nutrient solution concentration, specific plant requirements and seasonal variations.	To optimize the filtration process in aquaponics, it's recommended to flood the grow bed for 10-20 minutes every 45-60 minutes. This interval allows the grow bed to efficiently filter the fish waste, which is one of its primary functions within the aquaponic system.

Hydroponics	*Aquaponics*
Depth of grow beds	
The depth of grow bed should be sufficient to accommodate plant roots comfortably, taking into account the specific requirements of the system in use.	In aquaponics, the depth of growbed is crucial as it serves a dual function: providing a habitat for plants and acting as a bio-filter for fish waste. Unlike traditional hydroponic systems, aquaponic grow beds must be deep enough to effectively filter liquid waste and accommodate beneficial organisms like composting red worms and heterotrophic bacteria. Typically, aquaponic gardeners opt for grow beds that are around 30cm deep, filled with an inert media to support plant growth and facilitate the breakdown of fish waste. This depth ensures optimal conditions for both plant cultivation and waste filtration within the aquaponic system.
Change interval of water solution	
Regular replacement is necessary as the accumulation of salts and chemicals in the nutrient solution can affect its composition and efficacy over time.	Aquaponic systems require only occasional top-ups with fresh water.

20.2 Advantages of Aquaponic System

When considering aquaponics, there are numerous benefits to highlight:

1. **Self-sustaining:** Once established, aquaponic systems require minimal intervention. These self-circulate and self-sustain, eliminating the need for manual watering or fertilizing.
2. **Minimal effort and gardening problems:** Sealed greenhouse setups reduce gardening issues, making maintenance hassle-free.
3. **Low water requirement:** Aquaponics stands out as a highly water-efficient option, utilizing significantly less water compared to hydroponics and as little as one-tenth relative to aquaculture.
4. **Low power requirement:** Gravity flow setups require minimal power, contributing to reduced energy consumption and operational costs.
5. **Nutrient recycling:** The waste produced by fish serves as a natural fertilizer for the plants, reducing the need for synthetic fertilizers.
6. **Environmental-friendly:** Aquaponics eliminates the need for solution waste dumping or runoff thus promoting environmental sustainability. It also minimizes the use of chemicals, pesticides and fertilizers thus further reducing ecological impact.

7. **Diverse crop range:** Aquaponic systems can support a wide variety of crops including leafy greens, herbs, vegetables and even some fruits thus providing flexibility for growers.
8. **Nutrient-rich produce:** Aquaponically grown fish and vegetables are known for their superior taste and nutritional value.
9. **High yields:** Aquaponic systems can produce high yields of both fish and plants in a relatively small space, making them suitable for urban farming and areas with limited land availability.
10. **Year-round production:** With proper environmental control, aquaponic systems can operate year-round thus allowing for continuous harvests regardless of the season.
11. **Natural processes:** Beneficial organisms such as worms and bacteria along with hardy fish species and plants like duckweed work synergistically to create optimal growing conditions thus yielding abundant harvests at low cost.
12. **Educational opportunities:** Aquaponics serves as an excellent educational tool, teaching concepts of biology, ecology and sustainable agriculture to students of all ages.
13. **Community engagement:** Aquaponic systems can foster community engagement and social cohesion through shared learning experiences and local food production initiatives.

20.3 Considerations

Aquaponic systems offer numerous benefits along with several challenges that growers must navigate. One significant drawback is the initial setup cost which can be relatively high due to need for specialized equipment such as tanks, pumps and filtration systems. Moreover, the process of properly cycling a new system which can take up to a month in warm temperatures and even longer in colder climates is essential to establish the necessary beneficial bacteria and ensure water quality. Failure to cycle the system adequately can result in poor water quality, fish shock and even fish mortality.

Maintaining optimal conditions for all the three life forms ~ fish, plants and bacteria pose another challenge. Balancing fish and plant populations requires careful attention to prevent overstocking or overfeeding which can lead to nutrient imbalances and water quality issues. Reliance on electricity to power pumps and aeration systems increases operational costs and vulnerability to power outages. In the event of a power outage without backup systems, fish mortality can occur due to oxygen deprivation and ammonia buildup.

Furthermore, aquaponic systems demand diligent monitoring and management to prevent diseases in both plants and fish as well as to maintain optimal growing conditions. Growers must regularly monitor water quality parameters such as pH, oxygen levels, nitrate concentrations and temperature to ensure that these remain within ideal range for biological processes. pH levels must be carefully regulated to support nutrient uptake by plants and prevent stress on fish while oxygen levels need to be sufficient to support respiration and aerobic bacterial activity. Nitrate concentrations should be monitored to prevent toxicity to fish and promote healthy plant growth. Maintaining stable water temperatures is crucial to prevent stress and ensure optimal metabolic rates for both fish and plants. Regular testing and adjustments are necessary to address any fluctuations and maintain a stable aquatic environment conducive to the success of aquaponic system.

Finally, while aquaponics offers a sustainable approach to food production, it may not be suitable for all types of crops or climates thus limiting its applicability in certain regions or for specific agricultural needs.

20.4 Aquaponics Operation

Aquaponics systems operate on a dynamic input-process-output cycle wherein external inputs are processed by various components and life forms within the system to yield edible or ornamental plants and aquatic animals. Here is a breakdown of the key elements involved:

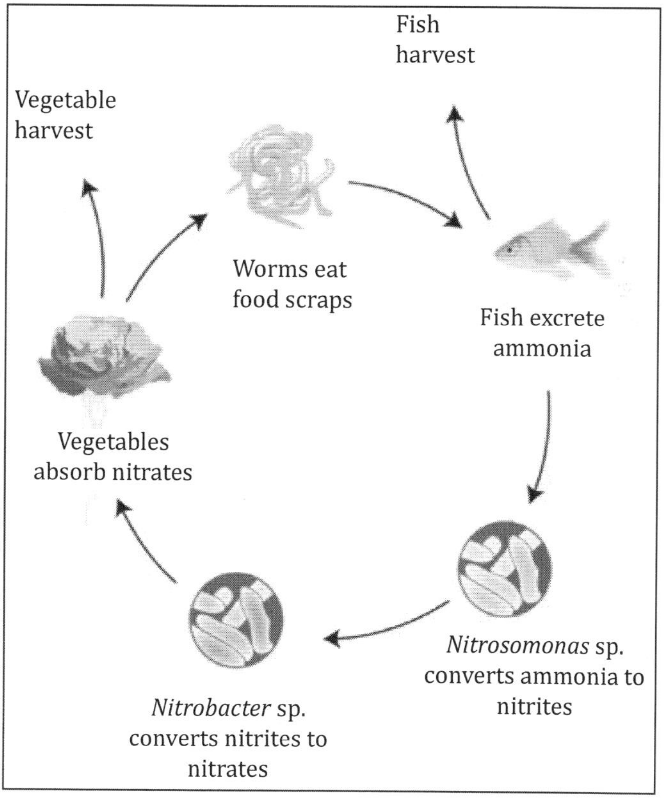

Aquaponics Cycle.

Inputs

- ☆ **Water**: It serves as the primary medium for nutrient transport and sustenance of aquatic life.
- ☆ **Fish feed**: It provides essential nutrients for aquatic animals and contributes to nutrient cycle.
- ☆ **Oxygen**: It is crucial for respiration of both fish and beneficial bacteria.
- ☆ **Additional minerals or nutrients**: Iron may be added to enhance plant growth.
- ☆ **pH adjustment solutions**: It is used to regulate acidity or alkalinity of water.

Process

- ☆ **Fish**: These produce waste continuously serving as the primary nutrient source for plants.
- ☆ **Filters**: These remove solid waste and uneaten fish feed from the water while also hosting beneficial bacteria essential for nutrient conversion.
- ☆ **Aeration/oxygenation**: It ensures adequate oxygen levels in the water ~ vital for fish respiration and bacterial activity.
- ☆ **Growing media**: These provide habitat for beneficial bacteria and worms as well as physical support for plant roots.
- ☆ **Aquarium heater**: It maintains optimal water temperature within the fish tank to support aquatic life.
- ☆ **Beneficial bacteria**: It is responsible for converting fish waste compounds (ammonia to nitrite and nitrite to nitrate) thus facilitating nutrient uptake by plants.
- ☆ **Plant roots**: These absorb nutrients from the water, effectively purifying it for fish and contributing to plant growth.

Output

- ☆ **Edible plants**: Vegetables, fruits or herbs cultivated hydroponically using the nutrient-rich water.
- ☆ **Aquatic species**: Such as fish or crustaceans which may also be harvested for consumption.

By utilizing natural processes efficiently and symbiotic relationships between aquatic organisms and plants, aquaponic systems offer a sustainable and integrated approach to food production.

20.5 Types of Aquaponics Systems

Aquaponic systems can be designed using solution culture (without media) or medium culture. The three primary aquaponic systems widely utilized are:

i) **Raft system (float/deep channel/deep flow system):** Plants in this system are cultivated on floating styrofoam boards within a separate tank from the fish. Though effective but raft systems are limited in the variety of plants these can support.

ii) **Nutrient film technique:** Plants are grown in elongated narrow channels through which a thin film of nutrient-rich water flows, providing essential nutrients directly to the roots.

iii) **Media-filled beds:** These beds consist of containers filled with a growing medium such as gravel, perlite or hydroton. Plant roots are embedded in the medium where a flood and drain sequence delivers nutrients. Maintaining a dry top layer of 5 to 6 cm prevents weed, fungus and harmful bacterial growth while also safeguarding worms from drowning. Ideal media bed depth ranges from 30 to 45cm or more depending on the plant type. The shell of the bed should extend 5cm above the media bed with an overflow pipe returning water to the sump tank in case of plumbing stoppage or directly to the fish tank if no sump tank is used.

While first two methods are prevalent in commercial-scale operations, media-filled beds are commonly found in backyard setups. Other aquaponic methods include static solution culture, wick systems, drip systems and more. Each method offers unique advantages and is chosen based on factors such as scale, plant variety and operational preferences.

20.6 Main Components and their Working

Aquaponic systems offer flexibility in design with various configurations tailored to specific needs. Despite this diversity these typically consist of three core subsystems ~ the rearing tank or fish tank, the bacteria colony housed in the grow bed, filters, gravels in rearing tank or in a separate mineralization unit and the hydroponics subsystem. Each subsystem plays a vital role in overall functionality of aquaponics system. 'Rearing tank' receives external inputs such as fish feed, water and oxygen. The output, comprising fish waste and uneaten feed, serves as the primary nutrient source for bacterial colony where beneficial bacteria convert ammonia from fish waste into nitrates. These nitrates are then absorbed by plants in the hydroponics subsystem thus purifying the water for fish effectively. The treated water is subsequently returned to the rearing tank to maintain optimal conditions for the aquatic species.

In addition to these primary subsystems, several other components or subsystems contribute to the effective operation of an aquaponics system. Common components include grow beds, filtration systems, water pumps, aeration equipment and monitoring devices. By integrating these components strategically, aquaponic systems can achieve efficient nutrient cycling, optimal plant growth and sustainable aquatic animal production.

20.6.1 Rearing Tank or Fish Tank

This crucial component serves as the habitat for raising and feeding the fish, forming the cornerstone of the aquaponics system. Fish tanks can vary in material, ranging from glass or fibreglass aquariums to plastic tubs, buckets or barrels provided these maintain water quality and ensure fish welfare. Optimal materials include fibreglass and polyethylene (PE) to minimize fish exposure to potential toxins though stainless steel or PE liners are also viable options. Fish tanks may adopt a single-tank configuration for hobby-scale systems or a multi-tank setup for commercial-scale operations. Buried fish tanks offer easier and more cost-effective water temperature control.

20.6.1.1 Water Movement and Optimization

Effective water movement within the fish tank is crucial for maintaining optimal water quality. Key parameters include flow rate, *i.e.* litres per hour and water turnover rate, *i.e.* volume exchanged per hour. Proper fluid dynamics optimization ensures efficient water circulation thus facilitating adequate gas exchange and nutrient distribution.

20.6.1.2 Size and Design

The size and design of fish tanks are influenced by various factors such as water flow, gas exchange, fish density and water quality requirements. Factors such as removal of solids and treatment of dissolved fish waste also influence tank design considerations. Achieving a high surface-to-volume ratio is essential to facilitate gas exchange, ensuring sufficient oxygen levels and carbon dioxide release within the water body. In the deep and narrow tanks where surface area is too low as compared to water volume, enough oxygen cannot enter the water body and enough carbon dioxide is not released from the water body. Therefore the surface area to volume ratios should be as high as possible.

20.6.1.3 Rearing Techniques

Aquaponics systems employ different rearing techniques including sequential rearing, stock splitting and multiple rearing units. Each method offers distinct advantages and challenges thus influencing fish growth and system management.

- ☆ **Sequential rearing:** This straightforward method involves introducing fish of different age groups into the same tank. As fish reach harvest size, these are removed, and the new smaller fish are introduced. While simple to implement, sequential rearing may induce stress in fish not yet ready for market when others are harvested thus complicating stock management.
- ☆ **Stock splitting:** In stock splitting, fish are divided into two tanks randomly once the first tank reaches its carrying capacity. While this method allows for better distribution of fish and reduces competition for resources,

the stress of transferring fish between tanks can negatively impact their growth and well-being.

- ☆ **Multiple rearing units:** This approach involves starting fish populations at different ages and transferring them to larger tanks as these grow. By segregating fish based on size and developmental stage, multiple rearing units optimize space utilization and minimize competition for resources thus promoting healthier growth and overall system efficiency.

20.6.1.4 Stocking Density

Fish stocking density expressed as the weight of fish per unit volume of water (kg/m^3) impacts system productivity and fish health significantly. Typically commercial growers maintain densities up to $50 kg/m^3$ in aerated systems and even higher in systems with direct oxygen injection while hobby growers usually keep densities below $20 kg/m^3$ to minimize stress and optimize growth.

20.6.1.5 Gravel Substrate

Gravel substrates in fish tanks provide a habitat for nitrifying bacteria which convert ammonia to nitrite and then to nitrate which is essential nutrient for plant growth. These substrates available in various colours should be thoroughly washed before introduction to prevent clouding of the tank water thus enhancing both aesthetic appeal and system functionality.

20.6.1.6 Enhancing Oxygen Levels in Aquatic Environments

The presence of oxygen in the tank water is crucial for well being of both ~ the fish and the plants in aquaponic systems. Depending on the method used to oxygenate the water in the fish tank, aquaponic systems can be classified into two categories ~ aerated aquaponic systems and aquaponic systems with direct oxygen injection.

- ☆ **Aerated aquaponic systems**: In aerated aquaponic systems, oxygenation is achieved through introduction of air into water. This is typically accomplished using an air pump which forces air through air stones or diffusers placed at the bottom of fish tank or within grow beds. As air bubbles rise to the surface these create turbulence and agitation thus facilitating transfer of oxygen from air into the water.
  - ❖ **Air pump:** The air pump is the central component responsible for generating airflow. It draws in air from environment and pushes it through tubing to air stones or diffusers.
  - ❖ **Air stones/diffusers:** These devices break the incoming air into tiny bubbles thus increasing surface area available for oxygen exchange. These are strategically placed at bottom of the fish tank or grow beds to ensure thorough oxygenation of the water.

- **Tubing:** Tubing connects the air pump to an air stone at the bottom of the tank.
- **Oxygen transfer:** As air bubbles rise through the water column, these release oxygen molecules into the surrounding water. This increases the dissolved oxygen levels which are crucial for health and well-being of fish, beneficial bacteria and plant roots.

☆ **Aquaponic systems with direct oxygen injection:** Aquaponic systems with direct oxygen injection utilize specialized equipment to introduce oxygen directly into the water in a more controlled manner. This is typically achieved using oxygen generators, diffusers or other oxygenation devices that deliver oxygen in a concentrated form.

- **Oxygen generator:** An oxygen generator produces oxygen gas (O_2) from ambient air using various methods such as pressure swing adsorption (PSA) or membrane separation. The generated oxygen is then delivered to aquaponic system through tubing or pipes.
- **Oxygen diffusers:** These devices are designed to disperse oxygen gas directly into water. These can be placed strategically within fish tank or grow beds to ensure optimal oxygen distribution.
- **Direct oxygen injection:** Unlike aerated systems where air is bubbled through water aquaponic systems with direct oxygen injection introduce oxygen gas directly into water column. This allows for more precise control over dissolved oxygen levels which can be crucial for maintaining optimal conditions in larger-scale or high-density aquaponic setups.

20.6.1.7 Filtration Systems

☆ **Biofilter:** This is where nitrification bacteria thrive ~ converting ammonia into nitrates.

☆ **Mechanical filtration (for solid waste removal):** A solid removal filter is employed if fish waste exceeds the capacity of biofilter and plants to control. The choice of solid removal system depends on the amount of organic waste generated in the system which may accumulate in grow beds or settle in the tank. Maintaining a balanced ratio of fish to plants helps minimize solid waste accumulation.

20.6.1.8 Mineralization Tank

If sophisticated solid removal filters are used to eliminate solids from the system, mineralization occurs separately in dedicated units. Such systems experience minimal or no competition for dissolved oxygen in fish tanks or grow beds thus formation of anaerobic zones is rare.

20.6.1.9 Temperature Regulation

Aquarium heater and shade cloth covers are suitable for small aquaponics systems to prevent sudden temperature shifts. In larger systems, a greenhouse provides similar temperature stabilization benefits. A larger water volume of 1,000 litres or more enhances system stability and mitigates abrupt temperature fluctuations.

20.6.1.10 Lightening for Fish Tanks

Lightening serves multiple purposes and this is typically employed in indoor systems such as aquariums or aquaponic fish tanks featuring ornamental fish. In most aquarium setups, fluorescent lights are utilized to illuminate the tank thus allowing for clear visibility of the fish and facilitating health monitoring. These lights enhance the aesthetic appeal of the aquarium so augmenting its overall beauty.

20.6.2 Grow Beds (Hydroponic subsystem of aquaponic system)

Grow beds serve as the primary hydroponic component in aquaponics systems providing medium for plant growth. These are typically filled with a growing medium which supports root systems of plants. These beds are where plants absorb nutrients ~ primarily nitrates from the water circulating through the system. Beneficial bacteria colonize the grow media converting ammonia from fish waste into forms usable by plants.

20.6.2.1 Growing Medium (for media-based systems)

Growing medium serves as a crucial element in media-based systems acting as a porous inert material that supports plant roots and retains moisture. These also provide habitat for beneficial bacteria crucial for nutrient conversion. Suitable media include perlite, expanded clay pebbles, peat moss, pea gravel and coconut coir. Each type of medium presents unique characteristics such as cost-effective gravel or non-calcium-based rock while baked clay which is lightweight tends to be more expensive. Furthermore, smaller-sized media facilitate increased surface area for colonization of beneficial bacteria thus aiding in nutrient conversion processes.

20.6.2.2 Planting Area

Within grow beds there are designated areas for planting various crops. These areas allow for cultivation of a wide range of plants such as leafy greens, herbs, vegetables and even certain fruit crops.

20.6.3 Water Distribution System

An intricate network comprising pipes, pumps and potentially gravity-fed channels facilitates the distribution of nutrient-rich water from the fish tank to the grow beds. Simultaneously, it returns filtered water back to the fish tank efficiently. This mechanism guarantees a consistent and uninterrupted supply of water and nutrients to plants thus fostering optimal growth conditions.

20.6.3.1 Water Pump and Tubing

A water pump is employed to facilitate water circulation within aquaponics system ~ typically directing water from the fish tank to the grow bed. The pump can either operate continuously or be timer-operated for controlled intervals. In a basic aquaponic setup, water is pumped into the grow bed and then allowed to drain back into the fish tank through gravity. Sufficient tubing is necessary to ensure proper water circulation extending from the pump outlet to the top of the grow bed and allowing for drainage back into the fish tank from the grow bed outlet.

20.6.3.2 Auto-siphons

Auto-siphons play a vital role in maintaining optimal water levels within the aquaponic system, ensuring conducive conditions to survival and thriving of both fish and plants. These devices function by regulating runoff water from the plants, effectively returning it to the main tank thus preventing water levels from dropping too low.

20.6.3.3 Plumbing

Plumbing an aquaponics system requires careful consideration of various factors including design of the system, available space and location. Typically, PVC or UPVC pipes are preferred choice for plumbing in both hydroponic and aquaponic setups. These pipes offer durability, affordability, wide availability and standardized sizes with a plethora of adapters and connectors globally.

While PVC and UPVC pipes are commonly used, alternative options such as agricultural pipes, flexi-pipes, bamboo hosepipes may also be suitable. However, it is crucial to ensure that any alternative pipes chosen are non-toxic to both fish and plants. Iron pipes are discouraged due to their susceptibility to corrosion while copper pipes pose toxicity risks to fish.

By selecting appropriate piping materials and considering factors such as durability, toxicity and compatibility with aquaponic environment gardeners can ensure efficient and safe plumbing for their systems.

20.6.4 Grow Lights

Light is essential for overall development of plants as these utilize carbon dioxide, water and light along with various minerals to produce food through photosynthesis. In outdoor hydroponic systems or indoor setups with ample sunlight artificial grow lights may not be necessary. However, if sufficient sunlight is lacking grow lights become indispensable. Selection of grow lights for a hydroponic system depends on several factors such as plant variety, size and growth stage, *vis-à-vis* the type and dimensions of hydroponic setup, indoor or outdoor placement and the season. Various types of grow lights are tailored specifically for indoor hydroponic systems including fluorescent, metal halide, high-pressure sodium and specialized LED lights. Each type has its unique advantages and is chosen based on specific needs of the plants and the conditions of the hydroponic environment.

20.6.5 Mineralization Tank

A mineralization tank is often necessary in raft or NFT systems particularly when employing solid removal filters. In its basic configuration, it consists of porous media, water and aeration typically provided by an air stone. Solid waste can manually be directed or automatically channelled into the mineralization tank for breakdown, converting it into mineralized nutrients that can be reintroduced into the aquaponics system either manually or automatically. Aerated mineralization tanks come in various configurations ~ some feature intricate baffle systems and screens to separate water containing dissolved nutrients from remaining fish waste solids while others incorporate automatic water and solids addition along with nutrient-rich water removal systems.

However, in media-filled beds where solids are trapped within the media a separate mineralization tank may not be necessary. Properly sized and designed media beds engineered to promote aerobic conditions can effectively serve as both solid filtration and mineralization devices.

20.6.6 Sump

In an aquaponics system the sump serves as the lowest point where water collects and is subsequently pumped back to rearing tanks. It acts as the central hub for water circulation within the system. The primary function of the sump is to maintain a consistent water level in the fish tank irrespective of varying levels of water in the grow beds. This stability helps regulate pH and temperature thus enhancing overall health of the system. The sump provides an opportune location to replenish lost water and can even accommodate the growth of small fish.

Traditionally situated below the grow beds the sump receives water drained from the grow beds. It can be positioned at a constant height with the fish tank or placed lower. When sizing a sump tank it is essential to account for the total water volume of all grow beds, factoring in the displacement caused by the grow media. For instance, hydroton displaces about 55 per cent of water while 1.8" gravel displaces approximately 62 per cent. Calculation should consider the minimum water volume required to cover the pump. This total volume determines the minimum capacity needed for the sump tank. If necessary, multiple sump tanks can be interconnected using a bridge siphon.

The versatility of sumps allows for various applications and configurations within aquaponic systems. Some common variations include:

- ☆ The water circulation within the aquaponic system follows a continuous cycle. Initially, water is pumped from the sump tank into the fish tank using a water pump. From the fish tank water flows into the grow beds providing nourishment to the plants. Subsequently, the water is drained from the grow beds back into the sump tank so completing the cycle.

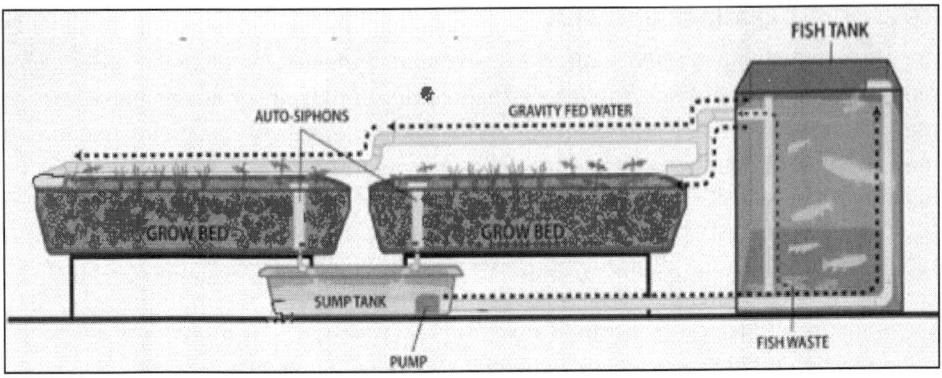

Aquaponics with One Sump, One Fish Tank and Two Grow Beds.

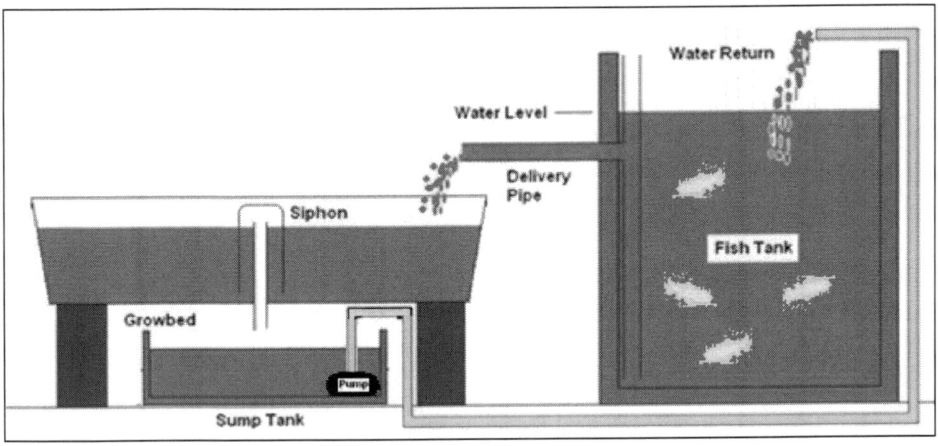

Aquaponics with One Sump, One Fish Tank and One Grow Bed.

☆ The water from the sump is pumped to both the grow beds and the fish tank with both systems designed to return the water back to the sump after use.

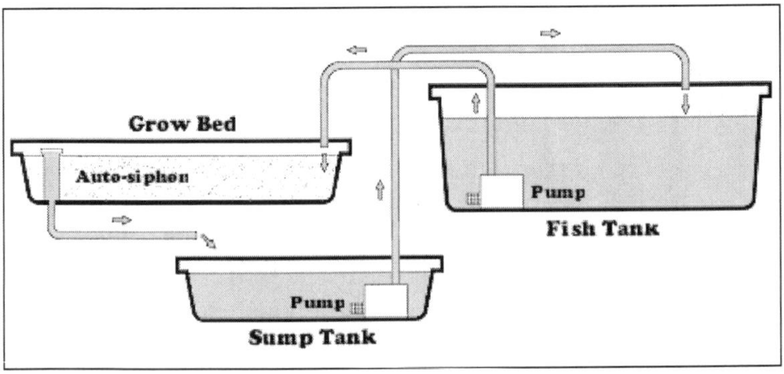

- ✰ The water is pumped from the sump tank to the grow beds. After passing through the grow beds the water drains into the fish tank. From the fish tank it is then directed back to the sump tank so completing the cycle.
- ✰ Water is pumped from sump to fish tank. Another pump in fish tank pumps water in fish tank to grow beds. Water from grow beds is drained by gravity to sump.

20.6.7 Backup System

Ensuring continuous power supply is crucial for sustainability of aquaponic systems as even a brief power outage can have detrimental effects. Despite meticulous system design and control, aquaponic gardeners must be prepared with backup power solutions to mitigate the risks associated with power failure.

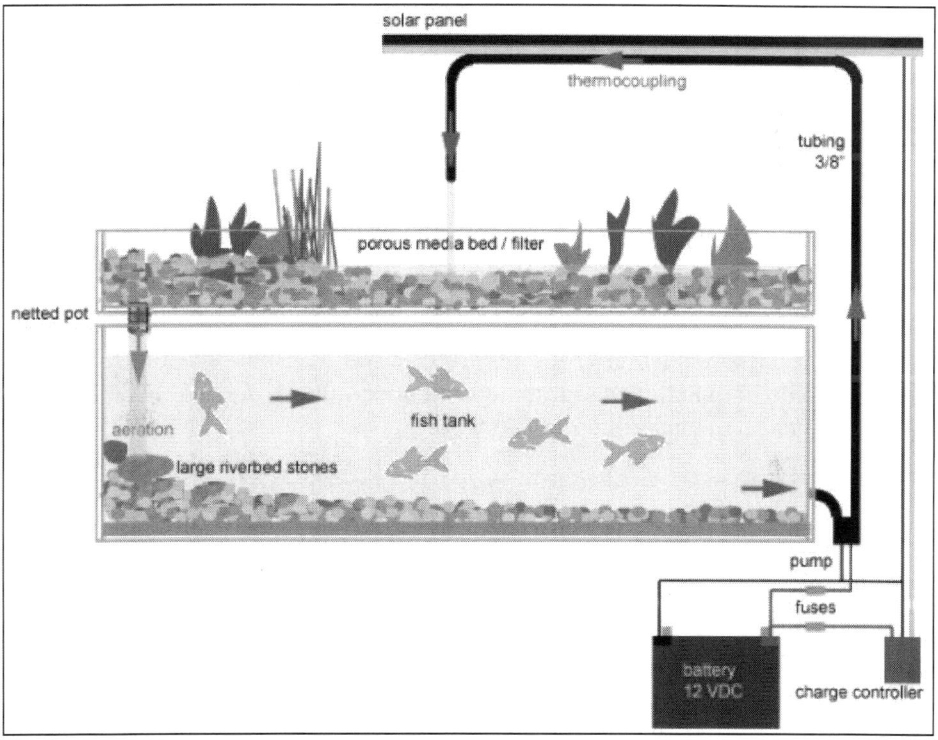

Aquaponics System with Backup Power Supply.

In the event of a power outage the consequences can be severe. Without electricity to run water and air pumps the fish may suffer from oxygen depletion so leading to mortality. Plants may wilt or perish due to lack of water and oxygen. Therefore, it is imperative for gardeners to have alternative power sources readily available.

While generators offer a viable backup option but these necessitate continuous human presence to operate and maintain thus making them less practical for long-term solutions. An ideal backup system is an inverter setup equipped with internal batteries that seamlessly switch over during power outages. Once power is restored the internal batteries recharge automatically thus ensuring uninterrupted operation of water and air pumps.

Another sustainable option is the installation of a solar power system to recharge batteries and power essential components of aquaponic system. This environment friendly solution harnesses renewable energy thus providing a reliable and sustainable backup power source.

Some growers opt for air pumps with built-in rechargeable batteries which operate both on mains power and internal batteries. During power outages, these pumps automatically switch to internal battery mode thus continuing to aerate the water and sustain fish life for up to 10 hours. Once power is restored the internal batteries recharge automatically ensuring continuous operation.

By implementing robust backup power solutions such as inverters with internal batteries, solar power systems or air pumps with rechargeable batteries, aquaponic gardeners can safeguard their systems against the risks posed by power interruptions thus ensuring the health and longevity of their fish and plants.

20.6.8 Monitoring and Control Systems

Hydroponic components may also include monitoring and control systems that regulate factors such as pH, temperature and nutrient levels in the water. These systems help maintain optimal growing conditions for plants and ensure the overall health of aquaponics ecosystem.

20.7 Reducing Electricity Costs in Aquaponics Systems

Electricity consumption in aquaponics systems, primarily for running aerators, pumps, artificial lighting and temperature maintenance can constitute a significant portion of operational expenses. However, employing strategic design approaches and carefully selecting fish and plant species can lead to substantial savings in electricity costs such as:

- ✰ Aerators can be replaced with strategically designed grow beds that allow water to fall from sufficient height thus creating a splash-back effect into the fish tank thereby aerating the water naturally.
- ✰ Mechanical filtration can be achieved through simple methods such as incorporating foam blocks underneath irrigation pipes entering grow beds. Regular cleaning of these blocks effectively traps large debris thus minimizing the need for energy-intensive filtration systems.
- ✰ Careful selection of fish and plant species can help optimize temperature requirements so reducing the need for energy-intensive temperature maintenance measures.

- ☆ Utilizing short tube lengths for water transfer minimizes power consumption by reducing the workload on water pumps.
- ☆ In smaller systems, manual irrigation methods can eliminate electricity costs entirely. These systems typically require:
  - ❖ A fish-holding tank such as a large plastic tub.
  - ❖ Elevated plant containers positioned to allow water to splash back into the fish tank ~ promoting aeration.
  - ❖ Simple watering devices.
  - ❖ To initiate such a system it is essential to introduce fish into the tank at least a week before. Before watering the plants the fish-rearing tank should be agitated and nutrient-rich water manually transferred to grow beds using a watering can. Regular maintenance includes periodic cleaning of fish-rearing tank and ensuring that containers are flooded at least three times a day. By implementing these measures, aquaponic practitioners can significantly reduce electricity costs while maintaining efficient system functionality and sustainability.

20.8 Success of Aquaponics System

The success of an aquaponics system hinges on several critical factors. Firstly, establishing nitrification and mineralization processes is essential facilitated by beneficial bacteria and denitrification mechanisms. Secondly, maintaining good water quality is paramount, encompassing parameters such as ammonia, nitrite and nitrate levels as well as water hardness, temperature, pH, alkalinity, carbon dioxide and dissolved oxygen. Regular water testing is necessary to monitor these parameters. Effective filtration is another key aspect involving use of biofilters to host nitrifying bacteria and solid removal systems to ensure water clarity. System size and ratios including appropriate fish stocking densities relative to the size of aquaponics system and plant requirements must be carefully considered. Proper water flow achieved through well-designed and sized pipes is crucial for adequate water circulation. The ratio of water volume to surface area in the fish tank plays a role in promoting oxygen exchange and maintaining water quality. Regular water changes are necessary to prevent nutrient buildup. Nutrient supplementation, fish feeding practices and stress management are vital for overall system health. Judicious use of salt, strategies for pest and disease management and selection of suitable fish and plant species are also significant considerations. Incorporating biological pest control methods can further enhance system resilience and productivity. By addressing and managing these factors comprehensively, aquaponic practitioners can optimize system performance and ensure long-term success.

20.9 Nitrification (System Cycling)

In an aquaponics system nitrification plays a crucial role as it involves the aerobic conversion of toxic ammonia into beneficial nitrates. Primarily ammonia released into the water through fish excretions and gills can become harmful to fish at higher concentrations. However, plants cannot directly utilize the nitrogen in ammonia for nutrition. Therefore, establishing nitrification is vital to reduce water toxicity for fish and to provide plants with essential nutrients in the form of nitrates.

Efficient nitrification ensures a healthy nitrogen cycle, facilitating the removal of toxic compounds while promoting nutrient uptake by plants. Over time, aquarium filtration systems have evolved from traditional box filters filled with charcoal and glass wool to more sophisticated methods like under-gravel filters, trickle filters and fluidized bed filters. Each advancement aims to enhance the effectiveness of biological filtration, thus optimizing the efficiency of the nitrogen cycle.

Establishing a well-functioning nitrogen cycle quickly and with minimal stress on the aquatic inhabitants is essential for overall success of an aquaponics system. When employing a fishless cycling technique it is advisable to introduce fish into the system once nitrates become detectable and the levels of both ammonia and nitrite have reached their peak and subsequently decreased to below 1ppm.

20.9.1 Nitrifying Bacteria

Several genera of bacteria play a crucial role in aquaponics by converting ammonia into nitrite and nitrite into nitrate thus transforming fish waste into a highly effective plant fertilizer. Among these, *Nitrosomonas* and *Nitrobacter* are the most common genera of autotrophic bacteria responsible for converting ammonia to nitrites and nitrites to nitrates, respectively. Healthy populations of these bacteria can further convert ammonia into other nitrogenous compounds.

Beneficial bacteria not only aid in organic material breakdown but also help prevent the proliferation of harmful bacteria within the aquaponic system. As the nitrification process tends to acidify the water, it may be necessary to add non-sodium bases such as potassium hydroxide or calcium hydroxide to neutralize the pH. Furthermore, supplementary minerals or nutrients such as iron can be introduced alongside fish waste thus serving as additional nutrients for plant growth.

These bacteria form a biofilm on all solid surfaces within the aquaponic system that are in constant contact with water. In systems without media such as floating raft systems, sufficient bacteria for nitrification can accumulate on submerged plant roots and aquarium filters. Many aquaponic setups include a biofiltering unit specifically designed to foster the growth of these microorganisms.

Nitrification thrives under conditions of high dissolved oxygen and low organic matter levels. However, it is important to note that nitrification process consumes alkalinity, generates acidity, and consequently lowers pH levels.

When initially setting up an aquaponic system, beneficial bacteria populations are typically absent. It may take approximately 20-30 days and sometimes even up to 8 weeks for these bacteria colonies to establish themselves naturally. To expedite this process it is advisable to introduce *Nitrosomonas* and *Nitrobacter* colonies before introducing fish and plants into the system. While these bacteria will eventually colonize the system naturally, these can also be introduced from established aquatic environments such as aquaria or ponds.

20.9.1.1 Establishing Beneficial Bacteria Build-up

Various methods can be employed to promote the buildup of beneficial bacteria, ensuring the efficient cycling of nutrients in the aquaponic system:

i) **Natural build-up:** Allowing a natural build-up of bacteria colonies is a slower but effective method.

ii) **Transferring from established systems:** Beneficial bacteria can be introduced into the system by transferring water, gravel or filters from already established disease-free aquaponic systems.

iii) **Manual addition:** Beneficial bacteria cultures can be manually added to the system which is advisable if immediate establishment of the aquaponic system is desired. These cultures can be obtained from reputable aquaponics stores.

iv) **Fish feed cycling:** Prior to introducing plants and fish, cycling the system using fish feed can kickstart the establishment of beneficial bacteria. As the feed breaks down it releases ammonia ~ providing a food source for the bacteria to thrive on.

v) **Additional substances for bacterial establishment:**

- **Cow dung or cow urine:** Aged manure (15-20 days old) from cows can be added to the system. Fresh dung is not recommended.
- **Urea fertilizers:** Using urea fertilizers offers a straightforward method of cycling the system.
- **Household ammonia:** Only food-grade ammonia should be used for this purpose.
- **Decaying organic matter:** Addition of decaying prawn or fish to the system provides a source of ammonia for the bacteria to metabolize.

The timeline for establishing a bacterial colony varies based on several factors. It is imperative to avoid overdosing any additives in the system.

20.9.1.2 Factors Affecting Nitrification

- **Water temperature:** The optimum temperature for nitrifying bacteria ranges between 25 to 30°C. Temperatures below 11°C or above 40.5°C are likely to result in death of nitrifying bacteria. *Nitrobacter* is generally less tolerant to low temperatures compared to *Nitrosomonas*.
- **pH range:** Nitrification is inhibited if the pH drops below 6.0. The optimal pH range for *Nitrosomonas* is 7.8 to 8.0 while for *Nitrobacter* it is 7.3 to 7.5. A compromise point between the needs of fish and plants lies between

6.8 to 7.0. Fluctuations in pH levels outside the system's acceptable range can halt the conversion of ammonia to nitrate.

- **Dissolved oxygen:** Nitrification is maximized when dissolved oxygen levels exceed 80 per cent saturation. Concentrations dropping to 2mg/l (ppm) or lower will inhibit nitrification with *Nitrobacter,* being more susceptible to low dissolved oxygen levels compared to *Nitrosomonas.*
- **Depth of grow bed:** An ideal depth of 30 to 35cm promotes better nitrification.
- **Light:** Nitrifying bacteria are sensitive to light, particularly blue and UV light. However, once their colony is established light sensitivity becomes less of an issue.
- **Chlorine:** Chlorine is harmful to both beneficial bacteria and fish. Water conditioners or chlorine neutralizers suitable for use with fish can be utilized to neutralize chlorine.
- **Other additives:** Avoid adding anything that may harm beneficial bacteria or fish to the aquaponics system such as pesticides and growth enhancers. Maintaining a chemical-free environment is crucial for health of the ecosystem.

20.10 Monitoring Water Parameter

Regular monitoring of key water parameters is essential for maintaining a healthy aquaponic system. During cycling phase which typically requires daily monitoring it is crucial to assess levels of ammonia, nitrite, nitrate, pH and water temperature.

Aquarium water test kits are indispensable tools for measuring ammonia, nitrite and nitrate levels in ppm as well as pH and temperature. It is advisable to commence daily testing once the ammonia source is introduced into the system. The cycling process is considered complete when nitrates are detected and both ammonia and nitrite concentrations have dropped to 0.5 ppm or lower.

The pH level and water temperature play significant roles in the cycling rate and overall health of fish and plants within the system. A pH test kit allows for regular assessment of pH levels while a submersible thermometer accurately measures water temperature.

By diligently monitoring these parameters aquaponic practitioners can ensure optimal conditions for biological cycling as well as the thriving of their aquatic inhabitants and plants.

20.11 Denitrification

Denitrification typically occurs in anaerobic conditions ~ devoid of oxygen where anaerobic bacteria thrive. In this process, nitrate may be converted back to nitrite which is toxic to fish. As denitrification progresses nitrogen within the aquaponics system transforms into various forms so ultimately culminating in nitrogen gas (N_2) which dissipates from the system. However, denitrification can also elevate the pH of water within the aquaponics setup.

Excessive denitrification resulting from an abundance of anaerobic zones can lead to a significant rise in pH levels ~ potentially reaching neutrality (pH 7) or even higher. This scenario is undesirable as it limits the ability to adjust pH using buffers to supplement crucial nutrients such as potassium and calcium essential for optimal plant growth. Hence, a well-designed and meticulously managed aquaponics system should ideally operate under aerobic conditions where pH tends to decrease gradually.

It is worth noting that reducing nitrate levels can be advantageous, particularly when cultivating fruiting plants like tomatoes as elevated nitrate levels tend to promote vegetative growth while hindering fruit set. Conversely, high nitrate concentrations are beneficial for stimulating growth of leafy green plants.

20.12 Earthworms (Red Worms)

Earthworms are valuable contributors to the health and balance of media-based aquaponic systems. These worms play a vital role in breaking down solid organic matter such as fish waste, leftover feed and other biodegradable materials into nutrient-rich substances known as worm castings or vermicompost. Not only do red worms facilitate the conversion of waste into accessible nutrients for plants these also contribute to the ecosystem by aiding in the management of plant diseases, harmful insects and nematodes.

Once the aquaponic system has completed its cycling process and fish have been introduced gardeners can introduce a handful of composting red worms into each grow bed. This addition ensures ongoing waste management and nutrient cycling within the system thus promoting healthier plant growth and overall system sustainability.

20.13 Ensuring Optimum Water Quality

Ensuring optimal water quality is paramount for health and vitality of both fish and plants in an aquaponics system. Several factors contribute to water quality including its source, pH levels, presence of ammonia, nitrite/nitrate levels, bacterial balance, temperature, oxygen content and alkalinity. To maintain high water quality, following measures are essential:

- ☆ **Use of de-chlorinated, disease-free water:** Only high-quality water preferably de-chlorinated and free from pests and diseases should be introduced into the aquaponics system. Reverse osmosis (RO) water is considered ideal for its purity.
- ☆ **Dissolved compounds:** Water intended for use in aquaponics systems may contain various dissolved compounds. Therefore, it is essential to first

identify the specific elements present in the water and take appropriate corrective measures as needed.

- **Regular monitoring and maintenance:** It requires continuous monitoring and adjusting water temperature *cum* pH levels to ensure that these remain within optimal ranges for species being cultivated.
- **Effective filtration techniques:** Reliable filtration methods should be employed to remove solid waste and to maintain water clarity. This helps prevent the accumulation of harmful substances and ensures a healthy aquatic environment.
- **Control of ammonia, nitrite, nitrate and carbon dioxide:** Proper control over the concentrations of ammonia, nitrite and nitrate should be maintained to prevent toxicity and promote overall water quality. Similarly, managing carbon dioxide levels is essential for sustaining healthy plant growth.
- **Aeration:** Proper aeration of the water should be ensured to enhance oxygenation and circulation. Adequate oxygen levels are vital for well-being of both fish and plants.
- **Proper feeding practices:** Responsible feeding practices should be implemented to avoid overfeeding and to minimize waste accumulation. This helps maintaining water quality by reducing release of excess nutrients into the system.

By adhering to these practices and maintaining vigilant oversight of water quality parameters, aquaponic practitioners can create and sustain an optimal environment for thriving growth of fish and plants alike.

20.14 Monitoring and Managing Ammonia, Nitrite and Nitrate

Monitoring and managing ammonia, nitrite and nitrate levels are crucial aspects of successful aquaponics. High concentrations of ammonia and nitrite can be toxic to fish therefore regular testing and maintenance are essential. For most fish species, levels below 0.02ppm of un-ionized ammonia are considered safe. Ammonia is less toxic at pH levels slightly below neutral (7.0). Ideally, ammonia levels should be kept below 0.5mg/l and nitrite levels below 0.2mg/l to prevent harming to fish. Nitrate serving as primary nutrient source for plants should be maintained between 10 to 150mg/l.

20.14.1 Ammonia

Sudden spikes in ammonia levels can occur in newly established or uncycled systems or if fish are overfed particularly in warmer-than-normal temperatures. Ammonia poisoning can lead to tissue damage especially in the gills and kidneys, physiological imbalances, reduced disease resistance, impaired growth and even mortality. Physical symptoms may include reddening of eyes of fish.

Therefore regular monitoring of ammonia levels especially in new systems or when fish are overfed is crucial for maintaining a healthy aquaponic environment. If ammonia levels exceed safe limits, replacing one-third of the tank's water with fresh de-chlorinated water at a similar temperature and pH level can help alleviate toxicity.

20.14.2 Nitrite

Nitrite poisoning also known as brown blood disease hampers oxygen-carrying capacity of red blood cells so posing a severe threat to fish health. Concentrations of nitrite exceeding 10ppm in the fish tank can lead to fatalities due to brown blood disease. In such critical situations immediate action is necessary to mitigate impact on fish.

To address high nitrite levels it is recommended to replace approximately one-third of the water in the fish tank with fresh de-chlorinated water. Sodium chloride salt should be introduced into the system at a ratio of about 1 part per thousand (1 kg of salt per 1000 litres of water) using non-iodized salt. While pool salt or water softener salt can suffice, it is essential to avoid table salt which may contain harmful anti-caking agents. Utilizing salt specifically formulated for fish is preferred.

Prior to adding salt to the system ensure complete dissolution by mixing it thoroughly in a separate container. This salt solution can then be gradually added to the fish tank. Salt aids in mitigating nitrites by binding chloride ions with nitrites thereby reducing toxic effects on fish.

During this treatment process, fish feeding should be ceased and aeration to maximum levels should be increased until nitrite levels decline below 1ppm. Diligent monitoring of water parameters and fish health is essential throughout the recovery process to ensure well-being of aquatic inhabitants.

20.14.3 Nitrate

While high nitrate levels typically won't directly harm the fish but these can impact their well-being once levels surpass 100ppm especially if sustained over time. Elevated nitrate levels induce stress in fish thus rendering them more vulnerable to diseases. Conversely, a low to negligible nitrate reading in fish tank indicates a well-balanced aquaponics system and thriving vegetation.

To address high nitrate levels in the fish tank, adjustments to the system may be necessary. Options include reducing the number of fish stocked or increasing the number of plants and/or grow beds to enhance nutrient absorption. The growth of plants serves as a reliable gauge of system health. Presence of algae in the tank often signals elevated nitrate levels thus prompting proactive measures to restore balance.

20.15 Water Temperature

Optimal temperature is crucial for thriving of both plants and fish within an aquaponic system. A temperature range of 21-29°C is considered ideal for their growth and well-being. Variations outside this range can induce stress and impacting their metabolism, growth rates and overall health. Water temperature plays a pivotal role influencing oxygen levels, ammonia concentration (particularly the proportion of unionized ammonia) and water salinity. Warmer water tends to hold less oxygen, higher levels of unionized ammonia and increased salinity compared to cooler water.

Careful consideration should be given in selecting the species that thrive within the same temperature range and are well-suited to local environment. Fish species exhibit temperature-dependent behaviours, each requiring specific temperature ranges for comfort and vitality. Depending on the climate, adjustments such as heating or cooling may be necessary to ensure the well-being of the fish. It is essential to minimize fluctuations in water temperature and conditions as sudden changes can stress both fish and plants particularly in the root zone. Ideally, temperature variations should not exceed 1.7°C per day. To achieve equilibrium it is advisable to allow new water added to the fish tank to reach the same temperature as the existing water. Different fish species may have slightly different temperature requirements while most plants thrive in temperatures range of 17 to 27°C. Achieving optimal conditions in aquaponics involves striking a balance between the needs of fish, plants and beneficial bacteria.

For temperature regulation, small-scale aquaponic systems can utilize aquarium heaters and shade cloth covers to prevent sudden temperature shifts. In larger setups, greenhouses offer effective temperature control. Larger water volumes of 1,000 litres or more provide stability and mitigate sudden temperature swings.

For enhanced temperature control, burying the fish tank in the ground can offer insulation and stability. It's crucial to note that fish exhibit stress responses, including decreased appetite, in response to extreme temperatures. By carefully managing and maintaining optimal temperature conditions, aquaponic practitioners can create a conducive environment for the thriving of both fish and plants within their systems.

20.15.1 Fish Species Based on Temperature Requirement

Depending on their preferred temperature ranges for optimal growth, fish species are commonly classified into three categories ~ warmwater, coolwater and coldwater. These categories help aquaponic enthusiasts select suitable fish species for their specific environmental conditions.

Fish			*Bacteria*	*Plants*
Coldwater Optimum 12.8-18.3°C	Coolwater Optimum 18.3-23.9°C	Warmwater Optimum 23.9-32.2°C	Optimum 25-30°C	Optimum for most plants 17-27°C
Example: Rainbow trout	Example: Walleye and yellow perch	Example: Tilapia and catfish	Will most probably die at temperature below 1.1°C or above 46.1°C. *Nitrobacter* is less tolerant to low temperatures than *Nitrosomonas*	

The temperature within the fish tank of an aquaponics system is meticulously maintained at compromised optimum levels to support health and vitality of fish, bacteria and plants; such as:

- Coldwater fish thrive at temperatures ranging from 15.6-18.3°C.
- Cool water fish prefer 21.1-23.9°C temperatures.
- Warm water fish flourish at 23.9-27°C temperatures.

These temperature ranges serve as a foundational guideline with actual compromised optimum temperatures varying based on the specific species of fish and plants cultivated within the system.

Extended exposure to temperatures outside the acceptable range can lead to various issues:

- Fish may exhibit poor food consumption and heightened stress levels so increasing their vulnerability to diseases.
- Plants may experience root diseases and sluggish growth so making them susceptible to foliage diseases and pest infestations.
- Bacterial activity responsible for ammonia conversion may slow down or cease entirely, resulting in potential losses of both fish and plants.

20.15.2 Growing Plants and Fish with Entirely different Temperature Requirements

When cultivating plants and fish with divergent temperature requirements within the same aquaponic system, special considerations are necessary to maintain optimal conditions for both. One approach involves routing water from the fish tank to a separate container where its temperature can be regulated for plant use before transferring it to the grow beds. Conversely, purified water from the grow beds can be channelled into a container where its temperature is adjusted to meet the needs of the fish before returning it to the tank. Implementing this method may necessitate additional containers, drainage systems or temperature adjustment mechanisms within the plumbing infrastructure, resulting in increased investment costs. Alternatively, to streamline operations and reduce complexity it is advisable to select fish and plants that thrive under similar environmental conditions thereby minimizing the need for extensive temperature management measures.

The figure illustrates temperature adjustments for both fish and plants.

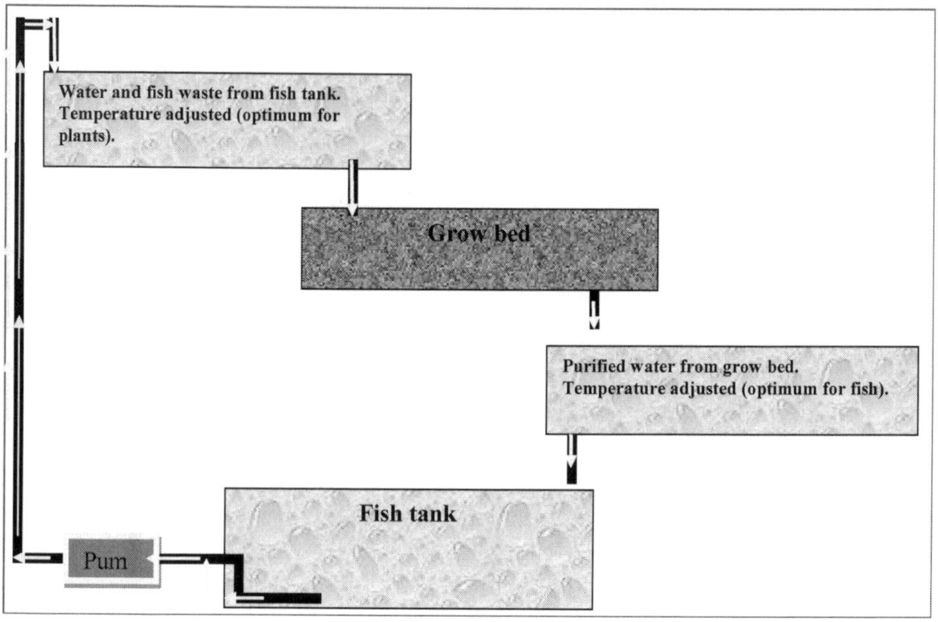

20.16 pH

The pH level in an aquaponic system influences various water quality parameters profoundly including the balance between ammonia (NH_3) and ammonium (NH_4) and the solubility of essential plant nutrients. Fluctuations in pH can significantly affect nutrient availability to plants and overall well-being of the fish. Therefore, it is imperative to regularly monitor and adjust pH levels to maintain a stable and conducive environment for both plants and fish.

In aquaponics, fish, beneficial microbes (such as bacteria) and plants coexist in a harmonious symbiotic relationship, mutually supporting growth of others in a natural and organic manner. To sustain this synergy it is crucial to maintain pH and temperature levels at optimal compromises that accommodate the needs of all components. In aquaponic systems, typically the pH range falls between 6.4 to 7.3 with our experience indicating favourable outcome within 6.8 to 7.0.

20.16.1 Optimum pH Levels for Fish, Bacteria and Plants

Fish (Freshwater)

- ☆ **Ideal range for most fish:** 6.5-7.5
- ☆ **Adaptable range:** 6.0-8.5

Bacteria

- *Nitrosomonas*: 7.8-8.0
- *Nitrobacter*: 7.3-7.5

Plants (Most varieties)

- **Usual range:** 5.5-6.5
- **Optimal range:** 5.8-6.3
- **Tolerated range:** 5.5-7.0
- **Optimum for vegetative growth:** 6.0-6.3
- **Optimum for flowering and fruiting:** 5.7-5.9

Compromised Ideal Level for Mature Aquaponics

- **Usual range:** 6.4-7.3
- **Effective range:** 6.8-7.0

While certain fish species may tolerate pH levels below 6.0, the majority thrive in slightly alkaline water, therefore it is advisable to maintain pH levels above this threshold. Dropping below a pH of 6.0 can inhibit nitrification processes crucial for the health of aquaponic ecosystem. Conversely, excessively high pH levels above 7.5 can exacerbate ammonia toxicity, posing a grave threat to fish health and survival. Generally, most freshwater fish flourish within a pH range of 6.0 to 8.5. It is important to note that while fish can adapt to different pH ranges, any adjustment should be made gradually to avoid stress or shock.

In aquaponics systems, maintaining pH within a specific range is critical for health and productivity of all organisms involved. While plants thrive in a pH range of 5.5 to 6.5, nitrifying bacteria essential for converting ammonia to nitrates perform optimally at a pH of 7.3 to 8.0. Balancing the pH to accommodate the preferences of both plants and bacteria necessitates maintaining it within a compromise range of 6.4 to 7.3, commonly around 6.8 in many aquaponics setups. This pH compromise is crucial for two main reasons: i) at slightly acidic pH levels below 7.0 ammonia becomes predominantly non-toxic which is beneficial if ammonia is present in the system, and ii) many plants thrive in slightly acidic conditions around pH 6.0.

20.16.2 Factors Affecting pH

It is important to note that pH levels in aquaponic systems can fluctuate due to various factors. During daytime, plants produce oxygen through photosynthesis but during nights these release carbon dioxide which can lower the pH. Fish respiration also releases carbon dioxide, potentially elevating carbon dioxide levels and lowering the pH. If the pH becomes too low anaerobic bacteria may proliferate in the media beds producing acids and creating pockets where air cannot reach which leads to plant and system health issues.

Conversely, if pH becomes too high it suggests that biofilters of system may not be processing ammonia production of fish adequately. In such cases, increasing the plant population can help address the issue by enhancing ammonia uptake and nutrient processing.

Maintaining optimal pH levels in aquaponics systems requires careful monitoring and adjustment to ensure the health and balance of the ecosystem thus facilitating thriving conditions for both plants and fish alike.

20.16.3 pH Instability and Prolonged Extremes

Prolonged fluctuations in pH levels and deviations beyond the acceptable range can trigger a cascade of issues within an aquaponics system. Firstly, such fluctuations can induce stress in fish populations thus rendering them more susceptible to diseases. Consequences are twofold ~ nutrient lockout where these are unable to absorb available nutrients effectively thus growth is stunted, *vis-a-vis* such condition increases their vulnerability to foliage diseases and pest infestations. Moreover, the disruption in pH levels can disrupt the crucial role of beneficial bacteria so halting conversion of ammonia to nitrate. This interruption not only impacts the health of the fish but also deprives plants of essential nutrients, potentially leading to losses in both fish and plant populations.

20.16.4 Adjusting pH

If the pH deviates from the optimal range, adjustments should be made gradually to avoid causing undue stress to the fish. It is recommended to shift the pH towards the target range slowly ~ ideally no more than 0.2 to 0.3 units per day. If necessary, regular testing and buffering of pH are advisable as minor fluctuations can occur even in established systems and may require only slight adjustments.

Various products are available for adjusting pH levels quickly and easily. Aquaponics water pH can be monitored using a pH meter, paper test strips and liquid test kits. When adjusting pH it is safest to use pH up or pH down adjusters following the instructions on their labels carefully. To raise pH levels in an aquaponics system, calcium hydroxide, potassium carbonate or bicarbonate, or potassium hydroxide can be used. It is important to note that carbonates can buffer pH making precise adjustments challenging. Avoid using citric acid as it can have antibacterial properties.

When cultivating plants with different pH requirements in the same grow bed, strategic placement can help accommodate their needs. Heavy feeding plants should be positioned where water flows into the grow tray thus favouring their pH requirements while light feeding plants can be placed at the end where water flows out. Consider growing blooming plants and non-blooming plants separately in dedicated grow trays to facilitate precise monitoring and adjustment of pH levels for each variety. This approach allows growers to optimize growing conditions for different types of plants and ensuring their success in the aquaponics system.

20.16. Stable pH

A stable pH condition resulting from denitrification in anaerobic zones often leads to increased alkalinity levels. However, prolonged periods of pH stability can be detrimental as it hinders the supplementation of essential nutrients such as calcium and potassium. To address this issue it is essential to implement regular maintenance practices. This includes cleaning the filter tanks regularly ideally twice a week and removing any organic matter deposits from the grow beds. By maintaining proper cleaning schedules and managing organic buildup, aquaponic systems can sustain optimal pH levels and nutrient balance.

20.16.6 Nitrification and pH

Nitrification and pH levels play crucial roles in the health and balance of aquaponic ecosystems. Nitrification ~ the biological process of converting ammonia into nitrates is most efficient within a pH range of 6.8 to 7.0 or slightly higher. Interestingly, this pH range aligns with optimal conditions for both fish and plant growth thus striking a balance between their respective needs. When pH falls below 7.0 nitrification activity slows down and at pH levels below 6.0 it nearly ceases altogether. This decline in nitrification efficiency is attributed to the generation of nitric acid during the process which subsequently lowers the pH of water. Fluctuations in pH may result from other factors such as water alkalinity, buffering capacity, overall water chemistry and temperature variations.

20.16.7 Alkalinity

Alkalinity serves as a buffer against acidity and is quantified by its equivalent concentration of calcium carbonate ($CaCO_3$). To elevate alkalinity, bases can be introduced into the system. In aquaponics, the acceptable range for alkalinity is quite wide typically falling between 59mg and 300mg per liter.

20.17 Carbon Dioxide

In aquaponics systems, it is crucial to maintain CO_2 levels below 20mg/litre to ensure optimal conditions for fish health. Elevated CO_2 levels can impede oxygen absorption through gills of the fish which may lead to respiratory issues. However, in systems employing diffused aeration, excess CO_2 is efficiently vented off into the atmosphere thus mitigating the risk of buildup. The use of pure oxygen is not recommended in aquaponics as it may further reduce the likelihood of CO_2 accumulation.

20.18 Dissolved Oxygen

Dissolved oxygen plays a crucial role in supporting optimal growth of all life forms within an aquaponics system including fish, plants and beneficial bacteria. While fish typically require higher levels of dissolved oxygen for respiration compared to plants, maintaining adequate oxygen levels benefits all components of the ecosystem. Utilizing diffused aeration methods such as air stones and porous hoses are recommended to ensure efficient oxygen distribution throughout the system. Monitoring dissolved oxygen levels is essential especially in new aquaponic systems to gauge health and functionality of the system. While frequent testing is advisable during initial stages, mature systems may require less frequent monitoring. This can be achieved using test kits or dissolved oxygen meters.

The recommended dissolved oxygen levels in an aquaponics system should ideally be maintained at 5mg/l or higher. Optimal levels approaching 80 per cent saturation translate to approximately 6-7 milligrams per litre. Oxygen levels below 3mg/l impose stress on aquatic life while levels below 1mg/l can be fatal.

20.18.1 Factors Affecting Dissolved Oxygen

It is important to note that oxygen-holding capacity of water is influenced by various factors where temperature is a primary determinant. As water temperature increases its ability to retain dissolved oxygen decreases. Therefore, measures to mitigate temperature fluctuations and ensuring adequate aeration are essential for maintaining optimal oxygen levels and promoting the health and vitality of the aquaponics ecosystem. Other factors influencing the dissolved oxygen content in the system include:

- **Fish feeding:** The oxygen levels in the fish tank experience a significant decrease shortly after feeding.
- **Time of day:** Oxygen levels typically peak in the late afternoon and reach their lowest point in the early morning.
- **Weather conditions:** Oxygen levels decrease during cloudy or rainy days.
- **Stocking density of fish:** Higher fish density leads to reduced oxygen availability.
- **Salinity:** Increased concentrations of dissolved salts result in decreased oxygen levels.
- **Use of air diffusers:** Utilizing air diffusers with smaller bubbles enhances oxygenation.
- **Fish waste:** Elevated levels of fish waste contribute to reduced oxygen levels.

20.18.2 Effect of Drop in Dissolved Oxygen

When dissolved oxygen levels in the fish tank drop below desired threshold fish behaviour can change. Signs such as decreased appetite, gasping at the water surface, congregating near inflow areas and increased susceptibility to diseases may manifest.

Insufficient aeration can lead to dire consequences with fish at risk of perishing within 45 minutes. Even if immediate death is avoided, permanent gill damage can occur leading to a decline in the fish population over time. Therefore, it is essential to have a backup power system in place to operate aerators and water circulation pumps during power outages.

While water aerators are available for purchase at aquarium supply stores alternative methods can also be employed to oxygenate fish tank. For instance, arranging the water flow from the grow beds to cascade back into the fish tank from a height can effectively aerate the water by introducing air through splashing.

20.18.3 Considerations to Achieve a Balanced DO Level

Here are some considerations to achieve a balanced DO level:

- ☆ **Optimal range:** One should aim to maintain the DO level within a range that is suitable for both fish and plants. Typically, a DO level of 5-8 mg/l is considered optimal for both aquatic organisms in aquaponic systems.
- ☆ **Aeration:** Aeration devices such as air stones, diffusers or surface aerators should be implemented to increase oxygen levels in the water. These devices help oxygenate the water thus benefiting both fish and plant roots.
- ☆ **Water movement:** Adequate water movement and circulation should be ensured throughout the system. Proper water flow helps distributing oxygen evenly and preventing stagnation thus ensuring oxygen-rich water reaching all the areas of the system.
- ☆ **Photosynthesis:** For performing proper photosynthesis in the plants sufficient light should run on the plants in the beds. During photosynthetic activities plants release oxygen into the water thus contributing overall oxygen levels in the system.
- ☆ **Monitoring and adjustment:** Regular monitoring of DO levels should be done using a dissolved oxygen meter. Aeration and water flow should be adjudted as needed to maintain desired DO range.

20.19 Environmental Condition

In aquaponics, creating and maintaining optimal environmental conditions for both plants and fish is crucial for their health and growth. Here is a breakdown of the key environmental factors:

- ☆ **Light:** Plants require adequate light for photosynthesis which is crucial for their growth and development. Different plant species have varying light requirements so it is essential to provide appropriate lighting conditions based on their needs. For indoor aquaponic systems artificial grow lights may be necessary to supplement natural sunlight especially in environments with limited light exposure. Also see "Environmental Conditions for Plants".

- ☆ **Humidity:** Plants thrive in environments with moderate humidity levels generally between 50 per cent to 70 per cent. Maintaining proper humidity levels helps prevent issues such as wilting, leaf scorching and fungal diseases. Also see "Environmental Conditions for Plants"
- ☆ **Temperature:** See "Environmental Conditions for Plants"
- ☆ **CO_2 levels:** See "Environmental Conditions for Plants"

20.19.1 Aquaponics Greenhouse

While not essential, a greenhouse offers significant protection to an aquaponics system compared to outdoor exposure to elements like rain, hail and wind. For small indoor setups, maintaining ideal light, temperature and humidity levels can be achieved through artificial lighting, ventilation, aquarium heaters and covers. In colder months, additional measures such as insulation, solar panels and aqua heaters may be employed to regulate fish tank temperatures.

In large-scale commercial aquaponics systems, providing optimal environment for all the organisms, *viz.* fish, plants and bacteria, is best achieved within a greenhouse or dedicated grow room. For more details on environmental conditions for plants, please see the Chapter "Environmental Conditions for Plants".

20.20 Filtration

Filtration in aquaponics is a crucial process that helps maintain water quality by removing solid waste, ammonia, nitrites and other contaminants from the system. Inadequate filtration whether it is through a small pump, the grow bed or an external bio-filter can lead to water clarity issues, especially with increased fish stocking densities. A foul odour emanating from water is a clear indication of trouble in the aquaponic environment. If water impurities reach critical levels it is imperative to halt fish feeding and perform a partial water change. Removing uneaten food from the bottom of tank is crucial to prevent its decomposition which can result in an ammonia buildup within the system.

20.20.1 Filteration Methods

Here is an overview of filtration methods commonly used in aquaponics:

20.20.1.1 Biofiltration

Biofiltration within aquaponics involves the use of biological agents ~ primarily aerobic bacteria to metabolize and remove harmful substances such as ammonia and nitrite from the system. Biofilters such as trickle filters or moving bed biofilters providing ample surface area for bacterial colonization are essential component of the aquaponic ecosystem.

This process transforms fish excreta and excess feed into forms readily absorbable by plants. In media-based systems such as those utilizing gravel or expanded clay, the medium provides ample surface area for bacterial colonization

~ serving as the primary biofilter. Conversely, in raft systems the combination of plant roots and raft surface area fulfills this function. Additional biofilters can be integrated into the system to enhance waste processing. These lightweight containers typically contain biomedia designed for maximum surface area which may facilitate bacterial colonization. Placed between the fish tank and grow beds, these biofilters supplement existing filtration mechanisms. While not always necessary in raft or media-filled systems due to sufficient surface area, these are crucial in NFT setups to ensure optimal bacterial colonization and waste processing.

The establishment of biofilter begins during initial introduction of ammonia into the system either through fish waste or synthetic means ~ a process known as 'cycling.' Cycling is considered complete when bacterial populations are sufficient to convert ammonia and nitrites to nitrates resulting in zero ammonia and nitrite levels. It is advantageous to establish the biofilter with ammonia compounds before adding fish to the system. Over time, the biofilter matures and becomes more efficient at converting fish waste into plant nutrients. This maturation process typically spans 2-4 weeks during which careful monitoring of water parameters is essential.

Adding earthworms to grow beds can aid in keeping the media clean as these thrive on fish waste. It is important to note that fish can tolerate higher levels of total ammonia nitrogen if pH remains below 7.0.

20.20.1.2 Solid Removal Filteration

The primary purpose of solid removal in aquaponics is to maintain water quality by preventing the accumulation of solid waste which can lead to clogs, nutrient imbalances and poor water clarity. Removing solid waste helps to reduce the organic load on the system thus minimizing the risk of ammonia spikes and other water quality issues.

Maintaining a balanced ratio of fish to plants helps minimize solid waste accumulation. In a fully cycled media-based aquaponics system with appropriate fish-to-plant ratios and well-established bacteria colonies *cum* earthworm populations, a solid removal filter may not be necessary. These beneficial organisms effectively decompose fish waste into plant nutrients thus obviating the need for additional solid removal filtration. The selection of a solid removal system depends on the organic waste production within the system which may accumulate in grow beds or settle in the tank. Here is a comprehensive overview:

20.20.1.2.1 Mechanical Filters

Mechanical filtration is integral to managing solid waste in aquaponic systems especially when fish waste surpasses the biofilter and plant capacity. Devices like mesh screens, filter pads and socks physically capture suspended solid particles as water passes through. Effectively these filters trap larger debris such as uneaten fish food and waste so preventing these from entering the grow beds.

One basic mechanical filtration method involves using a foam block beneath the irrigation pipe entering the grow beds. This straightforward approach efficiently traps sludge and food remnants thus necessitating regular cleaning. Suspended solids in filter tanks adhere to orchard netting and are cleaned out weekly. Filtered waste facilitates conversion of ammonia and other waste products before delivering nutrient-rich water to hydroponic vegetables. Rapid settling of solids reduces waste buildup in grow beds thus enhancing oxygen availability to plant roots. Repurposing filtered waste for worm beds, fruit tree fertilizer or ground plants maximize resource utilization.

Smaller systems often employ submersible aquarium filters to collect settled solids from the tank. Larger setups utilize clarifiers and netting with the latter installed post-clarifier to capture excess organic waste. Routine net cleaning weekly prevents solids from entering grow beds and creating anaerobic conditions that foster toxic substances.

20.20.1.2.2 Clarifier and Swirl Filter

More advanced mechanical filtration uses clarifiers and swirl filters that are indispensable components within aquaponic systems which are vital for maintaining water quality and ensuring efficient operation of the setup. These collect solids from all or part of the fish tank water which are manually removed and serve as valuable fertilizer for soil-based plants.

Clarifiers serve a crucial role in aquaponic systems by facilitating the settling of suspended solids from water column. This process involves allowing water to flow slowly through a large tank or chamber where gravity causes heavier particles to settle at the bottom. Typically, clarifiers feature a conical or sloped bottom design to aid in the collection and removal of settled solids. Aquaponics growers often rely on clarifiers employing various methods to remove solids effectively from the water column. Commonly used clarifier designs include conical clarifiers and settling basins which facilitate the settling of solids out of the water column. It is essential for clarifiers to have sufficient storage capacity to maintain required water volume and retention time. The water flow within a clarifier follows a specific path ~ entering from fish tank to central baffle then moving either through the discharge baffle or into the outlet leading to filter tanks or exiting through the sludge drain. Baffle obstacles strategically placed within the clarifier aid in trapping suspended particles as the water flows downward, upward and out through the tank. To remove settled solids from the clarifier growers can manually open the valve fitted at the end of the sludge drain pipe or utilize a manual siphon mechanism similar to a toilet siphon for extraction.

In systems utilizing raft and NFT the necessity for solid removal becomes particularly evident as solid waste can interfere with functionality of system components. While media-filled beds can trap solids within the media, clarifiers remain beneficial even in these systems especially when dealing with substantial solid waste accumulation.

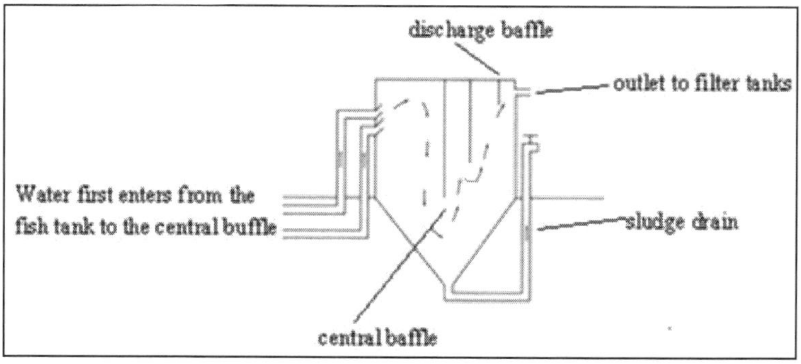

Clarifier

Swirl filter rotates water gently in a sweeping arc encouraging the gradual settling of solids. Positioned at the centre of the tank just below the water surface the outlet pipe captures and redirects clarified water efficiently. Over time, solid particles descend in a slow swirling motion, eventually accumulating at the tank bottom.

Constructed as a round tank with sufficient water volume for the required retention time the swirl filter's inward slope directs water tangentially along the tank wall upon entry. This tangential flow induces a circular motion within the tank thus facilitating sedimentation in the centre. The outlet at the centre of the bottom of tank enables easy removal of settled solids *via* hydraulic pressure ~ a valve can be opened to suction out solids with water flow.

Clarifiers and swirl filters employ sedimentation techniques harnessing gravity to settle solid particles at the tank bottom. Basic sedimentation devices such as rectangular tanks facilitate slow water flow allowing solids to settle before exit. Proper design and sizing of sedimentation devices hinge on two key factors: i) the flow rate of water through the device dictated by water exiting fish tank and device volume, and ii) the retention time which determines efficacy of solid particle separation.

While clarifiers and swirl filters remove a significant portion of solid matter, additional tanks equipped with netting capture remaining particles before water proceeds to vegetable-growing raft systems. Without regular cleaning, accumulated solids can coat tank pipes and fittings impeding system function.

20.20.1.2.3 Screen Filters

Screen filters are designed to separate solids from water by employing various materials that allow water to pass through while trapping solids. The effectiveness of solid removal depends on pore size of screen material. Several materials can be utilized for screens including mesh screens akin to those used in screen printing, plastic woven screens, filter mats and even cotton or foam.

Static screen filters feature stationary screens through which water flows. These filters are commonly used in smaller systems and can be easily constructed by housing the screen material for support. However, these require active management as screens must be regularly removed, cleaned and reinserted. One potential drawback is for frequent clogging particularly in systems with high fish numbers or biomass.

In contrast, moving screen filters utilize screens that are in motion allowing entire surface to come into contact with the water being filtered.

20.20.1.2.4 Drum Screen Filter and Belt Screen Filter

The drum screen filter stands out as the preferred choice for large commercial aquaponic systems due to its automatic and self-cleaning capabilities. In this design, a screen is affixed to a rotating drum allowing unfiltered water to enter the interior of the drum and pass through the screen to the exterior. Inside the drum, a sensor monitors water levels. Over time, as the screen becomes obstructed with solids, water flow diminishes causing water level within the drum to rise. Once a predetermined threshold is reached the sensor triggers the filter to initiate a 'back flush' cycle.

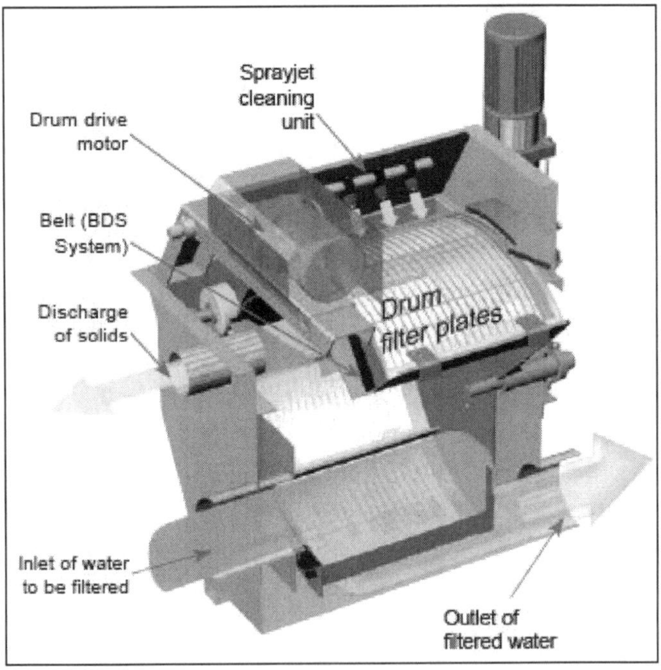

During this back flush operation a series of high-pressure sprays target the exterior of the drum dislodging and flushing accumulated solids from the screen. The dislodged solids are directed into a channel within the drum positioned opposite the high-pressure spray nozzles. This process effectively cleans the screen and guides the solids towards an outlet for disposal.

A similar alternative to the drum screen filter is the belt screen filter which functions on the same principle but utilizes a rotating belt instead of a drum. Both systems offer efficient and automated filtration ensuring the sustained cleanliness and functionality of aquaponic systems on a commercial scale.

20.20.1.2.5 Media Bed Filtration

Media bed filtration functions akin to screen filtration utilizing the porous structure of the media to facilitate water flow. The interstitial spaces between media particles serve as channels through which water passes. Larger solid particles such as fish waste are intercepted and separated from the water column if these exceed the pore size between media particles.

However, inadequate sizing of media beds to handle solid loads can lead to rapid clogging. In such cases water may bypass the media bed and flow out from the top instead of filtering through the media. The necessity for frequent cleaning of media bed filters demands significant labour and requires halting the system temporarily.

20.20.1.3 Chemical Filtration

While less commonly used in aquaponics, chemical filtration involves the use of specialized media such as activated carbon or zeolite to remove impurities or excess nutrients from water. Chemical filtration can help address specific water quality issues such as excess dissolved organic compounds or heavy metals but should be used judiciously to avoid disrupting the delicate balance of the aquaponic system.

20.20.1.4 Oxygenation

Oxygenation is an integral part of filtration in aquaponics as adequate oxygen levels are essential for fish health and beneficial bacteria. Aeration devices such as air stones or diffusers are often used to increase dissolved oxygen levels in the water, promoting efficient biological filtration and preventing anaerobic conditions that can lead to nutrient imbalances or water quality issues.

20.20.2 Mineralization Tank

Mineralization tanks in aquaponics are specialized components dedicated to further processing solid waste removed from the system. These tanks facilitate the conversion of organic waste such as uneaten fish food and fish excrement into mineral forms that are readily absorbed by plants. Here is a detailed overview of mineralization tanks in aquaponics:

i) **Purpose:** The primary purpose of mineralization tanks is to break down solid organic waste into mineralized forms such as nitrates, phosphates and other essential nutrients that can be readily absorbed by plants. By converting organic waste into plant-available nutrients mineralization tanks contribute to overall health and productivity of aquaponic system.

ii) **Process:** Solid waste collected from aquaponic system whether through mechanical filtration or other means is directed into mineralization tank. Within the mineralization tank beneficial bacteria facilitate decomposition of organic matter thus converting it into mineral forms that can be utilized by plants. Nitrifying bacteria play a crucial role in this process converting ammonia produced by fish waste into nitrites and then into nitrates which serve as a vital source of nitrogen for plants.

iii) **Components and design:** Mineralization tanks are typically constructed as large durable containers with sufficient volume to accommodate solid waste and provide adequate space for bacterial activity. Porous media such as gravel, lava rock or bio-media may be added to the tank to enhance surface area for bacterial colonization and facilitating waste breakdown. Aeration systems such as air stones or diffusers are often incorporated into mineralization tanks to ensure sufficient oxygenation for bacterial activity.

iv) **Benefits:** Effective mineralization helps maintaining water quality by reducing accumulation of organic matter in the system which can lead to ammonia spikes and other water quality issues. Mineralized nutrients produced in the tank are readily available for plant uptake, supporting healthy growth and productivity. By reducing reliance on external fertilizers, mineralization tanks contribute to the sustainability and self-sufficiency of the aquaponic system.

v) **Maintenance:** Regular maintenance of mineralization tanks is essential to ensure efficient nutrient cycling and preventing the buildup of organic matter. This may involve periodic cleaning of the tank and media, adjustment of water parameters to optimize bacterial activity and replenishment of bacterial populations as needed.

20.20.2.1 Importance of Mineralization/Nitrification Unit and Solid Removal Filters

Allowing solids to remain in the main aquaponics system for mineralization can lead to several challenges. Firstly, aerobic bacteria responsible for mineralization will compete for dissolved oxygen in both the grow beds and fish tank. As solids break down into organic macromolecules, these consume dissolved oxygen so leading potentially to oxygen depletion in critical areas of the system. These macromolecules may also disperse throughout the system contributing to unwanted mineralization in areas where oxygen competition is undesirable such as fish tanks or deep flow beds.

Accumulation of solids in certain parts of the system can create anaerobic zones where anaerobic mineralization occurs. This process can release harmful elements that may pose risks to fish and plants in the system.

In aquaponics systems where solids are removed using solid removal filters and mineralization is conducted separately in dedicated units, these challenges are mitigated. By separating solid removal and mineralization processes, competition for dissolved oxygen in fish tanks and grow beds is minimized thus reducing the likelihood of anaerobic zones developing. This approach promotes optimal oxygen levels throughout the system thus supporting health and productivity of both fish and plants.

20.21 Size of an Aquaponics System and Fish to Plant Ratios

The fish-to-plant ratio in aquaponics hinges on several factors chiefly the waste production of the fish and the nutrient demands of the plants. Understanding these dynamics is crucial for designing a balanced and thriving system:

i) **Fish waste production:** The quantity of waste generated by fish depends on factors such as species, feeding habits and feed quality. Different species produce varying amounts of waste. Growers must assess the approximate waste output of their chosen fish species to determine the appropriate stocking density.

ii) **Fish and tank size:** The number of fish to be raised is determined by factors like size of the fish and volume of the tank. As a rule of thumb aim for around 1 pound of fish per 5-7 gallons of tank water. This guideline ensures adequate space and water quality for the fish to thrive.

iii) **Plant nutrient needs:** The number of plants to be cultivated depends on the waste produced by the fish. Typically, it is recommended to have 1 pound of fish waste for every square foot of grow bed surface area assuming that beds are at least 30cm deep. Grow bed size and configuration are tailored to accommodate desired plant density and diversity.

iv) **System design factors:** Several factors influence system design, including:

- Number, type and size of fish species to be cultivated.
- Water volume requirements for well-being of the fish.
- Quantity and quality of fish feed necessary to sustain the fish population.
- Space and nutrient demands of the selected plants.
- Grow bed depth of at least 30cm to support a variety of plants and facilitate complete filtration and bacterial activity.

v) **Adjustments for young fish:** When raising young or small fish it is essential to adjust the number of plants accordingly to match their lower feed rates and waste production.

By considering these factors carefully and tailoring the system to meet the specific needs of the fish and plants, growers can create a well-balanced aquaponics system that maximizes efficiency and productivity.

20.21.1 Recommended Fish Stocking Densities

Stocking density in aquaponics refers to the amount of fish stocked in relation to the volume of water in the system. It is commonly expressed as the weight of fish (in kilograms) per unit volume of water (in litres) typically denoted as kg/m^3 (where $1m^3$ of water equals 1000 litres).

Commercial aquaponics operations often maintain fish at densities ranging up to nearly $50kg/m^3$ in aerated systems and up to approximately $150kg/m^3$ or higher in systems with direct oxygen injection. Conversely, hobbyist growers usually maintain fish at lower densities typically below $20kg/m^3$.

20.21.2 Flow Rate of Water

Proper water flow through fish tanks ensures timely removal of waste, effective biofiltration, adequate oxygen levels and uniform water parameters throughout the system. While hobby aquaponic setups with low fish stocking densities may have fewer concerns regarding water quality, optimizing water flow remains essential for overall system health and productivity.

The flow rate of water through the fish tank is influenced by fish density housed within it, *e.g.*:

- For low to medium fish densities (up to $15kg/m^3$) it is advisable to maintain a minimum exchange rate of half the fish tank volume per hour.
- For higher densities exceeding $15kg/m^3$ a minimum exchange rate of one full fish tank volume per hour is recommended to ensure adequate water circulation and oxygenation for the fish.

Several factors influence and maintain the desired flow rate in aquaponic systems:

i) **Volume of water to move:** Calculate the total volume of water to be moved per hour to maintain optimal water quality for the fish. This includes the combined volume of grow beds, fish tank and sump tank. Water volume in the fish tank should be exchanged at a rate that meets the water quality needs of the fish. The flow rate must strike a balance to avoid energy wastage for fish swimming against strong currents or health issues from low water flow.

ii) **Tank shape and size:** Round tanks are preferred for commercial aquaponic systems due to uninhibited water flow. Corners in rectangular tanks can hinder water flow and lead to areas with poor water exchange. In circular tanks, solids tend to settle at the centre thus facilitating easier removal. Non-circular tanks may pose challenges in determining optimal placement for outlets or pumps.

iii) **Water movement method:** Consider the height to which water needs to be pumped and leverage gravity where possible. For instance, if grow beds are positioned higher than the fish tank, water can be pumped from the tank to the beds and allowed to flow back by gravity.

iv) **Pump selection:** Choose a high-quality pump and decide whether it should run continuously or operate on a timer for a flood and drain system. Consider the pump's flow rate and head pressure, ensuring it meets the system's requirements efficiently. It is advisable to select a slightly larger pump for future system expansions.

v) **Pipe design:** Water flows faster through straight pipes and slower through those with bends. Ensure pipes are adequately sized to handle water volume and incorporate controls and safety measures. Joints and bends decrease water flow rates.

vi) **Pipe size:** Use smaller diameter pipes to deliver water to grow beds and larger ones to remove water, ensuring proper drainage and preventing overflow. Maintain slightly larger pipe diameters to minimize blockages and clean pipes regularly to prevent debris buildup.

vii) **Connectors:** Common connector types include threaded and slip connectors. Choose connectors based on system needs and compatibility with pipe materials.

20.21.3 Ratio of Water Volume to Surface Area of Fish Tank

The ratio of water volume to surface area in the fish tank is crucial for effective gas exchange including release of carbon dioxide and uptake of oxygen. In deep and narrow tanks with low surface areas relative to their volumes adequate oxygen may struggle to penetrate the water and carbon dioxide may not efficiently escape. Hence, maximizing the surface area to volume ratio is essential to ensure optimal gas exchange within the tank.

20.22 Water Change

It is advisable to avoid large water changes in an aquaponics system unless absolutely necessary as these can induce stress in the fish. Large and sudden changes in pH levels resulting from significant water changes can be detrimental to fish health. Instead, a gentler approach involves using a self-regulating ball valve to top up the water in the fish tank as it naturally evaporates or is transpired by the plants. When water changes are required it is recommended to change no more than one-third of the water volume in the system at a time. This gradual approach helps maintain stable water conditions and minimizes stress on the fish.

20.23 Nutritional Supplements for Plants and pH Control

In aquaponic systems, the majority of plant nutrients are supplied by fish waste and feed. However, certain deficiencies may arise in plants necessitating the addition of supplements. Ca, K, Fe and occasionally Mg are commonly required. It is crucial to use nutrient supplements judiciously as some may have adverse effects on fish health.

In a well-established system, the nitrification process can cause the pH to gradually decrease over time. Adjusting pH while adding essential nutrients like Ca and K can be achieved by alternating between potassium hydroxide and calcium hydroxide. Some prefer potassium carbonate and calcium carbonate but these carbonates can buffer pH so making precise adjustment challenging. Chelated iron is commonly used to supplement iron in the system.

To lower pH and provide phosphorus, phosphoric acid can be utilized as this benefits tomatoes or cucumbers that require additional nutrients for fruiting and flowering. Alternatively, rock phosphate can be added to supply phosphorus if pH adjustment is unnecessary.

Maintaining optimal pH levels is crucial for nutrient availability and overall system health. Suboptimal pH can lead to toxic material buildup, disrupt bacterial nutrient generation and hinder nutrient uptake by plant roots potentially harming fish. Ideal pH range for aquaponic systems is typically between 6.8 to 7.

In well-managed mature systems with proper fish feed, pH within the optimal range, well-oxygenated water and appropriate temperatures nutrient supplements are usually unnecessary. Under these controlled conditions bacteria within the system provide ample nutrients for plant growth.

20.23.1 Fish Feed and Feeding Guidelines for Aquarium Fish

The quality of fish feed is paramount for the success of an aquaponics system as it profoundly influences the health of both the fish and the plants. Maintaining proper feeding practices is crucial to ensure optimal nutrient availability for robust plant growth. Observing and documenting the feeding habits of fish can offer valuable insights for effective system management.

Commercial fish feed readily accessible at pet or aquarium shops stands as the most convenient option for feeding fish in an aquaponics system. These feeds boast a variety of formulations tailored to suit the dietary requirements of different fish species. Moreover, homemade or live fish foods such as worms, snails, black soldier fly larvae, red worms and duckweed can complement commercial feeds or serve as viable alternatives. A blend of these feed sources can provide a well-rounded diet for the fish.

It is imperative to prioritize the health and well-being of the fish by ensuring that these receive high-quality feed and inhabit a stress-free environment. Fish may encounter stress due to physical, chemical or biological factors which can lead to reduced appetite and compromised health. Regular monitoring of water quality, temperature and other environmental conditions is indispensable for mitigating stress and fostering fish health within the aquaponics system.

While hobby growers have the flexibility to choose fish food based on availability and suitability, commercial aquaponics systems typically benefit from professionally formulated feeds. Carnivorous fish such as perch and trout thrive on feeds rich in proteins while omnivorous species like tilapia, koi and catfish prefer feeds with lower protein levels. Younger fish have higher protein requirements compared to mature counterparts. Ensuring proper nutrition tailored to the specific needs of the fish species contributes to their overall health and vitality within the aquaponics system.

In aquaponics systems, proper feeding practices are essential to maintain health and well-being of aquarium fish. Here are some guidelines to ensure optimal feeding and prevent issues such as overeating and water contamination:

i) **Feed little and often:** It is generally better to feed aquarium fish little and often rather than large meals infrequently. Aim to feed them at least twice a day though ideally 3-4 times.

ii) **Monitor feeding amounts:** As a guideline, mature fish such as tilapia typically consume about 1 per cent of their body weight per day. Fingerlings may consume up to 7 per cent of their body weight.

iii) **Remove uneaten food:** Any uneaten food should be promptly removed from the system to prevent it from sinking and decomposing, which can consume oxygen and increase ammonia levels in the water.

iv) **Feed for a few minutes:** Feed the fish only as much as they can consume within a few minutes usually 5-6 minutes. This helps prevent overfeeding and minimizes waste.

v) **Use floating pellets:** Consider using floating pellets for feeding as these make it easier to monitor how much food is being consumed. This allows to adjust feeding amounts accordingly.

20.24 Understanding Fish Stress in Aquaponics Systems

Fish stress is a significant concern in aquaponics systems as it can compromise the health and well-being of fish, affecting their survival and overall system balance. Here is a detailed exploration of various factors contributing to fish stress and how to mitigate them effectively:

i) Physical and Psychological Stress

- **Environmental conditions:** Fish have specific temperature ranges within which they thrive. Sudden fluctuations in temperature can disrupt their metabolic processes and make them more susceptible to diseases.
- **Light exposure:** Abrupt exposure to bright light can startle fish causing them to panic and potentially injure themselves by colliding with tank walls.

- ☆ **Vibrations:** Fish can perceive vibrations in their environment and sudden disturbances such as tapping on the tank can induce stress.
- ☆ **Water velocity:** Some fish species prefer calm waters and may become stressed in tanks with high water flow rates.
- ☆ **Water changes:** Large and sudden changes in water parameters such as pH and temperature can stress fish. Gradual acclimation is essential when introducing new fish or altering tank conditions.
- ☆ **Transportation:** Handling and transportation can damage the protective slime coat of fish, weakening their immune system and making them more vulnerable to stress-related illnesses.
- ☆ **Handling and netting:** Excessive handling or netting of fish should be minimized to avoid unnecessary stress. Care should be taken to use proper techniques and equipment to prevent injury to the fish.
- ☆ **Social interactions:** Aggressive behaviour from tankmates such as territorial disputes or bullying can induce stress in fish.

ii) Chemical Stress

- ☆ **Water quality:** Poor water quality characterized by high levels of ammonia, nitrites or anaerobic zones can be detrimental to fish health.
- ☆ **pH imbalance:** Extreme pH levels, either too high or too low, can stress fish and compromise their physiological functions.
- ☆ **Insufficient filtration:** Inadequate filtration and low dissolved oxygen levels can lead to chemical imbalances in the water, exacerbating stress.
- ☆ **Overstocking:** Overcrowding the tank with fish can exacerbate water quality issues and increase stress levels.

iii) Biological Stress

- ☆ **Disease:** Infections caused by viruses, bacteria, fungi and parasites can induce biological stress in fish populations. Proper disease prevention measures and monitoring are essential for maintaining fish health.
- ☆ **Salt treatment:** While salt can be used to treat certain fish diseases, it should be applied cautiously to avoid harming other organisms in the system.

Overall, minimizing stress factors and promoting a healthy environment are crucial for supporting fish health and immune function in aquaponics systems. Regular monitoring of water parameters, careful handling practices and appropriate stocking densities can help mitigate stress and ensure the well-being of fish populations.

20.25 Salt Usage in Aquaponics Systems

In aquaponics, salt is commonly utilized for parasite control and stress reduction in fish. Here is a concise guide on salt usage in aquaponics systems:

i) **Purpose of salt:** Salt is used in aquaponics to eliminate fish parasites and alleviate stress. When fish are stressed, these release minerals into the water which can be fatal if prolonged. Adding salt helps increasing fish resistance to diseases and infections by affecting their mucous layer positively.

ii) **Recommended salinity level:** The normal recommended salinity level for dosing fish is typically in the range of 0.5-3 parts per thousand (ppt) or kg/1000 litres of water. It is important to note that some plants such as strawberries may not tolerate salt levels beyond 3 ppt.

iii) **Avoid overdosing:** Excessive salt dosage can harm all life forms in the aquaponics system including plants, fish and beneficial bacteria. Therefore, it is crucial to adhere to recommended salinity levels to prevent adverse effects.

iv) **Salt selection:** Table salt may contain potentially harmful anti-caking agents, making sea salt a preferable choice for aquaponics systems. Alternatively, using salt specifically formulated for fish is recommended to ensure the safety of all organisms in the system.

v) **Dissolving and application:** Prior to adding salt to the fish tank it should be completely dissolved in a separate container. This ensures even distribution and prevents clumping. Once dissolved, the salt solution can be gradually added to the fish tank.

By following these guidelines, aquaponics practitioners can effectively utilize salt to maintain fish health and promote overall system balance without risking harm to plants or beneficial bacteria.

20.26 Managing Insect-Pests and Diseases in Aquaponic Systems

Dealing with plant pests and diseases in aquaponics requires careful attention to maintain the health of both the plants and the aquatic ecosystem. Here are some effective methods for managing pests and diseases:

a) Physical Methods

i) **Water blast:** A strong blast of water from a hose can help kill and wash away pests from plant surfaces.

ii) **Hand-picking:** Remove pests such as caterpillars by hand and drop them into a solution of soapy water.

iii) **Shaking off:** Shaking plants infested with pests like potato beetles onto a sheet to remove them.

iv) **Sticky traps:** Bright sticky traps are effective for capturing and monitoring insect-pests.

v) **Beneficial insects:** Introduce beneficial insects such as 'ladybugs beetles', 'praying mantis', *Trichogramma* wasps, *Encarsia formosa, etc.* to control various pests and their larvae.

b) Organic pesticides

☆ Homemade or commercial organic pesticides can be used as an alternative method for pest control. However, precautions should be taken to prevent the pesticides from entering the aquaponics system.

☆ Covering the grow bed media and fish tank with a plastic sheet or other materials can help prevent excessive pesticide exposure to the system. Pesticides should be applied sparingly and in a targeted manner to minimize harm to beneficial bacteria and fish and to prevent disturbances in pH levels.

It is crucial to avoid using pesticides or any other substances that may harm the fish and beneficial bacteria in the aquaponics system. Any diseased plants should be promptly removed to prevent spread of the malady. Above all, any treatment methods used for plants should prioritize the well-being of the fish and minimize stress to the aquatic ecosystem.

See 'Pests and Diseases' under 'Troubleshooting Common Issues'.

20.27 Fish Species for Aquaponics

Fish play a crucial role in aquaponics systems serving as both a source of food and a provider of nutrients for plants. Through thoughtful selection and managing fish species in aquaponics systems growers can optimize productivity while fostering the health and well-being of both the fish and the plants. Here are some key considerations for managing fish in aquaponics:

i) **Diverse species selection:** Aquaponics growers have a wide array of freshwater fish species to choose from including both edible varieties like tilapia, barramundi and silver perch; *vis-a-vis* ornamental species such as koi and goldfish. Freshwater crayfish, mussels and prawns are occasionally incorporated into aquaponics setups.

ii) **Optimal conditions:** Each fish species has specific temperature and pH requirements for optimal health and growth. While maintaining ideal conditions is crucial, it is often necessary to strike a balance as their requirements may conflict with the needs of plants and beneficial bacteria. Typical conditions range from 21-23°C and a pH range of 6.4-7.0.

iii) **Temperature management:** Regulating the temperature of the fish tank can be achieved through the use of cooling and heating equipment. However, growers can also stabilize temperature and reduce energy costs by selecting fish species native to their region and employing insulation and aquarium cover cloth.

iv) **Species compatibility:** Different fish species with similar environmental requirements can coexist in the same aquaponics system. However, caution must be exercised to avoid mixing aggressive carnivores with omnivores or introducing fish of significantly different sizes. Fast-growing species should be kept separate from slower ones to ensure equitable access to food whereas larger and aggressive fish may pose a threat to smaller *cum* more passive species.

v) **System size consideration:** Larger aquaponics systems such as ponds offer greater flexibility in accommodating a variety of fish species due to ample space available. However, even in smaller systems, careful species selection and management can foster harmony among aquatic inhabitants.

20.27.1 Some Fish Species Suitable for an Aquaponics System

i) Tilapia (*Oreochromis* spp.) ~ Freshwater Edible Species

Growing tilapia in aquaponics can be a rewarding endeavour as these are hardy fish that thrive in a variety of environmental conditions. Here is a guide to cultivating tilapia successfully in an aquaponics system:

☆ **Variability in tilapia types:** Tilapia encompasses a diverse range of species each with unique characteristics. There are several species of tilapia used in aquaculture with Nile tilapia (*Oreochromis niloticus*) being one of the most commonly cultivated one. Other species like Mozambique tilapia (*Oreochromis mossambicus*) and blue tilapia (*Oreochromis aureus*) are also popular choices. Consider factors such as temperature tolerance, growth rate and local regulations when selecting the tilapia species for aquaponics system. Tilapia males exhibit faster growth rates as compared to females. To ensure uniform growth some farmers opt for hormone treatment of fingerlings to induce male development in all individuals.

- **Feeding habit:** Tilapias are omnivorous by nature displaying a broad dietary spectrum. Mature tilapias usually consume around 1 per cent of their body weight daily while fingerlings can exhibit higher consumption rates of up to 7 per cent. Their diet can include pellet fish food, duckweed, algae, worms and other organic matter.
- **Water parameters:**

  pH range: Tilapias thrive within a pH range of 7-8 though aquaponics systems typically maintain a slightly lower pH (around 6.8-7.0) to accommodate both fish and plant preferences. Gradual adjustments are key to ensuring fish health and acclimatization.

  Salinity: Tilapias can tolerate salinities ranging from 16 to 20 parts per thousand, making them adaptable to various water conditions. Adding salt to the water can deter pests and create thermal stratification in deep ponds during colder seasons.

  Temperature preferences: Tilapias exhibit diverse temperature tolerances depending on their species. While some breeds thrive in temperatures ranging from 12 to 36°C though others prefer narrower ranges such as 25 to 30°C or 31 to 36°C. Certain tilapia strains may endure short-term exposure to temperatures as low as 8 or 9°C.

  Dissolved oxygen (DO) levels: Tilapia have relatively low oxygen requirements with optimal DO levels ranging from 3 to 10mg/l. While these can survive in lower oxygen environments but maintaining adequate oxygenation is essential for optimal growth and health.
- **Tolerance to environmental changes:** Tilapias are renowned for their hardiness and resilience to fluctuating water conditions. These can withstand shifts in pH, temperature variations, elevated ammonia levels and low dissolved oxygen levels thus making these well-suited for aquaponic systems subject to dynamic environmental factors.
- **Tank size and stocking density:** For tilapias there should be adequate space to swim and grow so appropriately sized tanks or ponds should be selected in aquaponics system. Stock tilapia at a density that allows for efficient nutrient cycling and preventing overcrowding typically ranging from 10 to 20 kilograms of fish per cubic metre of water.
- **Harvesting tilapia:** Tilapia can reach market size within 6 to 9 months depending on factors such as water temperature, feeding rate and genetics. Harvest tilapia by netting or draining the tank once these reach the desired size, *i.e.* typically around 250 to 500 grams.

ii) Rainbow Trout (*Oncorhynchus mykiss*)

Rainbow trouts are freshwater edible species requiring clean and cold water. These are known for their vibrant colours and delicious taste. Raising rainbow trout in an aquaponics system can be a rewarding experience as these are prized for their vibrant colours and delicious flavour. Here is a guide to grow rainbow trout in an aquaponics setup successfully:

- ☆ **Selecting rainbow trout:** Rainbow trout are a popular choice for aquaculture due to their adaptability to various environmental conditions. When selecting rainbow trout for aquaponics system choose healthy fingerlings from a reputable hatchery or supplier.
- ☆ **Water temperature and quality:** Rainbow trouts thrive in cool and clean water with temperatures ranging from 10 to 18°C though these can tolerate temperatures up to 22°C for short periods. These are sensitive to changes in pH, water quality and temperature fluctuations. Ensuring water quality parameters such as pH (ideally between 6.5 to 8.0), dissolved oxygen levels (above 5 mg/l) and ammonia levels should be within acceptable ranges. DO should not drop below 5.5 mg/l. Above 8mg/l is preferable for trout culture. Provision of ample aeration and water circulation should be there to maintain oxygen levels and to prevent stratification.
- ☆ **Feeding rainbow trout:** Rainbow trout are carnivorous and require a diet rich in protein. These may prey on smaller fish in the system. Feed rainbow trout high-quality commercial fish pellets formulated specifically for trout as well as supplemental live or frozen foods such as insects, larvae and small fish. Monitor feeding rates based on fish size, water temperature and growth rates. Overfeeding should be avoided to prevent water quality issues.
- ☆ **Tank size and stocking density:** Provide rainbow trout with ample space to swim and grow by selecting appropriately sized tanks or raceways for the aquaponics system. Stock rainbow trout at a density that allows for efficient nutrient cycling and prevents overcrowding, typically ranging from 15 to 20 kilograms of fish per cubic meter of water.

- ☆ **Harvesting rainbow trout:** Rainbow trout can reach market size within 12 to 18 months depending on factors such as water temperature, feeding rate and genetics. Harvest rainbow trout by netting or draining the tank once these reach desired size, typically around 250 to 500 grams.

iii) Silver Perch (*Bidyanus bidyanus*)

Growing Silver Perch in aquaponics can be a rewarding endeavour as these are known for their adaptability, rapid growth and suitability for aquaculture. Here is a guide to raising Silver Perch successfully in an aquaponics system:

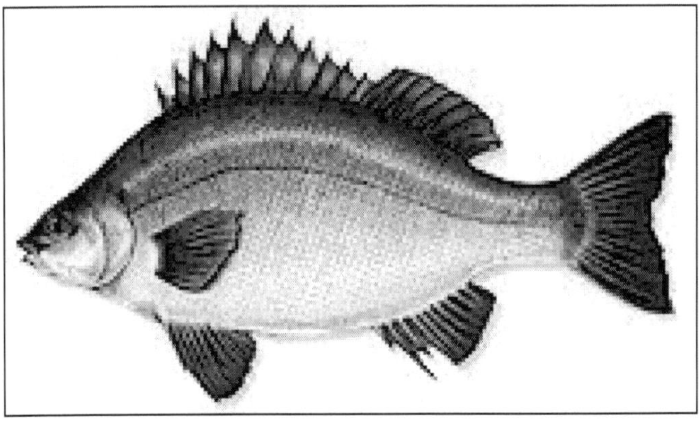

- ☆ **Tank setup:** Provide a spacious tank or pond with ample water volume to accommodate Silver Perch population. Ensure the tank has sufficient aeration to maintain adequate oxygen levels for the fish.
- ☆ **Water quality:** Monitor and maintain water parameters within suitable ranges, including pH (6.5-8.0), temperature (preferably 20-26°C), ammonia, nitrite and nitrate levels. Use a reliable filtration system to remove solid waste and maintain water clarity.
- ☆ **Feeding:** Offer a balanced diet consisting of commercial fish pellets appropriate for Silver Perch. Supplement their diet with live or frozen foods such as insects, worms and small crustaceans for variety and enrichment.
- ☆ **Stocking density:** Avoid overcrowding by maintaining appropriate stocking densities based on size of the tank and growth rate of the fish. Provide adequate space for each fish to swim freely and exhibit natural behaviours.
- ☆ **Compatibility with plants:** Silver Perch produce waste rich in nutrients that can benefit hydroponic plants in the aquaponics system. Ensure a balanced ratio of fish to plants to optimize nutrient uptake and minimize waste accumulation.

- ☆ **Harvesting:** Silver Perch are typically ready for harvest within 12-18 months depending on factors such as water temperature and feeding regimen. Harvest fish selectively based on size and growth rate to maintain a balanced population in the aquaponics system.

iv) Goldfish (*Carassius auratus*) and Koi (*Cyprinus carpio*) ~ Freshwater Ornamental Carps

Goldfish and koi are both popular ornamental fish species that can be raised in aquaponics systems. Here is a guide to growing goldfish and koi in aquaponics:

- ☆ **Tank setup:** Provide a tank or pond with ample space for goldfish and koi population, ensuring adequate water volume and filtration. Both goldfish and koi prefer water temperatures between 18 to 24°C though these can tolerate a wide range of temperatures (8-32°C). Although these have low oxygen needs but for best growth growers must ensure the tank has sufficient aeration to maintain high oxygen levels especially in warmer weather when oxygen levels may decrease.
- ☆ **Water quality:** Monitor and maintain water parameters within suitable ranges including pH (6.5-8.5), ammonia, nitrite, nitrate levels and dissolved oxygen. Both goldfish and koi produce waste that can contribute to ammonia buildup so it is important to have a robust filtration system to keep water clean and healthy. These are quite hardy and can tolerate a range of water quality parameters.
- ☆ **Feeding:** These omnivorous fish should be given a balanced diet of high-quality commercial fish pellets formulated specifically for goldfish and koi. Their diet supplemented with live or frozen foods such as insects, worms and vegetation to provide variety and enrichment.
- ☆ **Stocking density:** Avoid overcrowding by maintaining appropriate stocking densities based on the size of the tank or pond and the growth rate of the fish. Provide ample space for each fish to swim freely and exhibit natural behaviours.
- ☆ **Compatibility with plants:** Goldfish and koi produce nutrient-rich waste that serves as an excellent fertilizer for hydroponic plants in the aquaponics system. Ensure a balanced ratio of fish to plants to optimize nutrient uptake and minimize waste accumulation.
- ☆ **Harvesting:** While goldfish are typically kept as ornamental fish and not harvested, koi can be harvested selectively for sale or consumption based on size and colouration. If intending to harvest koi ensure these are grown in a separate system from goldfish to prevent hybridization.

v) Jade Perch (*Scortum barcoo*) ~ Warm Water Edible Fish

Jade perch also known as Barcoo grunter are freshwater fish species native to Australia. These are being raised increasingly in aquaponics systems due to their rapid growth, hardiness and suitability for aquaculture. Here is a guide to growing jade perch in aquaponics:

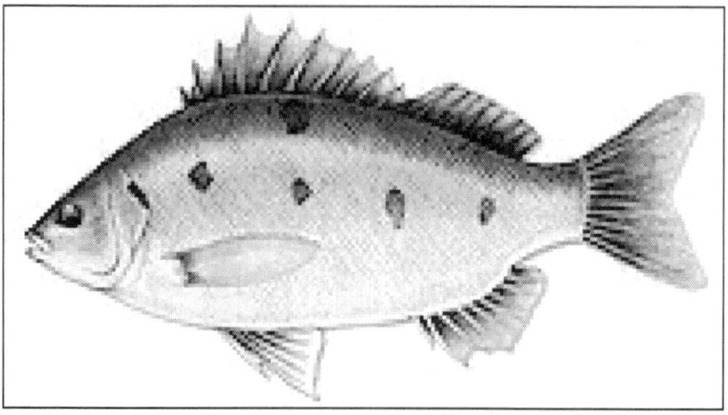

- **Tank setup**: Provide a tank or pond with ample space for the jade perch population, ensuring adequate water volume and filtration. Jade perch thrive in warm water temperatures above 24°C so consider using a heater or locating the system in a suitable climate. Aeration needed to increase when temperature exceeds 30°C. Growth rate declines rapidly when temperature falls below 20°C. Most farmers cease feeding at temperature below 16°C.
- **Water quality**: Monitor and maintain water parameters within suitable ranges including 6.5-8.0 pH range, ammonia, nitrite, nitrate levels and dissolved oxygen of 4 mg/l or greater for maximum growth and survival. Growers must ensure water quality remains stable to prevent stress and promote healthy growth.
- **Salinity**: Up to 5g/l of sodium chloride are acceptable for long term exposure.
- **Feeding** (Omnivorous): Jades readily accept pellets and do well on a low protein and high fibre diet. These are aggressive feeders and care should be taken to ensure that all the fish in the tank get fed not just the dominant fish in the population. Offer a balanced diet of high-quality commercial fish pellets formulated specifically for jade perch. Supplement their diet with live or frozen foods such as insects, worms and vegetation to provide variety and enrichment.
- **Stocking density**: Avoid overcrowding by maintaining appropriate stocking densities based on the size of the tank and the growth rate of the jade perch. Provide sufficient space for each fish to swim freely and exhibit natural behaviours.
- **Compatibility with plants**: Jade perch produces nutrient-rich waste that serves as an excellent fertilizer for hydroponic plants in the aquaponics system. Ensure a balanced ratio of fish to plants to optimize nutrient uptake and minimize waste accumulation.

- ☆ **Tolerance**: These are quite hardy but the biggest issue with jade perch is temperature. Jades stop metabolizing food at water temperatures below 18°C, start to show stress at 16°C and death at prolonged periods below 14°C.
- ☆ **Harvesting**: Jade perch can be harvested once these reach a desirable size typically within 12-18 months depending on factors such as water temperature and feeding regimen. Harvest fish selectively based on size and growth rate to maintain a balanced population in the aquaponics system.

References

1. Adania, F., G. Pierluigi, P. Zaccheoa and G. Zocchia, 1998. The effect of commercial humic acid on tomato plant growth and mineral nutrition. *J. Pl. Nutrit.*, **21**(3): 561–75.
2. Atkin, K. and M.A. Nichols, 2004. Organic Hydroponics. *Acta Hort.*, No. 648: 121–7. ISSN 0567-7572.
3. Benton, J.J. 2004. *Hydroponics: A Practical Guide for the Soilless Grower* (2nd ed.), pp. 29-70 and 225-9. Taylor and Francis, New York. ISBN 978-0-8493-3167-1.
4. Boylan, C. 2020. The future of farming: Hydroponics. *PSCI*. Retrieved Feb 9, 2024. https://psci.princeton.edu/tips/2020/11/9/the-future-of-farming-hydroponics.
5. Carter, W.A. 1942. *A method of growing plants in water vapor to facilitate examination of roots. Phytopath.*, **732**: 623–5.
6. Cooper, A. 1988. 1. The system. 2. Operation of the system. In: *The ABC of NFT. Nutrient Film Technique*, pp. 3-123. Grower Books (ed.), London, England.
7. Gericke, W.F. 1937. Hydroponics - crop production in liquid culture media. *Science*, **85**(2198): 177–8. Bibcode:1937Sci.85.177G.
8. Gericke, W.F. 1938. Crop production without soil. *Nature*, **141**(3569): 536–40. Bibcode:1938Natur.141.536G.
9. Gericke, W.F. 1945. The meaning of hydroponics. *Science*, **101**(2615): 142–3. Bibcode:1945Sci.101.142G. PMID 17800488.
10. Gericke, W.F. 1940. *The Complete Guide to Soilless Gardening* (1st ed.), pp. 9-10, 38 and 84. Putnam, London: ISBN 978-1-163-14049-9.

11. Hoagland, D.R. and D.I. Arnon, 1950. *The Water Cultural Method for Growing Plants without Soil* (Circular 347). California Agricultural Experiment Station, University of California, Berkeley, CA, USA.

12. Horn, M.E., S.L. Woodard and J.A. Howard, 2004. Plant molecular farming: systems and products. *Pl. Cell Rep.*, **22**(10): 711–20. PMC 7079917. PMID 14997337.

13. Jensen, M.H. and A.J. Malter, 1995. Protected Agriculture: A Global Review. *World Bank Technical Paper No. 253*, 157p. World Bank Publications, Washington DC, USA. http://ag.arizona.edu/hydroponictomatoes/nutritio.htm.

14. Kawamura-Aoyama, C, K. Fujiwara, M. Shinohara and M. Takano, 2014. Study on the hydroponic culture of lettuce with microbially degraded solid food waste as a nitrate source. *Japan agric. Res. Quart.*, **48**(1): 71–6. ISSN 0021-3551.

15. Lee, S. and J. Lee, 2015. Beneficial bacteria and fungi in hydroponic systems: Types and characteristics of hydroponic food production methods. *Scient. Hort.*, **195**: 206–15. ISSN 0304-4238.

16. Mc Near, D.H., Jr. 2013. The rhizosphere - roots, soil and everything in between. *Nature Educ.*, **4** (3): 1.

17. Mehandru, P., N.S Shekhawat, M.K. Rai, V. Kataria and H.S. Gehlot, 2014. Evaluation of aeroponics for clonal propagation of *Caralluma edulis*, *Leptadenia reticulata* and *Tylophora indica* – three threatened medicinal Asclepiads. *Physiol. and Molec. Biol. Pl.*, **20**(3): 365–73.

18. Moncada A., F. Vetrano and A. Miceli, 2020. Alleviation of salt stress by plant growth-promoting bacteria in hydroponic leaf lettuce. *Agronomy*, **10** (10): 1523. ISSN 2073-4395.

19. Murashige, T, and F. Skoog, 1962. A revised medium for rapid growth and bio assays with tobacco tissue cultures. *Physiol. Plantar.*, **15** (3): 473–97.

20. Nagel, K.A., B. Kastenholz, F. Gilmer, U. Schurr and A. Walter, 2010. Novel detection system for plant protein production of pharmaceuticals and impact on conformational diseases. *Protein and Peptide Letters*, **17**(6): 723–31. PMID 20015023.

21. NASA Spinoff, 2006. Progressive plant growing has business blooming. *Environ. and agric. Resour.*, pp. 68-72.

22. Nye, P.H. 1981. Changes of pH across the rhizosphere induced by roots. *Plant and Soil*, **61** (1–2): 7–26. S2CID 24813211.

23. Oakes, A.D., N.A. Kazcmar, C.A. Maynard and W.R. Argo, 2009. Vegetative propagation of American elm (*Ulmus americana*) varieties from softwood cuttings. *J. environ. Hort.*, **30**(2):73–6.

24. Peterson, B.J., S.E. Burnett and O. Sanchez, 2018. Submist is effective for propagation of Korean lilac and inkberry by stem cuttings. *HortTechnol.*, **28**(3): 378–81.

25. Phibunwatthanawong, T. and N. Riddech, 2019. Liquid organic fertilizer production for growing vegetables under hydroponic condition. *Intern. J. Recyc. Organic Waste in Agric.*, **8** (4): 369–80. Bibcode:2019IJROW.8.369P.

26. Santos, K.M. and P.R. Fisher, 2009. Stem versus foliar uptake during propagation of *Petunia* x *hybrida* vegetative cuttings. *HortSci.*, **44** (7): 1974–7.

27. Sholto Douglas, J. 1985. *Advanced Guide to Hydroponics: Soilless Cultivation*, pp. 169-87, 289-320 and 345-51. Pelham Books, London. ISBN 978-0-7207-1571-2.

28. ShuHua, K., H. Wei, H. ChiTe, F. Lo and L. Lur, 2015. Application of phototrophic bacterial inoculant to reduce nitrate content in hydroponic leafy vegetables. *Crop, Environ and Bioinform.*, **12**: 30–41.

29. Soffer, H. and D.W. Burger, 1988. Effects of dissolved oxygen concentration in aero-hydroponics on the formation and growth of adventitious roots. *J. Amer. Soc. Hort. Sci.*, **113** (2): 218–21.

30. Steiner, A.A. 1984. The universal nutrient solution. In: *Proc. 6th intern. Congr. Soil. Cult.*, pp. 633-50, held on April 29 to May 5, 1984. ISOSC, Wageningen, Holland.

31. Stoner, R.J. 1983. Aeroponics versus bed and hydroponic propagation. *Flor. Rev.*, **1**: 173.

32. Stutte, G.W., G. Newsham, R.M. Morrow, R.M. Wheeler, 2012. Concept for Sustained Plant Production on ISS using veggie capillary mat rooting system. 41st Intern. Conf. on Environm Syst., pp. 1-17, held on July 17–21, 2011 in Portland, Oregon, USA. hdl:2060/20110011606. ISBN 978-1-60086-948-8. S2CID 13847293.

33. Suryawanshi, Y. 2021. Hydroponic cultivation approaches to enhance the contents of the secondary metabolites in plants. Biotechnolgical Approaches to Enhance Plant Secondary Metabolites, pp. 71-88. CRC Press, USA. ISBN 978-1-003-03495-7. S2CID 239706318.

34. The Guardian, 2024 How much water is needed to produce food and how much do we waste?. *The Guardian*. Retrieved February 9, 2024. https://www.theguardian.com/news/datablog/2013/jan/10/how-much-water-food-production-waste.

35. Trejo-Téllez, L.I., F.C. Gómez-Merino and G.G. Alcántar, 2007. Elementos Benéficos, In: *Nutrición de Cultivos* (eds Alcántar, G and L. I. Trejo-Téllez), pp. 50-91. Mundi-Prensa, México, D. F.

36. Waldner, H. and K. Gunther, 1996. Characterization of low molecular weight zinc species in normal commercial vegetable foodstuffs. *Zeitschrift fur Lebensmittel-Untersuchung und Forschung*, 202 (3): 256–62.

37. Walker, T.S., H.P. Bais, E. Grotewold and J.M. Vivanco, 2003. Root exudation and rhizosphere biology. Pl. Physiol., 132(1): 44–51.

38. Williams, K.A, and J.S. Nelson, 2016. Challenges of using organic fertilizers in hydroponic production systems. Acta Hort., No. 1112, pp. 365–70.

39. Windsor, G. and M. Schwarz, 1990. Soilless Culture for Horticultural Crop Production (paper no. 101). FAO Plant Production and Protection. FAO, Roma, Italia.

40. Zeng, H, C. Xia, C. Zhang and L.Q. Chen, 2018. A simplified hydroponic culture of Arabidopsis. Bio-101. Retrieved February 9, 2024. https://bio-protocol.org/exchange/protocoldetail?id=3121 and type=1.

41. Zhang, H., A. Asutosh and W. Hu, 2018. Implementing vertical farming at University scale to promote sustainable communities: A feasibility analysis. Sustainability, 10(12): 4429. ISSN 2071-1050.

Index

B

D

E

F

G

I

O

P

Q

R

S

T

U

V

W

Z

COLOUR PLATES

Plants/Leaves Showing Nitrogen Deficiency. Pg: 298

Leaves Showing Phosphorus Deficiency. Pg: 299

Leaves Showing Potassium Deficiency. Pg: 300

Symptoms of Magnesium Deficiency Showing Yellowing in Older Leaves. Pg: 301

Symptoms of Magnesium Deficiency Showing Brown Spots on Leaves. Pg: 301

Leaves Showing Calcium Deficiency. Pg: 302

Calcium Deficiency Showing Blossom-End Rot of Tomato. Pg: 302

Leaves Showing Sulphur Deficiency. Pg: 303

Sulphur Deficiency in Corn and Soybean. Pg: 304

Leaves Showing Iron Deficiency. Pg: 305

Leaves Showing Chlorine Deficiency. Pg: 306

Leaves Showing Manganese Deficiency. Pg: 307

Leaves Showing Zinc Deficiency. Pg: 308

Leaves Showing Copper Deficiency. Pg: 310

Boron Deficiency Symptoms in Pepper Leaves. Pg: 311

Leaf Margins become Necrotic and may Curl Upward. Pg: 312

Without Surfactant

With Surfactant Pg: 316

Leaf Showing Sulphur Toxicity Pg: 320

Leaves Showing Boron Toxicity Pg: 321

Algae Pg: 322

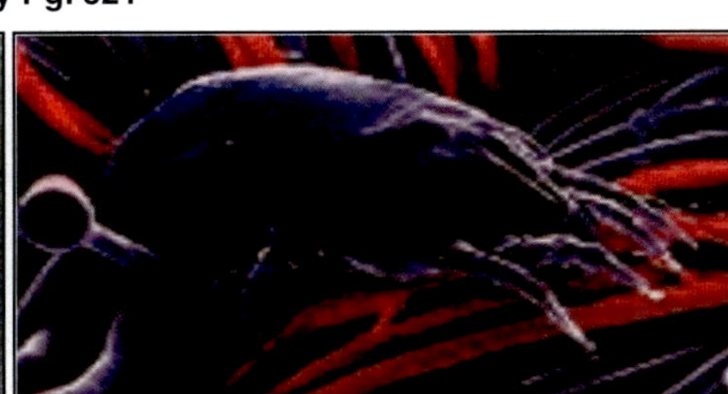

Pg: 340

Ladybug Eating Aphids Pg: 341

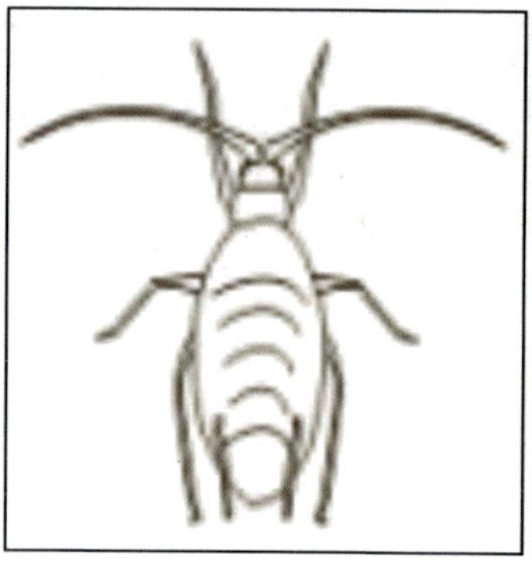

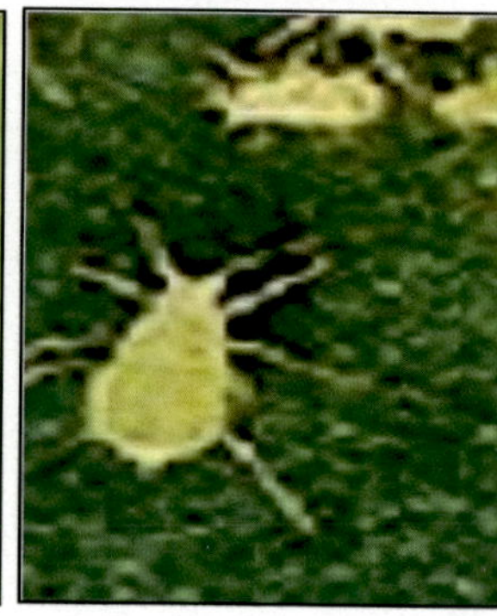

Pg: 341

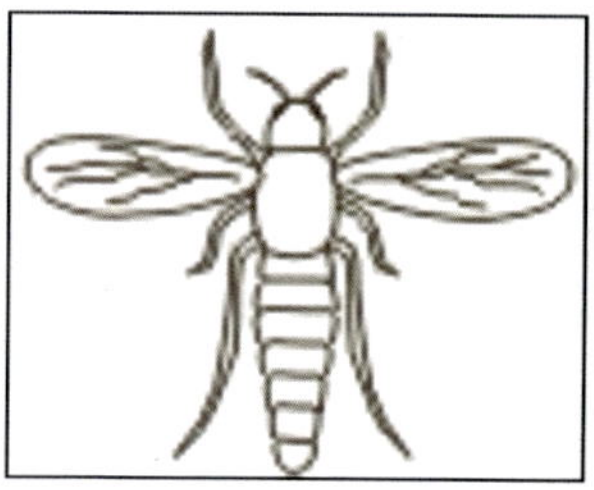

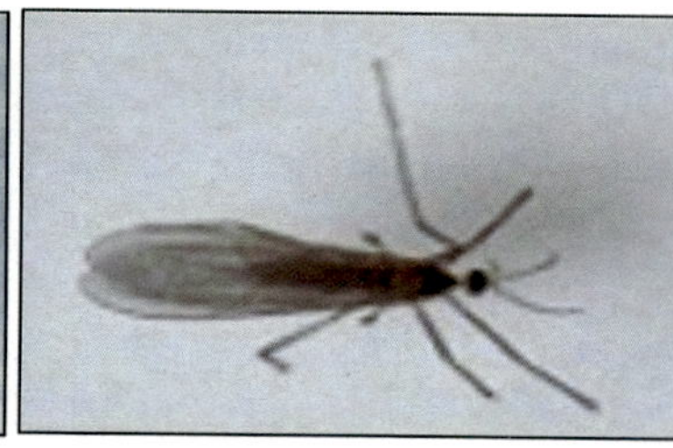

Fungus gnat Pg: 342

Pg: 343

Pg: 343

Flower Thrips Pg: 344

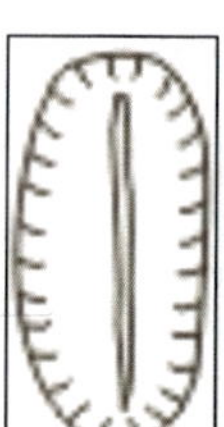

Pg: 344

Pg: 345

Earwigs Pg: 345

Cabbage worm Pg: 345

Mexican bean beetle Pg: 346

Caterpillars Pg: 346

European corn borers Pg: 346

Cutworms Pg: 346

Colorado potato beetle
Pg: 346

Tarnished plant bugs
Pg: 346

Striped cucumber beetle
Pg: 347

Root maggots Pg: 347

Leaf hoppers Pg: 347

Pg: 348

Pg: 349

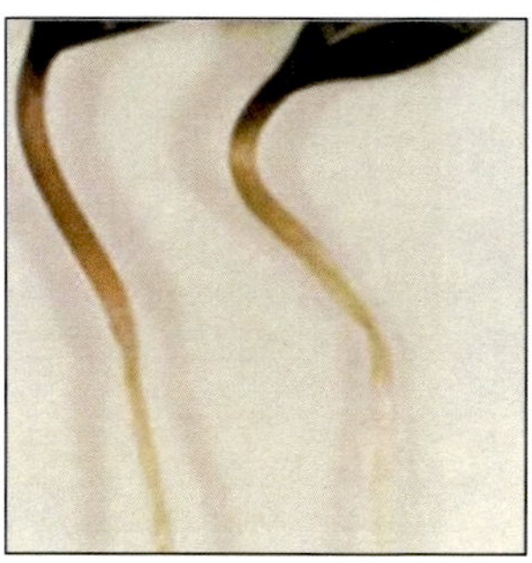

Pg: 349

Pg: 351

Pg: 352

Pg: 352

Pg: 353

Pg: 359

Pg: 355

Pg: 358

Pg: 356

Pg: 359

Pg: 360

Pg: 358

Pg: 358